Fundamentals and Applications of
Renewable Energy

Fundamentals and Applications of
Renewable Energy

Second Edition

MEHMET KANOĞLU
Alanya Alaaddin Keykubat University

YUNUS A. ÇENGEL
University of Nevada, Reno

JOHN M. CIMBALA
The Pennsylvania State University

New York   Chicago   San Francisco
Athens   London   Madrid
Mexico City   Milan   New Delhi
Singapore   Sydney   Toronto

Fundamentals and Applications of Renewable Energy, Second Edition

1 2 3 4 5 6 7 8 9   LWI   28 27 26 25 24 23

Library of Congress Control Number: 2023938138

ISBN 978-1-265-07965-9
MHID 1-265-07965-X

Sponsoring Editor
Robin Najar

Editorial Supervisor
Patty Mon

Project Manager
Tasneem Kauser,
KnowledgeWorks Global Ltd.

Acquisitions Coordinator
Olivia Higgins

Copy Editor
Susan McClung

Proofreader
Upendra Prasad

Indexer
Edwin Durbin

Production Supervisor
Lynn M. Messina

Composition
KnowledgeWorks Global Ltd.

Illustration
KnowledgeWorks Global Ltd.

Art Director, Cover
Jeff Weeks

About the Authors

Mehmet Kanoğlu is Professor of Mechanical Engineering at Alanya Alaaddin Keykubat University. He received his B.S. in mechanical engineering from Istanbul Technical University and his M.S. and Ph.D. in mechanical engineering from University of Nevada, Reno. His research areas include renewable energy systems, energy efficiency, refrigeration systems, gas liquefaction, hydrogen production and liquefaction, energy storage, geothermal energy, and cogeneration. He is the author or coauthor of dozens of journal and conference papers.

Dr. Kanoğlu has taught courses at University of Nevada, Reno, Ontario Tech University, American University of Sharjah, and University of Gaziantep. He is the coauthor of the books *Thermodynamics: An Engineering Approach* (10th ed., McGraw Hill, 2024), *Energy Efficiency and Management for Engineers* (McGraw Hill, 2020), and *Efficiency Evaluation of Energy Systems* (Springer, 2012).

Dr. Kanoğlu has served as an expert for certified energy manager training programs, the United Nations Development Programme (UNDP) for renewable energy and energy efficiency projects, and the EU's horizon projects. He instructed numerous training courses and gave lectures and presentations on renewable energy systems and energy efficiency. He has also served as advisor for state research funding organizations and industrial companies.

Yunus A. Çengel is Professor Emeritus of Mechanical Engineering at the University of Nevada, Reno. He received his B.S. in mechanical engineering from Istanbul Technical University and his M.S. and Ph.D. in mechanical engineering from North Carolina State University. His areas of interest are renewable energy, energy efficiency, energy policies, heat transfer enhancement, and engineering education. He served as the director of the Industrial Assessment Center (IAC) at the University of Nevada, Reno, from 1996 to 2000. He has led teams of engineering students to numerous manufacturing facilities in northern Nevada and California to perform industrial assessments, and has prepared energy conservation, waste minimization, and productivity enhancement reports. He has also served as an advisor for various government organizations and corporations.

Dr. Çengel is also the author or coauthor of the widely adopted textbooks *Thermodynamics: An Engineering Approach* (10th ed., 2024), *Heat and Mass Transfer: Fundamentals and Applications* (6th ed., 2020), *Fluid Mechanics: Fundamentals and Applications* (4th ed., 2018), *Fundamentals of Thermal-Fluid Sciences* (6th ed., 2022), *Differential Equations for Engineers and Scientists* (2013), and *Energy Efficiency and Management for Engineers* (2020), all published by McGraw Hill. Some of his textbooks have been translated into Chinese (long and short forms), Japanese, Korean, Spanish, French, Portuguese, Italian, Turkish, Greek, Tai, and Basq.

Dr. Çengel is the recipient of several outstanding teacher awards, and he received the ASEE Meriam/Wiley Distinguished Author Award for excellence in authorship in 1992 and again in 2000. He is a registered Professional Engineer in the State of Nevada, and is a member of the American Society of Mechanical Engineers (ASME) and the American Society for Engineering Education (ASEE).

John M. Cimbala is Professor of Mechanical Engineering at The Pennsylvania State University (Penn State), University Park, Pennsylvania. He received his B.S. in aerospace engineering from Penn State and his M.S. in aeronautics from the California Institute of Technology (CalTech). He received his Ph.D. in aeronautics from CalTech in 1984. His research areas include experimental and computational fluid mechanics and heat transfer, turbulence,

turbulence modeling, turbomachinery, indoor air quality, and air pollution control. Dr. Cimbala completed sabbatical leaves at NASA Langley Research Center (1993–1994), where he advanced his knowledge of computational fluid dynamics (CFD), and at Weir American Hydro (2010–2011), where he performed CFD analyses to assist in the design of hydro turbines.

Dr. Cimbala is the coauthor of four other textbooks: *Indoor Air Quality Engineering: Environmental Health and Control of Indoor Pollutants* (2003), published by Marcel Dekker; and *Fluid Mechanics: Fundamentals and Applications* (4th ed., 2018), *Essentials of Fluid Mechanics* (2008), and *Fundamentals of Thermal-Fluid Sciences* (6th ed., 2022), all published by McGraw Hill. He has also contributed to parts of other books, and is the author or coauthor of dozens of journal and conference papers. He has recently ventured into writing novels. More information can be found at www.me.psu.edu/cimbala.

Dr. Cimbala is the recipient of several outstanding teaching awards and views his book writing as an extension of his love of teaching. He is a Member and Fellow of the American Society of Mechanical Engineers (ASME). He is also a Member of the American Society for Engineering Education (ASEE) and the American Physical Society (APS).

Contents

Preface

BACKGROUND

The concern over the depletion of fossil fuels and pollutant and greenhouse emissions associated by their combustion can be tackled by essentially two methods: (1) using renewable energy sources to replace fossil fuels and (2) implementing energy efficiency practices in all aspects of energy production, distribution, and consumption so that less fuel is used while obtaining the same useful output. Energy efficiency can only reduce fossil fuel use while renewable energy can directly replace fossil fuels. The main renewable energy sources include solar, wind, hydropower, geothermal, and biomass. Wave and tidal energies are also renewable sources but they are currently not economical and the technologies are still in the developmental stage.

ABOUT THE BOOK

The study of renewable energy typically involves many different sciences including thermodynamics, heat transfer, fluid mechanics, physics, and chemistry. In this textbook, the primary emphasis is on thermodynamics, heat transfer, and fluid mechanics aspects of renewable energy systems and applications. This book provides an overview of common systems and applications for renewable energy sources. Systems are described and their fundamental analyses are provided.

The importance of renewable energy is relatively well-understood, and there are numerous books written on the subject. However, most of these books are concentrated on providing general information and practical guidance for practicing engineers and the public, and most books are not suitable as a textbook for classroom use. This book is primarily intended as a textbook for an upper-level undergraduate textbook for all relevant engineering majors. It may also be used as a convenient reference book for engineers, researchers, policy makers, and anyone else interested in the subject. This book provides insight into both the scientific foundations and the engineering practice of renewable energy systems. The thermodynamics, heat transfer, fluid mechanics, and thermochemistry background needed for the study of renewable energy is readily provided and thus the need for prerequisite courses is greatly minimized. This allows the use of this book for a variety of engineering majors since not all students may have backgrounds related to all thermal science courses. The book features both technical and economic analyses of renewable systems. It contains numerous practical examples and end-of-chapter problems and concept questions as well as multiple-choice questions.

OVERVIEW OF TOPICS

The first chapter covers a brief introduction to renewable energy systems, energy units, and a discussion of fossil fuels. Chapter 2 provides a comprehensive review of thermal-fluid sciences needed for studying renewable energy systems including thermodynamics, heat transfer, fluid mechanics, thermochemistry, power plants, and refrigeration systems. Chapter 3 is on fundamentals of solar energy, and Chap. 4 is on solar energy systems and applications. Chapter 5 is devoted to solar photovoltaic systems. Chapters 6, 7, 8, 9, and 10 cover wind, hydro, geothermal, biomass, and ocean (OTEC, wave, and tidal) energies, respectively. Hydrogen is introduced as an energy carrier and the principles of fuel cells are described in Chap. 11. Chapter 12 describes engineering economic analyses of renewable energy projects. Finally, environmental effects of energy are covered in Chap. 13.

KEY FEATURES

- A comprehensive review of thermodynamics, heat transfer, fluid mechanics, thermochemistry, power plants, and refrigeration systems
- Technical and economic analysis of renewable energy systems
- Rigorous descriptions and analyses of renewable energy systems and applications including concepts and formulations
- Approximately 100 worked-out example problems throughout the chapters
- Approximately 1100 end-of-chapter problems including conceptual and multiple-choice questions

UNIT SYSTEM

In recognition of the fact that English units are still widely used in some industries, both SI and English units are used in this text, with a primary emphasis on SI. The material in this text can be covered using combined SI/English units or SI units alone, depending on the preference of the instructor. The property tables in the appendices are presented in both units.

NEW IN THIS EDITION

This new edition of *Fundamentals and Applications of Renewable Energy* represents major improvements with the addition of a new chapter and several sections. The newly added coverage may be summarized as follows:

- In Chap. 5, solar photovoltaics is expanded greatly, and it is now an independent chapter: "Solar Photovoltaic Systems"
- In Chap. 1, a new section is added: "Dimensions and Units"
- In Chap. 3, a new section is added: "Estimation of Solar Radiation"
- In Chap. 4, the section "Flat-Plate Solar Collector" is expanded greatly, and a new section is added: "Evacuated Tube Collectors"
- In Chap. 6, three new sections are added: "Wind Turbine Operation and Aerodynamics," "Electricity Production from Wind Turbines," and "Wind Power Costing"
- In Chap. 9, six new sections are added: "Conversion of Biomass to Biofuel," "Anaerobic Digestion," "Thermal Gasification," "Ethanol," "Biodiesel," and "Liquid Fuels from Syngas, Pyrolysis Oil and Biocrude"

In addition to these major improvements, the number of examples and end-of-chapter problems in the text have increased significantly. Updates and changes for clarity and readability have been made throughout the text.

INSTRUCTOR RESOURCES

The following resources are available at https://www.mhprofessional.com/fare2e for instructors adapting the book for their courses.

- **Solutions Manual** The detailed solutions to all end-of-chapter problems are provided in PDF form.
- **PowerPoint Lecture Slides** PowerPoint presentation slides for all chapters in the text are available for use in lectures.

ACKNOWLEDGMENTS

The authors would like to acknowledge with appreciation the numerous and valuable comments, suggestions, constructive criticisms, and praise from several students, colleagues, and the users of the previous editions. We would like to express our appreciation to our family members for their continued patience, understanding, inspiration, and support throughout the preparation of this text.

Mehmet Kanoğlu
Yunus A. Çengel
John M. Cimbala

CHAPTER 1

Introduction to Renewable Energy

1-1 WHY RENEWABLE ENERGY?

Fossil fuels have been powering industrial development and the amenities of modern life since the 1700s, but this has not been without undesirable side effects. Pollutants emitted during the combustion of fossil fuels are responsible for smog, acid rain, and numerous other adverse effects on the environment. Environmental pollution has reached such high levels that it has become a serious threat to vegetation, wildlife, and human health. Air pollution has been the cause of numerous health problems, including asthma and cancer. But this fossil fuel–based economy is not sustainable since the estimated life of known reserves is limited. Therefore, the switch to renewable energy sources is inevitable.

Carbon dioxide (CO_2) is the greenhouse gas that makes the greatest contribution to global warming. Global climate change is widely regarded to be due to the excessive use of fossil fuels such as coal, petroleum products, and natural gas in electric power generation, transportation, buildings, and manufacturing, and it has been a concern in recent decades. The concentration of CO_2 in the atmosphere as of 2019 was about 410 ppm (or 0.41%). This is 20 percent higher than the level a century ago. Various scientific reports indicate that the earth has already warmed about 0.5°C during the last century, and it is estimated that the earth's temperature will rise another 2°C by the year 2100. A rise of this magnitude may cause severe changes in weather patterns with storms, heavy rains, and flooding in some parts and drought in others, major floods due to the melting of ice at the poles, loss of wetlands and coastal areas due to rising sea levels, variations in water supply, changes in the ecosystem due to the inability of some animal and plant species to adjust to the changes, increases in epidemic diseases due to the warmer temperatures, and adverse side effects on human health and socioeconomic conditions in some areas.

The combustion of fossil fuels produces the following undesirable emissions (Fig. 1-1):

- CO_2, primary greenhouse gas: contributes to global warming
- Nitrogen oxides (NO_x) and hydrocarbons (HC): cause smog
- Carbon monoxide (CO): toxic
- Sulfur dioxide (SO_2): causes acid rain
- Particulate matter (PM): causes adverse health effects

FOSSIL FUEL EMISSIONS
$CO_2 \Rightarrow$ global warming
NO_x and HC $\Rightarrow$ smog
CO $\Rightarrow$ toxic
$SO_2 \Rightarrow$ acid rain
PM $\Rightarrow$ adverse health effects

Figure 1-1   Effects of undesirable emissions from the combustion of fossil fuels.

Notice from this emissions list that CO_2 is different from the other emissions in that CO_2 is a greenhouse gas and a natural product of fossil fuel combustion, while other emissions are harmful air pollutants.

The concern over the depletion of fossil fuels and pollutant and greenhouse emissions associated with their combustion can be tackled by essentially two methods:

1. Using renewable energy sources such as solar, wind, hydroelectric, biomass, and geothermal to replace fossil fuels.

2. Implementing energy efficiency practices in all aspects of energy production, distribution, and consumption so that less fuel is used to obtain the same useful output.

The goal of *energy efficiency* is to reduce energy use to the minimum level, but to do so without reducing standards of living, production quality, and profitability. Energy efficiency is an expression for the most effective use of energy resources, and it results in energy conservation. Energy efficiency can only *reduce* fossil fuel use, while renewable energy can directly *replace* it.

Renewable Energy Sources

The main renewable energy sources include solar, wind, hydro, biomass, and geothermal (Fig. 1-2). Energy sources from the ocean, including ocean thermal energy conversion (OTEC), wave, and tidal, are also renewable sources, but they are currently not economical, and the technologies are still in the experimental and developmental stage.

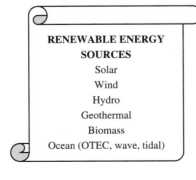

RENEWABLE ENERGY SOURCES
Solar
Wind
Hydro
Geothermal
Biomass
Ocean (OTEC, wave, tidal)

Figure 1-2   The switch from fossil fuels to renewable energy sources is inevitable.

Figure 1-3   Renewable energies such as solar water collectors are called *green energy* since they emit no pollutants or greenhouse gases.

An energy source is called *renewable* if it can be renewed and sustained without any depletion or significant effect on the environment. It is also called an *alternative, sustainable,* or *green* energy source (Fig. 1-3). Fossil fuels such as coal, oil, and natural gas, on the other hand, are not renewable, and they are depleted by use. They also emit harmful pollutants and greenhouse gases.

The best-known renewable source is *solar energy*. Although solar energy is sufficient to meet the energy needs of the entire world, currently it is not used as extensively as fossil fuels because of the *low concentration* of solar energy on earth and the *relatively high capital cost* of harnessing it. The conversion of the kinetic energy of wind into electricity via wind turbines represents *wind energy*, and it is one of the fastest-growing renewables as wind turbines are being installed all over the world. Collecting river water in large dams at some elevation and then directing the collected water into a hydraulic turbine is the common method of converting water energy into electricity. *Hydro* or *water energy* represents the greatest amount of renewable electricity production, and it supplies most of the electricity needs of some countries.

Geothermal energy refers to the heat of the earth. High-temperature underground geothermal fluid found in some locations is extracted, and the energy of the geothermal fluid is converted to electricity or heat. Geothermal energy conversion is one of the most mature renewable energy technologies. Geothermal energy is mostly used for electricity generation and district heating. Organic renewable energy is referred to as *biomass*, and a variety of sources (agriculture, forest, residues, crops, etc.) can be used to produce biomass energy. Biomass is becoming more popular with the help of the variety of available sources.

Wave and tidal energies are renewable energy sources, and they are usually considered to be part of ocean energy since they are available mostly in oceans. The thermal energy of oceans due to absorption of solar energy by ocean surfaces is also considered to be part of ocean energy, and this energy can be utilized using the OTEC system. Wave and tidal energies are mechanical forms of ocean energy since they represent the potential and kinetic energies of ocean water.

Hydrogen is an energy carrier that can be used to store renewable electricity. It is still a developing technology, and many research activities are under way to make it viable. *Fuel cells* convert the chemical energy of fuels (e.g., hydrogen) into electricity directly without a highly irreversible combustion process, and they are more efficient than combustion-based conversion to electricity.

All renewable energy sources can be used to produce useful energy in the form of electricity, and some renewables can also produce thermal energy for heating and cooling applications. Wind and water energies are converted to electricity only, while solar, biomass, and geothermal can be converted to both electricity and thermal energy (i.e., heat).

Electric cars (and other electricity-driven equipment) are often touted as "zero-emission" vehicles, and their widespread use is seen by some as the ultimate solution to the air pollution problem. It should be remembered, however, that the electricity used by electric cars is generated somewhere else, mostly by burning fossil fuels. Therefore, each time an electric car consumes 1 kWh of electricity, it bears the responsibility for the pollutants emitted to produce 1 kWh of electricity (plus the conversion and transmission losses) generated elsewhere. The electric cars can be claimed to be zero-emission vehicles only when the electricity they consume is generated by emission-free renewable resources such as hydroelectric, solar, wind, and geothermal energy. Therefore, the use of renewable energy should be encouraged worldwide, with incentives, as necessary, to make the earth a better place to live.

We should point out that what we call *renewable energy* is usually nothing more than the manifestation of solar energy in different forms. Such energy sources include wind energy, hydroelectric power, ocean thermal energy, ocean wave energy, and wood. For example, no hydroelectric power plant can generate electricity year after year unless the water evaporates by absorbing solar energy and comes back as rainfall to replenish the water source (Fig. 1-4).

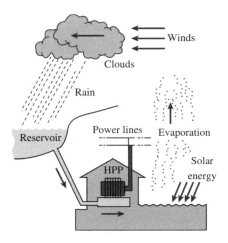

Figure 1-4 The cycle that water undergoes in a hydroelectric power plant (HPP).

1-2 DIMENSIONS AND UNITS

Any physical quantity can be characterized by *dimensions*. The magnitudes assigned to the dimensions are called *units*. Some basic dimensions such as mass (m), length (L), time (t), and temperature (T) are selected as *primary* or *fundamental dimensions*, while others such as velocity (V), energy (E), and volume (V) are expressed in terms of the primary dimensions and are called *secondary dimensions*, or *derived dimensions* (Çengel et al., 2019).

A number of unit systems have been developed over the years. Despite strong efforts in the scientific and engineering community to unify the world with a single unit system, two sets of units are still in common use today: the *English system*, which is also known as the *United States Customary System* (USCS), and the *metric SI* (from *Le Système International d' Unités*), which is also known as the *International System*.

The SI is a simple and logical system based on a decimal relationship between the various units, and it is being used for scientific and engineering work in most of the industrialized nations, including United Kingdom. The English system, however, has no apparent systematic numerical base, and various units in this system are related to each other rather arbitrarily (12 in = 1 ft, 1 mi = 5280 ft, 4 qt = 1 gal, etc.), which makes it confusing and difficult to learn.

The systematic efforts to develop a universally acceptable system of units date back to 1790 when the French National Assembly charged the French Academy of Sciences to come up with such a unit system. An early version of the metric system was soon developed in France, but it did not find universal acceptance until 1875 when *The Metric Convention Treaty* was prepared and signed by 17 nations, including the United States. In this international treaty, meter and gram were established as the metric units for length and mass, respectively, and a *General Conference of Weights and Measures* (CGPM) was established that was to meet every 6 years. In 1960, the CGPM approved the SI, which was based on six fundamental quantities, and their units adopted in 1954 at the Tenth General Conference of Weights and Measures: *meter* (m) for length, *kilogram* (kg) for mass, *second* (s) for time, *ampere* (A) for electric current, *degree Kelvin* (°K) for temperature, and *candela* (cd) for luminous intensity (amount of light). In 1971, the CGPM added a seventh fundamental quantity and unit *mole* (mol) for the amount of matter (Table 1-1).

Accurate and universal definitions of fundamental units have been challenging for the scientific community for many years. Recent new definitions of kilogram, mole, ampere, and kelvin are considered to be a historical milestone.

The kilogram unit represents the mass of one liter of pure water at 4°C. Previously, the kilogram was officially defined as the mass of a shiny metal cylinder that has been stored in

TABLE 1-1   The Seven Fundamental (or Primary) Dimensions and Their Units in SI

Dimension	Unit
Length	meter, m
Mass	kilogram, kg
Time	second, s
Temperature	kelvin, K
Electric current	ampere, A
Amount of light	candela, cd
Amount of matter	mole, mol

Paris since 1889. This International Prototype of Kilogram is an alloy of 90 percent platinum and 10 percent iridium, also known as *Le Grand K.*

On November 26, 2018, representatives from 60 countries gathered for the 26th General Conference on Weights and Measures in Versailles, France, and adopted a resolution to define the unit of mass in terms of the Planck constant *h*, which has a fixed value of $6.62607015 \times 10^{-34}$ m²·kg/s.

At the same conference, the approach of using fixed universal constants was also adopted for the new definitions of the mole, the kelvin, and the ampere. The mole is related to the value of Avogadro's constant and the ampere to the value of the elementary charge. The kelvin is related to the Boltzmann constant, whose value is fixed at 1.380649×10^{-23} J/K.

The standard meter unit was originally defined as 1/10,000,000 of the distance between the north pole and the equator. This distance was measured as accurately as possible at the time, and in the late 18th century a "master metre" stick of this length was made. All other meters were measured from this stick. Subsequent calculations of the pole-equator distance showed that the original measurement was inaccurate. In 1983, the meter is redefined as the distance traveled by light in a vacuum in 1/299,792,458 of a second.

Based on the notational scheme introduced in 1967, the degree symbol was officially dropped from the absolute temperature unit, and all unit names were to be written without capitalization even if they were derived from proper names. However, the abbreviation of a unit was to be capitalized if the unit was derived from a proper name. For example, the SI unit of force, which is named after Sir Isaac Newton (1647–1723), is *newton* (not Newton), and it is abbreviated as N. Also, the full name of a unit may be pluralized, but its abbreviation cannot. For example, the length of an object can be 5 m or 5 meters, *not* 5 ms or 5 meter. Finally, no period is to be used in unit abbreviations unless they appear at the end of a sentence. For example, the proper abbreviation of meter is m (not m.).

As pointed out, the SI is based on a decimal relationship between units. The prefixes used to express the multiples of the various units are listed in Table 1-2. They are standard for all units, and the student is encouraged to memorize them because of their widespread use (Fig. 1-5).

TABLE 1-2   Standard Prefixes in SI Units

Multiple	Prefix
10^{12}	tera, T
10^{9}	giga, G
10^{6}	mega, M
10^{3}	kilo, k
10^{2}	hecto, h
10^{1}	deka, da
10^{-1}	deci, d
10^{-2}	centi, c
10^{-3}	milli, m
10^{-6}	micro, μ
10^{-9}	nano, n

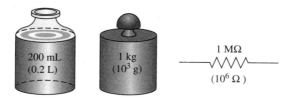

Figure 1-5 The SI unit prefixes are used in all branches of engineering.

In SI, the units of mass, length, and time are the kilogram (kg), meter (m), and second (s), respectively. The respective units in the English system are the pound-mass (lbm), foot (ft), and second (s). The pound symbol *lb* is actually the abbreviation of *libra*, which was the ancient Roman unit of weight. The English retained this symbol even after the end of the Roman occupation of Britain in 410. The mass and length units in the two systems are related to each other by

$$1 \text{ lbm} = 0.45356 \text{ kg}$$

$$1 \text{ ft} = 0.3048 \text{ m}$$

In the English system, force is usually considered to be one of the primary dimensions and is assigned a nonderived unit. This is a source of confusion and error that necessitates the use of a dimensional constant (g_c) in many formulas. To avoid this nuisance, we consider force to be a secondary dimension whose unit is derived from Newton's second law, that is,

$$\text{Force} = (\text{Mass})(\text{Acceleration})$$

or

$$F = ma \tag{1-1}$$

In SI, the force unit is the newton (N), and it is defined as the *force required to accelerate a mass of 1 kg at a rate of 1 m/s²*. In the English system, the force unit is the pound-force (lbf) and is defined as the *force required to accelerate a mass of 1 slug (32.174 lbm) at a rate of 1 ft/s²* (Fig. 1-6). That is,

$$1 \text{ N} = 1 \text{ kg·m/s}^2$$

$$1 \text{ lbf} = 32.174 \text{ lbm·ft/s}^2$$

Another force unit in common use in many European countries is the kilogram-force (kgf), which is the weight of 1 kg mass at sea level (1 kgf = 9.807 N).

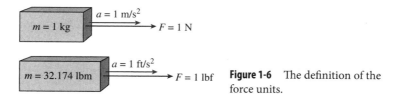

Figure 1-6 The definition of the force units.

The term *weight* is often incorrectly used to express mass, particularly by the "weight watchers." Unlike mass, weight (W) is a *force*. It is the gravitational force applied to a body, and its magnitude is determined from Newton's second law

$$W = mg \quad \text{(N)} \tag{1-2}$$

where m is the mass of the body, and g is the local gravitational acceleration (g is 9.807 m/s^2 or 32.174 ft/s^2 at sea level and 45° latitude). An ordinary bathroom scale measures the gravitational force acting on a body.

The mass of a body remains the same regardless of its location in the universe. Its weight, however, changes with a change in gravitational acceleration. A body weighs less on top of a mountain since g decreases with altitude. On the surface of the moon, an astronaut weighs about one-sixth of what they normally weigh on the earth.

At sea level, a mass of 1 kg weighs 9.807 N, as illustrated in Fig. 1-7. A mass of 1 lbm, however, weighs 1 lbf, which misleads people into believing that pound-mass and pound-force can be used interchangeably as pound (lb), which is a major source of error in the English system.

Pressure is defined as *a normal force exerted by a fluid per unit area*. Normally, we speak of pressure when we deal with a gas or a liquid. Since pressure is defined as force per unit area, it has the unit of newtons per square meter (N/m^2), which is called a *pascal* (Pa). That is,

$$1 \text{ Pa} = 1 \text{ N/m}^2$$

The pressure unit pascal is too small for most pressures encountered in practice. Therefore, its multiples *kilopascal* (1 kPa = 1000 Pa) and *megapascal* (1 MPa = 10^6 Pa) are commonly used. Two other pressure units commonly used in practice, especially in Europe, are *bar* and *standard atmosphere* (atm)

$$1 \text{ bar} = 10^5 \text{ Pa} = 0.1 \text{ MPa} = 100 \text{ kPa}$$

$$1 \text{ atm} = 101{,}325 \text{ Pa} = 101.325 \text{ kPa} = 1.01325 \text{ bar}$$

Work, which is a form of energy, can simply be defined as force times distance; therefore, it has the unit "newton-meter (N·m)," which is called a *joule* (J). That is,

$$1 \text{ J} = 1 \text{ N·m}$$

A more common unit for energy in SI is the kilojoule (1 kJ = 1000 J). In the English system, the energy unit is the *Btu* (British thermal unit), which is defined as the energy required to raise the temperature of 1 lbm of water at 68°F by 1°F. In the metric system, the amount

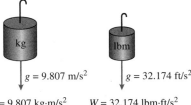

$g = 9.807$ m/s^2

$g = 32.174$ ft/s^2

$W = 9.807$ kg·m/s^2
    $= 9.807$ N
    $= 1$ kgf

$W = 32.174$ lbm·ft/s^2
    $= 1$ lbf

Figure 1-7 The weight of a unit mass at sea level.

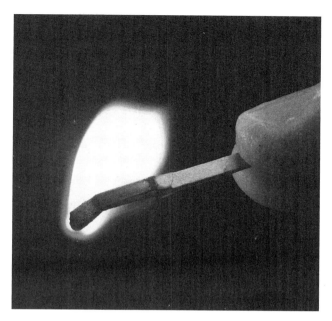

Figure 1-8 A typical match yields about 1 Btu (or 1 kJ) of energy
if completely burned. ©*John M. Cimbala*

of energy needed to raise the temperature of 1 g of water at 14.5°C by 1°C is defined as
1 *calorie* (cal), and 1 cal = 4.1868 J. The magnitudes of the kJ and Btu are almost identical (1 Btu = 1.0551 kJ). Here is a good way to get a feel for these units: If you light a typical
match and let it burn itself out, it yields approximately 1 Btu (or 1 kJ) of energy (Fig. 1-8).

The unit for amount of energy is kJ (or Btu in English units). The amount of energy per
unit time is called energy rate or the rate of energy. The unit for time rate of energy is joule
per second (J/s), which is called a *watt* (W):

$$\text{Energy rate (W)} = \frac{\text{Amount of energy (J)}}{\text{Time interval (s)}}$$

or

$$\dot{E} = \frac{E}{\Delta t} \tag{1-3}$$

In English units, a common unit for the rate of energy is Btu/h. In the case of work (such
as shaft work or electrical work), the time rate of energy is called *power*. A commonly used
unit of power is horsepower (hp), which is equivalent to 746 W. The amount of electrical
energy typically is expressed in the unit kilowatt-hour (kWh), which is equivalent to 3600 kJ.
An electric appliance with a rated power of 1 kW consumes 1 kWh of electricity when running
continuously for 1 hour.

When dealing with electric power generation, the units kW and kWh are often confused. Note that kW or kJ/s is a unit of power, whereas kWh is a unit of energy. Therefore,
statements like "the new wind turbine will generate 50 kW of electricity per year" are mean-
ingless and incorrect. A correct statement should be something like "the new wind turbine
with a rated power of 50 kW will generate 120,000 kWh of electricity per year."

EXAMPLE 1-1
Electric Power Generation
by a Wind Turbine

A school is paying $0.12/kWh for electric power. To reduce its power bill, the school installs a wind turbine (Fig. 1-9) with a rated power of 30 kW. If the turbine operates 2200 hours per year at the rated power, determine the amount of electric power generated by the wind turbine and the money saved by the school per year.

SOLUTION   The wind turbine generates electric energy at a rate of 30 kW or 30 kJ/s. Then the total amount of electric energy generated per year becomes

$$\text{Total energy} = (\text{Energy per unit time})(\text{Time interval}) = (30 \text{ kW})(2200 \text{ h/yr}) = \textbf{66,000 kWh/yr}$$

The money saved per year is the monetary value of this energy determined as

$$\text{Money saved} = (\text{Total energy})(\text{Unit cost of energy}) = (66,000 \text{ kWh/yr})(\$0.12/\text{kWh}) = \textbf{\$7920/yr}$$

The annual electric energy production also could be determined in kJ by unit manipulations as

$$\text{Total energy} = (30 \text{ kW})(2200 \text{ h/yr})\left(\frac{3600 \text{ s}}{1 \text{ h}}\right)\left(\frac{1 \text{ kJ/s}}{1 \text{ kW}}\right) = 2.38 \times 10^8 \text{ kJ/yr}$$

which is equivalent to 66,000 kWh/yr (1 kWh = 3600 kJ). ▲

Figure 1-9   A wind turbine.

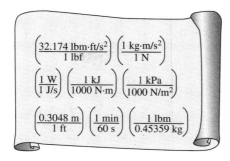

Figure 1-10 Every unity conversion ratio (as well as its inverse) is exactly equal to 1. Shown here are a few commonly used unity conversion ratios, each within its own set of parentheses.

Unity Conversion Ratio

In engineering, all equations must be *dimensionally homogeneous*. That is, every term in an equation must have the same unit. If, at some stage of an analysis, we find ourselves in a position to add two quantities that have different units, it is a clear indication that we have made an error at an earlier stage. So checking dimensions can serve as a valuable tool to spot errors.

We all know from experience that units can give terrible headaches if they are not used carefully in solving a problem. However, with some attention and skill, units can be used to our advantage. They can be used to check formulas; sometimes, they can even be used to *derive* formulas. You should keep in mind that a formula that is not dimensionally homogeneous is definitely wrong, but a dimensionally homogeneous formula is not necessarily right.

Just as all nonprimary dimensions can be formed by suitable combinations of primary dimensions, *all nonprimary units (secondary units) can be formed by combinations of primary units*. Force units, for example, can be expressed as

$$1 \text{ N} = 1 \text{ kg} \frac{\text{m}}{\text{s}^2} \qquad \text{and} \qquad 1 \text{ lbf} = 32.174 \text{ lbm} \frac{\text{ft}}{\text{s}^2}$$

They can also be expressed more conveniently as *unity conversion ratios* as

$$\frac{1 \text{ N}}{1 \text{ kg·m/s}^2} = 1 \qquad \text{and} \qquad \frac{1 \text{ lbf}}{32.174 \text{ lbm·ft/s}^2} = 1$$

Unity conversion ratios are identically equal to 1 and are unitless, and thus such ratios (or their inverses) can be inserted conveniently into any calculation to properly convert units (Fig. 1-10). You are encouraged to always use unity conversion ratios such as those given here when converting units.

EXAMPLE 1-2
Geothermal Power
Generation in Various
Units

Ton of oil equivalent (toe) is commonly used to express large amounts of energy. It represents the amount of energy released by burning 1 ton (1000 kg) of crude oil. One toe is equal to 41.868 GJ, sometimes rounded to 42 GJ. Another unit used in the United States for energy data is Quad, which is equal to one quadrillion Btu: 1 Quad $= 1 \times 10^{15}$ Btu.

The geothermal power generation in the world in 2020 was 89,300 gigawatt hours (GWh). Using unity conversion ratios, express the geothermal power generation in the world in kWh, toe, and Quad units.

SOLUTION   We note that 1 Quad $= 1 \times 10^{15}$ Btu and 1 toe $= 41.868$ GJ. Using these unity conversion ratios and the conversion factors available after the appendices, we express the global geothermal power generation in various units as follows:

$$\text{Geothermal power generation} = (89{,}300 \text{ GWh})\left(\frac{1 \times 10^6 \text{ kW}}{1 \text{ GW}}\right) = \mathbf{8.93 \times 10^{10} \text{ kWh}}$$

$$\text{Geothermal power generation} = (89{,}300 \text{ GWh})\left(\frac{1 \text{ GJ/s}}{1 \text{ GW}}\right)\left(\frac{3600 \text{ s}}{1 \text{ h}}\right)\left(\frac{1 \text{ toe}}{41.868 \text{ GJ}}\right) = \mathbf{7.678 \times 10^6 \text{ toe}}$$

$$\text{Geothermal power generation} = (89{,}300 \text{ GWh})\left(\frac{1 \text{ GJ/s}}{1 \text{ GW}}\right)\left(\frac{3600 \text{ s}}{1 \text{ h}}\right)\left(\frac{1 \times 10^6 \text{ kJ}}{1 \text{ GJ}}\right)\left(\frac{1 \text{ Btu}}{1.055 \text{ kJ}}\right)\left(\frac{1 \text{ Quad}}{1 \times 10^{15} \text{ Btu}}\right)$$
$$= \mathbf{0.3047 \text{ Quad}}$$

The quantity in large parentheses in these equations is a unity conversion ratio.   ▲

Capacity Factor

Capacity factor of a power plant is defined as the actual amount of power generated to the maximum possible power generation in a year. That is,

$$\text{Capacity factor} = \frac{\text{Actual amount of power generation}}{\text{Maximum amount of power generation}} \qquad (1\text{-}4)$$

The maximum amount of power generation is calculated assuming that the power plant operates nonstop 365 days a year and 24 hours a day at the rated power. Solar power generation has the lowest capacity factor among the renewable energy sources because solar energy is not available during nighttime as well as cloudy and rainy days. Storing solar energy daytime and using it nighttime is a viable method of increasing the capacity factor.

Wind power is dependent on the availability of sufficient wind velocity, which fluctuates in the seasons and even throughout the day. The capacity factor of wind power is usually under 50 percent. Hydropower is usually used at full load during peak hours. At other times, only some of the turbines are operated. As a result, their capacity factor is well below 100 percent. Geothermal power plants are known with their high-capacity factors because they are commonly used to meet base load requirement. Biomass-fueled power facilities also have high-capacity factors due to availability of biofuel much like fossil fuels.

When comparing renewable energy sources for the magnitude of their contribution, a more proper approach is to consider actual amount of annual power generation instead of their installed power capacity. For example, a geothermal power plant with a rated power capacity of 50 MW and a capacity factor of 90 percent generates more electricity than a wind farm with a total power rating of 100 MW and a capacity factor of 40 percent.

EXAMPLE 1-3
Capacity Factor of Solar
Power Installations

The global solar power capacity reached 970 gigawatt (GW) in 2021. On the other hand, the solar power generation in the world in 2021 was approximately 1000 terawatt hours (TWh). Using these values, determine the annual capacity factor of solar power installations in 2021.

SOLUTION   If the solar installations produced power 365 days a year and 24 hours a day at the rated power of 970 GW, the amount of power generation would be

$$\text{Maximum annual generation} = (970 \text{ GW})(365 \times 24 \text{ h}) = 8.497 \times 10^6 \text{ GWh}$$

But the actual generation is 1000 TWh or 1×10^6 GWh. Then, the annual capacity factor of solar power installations becomes

$$\text{Capacity factor} = \frac{1 \times 10^6 \text{ GWh}}{8.497 \times 10^6 \text{ GWh}} = 0.118 = \mathbf{11.8\%}$$

This is equivalent to operating solar power installations 11.8 percent of the time at a rated power of 970 GW. ▲

1-3 FOSSIL FUELS AND NUCLEAR ENERGY

The main energy sources include coal, oil, natural gas, nuclear energy, and renewable energy (Fig. 1-11). Among these, coal, oil, and natural gas are fossil fuels. Fossil fuels are responsible for more than 90 percent of global combustion-related CO_2 emissions with 37 gigatons (37,000 million tons) in 2017. The shares of fossil fuels in global CO_2 emissions are 45 percent for coal, 35 percent for oil, and 20 percent for natural gas (IEA, 2017). Here, we provide a short review of fossil fuels.

Coal

Coal is made of mostly carbon, and it also contains hydrogen, oxygen, nitrogen, sulfur, and ash (noncombustibles). The heating value of carbon is 32,800 kJ/kg. The percentages of carbon and other components vary depending on the production site. Energy content per unit mass (i.e., heating value) and sulfur content are among the important characteristics of coal. High energy content allows the extraction of more heat from coal, making the fuel more valuable. Low sulfur content is crucial to meet emission limits for sulfur compounds. Coal is used mostly for electricity production in steam power plants. It is also used for space heating, water heating, and steam generation.

There are four common types of coal with the following general characteristics.

Bituminous coal: It is also known as *soft coal*. It has high energy content but unfortunately also has high sulfur content. A representative composition (referred to in the industry as an "assay") of this coal by mass is 67 percent carbon, 5 percent hydrogen, 8.7 percent oxygen, 1.5 percent nitrogen, 1.5 percent sulfur, 9.8 percent ash, and 6.7 percent moisture. The higher heating value for this particular composition of coal is 28,400 kJ/kg. Bituminous coal is primarily used for electricity generation in power plants.

Subbituminous coal: It has lower energy content due to lower fractions of carbon and hydrogen but also lower sulfur content compared to bituminous coal. A representative composition of this coal by mass is 48.2 percent carbon, 3.3 percent

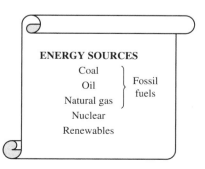

Figure 1-11 Main energy sources.

hydrogen, 11.9 percent oxygen, 0.7 percent nitrogen, 0.4 percent sulfur, 5.3 percent ash, and 30.2 percent moisture. The higher heating value for this composition of coal is 19,400 kJ/kg. Subbituminous coal is primarily used for electricity generation and heating applications.

Anthracite coal: It is also known as *hard coal*. It is far less common than bituminous and subbituminous coals. It is used mainly for residential and industrial heating applications. Few coal-fired plants burn it. It contains 80 to 95 percent carbon with low sulfur and nitrogen content. The ash content is between 10 and 20 percent and the moisture content is 5 to 15 percent. Its heating value is typically higher than 26,000 kJ/kg.

Lignite: It is also known as *brown coal*. It is the lowest-quality coal, with low energy content and high sulfur and moisture fractions. The carbon content is only 25 to 35 percent, with a heating value of less than 15,000 kJ/kg. The moisture and ash content can be as high as 75 and 20 percent, respectively. It is used mainly for electricity generation.

In the combustion of coal, hydrogen and sulfur burn first and carbon burns last. As a result, nearly all of the sulfur burns into SO_2 and nearly all of the hydrogen burns into H_2O by the following reactions:

$$S + O_2 \rightarrow SO_2$$
$$H_2 + \tfrac{1}{2}O_2 \rightarrow H_2O$$

Carbon burns according to the following reactions:

$$C + \tfrac{1}{2}O_2 \rightarrow CO$$
$$CO + \tfrac{1}{2}O_2 \rightarrow CO_2$$

If some CO cannot find sufficient oxygen to burn with by the time combustion is completed, some CO will be found in the combustion products. This represents a very undesirable emission as well as a waste of fuel because CO has energy content (the heating value of CO is 10,100 kJ/kg). This can happen even in the presence of stoichiometric or excess oxygen due to incomplete mixing and a short time for the combustion process.

Combustion of coal also causes pollutant emissions of unburned carbon particles, CO, unburned HC, SO_2, ash, and NO_x. The amount of CO_2 emission depends on the percentage of carbon in coal and the degree of completion of the combustion of carbon. Coal is considered to be a more polluting fossil fuel than liquid and gaseous fuels as well as being the largest contributor to global CO_2 emissions with about 40 percent.

EXAMPLE 1-4
Heating Value
of Coal

The assay of coal from Illinois is as follows by mass: 67.40 percent carbon (C), 5.31 percent hydrogen (H_2), 15.11 percent oxygen (O_2), 1.44 percent nitrogen (N_2), 2.36 percent sulfur (S), and 8.38 percent ash (noncombustibles). What are the higher and lower heating values of this coal? The heating value of sulfur is 9160 kJ/kg.

SOLUTION   The combustible constituents in the coal are carbon (C), hydrogen (H_2), and sulfur (S). The heating value of sulfur is given to be 9160 kJ/kg. The higher and lower heating values of hydrogen are 141,800 kJ/kg and 120,000 kJ/kg, respectively, and the heating value of carbon is 32,800 kJ/kg (Table A-7 in Appendix). Note that if the combustion of a fuel does not yield any water in the combustion gases, the higher and lower heating values are equivalent for that fuel.

Using their mass fractions (mf), the higher heating value of this particular coal is determined as

$$HHV = mf_C \times HHV_C + mf_{H_2} \times HHV_{H_2} + mf_S \times HHV_S$$
$$= (0.674)(32,800 \text{ kJ/kg}) + (0.0531)(141,800 \text{ kJ/kg}) + (0.0236)(9160 \text{ kJ/kg})$$
$$= \mathbf{29,850 \text{ kJ/kg}}$$

Similarly, the lower heating value of the coal is

$$LHV = mf_C \times LHV_C + mf_{H_2} \times LHV_{H_2} + mf_S \times LHV_S$$
$$= (0.674)(32,800 \text{ kJ/kg}) + (0.0531)(120,000 \text{ kJ/kg}) + (0.0236)(9160 \text{ kJ/kg})$$
$$= \mathbf{28,695 \text{ kJ/kg}}$$

The difference between the higher and lower heating values is about 4 percent.   ▲

Oil

Oil or petroleum is a mixture of a large number of hydrocarbons with different compositions. Crude oil has 83 to 87 percent carbon and 11 to 14 percent hydrogen with small amounts of other components such as sulfur, nitrogen, oxygen, ash, and moisture. End products such as gasoline, light diesel fuel, jet fuel, LPG (liquefied petroleum gas), natural gas, and heavy diesel fuel (fuel oil) are obtained by distillation and cracking in oil refinery plants (Fig. 1-12). Nonpetroleum liquid fuels may include ethanol, biodiesel, coal-to-liquids, natural gas liquids, and liquid hydrogen.

Gasoline and light diesel fuel are used in automobiles and can be approximated by C_8H_{15} and $C_{12}H_{22}$, respectively. Diesel fuel also includes some sulfur, but regulations in the United States and the European Union have reduced the sulfur limit from about 300 to 50 and then to 10 ppm (parts per million). The higher heating values of gasoline and light diesel fuel are 47,300 and 46,100 kJ/kg, respectively.

Oil is less commonly used for electricity generation than coal and natural gas. Two types are used in power plants and industrial heating applications:

Distillate oils: These are higher-quality oils that are highly refined. They contain much less sulfur than residual oils. A typical composition of distillate oils is 87.2 percent carbon, 12.5 percent hydrogen, and 0.3 percent sulfur. The higher heating value for this composition is 45,200 kJ/kg.

Residual oils: These oils undergo less refining. They are thicker with higher molecular mass, higher levels of impurities, and higher sulfur content. A typical composition is

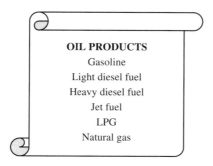

OIL PRODUCTS
Gasoline
Light diesel fuel
Heavy diesel fuel
Jet fuel
LPG
Natural gas

Figure 1-12   Main petroleum fuel products.

85.6 percent carbon, 9.7 percent hydrogen, 2.3 percent sulfur, 1.2 percent nitrogen, 0.8 percent oxygen, 0.1 percent ash, and 0.3 percent moisture. The higher heating value for this composition is 42,500 kJ/kg.

Natural Gas

Natural gas is mostly methane (CH_4), whose percentage varies between 60 and 98 percent. It also contains small amounts of ethane, propane, butane, nitrogen, oxygen, helium, CO_2, and other gases. It exists as a gas under atmospheric conditions and is stored as a gas under high pressure (15–25 MPa). It is mostly transported in gas phase by pipelines in and between cities and countries. When pipeline transportation is not feasible, it is first lique-fied to about −160°C using advanced refrigeration technologies before being carried in large insulated tanks in marine ships. Natural gas is used in boilers for space heating, hot water and steam generation, industrial furnaces, power plants for electricity production, and internal combustion engines.

The higher and lower heating values of methane are 55,530 kJ/kg and 50,050 kJ/kg, respectively. The heating value of natural gas depends mainly on the fraction of methane. The higher the methane fraction, the higher the heating value. Natural gas is commonly approximated as methane without much sacrifice in accuracy. The heating value of natural gas is usually expressed in kJ/m³, and the higher heating value ranges from 33,000 to 42,000 kJ/m³, depending on the resource. The lower heating value of natural gas is about 90 percent of its higher heating value. A comparison of higher heating values for various fuels is shown in Fig. 1-13.

Compared to coal and oil, natural gas is a cleaner fuel because it emits fewer pollutants. Air quality in certain cities improved dramatically when natural gas pipelines reached them and heating systems running on coal were replaced by their natural gas counterparts. Natural gas is used in public transportation (buses and taxis) to improve air quality in cities.

About 40 percent of the supply of natural gas is used by the industrial sector, while 33 percent is used for electricity generation in power plants (EIA, 2018). Residential and commercial applications account for the remaining use of natural gas. The supply of natural gas has recently risen substantially in the United States, Canada, and China. This is mostly due to exploitation of shale gas, which was made possible by horizontal drilling and hydraulic fracturing technologies.

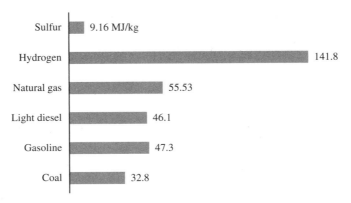

Figure 1-13   Higher heating values of various fuels, in MJ/kg. Coal is roughly approximated as carbon and natural gas as methane.

EXAMPLE 1-5
Higher Heating
Value of Methane
in Different Units

The lower heating value of methane (CH_4) is 50,050 kJ/kg. Determine its higher heating value in kJ/kg, kJ/m³, and therm/lbm units. The enthalpy of vaporization of water at 25°C is h_{fg} = 2442 kJ/kg. Assume natural gas is at 1 atm and 25°C.

SOLUTION The molar masses of CH_4 and H_2O are 16 and 18 kg/kmol, respectively. When 1 kmol of methane (CH_4) is burned with theoretical air, 2 kmol of water (H_2O) is formed. Then the mass of water formed when 1 kg of methane is burned is determined from

$$m_{H_2O} = \frac{N_{H_2O}M_{H_2O}}{N_{CH_4}M_{CH_4}} = \frac{(2 \text{ kmol})(18 \text{ kg/kmol})}{(1 \text{ kmol})(16 \text{ kg/kmol})} = 2.25 \text{ kg } H_2O/\text{kg } CH_4$$

The amount of heat released as 2.25 kg water is condensed is

$$Q_{latent} = m_{H_2O}h_{fg} = (2.25 \text{ kg } H_2O/\text{kg } CH_4)(2442 \text{ kJ/kg } H_2O) = 5495 \text{ kJ/kg } CH_4$$

Then the higher heating value of methane becomes

$$HHV = LHV + Q_{latent} = 50,050 \text{ kJ/kg} + 5495 \text{ kJ/kg} = \mathbf{55,545 \text{ kJ/kg}}$$

The gas constant of methane is R = 0.5182 kPa·m³/kg·K (Table A-1) and 1 atm = 101 kPa. The density of methane is determined from the ideal gas relation to be

$$\rho = \frac{P}{RT} = \frac{101 \text{ kPa}}{(0.5182 \text{ kPa·m}^3/\text{kg·K})(25 + 273 \text{ K})} = 0.6540 \text{ kg/m}^3$$

The higher heating value of methane in kg/m³ unit is

$$HHV = (55,545 \text{ kJ/kg})(0.6540 \text{ kg/m}^3) = \mathbf{36,330 \text{ kJ/m}^3}$$

Noting that 1 therm = 100,000 Btu = 105,500 kJ and 1 lbm = 0.4536 kg, the higher heating value of methane in therm/lbm unit is

$$HHV = (55,545 \text{ kJ/kg})\left(\frac{1 \text{ therm}}{105,500 \text{ kJ}}\right)\left(\frac{0.4536 \text{ kg}}{1 \text{ lbm}}\right) = \mathbf{0.2388 \text{ therm/lbm}} \quad \blacktriangle$$

Nuclear Energy

The tremendous amount of energy associated with the strong bonds within the nucleus of the atom is called *nuclear energy*. The most widely known fission reaction involves splitting the uranium atom (the U-235 isotope) into other elements, and it is commonly used to generate electricity in nuclear power plants, to power nuclear submarines, aircraft carriers, and even spacecraft, and as a component of nuclear bombs.

The first nuclear chain reaction was achieved by Enrico Fermi in 1942, and the first large-scale nuclear reactors were built in 1944 for the purpose of producing material for nuclear weapons. When a uranium-235 atom absorbs a neutron and splits during a fission process, it produces a cesium-140 atom, a rubidium-93 atom, 3 neutrons, and 3.2×10^{-11} J of energy. In practical terms, the complete fission of 1 kg of uranium-235 releases 6.73×10^{10} kJ of heat, which is more than the heat released when 3000 tons of coal are burned. Therefore, for the same amount of fuel, a nuclear fission reaction releases several million times more energy than a chemical reaction. The safe disposal of used nuclear fuel, however, remains a concern.

There are over 450 nuclear reactors operating worldwide with a total capacity of about 400,000 MW. Dozens of new reactors are under construction in 15 countries.

Nuclear energy by fusion is released when two small nuclei combine into a larger one. The huge amount of energy radiated by the sun and other stars originates from such a fusion process that involves the combination of two hydrogen atoms into a helium atom.

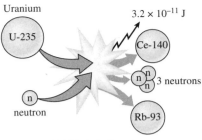

(a) Fission of uranium

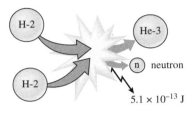

(b) Fusion of hydrogen

Figure 1-14 The fission of uranium and the fusion of hydrogen during nuclear reactions, and the release of nuclear energy.

When two heavy hydrogen (deuterium) nuclei combine during a fusion process, they produce a helium-3 atom, a free neutron, and 5.1×10^{-13} J of energy (Fig. 1-14).

Fusion reactions are much more difficult to achieve in practice because of the strong repulsion between the positively charged nuclei, called *Coulomb repulsion*. To overcome this repulsive force and to enable the two nuclei to fuse together, the energy level of the nuclei must be raised by heating them to about 100 million °C. But such high temperatures are found only in stars or in exploding atomic bombs (A-bombs). In fact, the uncontrolled fusion reaction in a hydrogen bomb (H-bomb) is initiated by a small atomic bomb. The first uncontrolled fusion reaction was achieved in the early 1950s, but all the efforts since then to achieve controlled fusion by massive lasers, powerful magnetic fields, and electric currents to generate power have failed.

Electricity

Electricity is the most valuable form of energy. Fuels cannot directly replace it because the vast majority of devices operate on electricity. Electricity is produced in power plants by burning coal, oil, and natural gas and in nuclear power stations. Renewable energy sources such as solar, wind, geothermal, and hydro are also used to produce electricity.

The contribution of renewable electricity is expected to increase in the coming years, but the incorporation of wind power and solar power into the grid involves some irregularities and uncertainties due to changing wind and solar conditions on an hourly, daily, and seasonal basis. This requires a more flexible electrical grid system than the existing conventional system in order to accommodate the inconsistent supply of renewable electricity. This new grid system is called a *smart grid*, which is an important area of research and development for electrical engineers.

Energy storage systems also help to deal with the irregularities of wind and solar electricity generation. Several techniques for storing energy have been suggested, but the two most common types are batteries and pumped storage. The former is well known since we all use rechargeable batteries in our cars, portable computers, cell phones, and other electronic devices. The latter involves pumping water "uphill" from a lower reservoir to a higher one when excess power is available, and then reversing the process ("turbining") when electricity demand is high. As more renewable energy sources go online, the need for such energy storage systems is also expected to grow significantly.

REFERENCES

Çengel YA, Boles MA, and Kanoğlu M. *Thermodynamics: An Engineering Approach*, 9th ed. New York: McGraw Hill, 2019.

EIA. 2018. U.S. Energy Information Administration, Annual Energy Outlook.

IEA. 2017. International Energy Agency, Key World Energy Statistics.

PROBLEMS

WHY RENEWABLE ENERGY?

1-1   Which undesirable emissions are produced by the combustion of fossil fuels? What adverse effects are produced by these emissions?

1-2   Is CO_2 an air pollutant? How does it differ from other emissions resulting from the combustion of fossil fuels?

1-3   What are the two main methods of tackling the concern over the depletion of fossil fuels and pollutant and greenhouse emissions associated by their combustion?

1-4   What is energy efficiency? How is it different from renewable energy use?

1-5   What are the main renewable energy sources? Why are ocean, wave, and tidal energies not considered main renewable sources?

1-6   How do you define a renewable energy source? Why are coal, oil, and natural gas not renewable energy sources?

1-7   Solar energy is the most widely available renewable energy source, and it is sufficient to meet the needs of the entire world. However, it is not used extensively. Why?

1-8   Which renewable sources are growing at the fastest rate? Which renewable source is used to produce the most electricity?

1-9   Which renewable energy sources are only used for electricity generation? Which renewable sources are converted to both electricity and thermal energy?

1-10   Some consider electric cars "zero-emission" vehicles and an ultimate solution to the air pollution problem. Do you agree? Explain.

1-11   Under what conditions can electric cars be considered "zero-emission" vehicles?

1-12   Most energy in the world is consumed by the _____ sector.
(*a*) Residential   (*b*) Commercial   (*c*) Industrial   (*d*) Transportation   (*e*) Service

1-13   The emission from fossil fuel combustion that is *not* an air pollutant is
(*a*) CO   (*b*) CO_2 (*c*) NO_x (*d*) SO_2 (*e*) PM

1-14   Which emission causes acid rain?
(*a*) CO   (*b*) CO_2 (*c*) NO_x (*d*) SO_2 (*e*) PM

1-15   Which source should not be considered a main renewable energy source?
(*a*) Wind   (*b*) Hydro   (*c*) Tidal   (*d*) Biomass   (*e*) Geothermal

1-16   The fastest-growing renewable energy sources in the world are
(*a*) Wind and solar   (*b*) Hydro and biomass   (*c*) Solar and hydro
(*d*) Biomass and wave   (*e*) Geothermal and biomass

1-17   Which renewable energy sources are only used for electricity generation?
(*a*) Wind and solar   (*b*) Hydro and solar   (*c*) Solar and geothermal
(*d*) Wind and hydro   (*e*) Hydro and geothermal

1-18   Which renewable energy source should not be considered to be the manifestation of solar energy in different forms?
(a) Wind   (b) Hydro   (c) Wave   (d) Biomass   (e) Geothermal

1-19   Reducing energy use to the minimum level, but to do so without reducing the standard of living, the production quality, and the profitability is called
(a) Energy renewability   (b) Energy efficiency   (c) Energy minimization
(d) Fuel savings   (e) Fuel conservation

DIMENSIONS AND UNITS

1-20   Why are the SI units simpler and preferred over English units? Explain.

1-21   What are the units of mass, length, time, energy, and power in the SI system and the English system?

1-22   What is the difference between kW and kWh units? Show that 1 kWh = 3600 kJ.

1-23   Show that $1 \text{ kJ} = 1 \text{ kPa·m}^3$.

1-24   Show that $1 \text{ kJ/kg} = 1000 \text{ m}^2/\text{s}^2$.

1-25   Total world delivered oil consumption by end-use sectors in 2017 was 188.8 Quad, while the renewable consumption was 19.4 Quad. Express these consumptions in Btu, GJ, and kWh.

1-26   Ton of oil equivalent (toe) is commonly used to express large amounts of energy. It represents the amount of energy released by burning 1 ton (1000 kg) of crude oil. One toe is equal to 41.868 GJ, sometimes rounded to 42 GJ. Total world delivered energy consumption in 2010 by fuel was 523.9 Quad, while that by end-use sector was 382.0 Quad. Express these values in toe units.

1-27   In 2013, 21.7 percent of global electricity was generated from natural gas–burning power plants. Total electricity generation in that year was 23,332 terawatt-hours (TWh) and it is estimated that 45 Quad natural gas was consumed to generate electricity. Determine the overall thermal efficiency of natural gas–burning power plants in 2013.

1-28   According to a 2007 report, 19,028 TWh electricity was produced in the world in 2006, and 55.3 percent of this production took place in OECD countries. The report also indicates that the average efficiency of thermal power plants in OECD countries in 2006 was 38 percent. Determine the amount of energy consumed in OECD countries in TWh, Quad, and toe.

1-29   The average annual electricity consumption by a household refrigerator has decreased from 1800 kWh in 1974 to 450 kWh today. Consider a country with 10 million households with a market penetration of 100 percent for modern refrigerators. If the refrigerators in this country were to continue to consume electric power at the 1974 levels, how much additional installed power would be needed to meet this extra demand? Assume that the load factor (average fraction of installed power load that is actually produced) of the power plants is 0.8 and the power plants operate 90 percent of the time on an annual basis.

1-30   A utility company decides to use kWh instead of therm for billing natural consumption of users. If the price of natural gas is $1.5/therm, express the natural gas price in $/kWh.

1-31   A 4-kW resistance heater in a water heater runs for 3 hours to raise the water temperature to the desired level. Determine the amount of electric energy used in both kWh and kJ.

1-32   The global wind power capacity was 743 GW in 2020. This was 6 percent of the global power capacity. The offshore wind power capacity increased by 15 percent in 2020 and reached 31.9 GW.
(a) Using unity conversion ratios, express the wind power capacity in the world in kW, MJ/h, and toe/h.
(b) Determine the percentage of offshore wind capacity with respect to total power capacity in the world.
(c) What is the approximate total power capacity in the world?

1-33   The global hydropower capacity reached 1.34 terawatt (TW) in 2021. On the other hand, the hydropower generation in the world in 2021 was 4370 terawatt hours (TWh).
(a) Using these values, determine the annual capacity factor of hydropower installations in 2021.
(b) Express the global amount of hydropower generation in exajoules.

1-34   The global nuclear power capacity reached 393 gigawatt (GW) in 2020. On the other hand, the nuclear power generation in the world in 2020 was 2553 terawatt hours (TWh). Using these values, determine the annual capacity factor of nuclear power installations.

1-35   The global geothermal power capacity was 15.6 GW in 2020. On the other hand, the geothermal power generation in the world in 2020 was 89,300 gigawatt hours (GWh). Using these values, determine the annual capacity factor of geothermal power plants in the world in 2020.

1-36   Consider a solar power plant with a rated power of 200 MW and a capacity factor of 25 percent and a wind farm with a total power rating of 110 MW and a capacity factor of 40 percent. Calculate annual power generation from each power installation.

1-37   Which of the following is a unit for the amount of energy?
(*a*) kW   (*b*) kJ/h   (*c*) J/min   (*d*) kWh   (*e*) Btu/h

1-38   Select the correct unit conversions.
  I. $1 kJ = 1 kPa \cdot m^3$
 II. $1 kWh = 3600 kJ$
III. $1 kW = 1 kJ/min$
(*a*) Only I   (*b*) I and II   (*c*) I and III   (*d*) II and III   (*e*) I, II, and III

1-39   Which of the following unit conversion is *incorrect*?
(*a*) $1 J/kg = 1 m^2/s^2$ (*b*) $1 kPa = 1 kN/m^2$ (*c*) $1 W = 1 J/s$
(*d*) $1 J = 1 N \cdot m$ (*e*) $1 GJ = 10^6 J$

1-40   A solar panel operates for a period of 10 hours at an average power rating of 1000 W. How much power is generated during this period?
(*a*) 10 kWh   (*b*) 10 kW   (*c*) 10 kJ   (*d*) 10,000 W   (*e*) 100 W/h

1-41   A wind turbine with an average power rating of 50 kW generates 600 kWh power in a given 24-hour period. What is the capacity factor of this turbine during this period?
(*a*) 0.1   (*b*) 0.25   (*c*) 0.5   (*d*) 0.6   (*e*) 1

1-42   Express 100 toe energy in kWh unit. 1 toe = 41.868 GJ.
(*a*) 1,163,000 kWh   (*b*) 1163 kWh   (*c*) 323 kWh
(*d*) 323,000 kWh   (*e*) 100,000 kWh

1-43   Express 999 GWh energy in toe unit. 1 toe = 41.868 GJ.
(*a*) 23.9 toe   (*b*) 23,860 toe   (*c*) 85.9 toe
(*d*) 85,899 toe   (*e*) 999 toe

FOSSIL FUELS AND NUCLEAR ENERGY

1-44   What are the main energy sources? What are the main fossil fuels?

1-45   What are the common coal types?

1-46   What causes CO emission in a combustion process? Will there be any CO emission when fuel is burned with stoichiometric or excess air? Explain.

1-47   What are the most common uses of coal and petroleum products?

1-48   What are the categories of oil used in power plants and industrial heating applications? Briefly describe their characteristics.

1-49   How is natural gas transported? Explain.

1-50   What are the common uses of natural gas?

1-51   What is nuclear energy? Briefly describe fission and fusion reactions.

1-52   Why is electricity the most valuable form of energy?

1-53   What is a smart grid? Explain.

1-54   The ultimate analysis of a coal from Colorado is as follows by mass: 79.61 percent carbon (C), 4.66 percent hydrogen (H_2), 4.76 percent oxygen (O_2), 1.83 percent nitrogen (N_2), 0.52 percent sulfur (S), and 8.62 percent ash (noncombustibles). What are the higher and lower heating values of this coal? The heating value of sulfur is 9160 kJ/kg.

1-55   Gasoline can be approximated by C_8H_{15}. Using this chemical formula, determine the higher and lower heating values of gasoline.

1-56   Light diesel fuel can be approximated by $C_{12}H_{22}$. Using this chemical formula, determine the higher and lower heating values of light diesel fuel.

1-57   The higher heating value of gasoline (approximated as C_8H_{15}) is 47,300 kJ/kg. Determine its lower heating value. The enthalpy of vaporization of water at 25°C is $h_{fg} = 2442$ kJ/kg.

1-58   In 2012, the United States produced 37.4 percent of its electricity in the amount of 1.51×10^{12} kWh from coal-fired power plants. Taking the average thermal efficiency to be 34 percent, determine the amount of coal consumed by these power plants. Take the heating value of coal to be 25,000 kJ/kg.

1-59   Which one cannot be considered a fossil fuel?
(*a*) Coal   (*b*) Natural gas   (*c*) Oil   (*d*) Hydrogen   (*e*) None of these

1-60   Which is not a fuel?
(*a*) Oil   (*b*) Natural gas   (*c*) Coal   (*d*) CO   (*e*) CO_2

1-61   Which is *not* a coal type?
(*a*) Bituminous coal   (*b*) Subbituminous coal   (*c*) Anthracite coal
(*d*) Lignite   (*e*) Green coal

1-62   Which coal type is of the lowest quality?
(*a*) Bituminous coal   (*b*) Subbituminous coal   (*c*) Anthracite coal
(*d*) Lignite   (*e*) Hard coal

1-63   Electricity is mostly produced from _____ burning power plants in the world.
(*a*) Coal   (*b*) Natural gas   (*c*) Oil   (*d*) Nuclear   (*e*) Solar

1-64   The most common use of petroleum products is in
(*a*) Motor vehicles   (*b*) Electricity generation   (*c*) Space heating
(*d*) Steam generation   (*e*) Industrial furnaces

1-65   Which fuel is the most polluting fuel and the largest contributor to global CO_2 emissions?
(*a*) Coal   (*b*) Natural gas   (*c*) Oil   (*d*) Nuclear   (*e*) Solar

1-66   Which fuel has the highest heating value?
(*a*) Coal   (*b*) Natural gas   (*c*) Oil   (*d*) Hydrogen   (*e*) Sulfur

1-67   Desirable characteristics of coal are
(*a*) High sulfur content, high heating value   (*b*) High sulfur content, low heating value
(*c*) Low sulfur content, high heating value   (*d*) Low sulfur content, low heating value

1-68   Order the following fuels from higher values of heating value to lower values.
  I. Coal
 II. Gasoline
III. Natural gas
(*a*) I, II, III   (*b*) I, III, II   (*c*) II, I, III   (*d*) II III, I   (*e*) III, II, I

1-69   In order to accommodate the inconsistent supply of renewable electricity, a more flexible electrical grid system than the existing conventional system is required. This is called
(*a*) Smart grid   (*b*) Flexible electricity   (*c*) Consistent grid supply
(*d*) Intelligent transmission   (*e*) None of these

1-70   The total amount of heat input to a coal-burning power plant is 5000 GJ. The heating value of coal is 25,000 kJ/kg. The amount of coal consumed per year is
(*a*) 5000 ton   (*b*) 500 ton   (*c*) 200 ton   (*d*) 200,000 ton   (*e*) 25,000 ton

CHAPTER 2

A Review of Thermal Sciences

2-1 THERMAL SCIENCES

The analysis of renewable energy systems requires a solid understanding of energy conversion processes, transformation between various forms of energy, and ways of defining efficiencies of energy producing and consuming systems. In this chapter, we review fundamental concepts of thermodynamics, heat transfer, fluid mechanics, thermochemistry, power plants, and refrigeration systems to form a solid and useful foundation for the renewable energy systems to be covered in the upcoming chapters.

The physical sciences that deal with energy and the transfer, transport, and conversion of energy are usually referred to as *thermal-fluid sciences* or just *thermal sciences*. Traditionally, the thermal-fluid sciences are studied under the subcategories of thermodynamics, heat transfer, and fluid mechanics (Çengel and Cimbala, 2018; Çengel et al., 2019; Çengel and Ghajar, 2020; Çengel et al., 2022).

The design and analysis of most thermal systems such as power plants, automotive engines, refrigerators, building heating and cooling systems, boilers, heat exchangers, and other energy conversion equipment involve all categories of thermal sciences. For example, designing a solar collector involves the determination of the amount of energy transfer from a knowledge of *thermodynamics,* the determination of the size of the heat exchanger using *heat transfer,* and the determination of the size and type of the pump using *fluid mechanics* (Fig. 2-1).

2-2 THERMODYNAMICS

Thermodynamics can be defined as the science of *energy*. Although everybody has a feeling for what energy is, it is difficult to give a precise definition for it. Energy can be viewed as the ability to cause changes. The name *thermodynamics* stems from the Greek words *therme* (heat) and *dynamis* (power), which is most descriptive of the early efforts to convert heat into power. Today the same name is broadly interpreted to include all aspects of energy and energy transformations, including power generation, refrigeration, and relationships among the properties of matter.

One of the most fundamental laws of nature is the *conservation of energy principle*. It simply states that during an interaction, energy can change from one form to another, but the total amount of energy remains constant. That is, energy cannot be created or destroyed. A rock falling off a cliff, for example, picks up speed as a result of its potential energy being converted to kinetic energy. The conservation of energy principle also forms the backbone of the diet industry: A person who has a greater energy input (food) than energy output

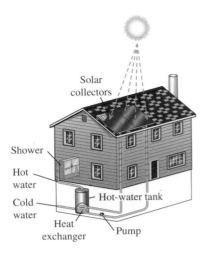

Figure 2-1   The design and analysis of renewable energy systems, such as this solar hot water system, involves thermal sciences.

(exercise) will gain weight (store energy in the form of fat), and a person who has a smaller energy input than output will lose weight. The change in the energy content of a body or any other system is equal to the difference between the energy input and the energy output, and the energy balance is expressed as $E_{in} - E_{out} = \Delta E_{system}$.

The *first law of thermodynamics* is simply an expression of the conservation of energy principle, and it asserts that *energy* is a thermodynamic property. The *second law of thermodynamics* asserts that energy has *quality* as well as *quantity,* and actual processes occur in the direction of decreasing quality of energy. For example, a cup of hot coffee left on a table eventually cools, but a cup of cool coffee in the same room never gets hot by itself. The high-temperature energy of the coffee is degraded (transformed into a less useful form at a lower temperature) once it is transferred to the surrounding air.

Heat and Other Forms of Energy

Energy can exist in numerous forms such as thermal, mechanical, kinetic, potential, electrical, magnetic, chemical, and nuclear, and their sum constitutes the *total energy E* (or *e* on a unit mass basis) of a system. The forms of energy related to the molecular structure of a system and the degree of the molecular activity are referred to as the *microscopic energy.* The sum of all microscopic forms of energy is called the *internal energy* of a system and is denoted by *U* (or *u* on a unit mass basis).

The international unit of energy is *joule* (J) or *kilojoule* (1 kJ = 1000 J). In the English system, the unit of energy is the *British thermal unit* (Btu), which is defined as the energy needed to raise the temperature of 1 lbm of water at 60°F by 1°F. The magnitudes of kJ and Btu are almost identical (1 Btu = 1.055056 kJ). Another well-known unit of energy is the *calorie* (1 cal = 4.1868 J), which is defined as the energy needed to raise the temperature of 1 g of water at 14.5°C by 1°C.

Internal energy may be viewed as the sum of the kinetic and potential energies of the molecules. The portion of the internal energy of a system associated with the kinetic energy of the molecules is called *sensible energy* or *sensible heat*. The average velocity and the degree of activity of the molecules are proportional to the temperature. Thus, at higher temperatures the molecules possess higher kinetic energy, and as a result, the system has a higher internal energy.

The internal energy is also associated with the intermolecular forces between the molecules of a system. These are the forces that bind the molecules to each other, and, as one would expect, they are strongest in solids and weakest in gases. If sufficient energy is added to the molecules of a solid or liquid, they will overcome these molecular forces and simply break away, turning the system to a gas. This is a *phase change* process, and because of this added energy, a system in the gas phase is at a higher internal energy level than it is in the solid or the liquid phase. The internal energy associated with the phase of system is called *latent energy* or *latent heat*.

The changes mentioned above can occur without a change in the chemical composition of a system. Most heat transfer problems fall into this category, and one does not need to pay any attention to the forces binding the atoms in a molecule together. The internal energy associated with the atomic bonds in a molecule is called *chemical* (or *bond*) *energy*, whereas the internal energy associated with the bonds within the nucleus of the atom itself is called *nuclear energy*. The chemical and nuclear energies are absorbed or released during chemical or nuclear reactions, respectively.

In the analysis of systems that involve fluid flow, we frequently encounter the combination of properties u and Pv. For the sake of simplicity and convenience, this combination is defined as *specific enthalpy h* or just *enthalpy*. We prefer the term *enthalpy* for convenience. That is, $h = u + Pv$, where the term Pv represents the *flow energy* of the fluid (also called the *flow work*), which is the energy needed to push a fluid and to maintain flow. In the energy analysis of flowing fluids, it is convenient to treat the flow energy as part of the energy of the fluid and to represent the microscopic energy of a fluid stream by enthalpy h (Fig. 2-2).

Specific Heats of Gases, Liquids, and Solids

An *ideal gas* is defined as a gas that obeys the relation

$$Pv = RT \qquad \text{or} \qquad P = \rho RT \qquad (2\text{-}1)$$

where P is the absolute pressure, v is the specific volume, T is the thermodynamic (or absolute) temperature, ρ is the density, and R is the gas constant. It has been experimentally observed that the ideal gas relation given above closely approximates the P-v-T behavior of real gases at low densities. At low pressures and high temperatures, the density of a gas decreases and the gas behaves like an ideal gas. In the range of practical interest, many familiar gases such as air, nitrogen, oxygen, hydrogen, helium, argon, neon, and krypton

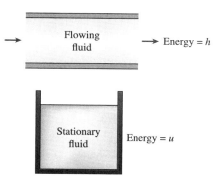

Figure 2-2   The *internal energy u* represents the microscopic energy of a nonflowing fluid, whereas *enthalpy h* represents the microscopic energy of a flowing fluid.

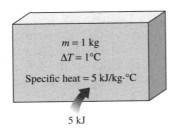

Figure 2-3 Specific heat is the energy required to raise the temperature of a unit mass of a substance by one degree in a specified way.

and even heavier gases such as carbon dioxide can be treated as ideal gases with negligible error (often less than 1%). Dense gases such as water vapor in steam power plants and refrigerant vapor in refrigerators, however, should not always be treated as ideal gases since they usually exist at a state near saturation.

Specific heat is defined as *the energy required to raise the temperature of a unit mass of a substance by one degree* (Fig. 2-3). In general, this energy depends on how the process is executed. We are usually interested in two kinds of specific heats: specific heat at constant volume c_v and specific heat at constant pressure c_p. The *specific heat at constant volume c_v* can be viewed as the energy required to raise the temperature of a unit mass of a substance by one degree as the volume is held constant. The energy required to do the same as the pressure is held constant is the specific heat at constant pressure c_p. The specific heat at constant pressure c_p is greater than c_v because at constant pressure the system is allowed to expand and the energy for this expansion work must also be supplied to the system. For ideal gases, these two specific heats are related to each other by $c_p = c_v + R$.

A common unit for specific heats is kJ/kg·°C or kJ/kg·K. Notice that these two units are identical since $\Delta T(°C) = \Delta T(K)$, and a 1°C change in temperature is equivalent to a change of 1 K. Also,

$$1 \text{ kJ/kg·°C} = 1 \text{ J/g·°C} = 1 \text{ kJ/kg·K} = 1 \text{ J/g·K}$$

and

$$1 \text{ Btu/lbm·°F} = 1 \text{ Btu/lbm·R}$$

The specific heat of a substance, in general, depends on two independent properties such as temperature and pressure. For an ideal gas, however, they depend on temperature only (Fig. 2-4). At low pressures all real gases approach ideal gas behavior, and therefore their specific heats depend on temperature only.

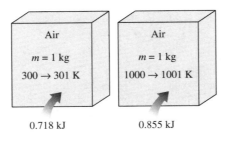

Figure 2-4 The specific heat of a substance changes with temperature.

The differential changes in the internal energy u and enthalpy h of an ideal gas can be expressed in terms of the specific heats as

$$du = c_v dT \qquad \text{and} \qquad dh = c_p dT \qquad \text{(2-2)}$$

The finite changes in the internal energy and enthalpy of an ideal gas during a process can be expressed approximately by using specific heat values at the average temperature as

$$\Delta u = c_{v,\text{avg}} \Delta T \qquad \text{and} \qquad \Delta h = c_{p,\text{avg}} \Delta T \qquad \text{(kJ/kg)} \qquad \text{(2-3)}$$

or

$$\Delta U = m c_{v,\text{avg}} \Delta T \qquad \text{and} \qquad \Delta H = m c_{p,\text{avg}} \Delta T \qquad \text{(kJ)} \qquad \text{(2-4)}$$

where m is the mass of the system.

A substance whose specific volume (or density) does not change with temperature or pressure is called an incompressible substance. The specific volumes of solids and liquids essentially remain constant during a process, and thus they can be approximated as incompressible substances without sacrificing much in accuracy.

The constant-volume and constant-pressure specific heats are identical for incompressible substances (Fig. 2-5). Therefore, for solids and liquids the subscripts on c_v and c_p can be dropped, and both specific heats can be represented by a single symbol, c. That is, $c_p = c_v = c$. This result could also be deduced from the physical definitions of constant-volume and constant-pressure specific heats. Specific heats of several common gases, liquids, and solids are given in the appendix.

The specific heats of incompressible substances depend on temperature only. Therefore, the change in the internal energy of solids and liquids can be expressed as

$$\Delta U = m c_{\text{avg}} \Delta T \qquad \text{(kJ)} \qquad \text{(2-5)}$$

where c_{avg} is the average specific heat evaluated at the average temperature. Note that the internal energy change of the systems that remain in a single phase (liquid, solid, or gas) during the process can be determined very easily using average specific heats.

Energy Transfer

Energy can be transferred to or from a given mass by two mechanisms: *heat transfer Q* and *work W*. An energy interaction is heat transfer if its driving force is a temperature difference. Otherwise, it is work. A rising piston, a rotating shaft, and an electrical wire crossing the system boundaries are all associated with work interactions. Work done *per unit time* is called *power* and is denoted by $\dot{W}$. The unit of power is kW or hp (1 hp = 0.746 kW).

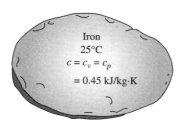

Figure 2-5 The c_v and c_p values of incompressible substances are identical and are denoted by c.

Car engines and hydraulic, steam, and gas turbines produce work; compressors, pumps, and mixers consume work. Notice that the energy of a system decreases as it does work and increases as work is done on it.

The amount of heat transferred during the process is denoted by Q. The amount of heat transferred per unit time is called *heat transfer rate* and is denoted by $\dot{Q}$. The overdot stands for the time derivative, or "per unit time." The heat transfer rate $\dot{Q}$ has the unit kJ/s (or Btu/h), which is equivalent to kW. In cooling applications, the rate of cooling provided by the cooling equipment (cooling capacity) is often expressed in "ton of refrigeration" units where 1 ton = 12,000 Btu/h.

When the *rate* of heat transfer $\dot{Q}$ is available, then the total amount of heat transfer Q during a time interval Δt can be determined from

$$Q = \int_0^{\Delta t} \dot{Q}\, dt \qquad (2\text{-}6)$$

provided that the variation of $\dot{Q}$ with time is known. For the special case of $\dot{Q}$ = constant, the equation above reduces to

$$Q = \dot{Q}\,\Delta t \qquad \text{(kJ)} \qquad (2\text{-}7)$$

The First Law of Thermodynamics

The *first law of thermodynamics*, also known as the *conservation of energy principle*, states that energy can neither be created nor be destroyed during a process; it can only change forms. Therefore, every bit of energy must be accounted for during a process. The conservation of energy principle (or the energy balance) for any system undergoing any process may be expressed as follows: The net change (increase or decrease) in the total energy of the system during a process is equal to the difference between the total energy entering and the total energy leaving the system during that process.

Noting that energy can be transferred to or from a system by *heat, work,* and *mass flow,* and that the total energy of a simple compressible system consists of internal, kinetic, and potential energies, the *energy balance* for any system undergoing any process can be expressed as

$$E_{\text{in}} - E_{\text{out}} = \Delta E_{\text{system}} \qquad \text{(kJ)} \qquad (2\text{-}8)$$

In the absence of significant electric, magnetic, motion, gravity, and surface tension effects (i.e., for stationary simple compressible systems), the change in the *total energy* of a system during a process is simply the change in its *internal energy*. That is, $\Delta E_{\text{system}} = \Delta U_{\text{system}}$.

Energy balance can be written in the *rate form* as

$$\dot{E}_{\text{in}} - \dot{E}_{\text{out}} = dE_{\text{system}}/dt \qquad \text{(kW)} \qquad (2\text{-}9)$$

Energy is a property, and the value of a property does not change unless the state of the system changes. Therefore, the energy change of a system is zero if the state of the system does not change during the process, that is, the process is steady. The energy balance in this case reduces to (Fig. 2-6)

$$\dot{E}_{\text{in}} = \dot{E}_{\text{out}} \qquad \text{(kW)} \qquad (2\text{-}10)$$

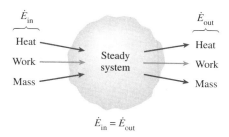

Figure 2-6   In steady operation, the rate of energy transfer to a system is equal to the rate of energy transfer from the system.

Energy Balance for Closed Systems

A closed system consists of a fixed mass. The total energy E for most systems encountered in practice consists of the internal energy U. This is especially the case for stationary systems since they do not involve any changes in their velocity or elevation during a process. The energy balance relation in that case reduces to

Stationary closed system:

$$E_{in} - E_{out} = \Delta U = mc_v\Delta T \qquad \text{(kJ)} \qquad (2\text{-}11)$$

where we expressed the internal energy change in terms of mass m, the specific heat at constant volume c_v, and the temperature change ΔT of the system. When the system involves heat transfer only and no work interactions across its boundary, the energy balance relation further reduces to (Fig. 2-7)

Stationary closed system, no work:

$$Q = mc_v\Delta T \qquad \text{(kJ)} \qquad (2\text{-}12)$$

where Q is the net amount of heat transfer to or from the system. This is the form of the energy balance relation we will use most often when dealing with a fixed mass.

Energy Balance for Steady-Flow Systems

A large number of engineering devices such as water heaters and car radiators involve mass flow in and out of a system and are modeled as *control volumes*. Most control volumes are analyzed under steady operating conditions. The term *steady* means *no change with time* at

Specific heat = c_v
Mass = m
Initial temp = T_1
Final temp = T_2

$Q = mc_v(T_1 - T_2)$

Figure 2-7   In the absence of any work interactions, the change in the energy content of a closed system is equal to the net heat transfer.

a specified location. The opposite of steady is *unsteady* or *transient*. Also, the term *uniform* implies *no change with position* throughout a surface or region at a specified time. These meanings are consistent with their everyday usage (steady job, uniform distribution, etc.). The total energy content of a control volume during a *steady-flow process* remains constant (E_{CV} = constant). That is, the change in the total energy of the control volume during such a process is zero ($\Delta E_{CV} = 0$). Thus the amount of energy entering a control volume in all forms (heat, work, mass transfer) for a steady-flow process must be equal to the amount of energy leaving it. In rate form, it is expressed as $\dot{E}_{in} = \dot{E}_{out}$.

The amount of mass flowing through a cross section of a flow device per unit time is called the *mass flow rate* and is denoted by $\dot{m}$. A fluid may flow in and out of a control volume through pipes or ducts. The mass flow rate of a fluid flowing in a pipe or duct is proportional to the cross-sectional area A_c of the pipe or duct, the density ρ, and the velocity V of the fluid. The flow of a fluid through a pipe or duct can often be approximated to be *one dimensional*. That is, the properties can be assumed to vary in one direction only (the direction of flow). As a result, all properties are assumed to be uniform at any cross section normal to the flow direction, and the properties are assumed to have *bulk average values* over the entire cross section. Under the one-dimensional flow approximation, the mass flow rate of a fluid flowing in a pipe or duct can be expressed as (Fig. 2-8)

$$\dot{m} = \rho V A_c \quad \text{(kg/s)} \tag{2-13}$$

The volume of a fluid flowing through a pipe or duct per unit time is called the *volume flow rate* $\dot{V}$ and is expressed as

$$\dot{V} = V A_c = \frac{\dot{m}}{\rho} \quad \text{(m}^3\text{/s)} \tag{2-14}$$

Note that the mass flow rate of a fluid through a pipe or duct remains constant during steady flow. This is not the case for the volume flow rate, however, unless the density of the fluid remains constant.

For a steady-flow system with one inlet and one exit, the rate of mass flow into the control volume must be equal to the rate of mass flow out of it. That is, $\dot{m}_{in} = \dot{m}_{out} = \dot{m}$. When the changes in kinetic and potential energies are negligible, which is usually the case, and there is no work interaction, the energy balance for such a steady-flow system reduces to (Fig. 2-9)

$$\dot{Q} = \dot{m}\Delta h = \dot{m}c_p\Delta T \quad \text{(kW)} \tag{2-15}$$

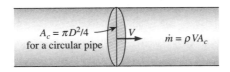

Figure 2-8 The mass flow rate of a fluid at a cross section is equal to the product of the fluid density, average fluid velocity, and cross-sectional area.

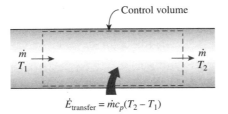

Figure 2-9 Under steady conditions, the net rate of energy transfer to a fluid in a control volume is equal to the rate of increase in the energy of the fluid stream flowing through the control volume.

where $\dot{Q}$ is the rate of net heat transfer into or out of the control volume. This is the form of the energy balance relation that we will use most often for steady-flow systems.

EXAMPLE 2-1
Electric Resistance Heating of a House

A house has an electric heating system that consists of a 250-W fan and an electric resistance heating element placed in a duct (Fig. 2-10). Air flows steadily through the duct at a rate of 0.5 kg/s and experiences a temperature rise of 7°C. The rate of heat loss from the air in the duct is estimated to be 400 W. Determine the power rating of the electric resistance heating element.

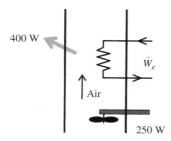

Figure 2-10   Schematic for Example 2-1.

SOLUTION   Air is considered to be an ideal gas, and constant specific heats at room temperature can be used for air. The specific heat of air at room temperature is $c_p = 1.005$ kJ/kg·°C (Table A-1). We take the heating duct as the system. This is a *control volume* since mass crosses the system boundary during the process. We observe that this is a steady-flow process since there is no change with time at any point, and thus $\Delta m_{CV} = 0$ and $\Delta E_{CV} = 0$. Also, there is only one inlet and one exit, and thus $\dot{m}_1 = \dot{m}_2 = \dot{m}$. The energy balance for this steady-flow system can be expressed in the rate form as

$$\dot{E}_{in} = \dot{E}_{out}$$

$$\dot{W}_{e,in} + \dot{W}_{fan,in} + \dot{m}h_1 = \dot{Q}_{out} + \dot{m}h_2$$

$$\dot{W}_{e,in} = \dot{Q}_{out} - \dot{W}_{fan,in} + \dot{m}(h_2 - h_1)$$

$$\dot{W}_{e,in} = \dot{Q}_{out} - \dot{W}_{fan,in} + \dot{m}c_p \Delta T$$

Substituting, the power rating of the heating element is determined to be

$$\dot{W}_{e,in} = (0.400 \text{ kW}) - (0.250 \text{ kW}) + (0.5 \text{ kg/s})(1.005 \text{ kJ/kg·°C})(7°C) = \textbf{3.67 kW} \quad \blacktriangle$$

Saturation Temperature and Saturation Pressure

Water starts to boil at 100°C. Strictly speaking, the statement "water boils at 100°C" is incorrect. The correct statement is "water boils at 100°C at 1 atm pressure." At 500 kPa pressure, water boils at 151.8°C. That is, the temperature at which water starts boiling depends on the pressure; therefore, if the pressure is fixed, so is the boiling temperature.

At a given pressure, the temperature at which a pure substance changes phase is called the saturation temperature T_{sat}. Likewise, at a given temperature, the pressure at which a pure substance changes phase is called the saturation pressure P_{sat}. At a pressure of 101.3 kPa, T_{sat} is 100°C. Conversely, at a temperature of 100°C, P_{sat} is 101.3 kPa.

Saturation tables that list the saturation pressure against the temperature (or the saturation temperature against the pressure) are available for practically all substances. A partial listing of such a table is given in Table 2-1 for water. This table indicates that the pressure of

TABLE 2-1   Saturation (Boiling) Pressure of Water at Various Temperatures

Temperature T, °C	Saturation Pressure P_{sat}, kPa
−10	0.260
−5	0.403
0	0.611
5	0.872
10	1.23
15	1.71
20	2.34
25	3.17
30	4.25
40	7.38
50	12.35
100	101.3 (1 atm)
150	475.8
200	1554
250	3973
300	8581

water changing phase (boiling or condensing) at 25°C must be 3.17 kPa, and the pressure of water must be maintained at 3976 kPa (about 40 atm) to have it boil at 250°C. Also, water can be frozen by dropping its pressure below 0.61 kPa.

It takes a large amount of energy to melt a solid or vaporize a liquid. The amount of energy absorbed or released during a phase-change process is called the *latent heat*. More specifically, the amount of energy absorbed during melting is called the *latent heat of fusion* and is equivalent to the amount of energy released during freezing. Similarly, the amount of energy absorbed during vaporization is called the *latent heat of vaporization* and is equivalent to the energy released during condensation. The magnitudes of the latent heats depend on the temperature or pressure at which the phase change occurs. At 1 atm pressure, the latent heat of fusion of water is 333.7 kJ/kg and the latent heat of vaporization is 2256.5 kJ/kg.

During a phase-change process, pressure and temperature are obviously dependent properties, and there is a definite relation between them, that is, $T_{sat} = f(P_{sat})$. A plot of T_{sat} versus P_{sat}, such as the one given for water in Fig. 2-11, is called a *liquid-vapor saturation curve*. A curve of this kind is the characteristic of all pure substances.

It is clear from Fig. 2-11 that T_{sat} increases with P_{sat}. Thus, a substance at higher pressures boils at higher temperatures. In the kitchen, higher boiling temperatures mean shorter cooking times and energy savings. A beef stew, for example, may take 1 to 2 h to cook in a regular pan that operates at 1 atm pressure, but only 20 min in a pressure cooker operating at 3 atm absolute pressure (corresponding boiling temperature: 134°C).

The atmospheric pressure, and thus the boiling temperature of water, decreases with elevation. Therefore, it takes longer to cook at higher altitudes than it does at sea level (unless a pressure cooker is used). For example, the standard atmospheric pressure at an

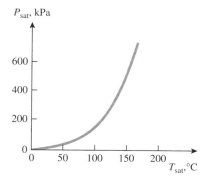

Figure 2-11 The liquid-vapor saturation curve of a pure substance (numerical values are for water).

elevation of 2000 m is 79.50 kPa, which corresponds to a boiling temperature of 93.3°C as opposed to 100°C at sea level (zero elevation). For each 1000 m increase in elevation, the boiling temperature drops by a little over 3°C.

2-3 HEAT TRANSFER

We define *heat* as the form of energy that can be transferred from one system to another as a result of temperature difference. A thermodynamic analysis is concerned with the *amount* of heat transfer that occurs as a system undergoes a process from one equilibrium state to another. The science that deals with the determination of the *rates* of such energy transfers is the *heat transfer*. The transfer of energy as heat is always from the higher-temperature medium to the lower-temperature one, and heat transfer stops when the two mediums reach the same temperature.

Heat can be transferred in three different modes: *conduction, convection,* and *radiation.* All modes of heat transfer require the existence of a temperature difference, and all modes are from the high-temperature medium to a lower-temperature one.

Conduction Heat Transfer

Conduction is the transfer of energy from the more energetic particles of a substance to the adjacent less energetic ones as a result of interactions between the particles. Conduction can take place in solids, liquids, or gases. In gases and liquids, conduction is due to the *collisions* and *diffusion* of the molecules during their random motion. In solids, it is due to the combination of *vibrations* of the molecules in a lattice and the energy transport by *free electrons.* A cold canned drink in a warm room, for example, eventually warms up to the room temperature as a result of heat transfer from the room to the drink through the aluminum can by conduction.

The *rate* of heat conduction through a medium depends on the *geometry* of the medium, its *thickness,* and the *material* of the medium, as well as the *temperature difference* across the medium. We know that wrapping a hot water tank with glass wool (an insulating material) reduces the rate of heat loss from the tank. The thicker the insulation, the smaller the heat loss. We also know that a hot water tank loses heat at a higher rate when the temperature of the room housing the tank is lowered. Further, the larger the tank, the larger the surface area and thus the rate of heat loss.

Consider steady heat conduction through a large plane wall of thickness $\Delta x = L$ and area A, as shown in Fig. 2-12. The temperature difference across the wall is $\Delta T = T_2 - T_1$. Experiments have shown that the rate of heat transfer $\dot{Q}$ through the wall is *doubled* when the

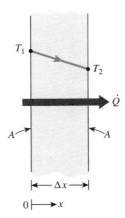

Figure 2-12 Heat conduction through a large plane wall.

temperature difference ΔT across the wall or the area A normal to the direction of heat transfer is doubled, but is *halved* when the wall thickness L is doubled. Thus we conclude that *the rate of heat conduction through a plane layer is proportional to the temperature difference across the layer and the heat transfer area, but it is inversely proportional to the thickness of the layer.* That is,

$$\dot{Q}_{\text{cond}} = kA\frac{T_1 - T_2}{\Delta x} = -kA\frac{\Delta T}{\Delta x} \qquad \text{(kW)} \qquad (2\text{-}16)$$

where the constant of proportionality k is the *thermal conductivity* of the material, which is a *measure of the ability of a material to conduct heat* (Fig. 2-13). In the limiting case of $\Delta x \to 0$, the equation above reduces to the differential form

$$\dot{Q}_{\text{cond}} = -kA\frac{dT}{dx} \qquad \text{(kW)} \qquad (2\text{-}17)$$

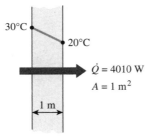

(a) Copper ($k = 401$ W/m·K)

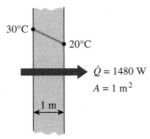

(b) Silicon ($k = 148$ W/m·K)

Figure 2-13 The rate of heat conduction through a solid is directly proportional to its thermal conductivity.

which is called *Fourier's law of heat conduction* after J. Fourier, who expressed it first in his heat transfer text in 1822. Here dT/dx is the *temperature gradient*, which is the slope of the temperature curve on a *T-x* diagram (the rate of change of *T* with *x*), at location *x*. The preceding relation indicates that the rate of heat conduction in a given direction is proportional to the temperature gradient in that direction. Heat is conducted in the direction of decreasing temperature, and the temperature gradient becomes negative when temperature decreases with increasing *x*. The *negative sign* in Eq. (2-17) ensures that heat transfer in the positive *x* direction is a positive quantity.

The heat transfer area *A* is always *normal* to the direction of heat transfer. For heat loss through a 5-m-long, 3-m-high, and 25-cm-thick wall, for example, the heat transfer area is $A = 15$ m². Note that the thickness of the wall has no effect on *A*.

Thermal Conductivity

We can define specific heat c_p as a measure of a material's ability to store thermal energy. For example, $c_p = 4.18$ kJ/kg·°C for water and $c_p = 0.45$ kJ/kg·°C for iron at room temperature, which indicates that water can store almost 10 times the energy that iron can per unit mass. Likewise, the thermal conductivity *k* is a measure of a material's ability to conduct heat. For example, $k = 0.607$ W/m·K for water and $k = 80.2$ W/m·K for iron at room temperature, which indicates that iron conducts heat more than 100 times faster than water can. Thus we say that water is a poor heat conductor relative to iron, although water is an excellent medium to store thermal energy.

Equation (2-16) for the rate of conduction heat transfer under steady conditions can also be viewed as the defining equation for thermal conductivity. Thus the *thermal conductivity* of a material can be defined as *the rate of heat transfer through a unit thickness of the material per unit area per unit temperature difference.* The thermal conductivity of a material is a measure of the ability of the material to conduct heat. A high value for thermal conductivity indicates that the material is a good heat conductor, and a low value indicates that the material is a poor heat conductor or *insulator*. The thermal conductivities of some common materials at room temperature are given in Table 2-2. The thermal conductivity of pure copper at room temperature is $k = 401$ W/m·K, which indicates that a 1-m-thick copper wall will conduct heat at a rate of 401 W per m² area per K temperature difference across the wall. Note that materials such as copper and silver that are good electric conductors are also good heat conductors, and they have high values of thermal conductivity. Materials such as rubber, wood, and styrofoam are poor conductors of heat and have low conductivity values.

A layer of material of known thickness and area can be heated from one side by an electric resistance heater of known output. If the outer surfaces of the heater are well insulated, all the heat generated by the resistance heater will be transferred through the material whose conductivity is to be determined. Then measuring the two surface temperatures of the material when steady heat transfer is reached and substituting them into Eq. (2-16) together with other known quantities give the thermal conductivity (Fig. 2-14).

The thermal conductivities of materials vary over a wide range, as shown in Fig. 2-15. The thermal conductivities of gases such as air vary by a factor of 10^4 from those of pure metals such as copper. Note that pure crystals and metals have the highest thermal conductivities, and gases and insulating materials the lowest.

Temperature is a measure of the kinetic energies of the particles such as the molecules or atoms of a substance. In a liquid or gas, the kinetic energy of the molecules is due to their random translational motion as well as their vibrational and rotational motions. When two molecules possessing different kinetic energies collide, part of the kinetic energy of the more energetic (higher-temperature) molecule is transferred to the less energetic

TABLE 2-2   Thermal Conductivities of Some Materials at Room Temperature

Material	Thermal Conductivity k, W/m·K*
Diamond	2300
Silver	429
Copper	401
Gold	317
Aluminum	237
Iron	80.2
Mercury (*l*)	8.54
Glass	0.78
Brick	0.72
Water (*l*)	0.607
Human skin	0.37
Wood (oak)	0.17
Helium (*g*)	0.152
Soft rubber	0.13
Glass fiber	0.043
Air (*g*)	0.026
Urethane, rigid foam	0.026

*Multiply by 0.5778 to convert to Btu/h·ft·°F.

(lower-temperature) molecule, much the same as when two elastic balls of the same mass at different velocities collide, part of the kinetic energy of the faster ball is transferred to the slower one. The higher the temperature, the faster the molecules move and the higher the number of such collisions, and the better the heat transfer.

The *kinetic theory* of gases predicts and the experiments confirm that the thermal conductivity of gases is proportional to the *square root of the thermodynamic temperature T,*

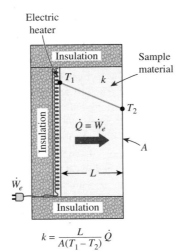

$$k = \frac{L}{A(T_1 - T_2)} \dot{Q}$$

Figure 2-14 A simple experimental setup to determine the thermal conductivity of a material.

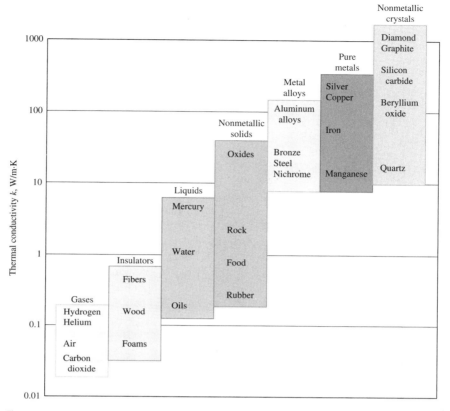

Figure 2-15   The range of thermal conductivity of various materials at room temperature.

and inversely proportional to the *square root of the molar mass M*. Therefore, for a particular gas (fixed *M*), the thermal conductivity increases with increasing temperature, and at a fixed temperature the thermal conductivity decreases with increasing *M*. For example, at a fixed temperature of 1000 K, the thermal conductivity of helium (*M* = 4) is 0.343 W/m·K and that of air (*M* = 29) is 0.0667 W/m·K, which is much lower than that of helium.

The mechanism of heat conduction in a *liquid* is complicated by the fact that the molecules are more closely spaced, and they exert a stronger intermolecular force field. The thermal conductivities of liquids usually lie between those of solids and gases. The thermal conductivity of a substance is normally highest in the solid phase and lowest in the gas phase. The thermal conductivity of liquids is generally insensitive to pressure except near the thermodynamic critical point. Unlike gases, the thermal conductivities of most liquids decrease with increasing temperature, with water being a notable exception. Like gases, the conductivity of liquids decreases with increasing molar mass. Liquid metals such as mercury and sodium have high thermal conductivities and are very suitable for use in applications where a high heat transfer rate to a liquid is desired, as in nuclear power plants.

In *solids*, heat conduction is due to two effects: the *lattice vibrational waves* induced by the vibrational motions of the molecules positioned at relatively fixed positions in a periodic manner called a lattice, and the energy transported via the *free flow of electrons* in the solid. The thermal conductivity of a solid is obtained by adding the lattice and electronic components. The relatively high thermal conductivities of pure metals are primarily due to the electronic component. The lattice component of thermal conductivity strongly depends on the way the molecules are arranged. For example, diamond, which is a highly ordered crystalline solid, has the highest known thermal conductivity at room temperature.

Unlike metals, which are good electrical and heat conductors, *crystalline solids* such as diamond and semiconductors such as silicon are good heat conductors but poor electrical conductors. As a result, such materials find widespread use in the electronics industry. Despite their higher price, diamond heat sinks are used in the cooling of sensitive electronic components because of the excellent thermal conductivity of diamond. Silicon oils and gaskets are commonly used in the packaging of electronic components because they provide both good thermal contact and good electrical insulation.

Pure metals have high thermal conductivities, and one would think that *metal alloys* should also have high conductivities. One would expect an alloy made of two metals of thermal conductivities k_1 and k_2 to have a conductivity k between k_1 and k_2. But this turns out not to be the case. Even small amounts in a pure metal of "foreign" molecules that are good conductors themselves seriously disrupt the transfer of heat in that metal. For example, the thermal conductivity of steel containing just 1 percent chrome is 62 W/m·K, while the thermal conductivities of iron and chromium are 83 and 95 W/m·K, respectively.

The thermal conductivities of materials vary with temperature. The variation of thermal conductivity over certain temperature ranges is negligible for some materials but significant for others, as shown in Fig. 2-16. The thermal conductivities of certain solids

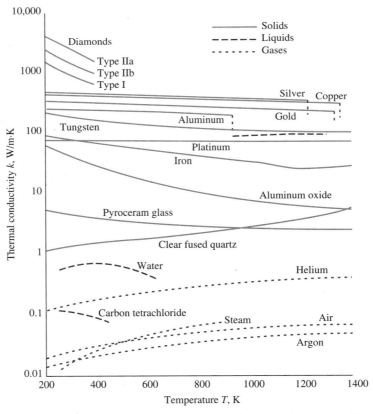

Figure 2-16 The variation of the thermal conductivity of various solids, liquids, and gases with temperature.

exhibit dramatic increases at temperatures near absolute zero, when these solids become *superconductors*. For example, the conductivity of copper reaches a maximum value of about 20,000 W/m·K at 20 K, which is about 50 times the conductivity at room temperature.

The temperature dependence of thermal conductivity causes considerable complexity in conduction analysis. Therefore, it is common practice to evaluate the thermal conductivity k at the *average temperature* and treat it as a *constant* in calculations. In heat transfer analysis, a material is normally assumed to be *isotropic;* that is, to have uniform properties in all directions. This assumption is realistic for most materials, except those that exhibit different structural characteristics in different directions, such as laminated composite materials and wood. The thermal conductivity of wood across the grain, for example, is different than that parallel to the grain.

EXAMPLE 2-2
Heat Loss Through a Wall

The east wall of an electrically heated home is 15 ft long, 8 ft high, and 1 ft thick and is made of brick whose thermal conductivity is $k = 0.42$ Btu/h·ft·°F. On a certain winter night, the temperatures of the inner and the outer surfaces of the wall are measured to be at about 65°F and 33°F, respectively, for a period of 10 h (Fig. 2-17). Determine (*a*) the rate of heat loss through the wall that night and (*b*) the cost of that heat loss to the homeowner if the cost of electricity is $0.12/kWh.

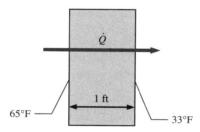

Figure 2-17   Schematic for Example 2-2.

SOLUTION   (*a*) Noting that the heat transfer through the wall is by conduction and the surface area of the wall is $A = 15$ ft × 8 ft = 120 ft², the steady rate of heat transfer through the wall can be determined from

$$\dot{Q} = kA\frac{T_1 - T_2}{L} = (0.42 \text{ Btu/h·ft·°F})(120 \text{ ft}^2)\frac{(65 - 33)°F}{1 \text{ ft}} = \textbf{1613 Btu/h}$$

or 0.473 kW since 1 kW = 3412 Btu/h.
(*b*) The amount of heat lost during a 10-h period and its cost are

$$Q = \dot{Q}\Delta t = (0.473 \text{ kW})(10 \text{ h}) = 4.73 \text{ kWh}$$

$$\text{Cost} = \text{Amount of energy} \times \text{Unit cost of energy} = (4.73 \text{ kWh})(\$0.12/\text{kWh}) = \textbf{\$0.57}$$

Therefore, the cost of the heat loss through the wall to the homeowner that night is $0.57.   ▲

Convection Heat Transfer

Convection is the mode of energy transfer between a solid surface and the adjacent liquid or gas that is in motion, and it involves the combined effects of *conduction* and *fluid motion*. The faster the fluid motion, the greater the convection heat transfer. In the absence of any bulk fluid motion, heat transfer between a solid surface and the adjacent

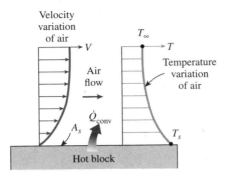

Figure 2-18   Heat transfer from a hot surface to air by convection.

fluid is by pure conduction. The presence of bulk motion of the fluid enhances the heat transfer between the solid surface and the fluid, but it also complicates the determination of heat transfer rates.

Consider the cooling of a hot block by blowing cool air over its top surface (Fig. 2-18). Heat is first transferred to the air layer adjacent to the block by conduction. This heat is then carried away from the surface by convection, that is, by the combined effects of conduction within the air that is due to random motion of air molecules and the bulk or macroscopic motion of the air that removes the heated air near the surface and replaces it with the cooler air.

Convection is called *forced convection* if the fluid is forced to flow over the surface by external means such as a fan, pump, or the wind. In contrast, convection is called *natural* (or *free*) *convection* if the fluid motion is caused by buoyancy forces that are induced by density differences due to the variation of temperature in the fluid (Fig. 2-19). For example, in the absence of a fan, heat transfer from the surface of the hot block in Fig. 2-18 is by natural convection since any motion in the air in this case is due to the rise of the warmer (and thus lighter) air near the surface and the fall of the cooler (and thus heavier) air to fill its place. Heat transfer between the block and the surrounding air is by conduction if the temperature difference between the air and the block is not large enough to overcome the resistance of air to movement and thus to initiate natural convection currents.

Heat transfer processes that involve *change of phase* of a fluid are also considered to be convection because of the fluid motion induced during the process, such as the rise of the vapor bubbles during boiling or the fall of the liquid droplets during condensation.

Despite the complexity of convection, the rate of *convection heat transfer* is observed to be proportional to the temperature difference, and is conveniently expressed by *Newton's law of cooling* as

$$\dot{Q}_{conv} = hA_s(T_s - T_\infty) \qquad \text{(kW)} \qquad (2\text{-}18)$$

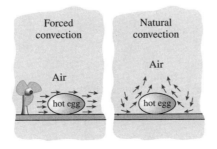

Figure 2-19   The cooling of a boiled egg by forced and natural convection.

TABLE 2-3 Typical Values of Convection Heat Transfer Coefficient

Type of Convection	Heat Transfer Coefficient h, W/m²·K*
Free convection of gases	2–25
Free convection of liquids	10–1000
Forced convection of gases	25–250
Forced convection of liquids	50–20,000
Boiling and condensation	2500–100,000

*Multiply by 0.176 to convert to Btu/h·ft²·°F.

where h is the *convection heat transfer coefficient* in W/m²·K or Btu/h·ft²·°F, A_s is the surface area through which convection heat transfer takes place, T_s is the surface temperature, and T_∞ is the temperature of the fluid sufficiently far from the surface. Note that at the surface, the fluid temperature equals the surface temperature of the solid.

The convection heat transfer coefficient h is not a property of the fluid. It is an experimentally determined parameter whose value depends on all the variables influencing convection such as the surface geometry, the nature of fluid motion, the properties of the fluid, and the bulk fluid velocity. Typical values of h are given in Table 2-3.

Radiation Heat Transfer

Radiation is the energy emitted by matter in the form of electromagnetic waves (or photons) as a result of the changes in the electronic configurations of the atoms or molecules. Unlike conduction and convection, the transfer of heat by radiation does not require the presence of an intervening medium. In fact, heat transfer by radiation is fastest (at the speed of light) and it suffers no attenuation in a vacuum. This is how the energy of the sun reaches the earth.

In heat transfer studies, we are interested in thermal radiation, which is the form of radiation emitted by bodies because of their temperature. It differs from other forms of electromagnetic radiation such as x-rays, gamma rays, microwaves, radio waves, and television waves that are not related to temperature. All bodies at a temperature above absolute zero emit thermal radiation.

Radiation is a volumetric phenomenon, and all solids, liquids, and gases emit, absorb, or transmit radiation to varying degrees. However, radiation is usually considered to be a *surface phenomenon* for solids that are opaque to thermal radiation such as metals, wood, and rocks since the radiation emitted by the interior regions of such material can never reach the surface, and the radiation incident on such bodies is usually absorbed within a few micrometers from the surface.

The maximum rate of radiation that can be emitted from a surface at a thermodynamic temperature T_s (in K or R) is given by the *Stefan-Boltzmann law* as

$$\dot{Q}_{\text{emit,max}} = \sigma A_s T_s^4 \qquad \text{(kW)} \qquad (2\text{-}19)$$

where $\sigma = 5.670 \times 10^{-8}$ W/m²·K⁴ or 0.1714×10^{-8} Btu/h·ft²·R⁴ is the *Stefan-Boltzman constant*. The idealized surface that emits radiation at this maximum rate is called a *blackbody*, and the radiation emitted by a blackbody is called *blackbody radiation* (Fig. 2-20). The radiation emitted by all real surfaces is less than the radiation emitted by a blackbody at the same temperature, and is expressed as

$$\dot{Q}_{\text{emit}} = \varepsilon \sigma A_s T_s^4 \qquad \text{(kW)} \qquad (2\text{-}20)$$

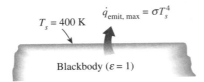

$T_s = 400\ \text{K}$

$\dot{q}_{\text{emit, max}} = \sigma T_s^4$

Blackbody ($\varepsilon = 1$)

Figure 2-20 Blackbody radiation represents the maximum amount of radiation that can be emitted from a surface at a specified temperature.

where ε is the *emissivity* of the surface. The property emissivity, whose value is in the range $0 < \varepsilon < 1$, is a measure of how closely a surface approximates a blackbody for which $\varepsilon = 1$. The emissivities of some surfaces are given in Table 2-4.

Another important radiation property of a surface is its *absorptivity* α, which is the fraction of the radiation energy incident on a surface that is absorbed by the surface. Like emissivity, its value is in the range $0 < \alpha < 1$. A blackbody absorbs all radiation incident on it. That is, a blackbody is a perfect absorber ($\alpha = 1$) just as it is a perfect emitter.

In general, both ε and α of a surface depend on the temperature and the wavelength of the radiation. *Kirchhoff's law* of radiation states that the emissivity and the absorptivity of a surface at a given temperature and wavelength are equal. In many practical applications, the surface temperature and the temperature of the source of incident radiation are of the same order of magnitude, and the average absorptivity of a surface is taken to be equal to its average emissivity. The rate at which a surface absorbs radiation is determined from (Fig. 2-21)

$$\dot{Q}_{\text{absorbed}} = \alpha \dot{Q}_{\text{incident}} \qquad (\text{kW}) \qquad (2\text{-}21)$$

TABLE 2-4   Emissivities of Some Materials at 300 K

Material	Emissivity ε
Aluminum foil	0.07
Anodized aluminum	0.82
Polished copper	0.03
Polished gold	0.03
Polished silver	0.02
Polished stainless steel	0.17
Black paint	0.98
White paint	0.90
White paper	0.92–0.97
Asphalt pavement	0.85–0.93
Red brick	0.93–0.96
Human skin	0.95
Wood	0.82–0.92
Soil	0.93–0.96
Water	0.96
Vegetation	0.92–0.96

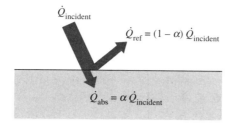

Figure 2-21   The absorption of radiation incident on an opaque surface of absorptivity α.

where $\dot{Q}_{\text{incident}}$ is the rate at which radiation is incident on the surface and α is the absorptivity of the surface. For opaque (nontransparent) surfaces, the portion of incident radiation not absorbed by the surface is reflected back.

The difference between the rates of radiation emitted by the surface and the radiation absorbed is the *net* radiation heat transfer. If the rate of radiation absorption is greater than the rate of radiation emission, the surface is said to be *gaining* energy by radiation. Otherwise, the surface is said to be *losing* energy by radiation. In general, the determination of the net rate of heat transfer by radiation between two surfaces is a complicated matter since it depends on the properties of the surfaces, their orientation relative to each other, and the interaction of the medium between the surfaces with radiation.

When a surface of emissivity ε and surface area A_s at a *thermodynamic temperature* T_s are *completely enclosed* by a much larger (or black) surface at thermodynamic temperature T_{surr} separated by a gas (such as air) that does not intervene with radiation, the net rate of radiation heat transfer between these two surfaces is given by (Fig. 2-22)

$$\dot{Q}_{\text{rad}} = \varepsilon \sigma A_s (T_s^4 - T_{\text{surr}}^4) \qquad \text{(kW)} \qquad (2\text{-}22)$$

In this special case, the emissivity and the surface area of the surrounding surface do not have any effect on the net radiation heat transfer.

Radiation heat transfer to or from a surface surrounded by a gas such as air occurs *parallel* to conduction (or convection, if there is bulk gas motion) between the surface and the gas. Thus the total heat transfer is determined by *adding* the contributions of both heat transfer mechanisms. For simplicity and convenience, this is often done by defining a combined heat transfer coefficient h_{combined} that includes the effects of both convection

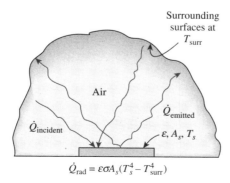

Figure 2-22   Radiation heat transfer between a surface and the surfaces surrounding it.

and radiation. Then the *total* heat transfer rate to or from a surface by convection and radiation is expressed as

$$\dot{Q}_{total} = \dot{Q}_{conv} + \dot{Q}_{rad} = h_{conv} A_s (T_s - T_\infty) + \varepsilon\sigma A_s (T_s^4 - T_{surr}^4)$$
$$\dot{Q}_{total} = h_{combined} A_s (T_s - T_\infty) \tag{2-23}$$
$$h_{combined} = h_{conv} + h_{rad} = h_{conv} + \varepsilon\sigma(T_s + T_{surr})(T_s^2 + T_{surr}^2)$$

Note that the combined heat transfer coefficient is essentially a convection heat transfer coefficient modified to include the effects of radiation.

Radiation is usually significant relative to conduction or natural convection, but negligible relative to forced convection. Thus radiation in forced convection applications is usually disregarded, especially when the surfaces involved have low emissivities and low to moderate temperatures.

EXAMPLE 2-3
Heat Transfer Between
Two Plates

Consider steady heat transfer between two large parallel plates at constant temperatures of $T_1 = 320$ K and $T_2 = 276$ K that are $L = 3$ cm apart (Fig. 2-23). Assuming the surfaces to be black (emissivity $\varepsilon = 1$), determine the rate of heat transfer between the plates per unit surface area assuming the gap between the plates is (*a*) filled with atmospheric air ($k_{air} = 0.02551$ W/m·K), (*b*) evacuated, (*c*) filled with fiberglass insulation ($k_{ins} = 0.036$ W/m·K), and (*d*) filled with superinsulation having an apparent thermal conductivity of 0.00015 W/m·K.

SOLUTION   (*a*) Disregarding any natural convection currents, the rates of conduction and radiation heat transfer are

$$\dot{Q}_{cond} = kA\frac{T_1 - T_2}{L} = (0.02551 \text{ W/m}^2\cdot\text{K})(1 \text{ m}^2)\frac{(320 - 276) \text{ K}}{0.03 \text{ m}} = 37.4 \text{ W}$$

$$\dot{Q}_{rad} = \varepsilon\sigma A_s (T_1^4 - T_2^4)$$

$$= 1(5.67\times10^{-8} \text{ W/m}^2\cdot\text{K}^4)(1 \text{ m}^2)[(320 \text{ K})^4 - (276 \text{ K})^4]$$

$$= 265.5 \text{ W}$$

$$\dot{Q}_{total} = \dot{Q}_{cond} + \dot{Q}_{rad} = 37.4 + 265.5 = \textbf{303 W}$$

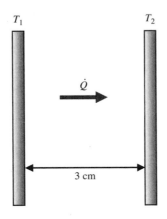

Figure 2-23   Schematic for Example 2-3.

(*b*) When the air space between the plates is evacuated, there will be radiation heat transfer only. Therefore,

$$\dot{Q}_{\text{total}} = \dot{Q}_{\text{rad}} = \textbf{266 W}$$

(*c*) In this case, there will be conduction heat transfer through the fiberglass insulation only,

$$\dot{Q}_{\text{total}} = \dot{Q}_{\text{cond}} = kA\frac{T_1 - T_2}{L} = (0.036 \text{ W/m·K})(1 \text{ m}^2)\frac{(320 - 276) \text{ K}}{0.03 \text{ m}} = \textbf{52.8 W}$$

(*d*) In the case of superinsulation, the rate of heat transfer will be

$$\dot{Q}_{\text{total}} = \dot{Q}_{\text{cond}} = kA\frac{T_1 - T_2}{L} = (0.00015 \text{ W/m·K})(1 \text{ m}^2)\frac{(320 - 276) \text{ K}}{0.03 \text{ m}} = \textbf{0.22 W}$$

Note that superinsulators are very effective in reducing heat transfer between the plates. ▲

2-4 FLUID MECHANICS

Mechanics is the oldest physical science that deals with both stationary and moving bodies under the influence of forces. The branch of mechanics that deals with bodies at rest is called statics, while the branch that deals with bodies in motion is called dynamics. The subcategory fluid mechanics is defined as the science that deals with the behavior of fluids at rest (fluid statics) or in motion (fluid dynamics), and the interaction of fluids with solids or other fluids at the boundaries.

Fluid mechanics itself is also divided into several categories. The study of the motion of fluids that can be approximated as incompressible (such as liquids, especially water, and gases at low speeds) is usually referred to as hydrodynamics. A subcategory of hydrodynamics is hydraulics, which deals with liquid flows in pipes and open channels. Gas dynamics deals with the flow of fluids that undergo significant density changes, such as the flow of gases through nozzles at high speeds. The category aerodynamics deals with the flow of gases (especially air) over bodies such as aircraft, rockets, and automobiles at high or low speeds. Some other specialized categories such as meteorology, oceanography, and hydrology deal with naturally occurring flows.

You will recall from physics that a substance exists in three primary phases: solid, liquid, and gas. (At very high temperatures, it also exists as plasma.) A substance in the liquid or gas phase is referred to as a *fluid*. Distinction between a solid and a fluid is made on the basis of the substance's ability to resist an applied shear (or tangential) stress that tends to change its shape. A solid can resist an applied shear stress by deforming, whereas a *fluid deforms continuously under the influence of a shear stress*, no matter how small. In solids, stress is proportional to *strain*, but in fluids, stress is proportional to *strain rate*. When a constant shear force is applied, a solid eventually stops deforming at some fixed strain angle, whereas a fluid never stops deforming and approaches a constant *rate* of strain.

In a liquid, groups of molecules can move relative to each other, but the volume remains relatively constant because of the strong cohesive forces between the molecules. As a result, a liquid takes the shape of the container it is in, and it forms a free surface in a larger container in a gravitational field. A gas, on the other hand, expands until it encounters the walls of the container and fills the entire available space. This is because the gas molecules are widely spaced, and the cohesive forces between them are very small. Unlike liquids, a gas in an open container cannot form a free surface (Fig. 2-24).

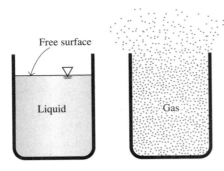

Figure 2-24   Unlike a liquid, a gas does not form a free surface, and it expands to fill the entire available space.

Viscosity

When two solid bodies in contact move relative to each other, a friction force develops at the contact surface in the direction opposite to motion. To move a table on the floor, for example, we have to apply a force to the table in the horizontal direction large enough to overcome the friction force. The magnitude of the force needed to move the table depends on the *friction coefficient* between the table legs and the floor.

The situation is similar when a fluid moves relative to a solid or when two fluids move relative to each other. We move with relative ease in air, but not so in water. Moving in oil would be even more difficult, as can be observed by the slower downward motion of a glass ball dropped in a tube filled with oil. It appears that there is a property that represents the internal resistance of a fluid to motion or the "fluidity," and that property is the *viscosity*. The force a flowing fluid exerts on a body in the flow direction is called the *drag force*, and the magnitude of this force depends, in part, on viscosity (Fig. 2-25).

Fluids for which the rate of deformation is linearly proportional to the shear stress are called *Newtonian fluids* after Sir Isaac Newton, who expressed it first in 1687. Most common fluids such as water, air, gasoline, and oils are Newtonian fluids. Blood and liquid plastics are examples of non-Newtonian fluids.

In one-dimensional shear flow of Newtonian fluids, shear stress can be expressed by the linear relationship

$$\tau = \mu \frac{du}{dy} \qquad (\text{N/m}^2) \qquad (2\text{-}24)$$

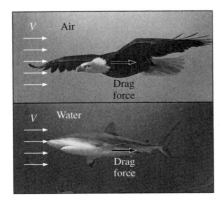

Figure 2-25   A fluid moving relative to a body exerts a drag force on the body, partly because of friction caused by viscosity. (© *Getty RF.*)

where the constant of proportionality μ is called the *coefficient of viscosity* or the *dynamic* (or *absolute*) *viscosity* of the fluid, whose unit is kg/m·s, or equivalently, N·s/m² (or Pa·s where Pa is the pressure unit pascal). A common viscosity unit is *poise*, which is equivalent to 0.1 Pa·s (or *centipoise*, which is one-hundredth of a poise). The viscosity of water at 20°C is 1.002 centipoise, and thus the unit centipoise serves as a useful reference.

In fluid mechanics and heat transfer, the ratio of dynamic viscosity to density appears frequently. For convenience, this ratio is given the name kinematic viscosity v and is expressed as $v = \mu/\rho$. Two common units of kinematic viscosity are m²/s and *stoke* (1 stoke = 1 cm²/s = 0.0001 m²/s).

In general, the viscosity of a fluid depends on both temperature and pressure, although the dependence on pressure is rather weak. For *liquids*, both the dynamic and kinematic viscosities are practically independent of pressure, and any small variation with pressure is usually disregarded, except at extremely high pressures. For *gases*, this is also the case for dynamic viscosity (at low to moderate pressures), but not for kinematic viscosity since the density of a gas is proportional to its pressure.

The viscosity of a fluid is a measure of its "resistance to deformation." Viscosity is due to the internal frictional force that develops between different layers of fluids as they are forced to move relative to each other.

The viscosity of a fluid is directly related to the pumping power needed to transport a fluid in a pipe or to move a body (such as a car in air or a submarine in the sea) through a fluid. Viscosity is caused by the cohesive forces between the molecules in liquids and by the molecular collisions in gases, and it varies greatly with temperature. The viscosity of liquids decreases with temperature, whereas the viscosity of gases increases with temperature (Fig. 2-26). This is because in a liquid the molecules possess more energy at higher temperatures, and they can oppose the large cohesive intermolecular forces more strongly. As a result, the energized liquid molecules can move more freely.

In a gas, on the other hand, the intermolecular forces are negligible, and the gas molecules at high temperatures move randomly at higher velocities. This results in more molecular collisions per unit volume per unit time and therefore in greater resistance to flow. The kinetic theory of gases predicts the viscosity of gases to be proportional to the square root of temperature.

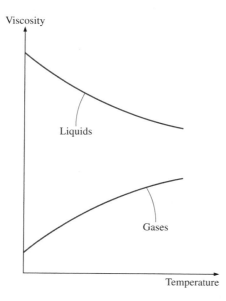

Figure 2-26 The viscosity of liquids decreases and the viscosity of gases increases with temperature.

TABLE 2-5 **Dynamic Viscosity of Some Fluids at 1 atm and 20°C** (Unless Otherwise Stated)

Fluid	Dynamic Viscosity μ, kg/m·s
Glycerin:	
−20°C	134.0
0°C	10.5
20°C	1.52
40°C	0.31
Engine oil:	
SAE 10 W	0.10
SAE 10 W30	0.17
SAE 30	0.29
SAE 50	0.86
Mercury	0.0015
Ethyl alcohol	0.0012
Water:	
0°C	0.0018
20°C	0.0010
100°C (liquid)	0.00028
100°C (vapor)	0.000012
Blood, 37°C	0.00040
Gasoline	0.00029
Ammonia	0.00015
Air	0.000018
Hydrogen, 0°C	0.0000088

The viscosities of some fluids at room temperature are listed in Table 2-5. They are plotted against temperature in Fig. 2-27. Note that the viscosities of different fluids differ by several orders of magnitude. Also note that it is more difficult to move an object in a higher-viscosity fluid such as engine oil than it is in a lower-viscosity fluid such as water. Liquids, in general, are much more viscous than gases.

Pressure Drop in Fluid Flow in Pipes

The *pressure loss* and *head loss* for all types of internal flows (laminar or turbulent, in circular or noncircular pipes, smooth or rough surfaces) are expressed as (Fig. 2-28)

$$\Delta P_L = f \frac{L}{D} \frac{\rho V^2}{2} \tag{2-25}$$

$$h_L = \frac{\Delta P_L}{\rho g} = f \frac{L}{D} \frac{V^2}{2g} \tag{2-26}$$

where ρ is the density, V is average velocity of fluid, L is the pipe length, g is gravitational acceleration, $\rho V^2/2$ is the *dynamic pressure*, and the dimensionless quantity f is the *friction factor*.

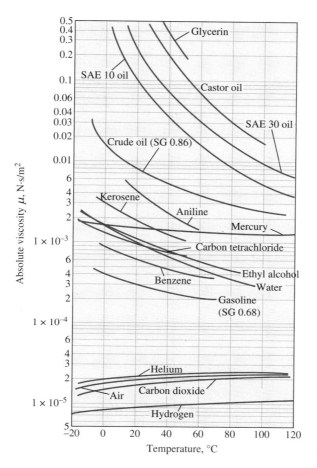

Figure 2-27   The variation of dynamic (absolute) viscosity of common fluids with temperature at 1 atm (1 N·s/m² = 1 kg/m·s = 0.020886 lbf·s/ft²) (White, 2011).

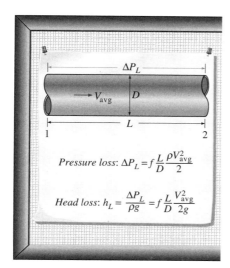

Pressure loss: $\Delta P_L = f \dfrac{L}{D} \dfrac{\rho V_{avg}^2}{2}$

Head loss: $h_L = \dfrac{\Delta P_L}{\rho g} = f \dfrac{L}{D} \dfrac{V_{avg}^2}{2g}$

Figure 2-28   The relation for pressure loss (and head loss) is one of the most general relations in fluid mechanics, and it is valid for laminar or turbulent flows, circular or noncircular pipes, and pipes with smooth or rough surfaces.

For fully developed laminar flow in a round pipe, the friction factor is $f = 64/\text{Re}$, where Re is Reynolds number $\text{Re} = VD/\nu$. Here, ν is the kinematic viscosity. Note that the flow is laminar for Reynolds numbers smaller than 2300 and turbulent for Reynolds numbers greater than 4000.

The head loss represents *the additional height that the fluid needs to be raised by a pump in order to overcome the frictional losses in the pipe*. The head loss is caused by viscosity, and it is directly related to the wall shear stress.

The *pressure drop* and the *volume flow rate* for laminar flow in a horizontal pipe are

$$\Delta P = \frac{32 \mu L V_{avg}}{D^2} \tag{2-27}$$

$$\dot{V} = V_{avg} A_c = \frac{\Delta P \pi D^4}{128 \mu L} \tag{2-28}$$

This equation is known as *Poiseuille's law*, and this flow is called *Hagen-Poiseuille flow* in honor of the works of G. Hagen (1797–1884) and J. Poiseuille (1799–1869) on the subject. Note that *for a specified flow rate, the pressure drop and thus the required pumping power is proportional to the length of the pipe and the viscosity of the fluid, but it is inversely proportional to the fourth power of the radius (or diameter) of the pipe.* Therefore, the pumping power requirement for a laminar-flow piping system can be reduced by a factor of 16 by doubling the pipe diameter (Fig. 2-29). Of course, the benefits of the reduction in the energy costs must be weighed against the increased cost of construction due to using a larger-diameter pipe.

For noncircular pipes, the diameter in the previous relations is replaced by the *hydraulic diameter* defined as $D_h = 4A_c/p$, where A_c is the cross-sectional area of the pipe and p is its wetted perimeter.

In fully developed turbulent flow, the friction factor depends on the Reynolds number and the *relative roughness ε/D*. The friction factor in turbulent flow is given by the *Colebrook equation*, expressed as

$$\frac{1}{\sqrt{f}} = -2.0 \log\left(\frac{\varepsilon/D}{3.7} + \frac{2.51}{\text{Re}\sqrt{f}}\right) \tag{2-29}$$

The plot of this formula is known as the *Moody chart*.

We routinely use the Colebrook equation to calculate the friction factor f for fully developed turbulent pipe flow. Indeed, the Moody chart is *created* using the Colebrook equation. However, in addition to being implicit, the Colebrook equation is valid only for *turbulent* pipe flow (when the flow is laminar, $f = 64/\text{Re}$). Thus we need to verify that the

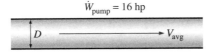

$\dot{W}_{pump} = 16$ hp

D ——————→ V_{avg}

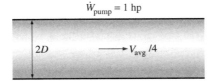

$\dot{W}_{pump} = 1$ hp

$2D$ ——→ $V_{avg}/4$

Figure 2-29   The pumping power requirement for a laminar-flow piping system can be reduced by a factor of 16 by doubling the pipe diameter.

Reynolds number is in the turbulent range. An equation was generated by Churchill (1977) that is not only explicit but is also useful for *any* Re and *any* roughness, even for laminar flow, and even in the fuzzy transitional region between laminar and turbulent flow. The *Churchill equation* is

$$f = 8\left[\left(\frac{8}{\text{Re}}\right)^{12} + (A+B)^{-1.5}\right]^{\frac{1}{12}}$$

(2-30)

where

$$A = \left\{-2.457 \cdot \ln\left[\left(\frac{7}{\text{Re}}\right)^{0.9} + 0.27\frac{\varepsilon}{D}\right]\right\}^{16}$$

The difference between the Colebrook and Churchill equations is less than 1 percent. Because it is explicit and valid over the entire range of Reynolds numbers and roughnesses, it is recommended that the Churchill equation be used to determine friction factor *f*.

Commercially available pipes differ from those used in the experiments in that the roughness of pipes in the market is not uniform, and it is difficult to give a precise description of it. Equivalent roughness values for some commercial pipes are given in Table 2-6 as well as on the Moody chart. But it should be kept in mind that these values are for new pipes, and the relative roughness of pipes may increase with use as a result of corrosion, scale buildup, and precipitation. As a result, the friction factor may increase by a multiple of 5 to 10. Actual operating conditions must be considered in the design of piping systems.

Once the pressure loss (or head loss) is known, the required pumping power *to overcome the pressure loss* is determined from

$$\dot{W}_{\text{pump},L} = \dot{V}\Delta P_L = \dot{V}\rho g h_L = \dot{m}g h_L$$

(2-31)

where $\dot{V}$ is the volume flow rate and $\dot{m}$ is the mass flow rate.

TABLE 2-6 Equivalent Roughness Values for New Commercial Pipes[*]

Material	Roughness, ε	
	ft	mm
Glass, plastic	0 (smooth)	
Concrete	0.003–0.03	0.9–9
Wood stave	0.0016	0.5
Rubber, smoothed	0.000033	0.01
Copper or brass tubing	0.000005	0.0015
Cast iron	0.00085	0.26
Galvanized iron	0.0005	0.15
Wrought iron	0.00015	0.046
Stainless steel	0.000007	0.002
Commercial steel	0.00015	0.045

[*]The uncertainty in these values can be as much as ±60 percent.

EXAMPLE 2-4
Pressure Drop and
Pumping Power
Requirement for Water
Flow in a Pipe

Water at 15°C (ρ = 999.1 kg/m³ and μ = 1.138 × 10⁻³ kg/m·s) is flowing steadily in a 30-m-long and 5-cm-internal-diameter horizontal pipe made of stainless steel at a rate of 9 L/s (Fig. 2-30). Determine the pressure drop, the head loss, and the pumping power requirement to overcome this pressure drop.

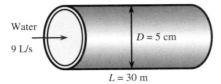

Water

9 L/s

$D = 5$ cm

$L = 30$ m

Figure 2-30   Schematic for Example 2-4.

SOLUTION   The density and dynamic viscosity of water are given to be ρ = 999.1 kg/m³ and μ = 1.138 × 10⁻³ kg/m·s, respectively. The roughness of stainless steel is 0.002 mm (Table 2-6). First we calculate the average velocity and the Reynolds number to determine the flow regime:

$$V = \frac{\dot{V}}{A_c} = \frac{\dot{V}}{\pi D^2/4} = \frac{0.009 \text{ m}^3/\text{s}}{\pi(0.05 \text{ m})^2/4} = 4.584 \text{ m/s}$$

$$\text{Re} = \frac{\rho V D}{\mu} = \frac{(999.1 \text{ kg/m}^3)(4.584 \text{ m/s})(0.05 \text{ m})}{1.138 \times 10^{-3} \text{ kg/m} \cdot \text{s}} = 2.012 \times 10^5$$

which is greater than 4000. Therefore, the flow is turbulent. The relative roughness of the pipe is

$$\varepsilon/D = \frac{2 \times 10^{-6} \text{ m}}{0.05 \text{ m}} = 4 \times 10^{-5}$$

The friction factor can be determined from the Moody chart, but to avoid a reading error, we determine it from the Colebrook equation using an equation solver (or an iterative scheme),

$$\frac{1}{\sqrt{f}} = -2.0\log\left(\frac{\varepsilon/D}{3.7} + \frac{2.51}{\text{Re}\sqrt{f}}\right) \quad \rightarrow \quad \frac{1}{\sqrt{f}} = -2.0\log\left(\frac{4 \times 10^{-5}}{3.7} + \frac{2.51}{2.012 \times 10^5 \sqrt{f}}\right)$$

It gives f = 0.01594. Then the pressure drop, head loss, and the required power input become

$$\Delta P = \Delta P_L = f\frac{L}{D}\frac{\rho V^2}{2}$$

$$= 0.01594\frac{30 \text{ m}}{0.05 \text{ m}}\frac{(999.1 \text{ kg/m}^3)(4.584 \text{ m/s})^2}{2}\left(\frac{1 \text{ kN}}{1000 \text{ kg·m/s}}\right)\left(\frac{1 \text{ kPa}}{1 \text{ kN/m}^2}\right)$$

$$= 100.4 \text{ kPa} \cong \textbf{100 kPa}$$

$$h_L = \frac{\Delta P_L}{\rho g} = f\frac{L}{D}\frac{V^2}{2g} = 0.01594\frac{30 \text{ m}}{0.05 \text{ m}}\frac{(4.584 \text{ m/s})^2}{2(9.81 \text{ m/s}^2)} = \textbf{10.2 m}$$

$$\dot{W}_{\text{pump}} = \dot{V}\Delta P = (0.009 \text{ m}^3/\text{s})(100.4 \text{ kPa})\left(\frac{1 \text{ kW}}{1 \text{ kPa·m}^3/\text{s}}\right) = \textbf{0.904 kW}$$

Therefore, useful power input in the amount of 0.904 kW is needed to overcome the frictional losses in the pipe. The power input determined is the mechanical power that needs to be imparted to the fluid. The shaft power will be more than this due to pump inefficiency; the electrical power input will be even more due to motor inefficiency.   ▲

2-5 THERMOCHEMISTRY

In this section, we review basic concepts and principles of thermochemistry, which is the thermodynamic study of chemical reactions, particularly combustion reactions.

Fuels and Combustion

Any material that can be burned to release thermal energy is called a *fuel*. Most familiar fuels consist primarily of hydrogen and carbon. They are called *hydrocarbon fuels* and are denoted by the general formula C_nH_m. Hydrocarbon fuels exist in all phases, some examples being coal, gasoline, and natural gas.

The main constituent of coal is carbon. Coal also contains varying amounts of oxygen, hydrogen, nitrogen, sulfur, moisture, and ash. It is difficult to give an exact mass analysis for coal since its composition varies considerably from one geographical area to the next and even within the same geographical location. Most liquid hydrocarbon fuels are a mixture of numerous hydrocarbons and are obtained from crude oil by distillation. The most volatile hydrocarbons vaporize first, forming what we know as gasoline. The less volatile fuels obtained during distillation are kerosene, diesel fuel, and fuel oil. The composition of a particular fuel depends on the source of the crude oil as well as on the refinery.

Although liquid hydrocarbon fuels are mixtures of many different hydrocarbons, they are usually considered to be a single hydrocarbon for convenience. For example, gasoline is treated as *octane*, C_8H_{18}, and the diesel fuel as *dodecane*, $C_{12}H_{26}$. Another common liquid hydrocarbon fuel is *methyl alcohol*, CH_3OH, which is also called *methanol* and is used in some gasoline blends. The gaseous hydrocarbon fuel natural gas, which is a mixture of methane and smaller amounts of other gases, is often treated as *methane*, CH_4, for simplicity.

Natural gas is produced from gas wells or oil wells rich in natural gas. It is composed mainly of methane, but it also contains small amounts of ethane, propane, hydrogen, helium, carbon dioxide, nitrogen, hydrogen sulfate, and water vapor. On vehicles, it is stored either in the gas phase at pressures of 150 to 250 atm as CNG (compressed natural gas), or in the liquid phase at $-162°C$ as LNG (liquefied natural gas). Over a million vehicles in the world, mostly buses, run on natural gas. LPG (liquefied petroleum gas) is a by-product of natural gas processing or crude oil refining. It consists mainly of propane, and thus LPG is usually referred to as propane. However, it also contains varying amounts of butane, propylene, and butylenes. Propane is commonly used in fleet vehicles, taxis, school buses, and private cars. Ethanol is obtained from corn, grains, and organic waste. Methanol is produced mostly from natural gas, but it can also be obtained from coal and biomass. Both alcohols are commonly used as additives in oxygenated gasoline and reformulated fuels to reduce air pollution.

Vehicles are a major source of air pollutants such as nitric oxides, carbon monoxide, and hydrocarbons, as well as the greenhouse gas carbon dioxide, and thus there is a growing shift in the transportation industry from the traditional petroleum-based fuels such as gasoline and diesel fuel to the cleaner burning *alternative fuels* friendlier to the environment, such as natural gas, alcohols (ethanol and methanol), LPG, and hydrogen. The use of electric and hybrid cars is also on the rise. Note that the energy contents of alternative fuels per unit volume are lower than that of gasoline or diesel fuel, and thus the driving range of a vehicle on a full tank is lower when running on an alternative fuel.

A chemical reaction during which a fuel is oxidized and a large quantity of energy is released is called *combustion*. The oxidizer most often used in combustion processes is air, for obvious reasons—it is free and readily available. Pure oxygen O_2 is used as an oxidizer only in some specialized applications, such as cutting and welding, where air cannot be used. Therefore, a few words about the composition of air are in order. On a mole or a volume basis, dry air is composed of 20.9 percent oxygen, 78.1 percent nitrogen, 0.9 percent argon, and small amounts of carbon dioxide, helium, neon, and hydrogen. In the analysis of combustion processes, the argon in the air is treated as nitrogen, and the gases that exist in trace amounts are disregarded. Then dry air can be approximated as 21 percent oxygen and 79 percent nitrogen by mole numbers. Therefore, each mole of oxygen entering a combustion chamber is accompanied by 0.79/0.21 = 3.76 mol of nitrogen. That is,

$$1 \text{ kmol O}_2 + 3.76 \text{ kmol N}_2 = 4.76 \text{ kmol air} \tag{2-32}$$

During combustion, nitrogen behaves as an inert gas and does not react with other elements, other than forming a very small amount of nitric oxides. However, even then the presence of nitrogen greatly affects the outcome of a combustion process since nitrogen usually enters a combustion chamber in large quantities at low temperatures and exits at considerably higher temperatures, absorbing a large proportion of the chemical energy released during combustion. Throughout this chapter, nitrogen is assumed to remain perfectly inert. Keep in mind, however, that at very high temperatures, such as those encountered in internal combustion engines, a small fraction of nitrogen reacts with oxygen, forming hazardous gases such as nitric oxide.

Air that enters a combustion chamber normally contains some water vapor (or moisture), which also deserves consideration. For most combustion processes, the moisture in the air and the H_2O that forms during combustion can also be treated as an inert gas, like nitrogen. At very high temperatures, however, some water vapor dissociates into H_2 and O_2 as well as into H, O, and OH. When the combustion gases are cooled below the dew point temperature of the water vapor, some moisture condenses. It is important to be able to predict the dew point temperature since the water droplets often combine with the sulfur dioxide that may be present in the combustion gases, forming sulfuric acid, which is highly corrosive.

During a combustion process, the components that exist before the reaction are called *reactants* and the components that exist after the reaction are called *products*. Consider, for example, the combustion of 1 kmol of carbon with 1 kmol of pure oxygen, forming carbon dioxide,

$$C + O_2 \rightarrow CO_2 \tag{2-33}$$

Here C and O_2 are the reactants since they exist before combustion, and CO_2 is the product since it exists after combustion. Note that a reactant does not have to react chemically in the combustion chamber. For example, if carbon is burned with air instead of pure oxygen, both sides of the combustion equation will include N_2. That is, the N_2 will appear both as a reactant and as a product.

We should also mention that bringing a fuel into intimate contact with oxygen is not sufficient to start a combustion process. (Thank goodness it is not. Otherwise, the whole world would be on fire now.) The fuel must be brought above its *ignition temperature* to start the combustion. The minimum ignition temperatures of various substances in atmospheric air are approximately 260°C for gasoline, 400°C for carbon, 580°C for hydrogen, 610°C for carbon monoxide, and 630°C for methane. Moreover, the proportions of the fuel and air must be in the proper range for combustion to begin. For example, natural gas does not burn in air in concentrations less than 5 percent or greater than about 15 percent.

As you may recall from your chemistry courses, chemical equations are balanced on the basis of the *conservation of mass principle* (or the *mass balance*), which can be stated as follows: *The total mass of each element is conserved during a chemical reaction.* That is, the total mass of each element on the right-hand side of the reaction equation (the products) must be equal to the total mass of that element on the left-hand side (the reactants), even though the elements exist in different chemical compounds in the reactants and products. Also, the total number of atoms of each element is conserved during a chemical reaction since the total number of atoms is equal to the total mass of the element divided by its atomic mass.

For example, both sides of Eq. (2-32) contain 12 kg of carbon and 32 kg of oxygen, even though the carbon and the oxygen exist as elements in the reactants and as a compound in the product. Also, the total mass of reactants is equal to the total mass of products, each being 44 kg. (It is common practice to round the molar masses to the nearest integer if great accuracy is not required.) However, notice that the total mole number of the reactants (2 kmol) is not equal to the total mole number of the products (1 kmol). That is, *the total number of moles is not conserved during a chemical reaction.*

A frequently used quantity in the analysis of combustion processes to quantify the amounts of fuel and air is the *air-fuel ratio* AF. It is usually expressed on a mass basis and is defined as *the ratio of the mass of air to the mass of fuel* for a combustion process (Fig. 2-31). That is,

$$AF = \frac{m_{\text{air}}}{m_{\text{fuel}}} \tag{2-34}$$

The mass m of a substance is related to the number of moles N through the relation $m = NM$, where M is the molar mass.

The air-fuel ratio can also be expressed on a mole basis as the ratio of the mole numbers of air to the mole numbers of fuel. But we will use the former definition. The reciprocal of the air-fuel ratio is called the *fuel-air ratio.*

Theoretical and Actual Combustion Processes

It is often instructive to study the combustion of a fuel by assuming that the combustion is complete. A combustion process is *complete* if all the carbon in the fuel burns to CO_2, all the hydrogen burns to H_2O, and all the sulfur (if any) burns to SO_2. That is, all the combustible components of a fuel are burned to completion during a complete combustion process. Conversely, the combustion process is *incomplete* if the combustion products contain any unburned fuel or components such as C, H_2, CO, or OH.

Insufficient oxygen is an obvious reason for incomplete combustion, but it is not the only one. Incomplete combustion occurs even when more oxygen is present in the combustion chamber than is needed for complete combustion. This may be attributed to insufficient mixing in the combustion chamber during the limited time that the fuel and the oxygen are in contact. Another cause of incomplete combustion is *dissociation,* which becomes important at high temperatures.

Oxygen has a much greater tendency to combine with hydrogen than it does with carbon. Therefore, the hydrogen in the fuel normally burns to completion, forming H_2O, even when there is less oxygen than needed for complete combustion. Some of the carbon, however, ends up as CO or just as plain C particles (soot) in the products.

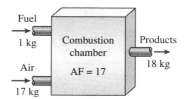

Figure 2-31 The air-fuel ratio (AF) represents the amount of air used per unit mass of fuel during a combustion process.

The minimum amount of air needed for the complete combustion of a fuel is called the *stoichiometric* or *theoretical air*. Thus, when a fuel is completely burned with theoretical air, no uncombined oxygen is present in the product gases. The theoretical air is also referred to as the *chemically correct amount of air,* or *100 percent theoretical air*. A combustion process with less than the theoretical air is bound to be incomplete. The ideal combustion process during which a fuel is burned completely with theoretical air is called the stoichiometric or *theoretical combustion* of that fuel. For example, the theoretical combustion of methane is

$$CH_4 + 2(O_2 + 3.76\ N_2) \rightarrow CO_2 + 2\ H_2O + 7.52\ N_2$$

Notice that the products of the theoretical combustion contain no unburned methane and no C, H_2, CO, OH, or free O_2.

In actual combustion processes, it is common practice to use more air than the stoichiometric amount to increase the chances of complete combustion or to control the temperature of the combustion chamber. The amount of air in excess of the stoichiometric amount is called *excess air*. The amount of excess air is usually expressed in terms of the stoichiometric air as *percent excess air* or *percent theoretical air*. For example, 50 percent excess air is equivalent to 150 percent theoretical air, and 200 percent excess air is equivalent to 300 percent theoretical air. Of course, the stoichiometric air can be expressed as 0 percent excess air or 100 percent theoretical air. Amounts of air less than the stoichiometric amount are called *deficiency of air* and are often expressed as *percent deficiency of air*. For example, 90 percent theoretical air is equivalent to 10 percent deficiency of air. The amount of air used in combustion processes is also expressed in terms of the *equivalence ratio*, which is the ratio of the actual fuel-air ratio to the stoichiometric fuel-air ratio.

Predicting the composition of the products is relatively easy when the combustion process is assumed to be complete and the exact amounts of the fuel and air used are known. All one needs to do in this case is simply apply the mass balance to each element that appears in the combustion equation; there is no need to take any measurements. Things are not so simple, however, when one is dealing with actual combustion processes. For one thing, actual combustion processes are hardly ever complete, even in the presence of excess air. Therefore, it is impossible to predict the composition of the products on the basis of the mass balance alone. Then the only alternative we have is to measure the amount of each component in the products directly.

Enthalpy of Formation and Enthalpy of Combustion

During a chemical reaction, some chemical bonds that bind the atoms into molecules are broken, and new ones are formed. The chemical energy associated with these bonds, in general, is different for the reactants and the products. Therefore, a process that involves chemical reactions involves changes in chemical energies, which must be accounted for in an energy balance. Assuming the atoms of each reactant remain intact (nonnuclear reactions) and disregarding any changes in kinetic and potential energies, the energy change of a system during a chemical reaction is due to a change in state and a change in chemical composition. That is,

$$\Delta E_{sys} = \Delta E_{state} + \Delta E_{chem} \tag{2-35}$$

Therefore, when the products formed during a chemical reaction exit the reaction chamber at the inlet state of the reactants, we have $\Delta E_{state} = 0$, and the energy change of the system in this case is due only to the changes in its chemical composition.

In thermodynamics, we are concerned with the *changes* in the energy of a system during a process, and not the energy values at the particular states. Therefore, we can choose any state as the reference state and assign a value of zero to the internal energy or enthalpy of a substance at that state. When a process involves no changes in chemical composition, the reference state chosen has no effect on the results. When the process involves chemical reactions, however, the composition of the system at the end of a process is no longer the same as that at the beginning of the process. In this case, it becomes necessary to have a common reference state for all substances. The chosen reference state is 25°C (77°F) and 1 atm, which is known as the *standard reference state*. Property values at the standard reference state are indicated by "°" such as $h°$ and $u°$.

When analyzing reacting systems, we must use property values relative to the standard reference state. However, it is not necessary to prepare a new set of property tables for this purpose. We can use the existing tables by subtracting the property values at the standard reference state from the values at the specified state. The ideal-gas enthalpy of N_2 at 500 K relative to the standard reference state, for example, is $\bar{h}_{500\,K} - \bar{h}° = 14{,}581 - 8669 = 5912$ kJ/kmol.

Consider the formation of CO_2 from its elements, carbon and oxygen, during a steady-flow combustion process. Both the carbon and the oxygen enter the combustion chamber at 25°C and 1 atm. The CO_2 formed during this process also leaves the combustion chamber at 25°C and 1 atm. The combustion of carbon is an *exothermic reaction* (a reaction during which chemical energy is released in the form of heat). Therefore, some heat is transferred from the combustion chamber to the surroundings during this process, which is 393,520 kJ/kmol CO_2 formed. (When one is dealing with chemical reactions, it is more convenient to work with quantities per unit mole than per unit time, even for steady-flow processes.)

The process described above involves no work interactions. Therefore, from the steady-flow energy balance relation, the heat transfer during this process must be equal to the difference between the enthalpy of the products and the enthalpy of the reactants. That is,

$$Q = H_{prod} - H_{react} = -393{,}520 \text{ kJ/kmol} \qquad (2\text{-}36)$$

Since both the reactants and the products are at the same state, the enthalpy change during this process is solely due to the changes in the chemical composition of the system. This enthalpy change is different for different reactions, and it is very useful to have a property to represent the changes in chemical energy during a reaction. This property is the *enthalpy of reaction* h_R, which is defined as *the difference between the enthalpy of the products at a specified state and the enthalpy of the reactants at the same state for a complete reaction.*

For combustion processes, the enthalpy of reaction is usually referred to as the *enthalpy of combustion* h_C, which represents the amount of heat released during a steady-flow combustion process when 1 kmol (or 1 kg) of fuel is burned completely at a specified temperature and pressure (Fig. 2-32). It is expressed as

$$h_R = h_C = H_{prod} - H_{react} \qquad (2\text{-}37)$$

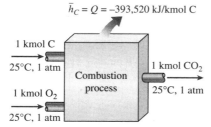

$\bar{h}_C = Q = -393{,}520$ kJ/kmol C

1 kmol C
25°C, 1 atm

Combustion process

1 kmol CO_2
25°C, 1 atm

1 kmol O_2
25°C, 1 atm

Figure 2-32   The enthalpy of combustion represents the amount of energy released as a fuel is burned during a steady-flow process at a specified state.

which is −393,520 kJ/kmol for carbon at the standard reference state. The enthalpy of combustion of a particular fuel is different at different temperatures and pressures.

The enthalpy of combustion is obviously a very useful property for analyzing the combustion processes of fuels. However, there are so many different fuels and fuel mixtures that it is not practical to list h_C values for all possible cases. Besides, the enthalpy of combustion is not of much use when the combustion is incomplete. Therefore, a more practical approach would be to have a more fundamental property to represent the chemical energy of an element or a compound at some reference state. This property is the *enthalpy of formation* $\bar{h}_f^\circ$, which can be viewed as *the enthalpy of a substance at a specified state due to its chemical composition.*

To establish a starting point, we assign the enthalpy of formation of all stable elements (such as O_2, N_2, H_2, and C) a value of zero at the standard reference state of 25°C and 1 atm. That is, $\bar{h}_f^\circ = 0$ for all stable elements. (This is no different from assigning the internal energy of saturated liquid water a value of zero at 0.01°C.) Perhaps we should clarify what we mean by *stable*. The stable form of an element is simply the chemically stable form of that element at 25°C and 1 atm. Nitrogen, for example, exists in diatomic form (N_2) at 25°C and 1 atm. Therefore, the stable form of nitrogen at the standard reference state is diatomic nitrogen N_2, not monatomic nitrogen N. If an element exists in more than one stable form at 25°C and 1 atm, one of the forms should be specified as the stable form. For carbon, for example, the stable form is assumed to be graphite, not diamond.

Now reconsider the formation of CO_2 (a compound) from its elements C and O_2 at 25°C and 1 atm during a steady-flow process. The enthalpy change during this process was determined to be −393,520 kJ/kmol. However, $H_{react} = 0$ since both reactants are elements at the standard reference state, and the products consist of 1 kmol of CO_2 at the same state. Therefore, the enthalpy of formation of CO_2 at the standard reference state is −393,520 kJ/kmol.

The negative sign is due to the fact that the enthalpy of 1 kmol of CO_2 at 25°C and 1 atm is 393,520 kJ less than the enthalpy of 1 kmol of C and 1 kmol of O_2 at the same state. In other words, 393,520 kJ of chemical energy is released (leaving the system as heat) when C and O_2 combine to form 1 kmol of CO_2. Therefore, a negative enthalpy of formation for a compound indicates that heat is released during the formation of that compound from its stable elements. A positive value indicates heat is absorbed.

You will notice that two $\bar{h}_f^\circ$ values are given for H_2O in Table A-6 in the appendix, one for liquid water and the other for water vapor. This is because both phases of H_2O are encountered at 25°C, and the effect of pressure on the enthalpy of formation is small. (Note that under equilibrium conditions, water exists only as a liquid at 25°C and 1 atm.) The difference between the two enthalpies of formation is equal to the h_{fg} of water at 25°C, which is 2441.7 kJ/kg or 44,000 kJ/kmol.

Another term commonly used in conjunction with the combustion of fuels is the *heating value* of the fuel, which is defined as the amount of heat released when a fuel is burned completely in a steady-flow process and the products are returned to the state of the reactants. In other words, the heating value of a fuel is equal to the absolute value of the enthalpy of combustion of the fuel. That is,

$$\text{Heating value} = |h_C| \qquad \text{(kJ/kg fuel)}$$

The heating value depends on the *phase* of the H_2O in the products. The heating value is called the *higher heating value* (HHV) when the H_2O in the products is in the liquid form, and it is called the *lower heating value* (LHV) when the H_2O in the products is in the vapor form (Fig. 2-33). The two heating values are related by

$$\text{HHV} = \text{LHV} + (mh_{fg})_{H_2O} \qquad \text{(kJ/kg fuel)} \qquad (2\text{-}38)$$

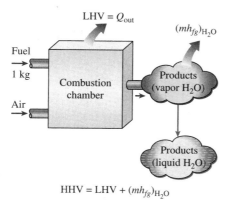

Figure 2-33 The higher heating value of a fuel is equal to the sum of the lower heating value of the fuel and the latent heat of vaporization of the H_2O in the products.

where m is the mass of H_2O in the products per unit mass of fuel and h_{fg} is the enthalpy of vaporization of water at the specified temperature. Higher and lower heating values of common fuels are given in Table A-7 in the appendix. The heating value or enthalpy of combustion of a fuel can be determined from a knowledge of the enthalpy of formation for the compounds involved.

When the exact composition of the fuel is known, the *enthalpy of combustion* of that fuel can be determined using enthalpy of formation data. However, for fuels that exhibit considerable variations in composition depending on the source, such as coal, natural gas, and fuel oil, it is more practical to determine their enthalpy of combustion experimentally by burning them directly in a bomb calorimeter at constant volume or in a steady-flow device.

First-Law Analysis of Reacting Systems

The energy balance (or the first-law) relations developed earlier are applicable to both reacting and nonreacting systems. However, chemically reacting systems involve changes in their chemical energy, and thus it is more convenient to rewrite the energy balance relations so that the changes in chemical energies are explicitly expressed. We do this for steady-flow systems.

Before writing the energy balance relation, we need to express the enthalpy of a component in a form suitable for use for reacting systems. That is, we need to express the enthalpy such that it is relative to the standard reference state and the chemical energy term appears explicitly. When expressed properly, the enthalpy term should reduce to the enthalpy of formation at the standard reference state. With this in mind, we express the enthalpy of a component on a unit mole basis as

$$\text{Enthalpy} = \bar{h}_f^\circ + (\bar{h} - \bar{h}^\circ) \qquad \text{(kJ/kmol)}$$

where the term in the parentheses represents the sensible enthalpy relative to the standard reference state, which is the difference between (the sensible enthalpy at the specified state) and (the sensible enthalpy at the standard reference state of 25°C and 1 atm). This definition enables us to use enthalpy values from tables regardless of the reference state used in their construction. The sensible enthalpy can be approximated as $(\bar{h} - \bar{h}^\circ) \cong \bar{c}_p \Delta T$, where $\bar{c}_p$ is the specific heat on a molar basis and ΔT is the difference in product temperature and reference state temperature.

When the changes in kinetic and potential energies are negligible, the steady-flow energy balance relation $\dot{E}_{in} = \dot{E}_{out}$ can be expressed for a *chemically reacting steady-flow system* more explicitly as

$$\dot{Q}_{in} + \dot{W}_{in} + \sum \dot{n}_r (\bar{h}_f^\circ + \bar{h} - \bar{h}^\circ)_r = \dot{Q}_{out} + \dot{W}_{out} + \sum \dot{n}_p (\bar{h}_f^\circ + \bar{h} - \bar{h}^\circ)_p \qquad (2\text{-}39)$$

where $\dot{n}_p$ and $\dot{n}_r$ represent the molal flow rates of the product p and the reactant r, respectively.

In combustion analysis, it is more convenient to work with quantities expressed *per mole of fuel*. Such a relation is obtained by dividing each term of the equation above by the molal flow rate of the fuel, yielding

$$Q_{in} + W_{in} + \sum N_r (\bar{h}_f^\circ + \bar{h} - \bar{h}^\circ)_r = Q_{out} + W_{out} + \sum N_p (\bar{h}_f^\circ + \bar{h} - \bar{h}^\circ)_p \qquad (2\text{-}40)$$

where N_r and N_p represent the number of moles of the reactant r and the product p, respectively, per mole of fuel. Note that $N_r = 1$ for the fuel, and the other N_r and N_p values can be picked directly from the balanced combustion equation. The energy balance relations above are sometimes written without the work term since most steady-flow combustion processes do not involve any work interactions.

A combustion chamber normally involves heat output but no heat input. Then the energy balance for a *typical steady-flow combustion process* becomes

$$Q_{out} = \sum N_r (\bar{h}_f^\circ + \bar{h} - \bar{h}^\circ)_r - \sum N_p (\bar{h}_f^\circ + \bar{h} - \bar{h}^\circ)_p \qquad (2\text{-}41)$$

It expresses that the heat output during a combustion process is simply the difference between the energy of the reactants entering and the energy of the products leaving the combustion chamber.

EXAMPLE 2-5
Analysis of a Combustion Process

A gaseous fuel mixture that is 40 percent propane (C_3H_8) and 60 percent methane (CH_4) by volume is burned at a rate of 3.75 kg/h with 50 percent excess dry air (Fig. 2-34) in the combustion chamber of a device producing heat (such as a boiler). Both the fuel and air enter the combustion chamber at 25°C and 100 kPa and undergo a complete combustion process. Combustion products leave the combustion chamber at 125°C. Determine (*a*) the balanced combustion equation, (*b*) the air-fuel ratio, (*c*) the dew point temperature of product gases, (*d*) the mass of CO_2 formed per unit mass of the fuel, and (*e*) the rate of heat produced in the equipment. Enthalpy of formation, molar mass, and specific heat data for the species are given in the following table.

	$\bar{h}_f^\circ$, kJ/kmol	M, kg/kmol	$\bar{c}_p$, kJ/kmol·K
C_3H_8 (*g*)	−103,850	44	
CH_4 (*g*)	−74,850	16	
CO_2	−393,520	44	41.16
CO	−110,530	28	29.21
H_2O (*g*)	−241,820	18	34.28
H_2O (*l*)	−285,830	18	75.24
O_2		32	30.14
N_2		28	29.27

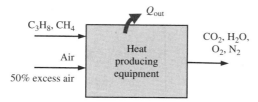

Figure 2-34   Schematic for Example 2-5.

SOLUTION   (*a*) The process is shown in Fig. 2-34. The balanced reaction equation with the stoichiometric air is

$$0.4 \, C_3H_8 + 0.6 \, CH_4 + a_{th}[O_2 + 3.76 \, N_2] \rightarrow 1.8 \, CO_2 + 2.8 \, H_2O + a_{th} \times 3.76 \, N_2$$

The stoichiometric coefficient a_{th} is determined from an O_2 balance:

$$a_{th} = 1.8 + 1.4 = 3.2$$

Substituting,

$$0.4 \, C_3H_8 + 0.6 \, CH_4 + 3.2 \, [O_2 + 3.76 \, N_2] \rightarrow 1.8 \, CO_2 + 2.8 \, H_2O + 12.032 \, N_2$$

There will be extra O_2 formed in the products when the reaction takes place with 50 percent excess air. Also, more N_2 will be formed:

$$0.4 \, C_3H_8 + 0.6 \, CH_4 + 1.5 \times 3.2 \, [O_2 + 3.76 \, N_2] \rightarrow 1.8 \, CO_2 + 2.8 \, H_2O + 1.5 \times 12.032 \, N_2 + 0.5 \times 3.2 \, O_2$$

That is,

$$0.4 \, C_3H_8 + 0.6 \, CH_4 + 4.8 \, [O_2 + 3.76 \, N_2] \rightarrow 1.8 \, CO_2 + 2.8 \, H_2O + 18.05 \, N_2 + 1.6 \, O_2$$

(*b*) The air-fuel ratio is

$$AF = \frac{m_{air}}{m_{fuel}} = \frac{(4.8 \times 4.76 \times 29) \, kg}{(0.4 \times 44 + 0.6 \times 16) \, kg} = \textbf{24.36 kg air/kg fuel}$$

(*c*) The partial pressure of water vapor is determined by multiplying the molar fraction of water in the products ($y_{H_2O} = N_{H_2O}/N_{total}$) by the total pressure of the product gases:

$$P_v = y_{H_2O} P_{total} = \frac{N_{H_2O}}{N_{total}} P_{total} = \frac{2.8}{1.8 + 2.8 + 18.05 + 1.6} (100 \text{ kPa}) = \frac{2.8 \text{ kmol}}{24.25 \text{ kmol}} (100 \text{ kPa}) = 11.55 \text{ kPa}$$

The dew point temperature of the product gases is the saturation temperature of water at this pressure:

$$T_{dp} = T_{sat@ \, 11.55 \, kPa} = \textbf{48.7°C} \qquad \text{(Table A-4)}$$

Since the temperature of the product gases is at 398 K (125°C), which is greater than the dew point temperature, there will be no condensation of water vapor in the product gases. If the temperature of exhaust gases drops below this temperature as they flow in piping and chimney, water vapor will start condensing.

(*d*) The mass of CO_2 formed per unit mass of the fuel is

$$\frac{m_{CO_2}}{m_{\text{fuel}}} = \frac{N_{CO_2} M_{CO_2}}{(N_{C_3H_8} M_{C_3H_8}) + (N_{CH_4} M_{CH_4})}$$

$$= \frac{(1.8 \text{ kmol})(44 \text{ kg/kmol})}{(0.4 \text{ kmol})(44 \text{ kg/kmol}) + (0.6 \text{ kmol})(16 \text{ kg/kmol})}$$

$$= \frac{79.2 \text{ kg}}{27.2 \text{ kg}} = \textbf{2.92 kg } CO_2 \textbf{/kg fuel}$$

When 1 kg of this fuel mixture is burned, 2.92 kg of CO_2 gas (a major greenhouse gas causing global warming) is formed in the product gases. Note that the amount of CO_2 production is the same whether the combustion is with theoretical air or with excess air as long as the combustion is complete. That is, using excess air does not affect the amount of CO_2 production.

(*e*) The heat transfer for this combustion process is determined from the energy balance $E_{\text{in}} - E_{\text{out}} = \Delta E_{\text{system}}$ applied on the combustion chamber with $W = 0$. It reduces to

$$-Q_{\text{out}} = \sum N_p \left(\bar{h}_f^{\circ} + \bar{h} - \bar{h}^{\circ} \right)_p - \sum N_r \left(\bar{h}_f^{\circ} + \bar{h} - \bar{h}^{\circ} \right)_r$$

The products are at 125°C, and the enthalpy of products can be expressed as

$$\bar{h} - \bar{h}^{\circ} = \bar{c}_p \Delta T$$

where $\Delta T = 125 - 25 = 100°C = 100$ K. Then, using the values given in the table,

$$-Q_{\text{out}} = (1.8)(-393,520 + 41.16 \times 100) + (2.8)(-241,820 + 34.28 \times 100)$$

$$+ (18.05)(0 + 29.27 \times 100) + (1.6)(0 + 30.14 \times 100) - (0.4)(-103,850) - (0.6)(-74,850)$$

$$= -1,224,320 \text{ kJ/kmol fuel}$$

or $Q_{\text{out}} = 1,224,320$ kJ/kmol fuel

The rate of heat output from the equipment for a fuel consumption rate of 3.75 kg/h is

$$\dot{Q}_{\text{out}} = \dot{m}_{\text{fuel}} \frac{Q_{\text{out}}}{M_{\text{fuel}}} = (3.75 \text{ kg fuel/h}) \frac{1,224,320 \text{ kJ/kmol fuel}}{(0.4 \times 44 + 0.6 \times 16) \text{ kg/kmol}} = 168,794 \text{ kJ/h} = \textbf{46.9 kW}$$

That is, this equipment supplies heat at a rate of 46.9 kW. If this is a boiler generating hot water, the value 46.9 kW represents the rate of heat transferred to the water in the boiler. ▲

2-6 HEAT ENGINES AND POWER PLANTS

Work can easily be converted to other forms of energy, but converting other forms of energy to work is not that easy. Work can be converted to heat directly and completely, but converting heat to work requires the use of some special devices. These devices are called *heat engines*. Heat engines differ considerably from one another, but all can be characterized by the following (Fig. 2-35):

1.   They receive heat from a high-temperature source (solar energy, oil furnace, nuclear reactor, etc.).

2.   They convert part of this heat to work (usually in the form of a rotating shaft).

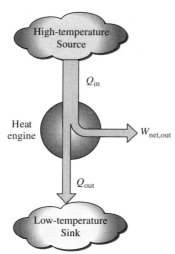

Figure 2-35 Part of the heat received by a heat engine is converted to work, while the rest is rejected to a sink.

3.   They reject the remaining waste heat to a low-temperature sink (the atmosphere, rivers, etc.).

4.   They operate on a cycle.

Heat engines and other cyclic devices usually involve a fluid to and from which heat is transferred while undergoing a cycle. This fluid is called the *working fluid*.

The term *heat engine* is often used in a broader sense to include work-producing devices that do not operate in a thermodynamic cycle. Engines that involve internal combustion such as gas turbines and car engines fall into this category. These devices operate in a mechanical cycle but not in a thermodynamic cycle since the working fluid (the combustion gases) does not undergo a complete cycle. Instead of being cooled to the initial temperature, the exhaust gases are purged and replaced by fresh air-and-fuel mixture at the end of the cycle.

The work-producing device that best fits into the definition of a heat engine is the *steam power plant,* which is an external-combustion engine. That is, combustion takes place outside the engine, and the thermal energy released during this process is transferred to the steam as heat. The schematic of a basic steam power plant is shown in Fig. 2-36. The various quantities shown in this figure are as follows:

Q_{in} = amount of heat supplied to steam in boiler from a high-temperature source (furnace)

Q_{out} = amount of heat rejected from steam in condenser to a low-temperature sink (the atmosphere, a river, etc.)

W_{out} = amount of work delivered by steam as it expands in turbine

W_{in} = amount of work required to compress water to boiler pressure

Notice that the directions of the heat and work interactions are indicated by the subscripts *in* and *out*. Therefore, all four of the described quantities are always *positive*.

The network output of this power plant is simply the difference between the total work output of the plant and the total work input (Fig. 2-37):

$$W_{net,out} = W_{out} - W_{in} \qquad (kJ) \qquad\qquad (2\text{-}42)$$

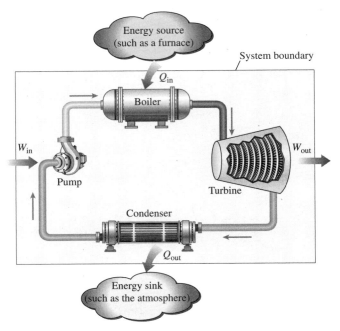

Figure 2-36 Schematic of a steam power plant.

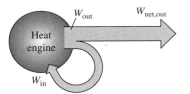

Figure 2-37 A portion of the work output of a heat engine is consumed internally to maintain continuous operation.

The network can also be determined from the heat transfer data alone. The four components of the steam power plant involve mass flow in and out, and therefore should be treated as open systems. These components, together with the connecting pipes, however, always contain the same fluid (not counting the steam that may leak out, of course). No mass enters or leaves this combination system, which is indicated by the shaded area on Fig. 2-36; thus, it can be analyzed as a closed system. Recall that for a closed system undergoing a cycle, the change in internal energy ΔU is zero, and therefore the net work output of the system is also equal to the net heat transfer to the system:

$$W_{\text{net,out}} = Q_{\text{in}} - Q_{\text{out}} \qquad \text{(kJ)} \qquad (2\text{-}43)$$

Thermal Efficiency

In Eq. (2-43), Q_{out} represents the magnitude of the energy wasted in order to complete the cycle. But Q_{out} is never zero; thus, the net work output of a heat engine is always less than the amount of heat input. That is, only part of the heat transferred to the heat engine is converted to work. The fraction of the heat input that is converted to net work output is a measure of the performance of a heat engine and is called the *thermal efficiency* η_{th} (Fig. 2-38).

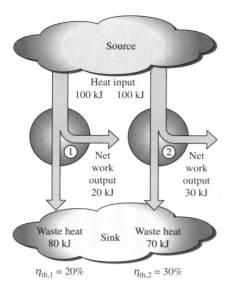

Figure 2-38 Some heat engines perform better than others (convert more of the heat they receive to work).

For heat engines, the desired output is the net work output, and the required input is the amount of heat supplied to the working fluid. Then the thermal efficiency of a heat engine can be expressed as

$$\eta_{th} = \frac{\text{Net work output}}{\text{Total heat input}} = \frac{W_{net,out}}{Q_{in}} \tag{2-44}$$

Using net power and rate of heat input,

$$\eta_{th} = \frac{\text{Net power output}}{\text{Total rate of heat input}} = \frac{\dot{W}_{net,out}}{\dot{Q}_{in}} \tag{2-45}$$

It can also be expressed as

$$\eta_{th} = 1 - \frac{Q_{out}}{Q_{in}} \tag{2-46}$$

since $W_{net,out} = Q_{in} - Q_{out}$.

Thermal efficiency is a measure of how efficiently a heat engine converts the heat that it receives to work. Heat engines are built for the purpose of converting heat to work, and engineers are constantly trying to improve the efficiencies of these devices since increased efficiency means less fuel consumption and thus lower fuel bills and less pollution.

The most efficient heat engine cycle is referred to as the Carnot cycle, which consists of totally reversible processes. The thermal efficiency of the Carnot cycle can be expressed in terms of temperatures as

$$\eta_{th,Carnot} = 1 - \frac{T_L}{T_H} \tag{2-47}$$

where T_L and T_H are the temperatures of the low-temperature and high-temperature reservoirs (i.e., sink and source temperatures), respectively. Notice that the thermal efficiency of a Carnot heat engine increases with increasing T_H and decreasing T_L values. The Carnot cycle is the *most efficient* heat engine cycle operating between two specified temperature levels.

Overall Plant Efficiency

In electricity-generating plants, heat is usually supplied by burning fuels such as coal, oil, and natural gas. For example, in a steam power plant, the chemical energy of fuel is converted to heat in a combustion process. Most of the produced heat is transferred to water to turn it into steam, which runs through a turbine to produce power. The remaining heat is lost to the environment with the exhaust gases of the combustion process.

We may define overall plant efficiency based on the total energy of the fuel consumed in the plant:

$$\eta_{plant} = \frac{\text{Net work output}}{\text{Total amount of heat input by fuel}} = \frac{W_{net,out}}{m_{fuel} \times HV_{fuel}} \qquad (2\text{-}48)$$

where $W_{net,out}$ (kJ) is the net work output from the plant, m_{fuel} (kg) is the amount of fuel consumed, and HV_{fuel} (kJ/kg) is the heating value of the fuel. It may also be expressed on a rate basis as

$$\eta_{plant} = \frac{\text{Net power output}}{\text{Total rate of heat input by fuel}} = \frac{\dot{W}_{net,out}}{\dot{m}_{fuel} \times HV_{fuel}} \qquad (2\text{-}49)$$

where $\dot{W}_{net,out}$ (kW) is the net power output from the plant and $\dot{m}_{fuel}$ (kg/s) is the mass rate of fuel consumed. The overall plant efficiency can be used to express the performance of all power plants which burn a fuel, such as steam power plants, gas-turbine power plants, and automobile engines. Note that the thermal efficiency and overall plant efficiency are sometimes used interchangeably.

The overall plant efficiencies of power plants are relatively low. Ordinary spark-ignition automobile engines have an efficiency of about 25 percent. That is, an automobile engine converts about 25 percent of the chemical energy of the gasoline to mechanical work. This number is as high as 40 percent for diesel engines and large gas-turbine plants and as high as 60 percent for large combined gas-steam power plants. Thus, even with the most efficient heat engines available today, almost one-half of the energy supplied ends up in the rivers, lakes, or the atmosphere as waste or useless energy.

EXAMPLE 2-6
Coal Consumption of a
Power Plant

A coal-fired power plant generates 1100 MW of power with an overall plant efficiency of 44 percent. The plant operates 75 percent of the time in a year. The coal used in the plant has a heating value of 26,000 kJ/kg. Determine the amount of coal consumed in the plant per year.

SOLUTION   The plant operates 75 percent of the time. Then, the total for annual operating hours is

$$\text{Operating days} = (0.75)(365 \times 24 \text{ h/yr}) = 6570 \text{ h/yr}$$

The rate of heat input to the plant is determined using plant efficiency as

$$\eta_{plant} = \frac{\dot{W}_{electric}}{\dot{Q}_{in}} \quad \rightarrow \quad \dot{Q}_{in} = \frac{\dot{W}_{electric}}{\eta_{plant}} = \frac{1100 \text{ MW}}{0.44} = 2500 \text{ MW}$$

The amount of heat input to the plant per year is

$$Q_{in} = \dot{Q}_{in} \times \text{Operating days} = (2500 \text{ MJ/s})(6570 \text{ h/yr})\left(\frac{3600 \text{ s}}{1 \text{ h}}\right)\left(\frac{1000 \text{ kJ}}{1 \text{ MJ}}\right) = 5.913 \times 10^{13} \text{ kJ/yr}$$

The corresponding coal consumption of the plant per year is

$$Q_{in} = m_{coal}HV_{coal} \rightarrow m_{coal} = \frac{Q_{in}}{HV_{coal}} = \frac{5.913 \times 10^{13} \text{ kJ/yr}}{26{,}000 \text{ kJ/kg}} = \mathbf{2.274 \times 10^9 \text{ kg/yr}} \quad \blacktriangle$$

2-7 REFRIGERATORS AND HEAT PUMPS

We all know from experience that heat flows in the direction of decreasing temperature, that is, from high-temperature regions to low-temperature ones. This heat-transfer process occurs in nature without requiring any devices. The reverse process, however, cannot occur by itself. The transfer of heat from a low-temperature region to a high-temperature one requires special devices called *refrigerators*. A refrigerator is a device used to keep a space cool, and it is called an *air conditioner* when indoor spaces of a building are kept at a lower temperature than the ambient air in summer.

Refrigerators are cyclic devices, and the working fluids used in the refrigeration cycles are called *refrigerants*. A refrigerator is shown schematically in Fig. 2-39*a*. Here Q_L is the magnitude of the heat removed from the refrigerated space at temperature T_L, Q_H is the magnitude of the heat rejected to the warm space at temperature T_H, and $W_{net,in}$ is the net work input to the refrigerator. Note that Q_L and Q_H represent magnitudes and thus are positive quantities.

Another device that transfers heat from a low-temperature medium to a high-temperature one is the *heat pump*. Refrigerators and heat pumps are essentially the same devices; they differ in their objectives only. The objective of a refrigerator is to maintain the refrigerated space at a low temperature by removing heat from it. Discharging this heat to a higher-temperature medium is merely a necessary part of the operation, not the purpose.

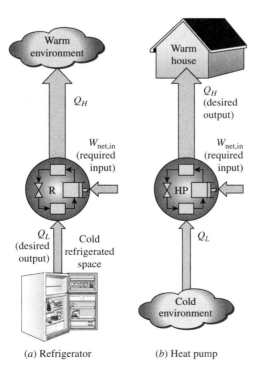

(*a*) Refrigerator   (*b*) Heat pump

Figure 2-39 The objective of a refrigerator is to remove heat (Q_L) from the cold medium; the objective of a heat pump is to supply heat (Q_H) to a warm medium.

The objective of a heat pump, however, is to maintain a heated space at a high temperature. This is accomplished by absorbing heat from a low-temperature source, such as well water or cold outside air in winter, and supplying this heat to a warmer medium such as a house (Fig. 2-39b).

The performance of refrigerators and heat pumps is expressed in terms of the *coefficient of performance* (COP), defined as

$$\text{COP}_R = \frac{\text{Cooling effect}}{\text{Work input}} = \frac{Q_L}{W_{\text{net,in}}} \tag{2-50}$$

$$\text{COP}_{HP} = \frac{\text{Heating effect}}{\text{Work input}} = \frac{Q_H}{W_{\text{net,in}}} \tag{2-51}$$

or in rate form

$$\text{COP}_R = \frac{\dot{Q}_L}{\dot{W}_{\text{net,in}}} \tag{2-52}$$

$$\text{COP}_{HP} = \frac{\dot{Q}_H}{\dot{W}_{\text{net,in}}} \tag{2-53}$$

Notice that both COP_R and COP_{HP} can be greater than 1. A comparison of Eqs. (2-52) and (2-53) reveals that

$$\text{COP}_{HP} = \text{COP}_R + 1 \tag{2-54}$$

for fixed values of Q_L and Q_H. This relation implies that $\text{COP}_{HP} \geq 1$ since COP_R is a positive quantity. That is, a heat pump functions, at worst, as a resistance heater, supplying as much energy to the house as it consumes. In reality, however, part of Q_H is lost to the outside air through piping and other devices, and COP_{HP} may drop below unity when the outside air temperature is too low. When this happens, the system normally switches to the fuel (natural gas, propane, oil, etc.) or resistance-heating mode.

The *cooling capacity* of a refrigeration system—that is, the rate of heat removal from the refrigerated space—is often expressed in terms of *tons of refrigeration*. The capacity of a refrigeration system that can freeze 1 ton (2000 lbm) of liquid water at 0°C (32°F) into ice at 0°C in 24 h is said to be 1 ton. One ton of refrigeration is equivalent to 211 kJ/min or 200 Btu/min or 12,000 Btu/h. The cooling load of a typical 200-m² residence is in the 3-ton (10-kW) range.

EXAMPLE 2-7
COP of a Heat Pump

A heat pump is used to heat water from 80 to 125°F. The heat absorbed in the evaporator is 32,000 Btu/h and the power input is 4.3 kW. Determine the rate at which water is heated in gal/min and the COP of the heat pump.

SOLUTION The rate of heat transfer in the condenser is

$$\dot{Q}_H = \dot{Q}_L + \dot{W}_{\text{in}} = 32{,}000 \text{ Btu/h} + (4.3 \text{ kW})\left(\frac{3412 \text{ Btu/h}}{1 \text{ kW}}\right) = 46{,}672 \text{ Btu/h}$$

Taking the specific heat of water to be 1.0 Btu/lbm·°F, the mass flow rate of water is determined to be

$$\dot{Q}_H = \dot{m}c_p(T_2 - T_1) \rightarrow \dot{m} = \frac{\dot{Q}_H}{c_p(T_2 - T_1)} = \frac{46{,}672 \text{ Btu/h}}{(1.0 \text{ Btu/lbm·°F})(125 - 80)°F} = 1037 \text{ lbm/h}$$

Taking the density of water to be 62.1 lbm/ft^3, the volume flow rate of water is determined to be

$$\dot{V} = \frac{\dot{m}}{\rho} = \frac{1037 \text{ lbm/h}}{62.1 \text{ lbm/ft}^3}\left(\frac{264.17 \text{ gal}}{35.315 \text{ ft}^3}\right)\left(\frac{1 \text{ h}}{60 \text{ min}}\right) = \textbf{2.08 gal/min}$$

Finally, the COP of the heat pump is determined from its definition to be

$$\text{COP}_{\text{HP}} = \frac{\dot{Q}_H}{\dot{W}_{\text{in}}} = \frac{46{,}672 \text{ Btu/h}}{(4.3 \text{ kW})\left(\dfrac{3412 \text{ Btu/h}}{1 \text{ kW}}\right)} = \textbf{3.18} \quad \blacktriangle$$

The most efficient refrigeration and heat pump cycles are referred to as Carnot refrigerators and heat pumps, which consist of totally reversible processes. The coefficients of performance of Carnot refrigerators and heat pumps can be expressed in terms of temperatures as

$$\text{COP}_{\text{R,Carnot}} = \frac{1}{T_H/T_L - 1} = \frac{T_L}{T_H - T_L} \tag{2-55}$$

and

$$\text{COP}_{\text{HP,Carnot}} = \frac{1}{1 - T_L/T_H} = \frac{T_H}{T_H - T_L} \tag{2-56}$$

where T_L and T_H are the temperatures of the cold and warm mediums, respectively. Notice that both COPs increase as the difference between the two temperatures decreases, that is, as T_L rises or T_H falls. This conclusion is also applicable to actual refrigerators and heat pumps. The reversed Carnot cycle is the *most efficient* refrigeration cycle operating between two specified temperature levels. The reversed Carnot cycle cannot be approximated in actual devices and is not a realistic model for refrigeration cycles. However, the reversed Carnot cycle can serve as a standard against which actual refrigeration cycles are compared.

REFERENCES

Çengel YA and Cimbala JM. 2018. *Fluid Mechanics: Fundamentals and Applications*, 4th ed. New York: McGraw Hill.

Çengel YA and Ghajar AJ. 2020. *Heat and Mass Transfer: Fundamentals and Applications*, 6th ed. New York: McGraw Hill.

Çengel, YA, Boles MA, and Kanoğlu M. 2019. *Thermodynamics: An Engineering Approach*, 9th ed. New York: McGraw Hill.

Çengel YA, Cimbala JM, and Ghajar AJ. 2022. *Fundamentals of Thermal-Fluid Sciences*, 6th ed. New York: McGraw Hill.

Churchill SW. "Friction Factor Equation Spans all Fluid-Flow Regimes," *Chemical Engineering*, 7 (1977), pp. 91–92.

White FM. 2011. *Fluid Mechanics*, 7th ed. New York: McGraw Hill.

PROBLEMS

THERMAL SCIENCES

2-1 What is the scope of the physical sciences called thermal sciences?

2-2 What are the subcategories of thermal-fluid sciences?

2-3 How does the design of a car radiator involve thermodynamics, heat transfer, and fluid mechanics? Explain.

2-4   Which science is not a subcategory of thermal-fluid sciences?
(*a*) Thermodynamics   (*b*) Statics   (*c*) Heat transfer
(*d*) Fluid mechanics   (*e*) Thermochemistry

2-5   Which is not a subcategory or application of thermal-fluid sciences?
(*a*) Refrigeration   (*b*) Turbomachinery   (*c*) Power plant design
(*d*) Mechatronics   (*e*) Internal combustion engines

THERMODYNAMICS

2-6   Express the first law of thermodynamics. Give an example.

2-7   Express the second law of thermodynamics. Give an example.

2-8   Describe two hypothetical processes, one violating the first law of thermodynamics and the other violating the second law of thermodynamics.

2-9   What does the total energy of a system consist of?

2-10   What is the difference between sensible and latent energies?

2-11   What is the difference between chemical and nuclear energies?

2-12   What is the difference between thermal energy and heat?

2-13   Define enthalpy and flow energy.

2-14   Define specific heat.

2-15   What is the difference between constant-volume specific heat and constant-pressure specific heat? Is one necessarily greater than the other?

2-16   What are the mechanisms of energy transfer to or from a given mass? How do you differentiate them? Give examples.

2-17   What is the difference between the amount and the rate of heat transfer? Give three proper units for each term.

2-18   Express general energy balances for a stationary closed system and for a steady-flow system. Simplify these relations when there is a heat input to the system with no work interaction and no changes in kinetic and potential energies. Assume that the fluid involved in both cases is an ideal gas.

2-19   What is the difference between mass flow rate and volume flow rate? What are their units?

2-20   Under what conditions do both the mass and volume flow rates of a fluid through a pipe remain constant?

2-21   How does a pressure cooker cook faster than an ordinary cooker?

2-22   How can water vapor in atmospheric air be condensed? Explain.

2-23   When the pressure is fixed during a phase change process, is the temperature automatically fixed?

2-24   What are the latent heat, latent heat of fusion, and latent heat of vaporization?

2-25   At which temperature, does it take more energy to vaporize a unit mass of water: 100°C or 150°C?

2-26   Which process requires less energy input: melting 1 kg of ice at 0°C or vaporizing 1 kg of water at 100°C?

2-27   Consider a house with a floor space of 200 m^2 and an average height of 3 m at sea level, where the standard atmospheric pressure is 101.3 kPa. Initially the house is at a uniform temperature of 13°C. Now the electric heater is turned on, and the heater runs until the air temperature in the house rises to an average value of 21°C. Determine how much heat is absorbed by the air, assuming some air escapes through the cracks as the heated air in the house expands at constant pressure. Also, determine the cost of this heat if the unit cost of electricity in that area is $0.095/kWh.

2-28   An 800-W iron is left on the ironing board with its base exposed to the air. About 85 percent of the heat generated in the iron is dissipated through its base, whose surface area is 0.15 ft^2, and the remaining 15 percent through other surfaces. Assuming the heat transfer from the surface to be uniform, determine (*a*) the amount of heat the iron dissipates during a 2-h period, in kWh, (*b*) the heat flux on the surface of the iron base, in W/ft², and (*c*) the total cost of the electrical energy consumed during this 2-h period. Take the unit cost of electricity to be $0.11/kWh.

2-29   Infiltration of cold air into a warm house during winter through the cracks around doors, windows, and other openings is a major source of energy loss since the cold air that enters needs to be heated to the room temperature. The infiltration is often expressed in terms of ACH (air changes per hour). An ACH of 2 indicates that all air in the house is replaced twice every hour by the cold air outside.

Consider an electrically heated house that has a floor space of 180 m² and an average height of 3 m at 1000 m elevation, where the standard atmospheric pressure is 89.6 kPa. The house is maintained at a temperature of 26°C, and the infiltration losses are estimated to amount to 0.5 ACH. Assuming the pressure and the temperature in the house remain constant, determine the amount of energy loss from the house due to infiltration for a day during which the average outdoor temperature is 9°C. Also, determine the cost of this energy loss for that day if the unit cost of electricity in that area is $0.088/kWh.

2-30   Water is heated in an insulated, constant-diameter tube by a 7-kW electric resistance heater. If the water enters the heater steadily at 10°C and leaves at 75°C, determine the mass flow rate of water.

2-31   A 2-m × 5-m × 6-m room is to be heated by a baseboard resistance heater. It is desired that the resistance heater be able to raise the air temperature in the room from 7 to 25°C within 25 min. Assuming no heat losses from the room and an atmospheric pressure of 100 kPa, determine the required power rating of the resistance heater. Assume constant specific heats at room temperature.

2-32   A 5-m × 6-m × 8-m room is to be heated by an electrical resistance heater placed in a short duct in the room. Initially, the room is at 15°C, and the local atmospheric pressure is 98 kPa. The room is losing heat steadily to the outside at a rate of 200 kJ/min. A 300-W fan circulates the air steadily through the duct and the electric heater at an average mass flow rate of 50 kg/min. The duct can be assumed to be adiabatic, and there is no air leaking in or out of the room. If it takes 13 min for the room air to reach an average temperature of 25°C, find (*a*) the power rating of the electric heater and (*b*) the temperature rise that the air experiences each time it passes through the heater.

2-33   A house has an electric heating system that consists of a 400-W fan and an electric resistance heating element placed in a duct. Air flows steadily through the duct at a rate of 0.3 lbm/s and experiences a temperature rise of 8°F. The rate of heat loss from the air in the duct is estimated to be 850 Btu/h. Determine the power rating of the electric resistance heating element.

2-34   The ducts of an air heating system pass through an unheated area. As a result of heat losses, the temperature of the air in the duct drops by 2°C. If the mass flow rate of air is 130 kg/min, determine the rate of heat loss from the air to the cold environment.

2-35   Air enters the duct of an air-conditioning system at 15 psia and 50°F at a volume flow rate of 450 ft³/min. The diameter of the duct is 10 in, and heat is transferred to the air in the duct from the surroundings at a rate of 2 Btu/s. Determine (*a*) the velocity of the air at the duct inlet and (*b*) the temperature of the air at the exit.

2-36   Air flows in a 2-in-diameter pipe at 15 psia and 80°F with a velocity of 12 ft/s. Determine the volume and mass flow rates of air.

2-37   Water flows in a 10-cm-diameter pipe at 15°C at a rate of 7 kg/s. Determine the velocity and volume flow rate of the water.

2-38   A boiler is used to generate saturated water vapor at 200°C. What is the pressure of this water vapor?

2-39   The minimum pressure that can be used in the condenser of a steam power plant is about 10 kPa. If a condenser operates at this pressure, what is the condensation temperature of steam?

2-40   Water is boiled at 1 atm pressure in a coffee maker equipped with an immersion-type electric heating element. The coffee maker initially contains 1 kg of water. Once boiling starts, it is observed that half of the water in the coffee maker evaporates in 12 min. If the heat loss from the coffee maker is negligible, what is the power rating of the heating element?

2-41   Saturated steam at 150°C is obtained from a boiler by transferring heat to water at a rate of 750 kW. Determine the rate at which steam is generated, in kg/min.

2-42   Water at 15°C is cooled and frozen at 1 atm at a rate of 50 kg/h. Determine the rate at which heat must be removed from the water. The heat of fusion of water at 1 atm is 334 kJ/kg.

2-43   Saturated steam coming off the turbine of a steam power plant at 40°C condenses on the outside of a 2-cm-outer-diameter, 60-m-long tube at a rate of 140 kg/h. Determine the rate of heat transfer from the steam to the cooling water flowing through the pipe.

2-44   The increase in which type of energy manifests itself as a rise in temperature?
(*a*) Sensible   (*b*) Latent   (*c*) Kinetic   (*d*) Chemical   (*e*) Potential

2-45   Which energy is the sum of sensible and latent energies?
(*a*) Nuclear   (*b*) Potential   (*c*) Chemical   (*d*) Internal   (*e*) Thermal

2-46   Consider the boiling of water in an open pan on an oven. In what form is energy added to water?
(*a*) Sensible   (*b*) Latent   (*c*) Kinetic   (*d*) Chemical   (*e*) Potential

2-47   Which energy is needed to push a fluid and to maintain flow?
(*a*) Sensible   (*b*) Latent   (*c*) Flow   (*d*) Kinetic   (*e*) Pressure

2-48   The microscopic energy of a flowing fluid is called
(*a*) Flow work   (*b*) Latent energy   (*c*) Chemical energy
(*d*) Enthalpy   (*e*) Internal energy

2-49   The specific heat of water is given to be 4.18 kJ/kg·°C. This is equivalent to
(*a*) 4.18 kJ/kg·K   (*b*) 7.52 kJ/kg·°F   (*c*) 0.0153 kJ/kg·K
(*d*) 17.5 Btu/lbm·°F   (*e*) 277.2 kJ/kg·K

2-50   Choose the incorrect statement.
(*a*) For liquids, the constant-volume specific heat is approximately equal to constant-pressure specific heat.
(*b*) For solids, the constant-volume specific heat is approximately equal to constant-pressure specific heat.
(*c*) The constant-volume specific heat is smaller than constant-pressure specific heat.
(*d*) For ideal gases, $c_p = c_v + R$.
(*e*) The specific heat of a substance depends on temperature only.

2-51   Choose the incorrect statement.
(*a*) Energy can be transferred to or from a closed system by heat, work, and mass.
(*b*) Heat is energy transfer due to temperature difference.
(*c*) Work per unit time is called power.
(*d*) The amount of heat transferred per unit time is called the heat transfer rate.
(*e*) A rotating shaft is an example for work interaction.

2-52   Which one is not a unit for an amount of energy?
(*a*) kWh   (*b*) kJ   (*c*) kcal   (*d*) hp   (*e*) Btu

2-53   Which one is not a unit for a rate of energy?
(*a*) Ton of refrigeration   (*b*) kW   (*c*) kcal   (*d*) hp   (*e*) Btu/s

2-54   Choose the incorrect statement.
(*a*) When the density, velocity, and cross-sectional area do not change, the mass flow rate does not change, either.
(*b*) When the velocity and cross-sectional area do not change, the volume flow rate does not change either.
(*c*) Volume flow rate is equal to mass flow rate divided by density.
(*d*) When the mass flow rate remains constant during a process for an ideal gas, so does the volume flow rate.
(*e*) The product of velocity and cross-sectional area is equal to volume flow rate.

2-55   Which process requires the least amount of energy transfer per unit mass of water?
(*a*) Vaporizing at 100°C   (*b*) Condensing at 110°C   (*c*) Vaporizing at 120°C
(*d*) Condensing at 130°C   (*e*) Melting at 0°C

2-56   Which process requires the greatest amount of energy transfer per unit mass of water?
(*a*) Boiling at sea level   (*b*) Boiling at 2000 m elevation   (*c*) Boiling at 500 kPa
(*d*) Condensing at 150°C   (*e*) Melting at 0°C

HEAT TRANSFER

2-57   What are the modes of heat transfer? What is the driving force for heat transfer? What is the direction of heat transfer?

2-58   What is conduction? What are the mechanisms of conduction in solids, liquids, and gases?

2-59   What does heat conduction through a medium depend on?

2-60   What is thermal conductivity? What does a low value of thermal conductivity represent?

2-61   What is convection? What is the difference between forced and natural convection?

2-62   What is radiation? How does radiation differ from conduction and convection?

2-63   What are the Stefan-Boltzmann law, Stefan-Boltzmann constant, blackbody, blackbody radiation, and emissivity?

2-64   The inner and outer surfaces of a 5-m × 6-m brick wall of thickness 30 cm and thermal conductivity 0.69 W/m·°C are maintained at temperatures of 18 and 9°C, respectively. Determine the rate of heat transfer through the wall, in W.

2-65   An aluminum pan whose thermal conductivity is 237 W/m·°C has a flat bottom whose diameter is 20 cm and thickness 0.6 cm. Heat is transferred steadily to boiling water in the pan through its bottom at a rate of 750 W. If the inner surface of the bottom of the pan is 108°C, determine the temperature of the outer surface of the bottom of the pan.

2-66   Two surfaces of a 1-in-thick plate are maintained at 32 and 212°F, respectively. If it is determined that heat is transferred through the plate at a rate of 180 Btu/h·ft², determine its thermal conductivity.

2-67   For heat transfer purposes, a standing man can be modeled as a 30-cm-diameter, 175-cm-long vertical cylinder with both the top and bottom surfaces insulated and with the side surface at an average temperature of 34°C. For a convection heat transfer coefficient of 12 W/m²·°C, determine the rate of heat loss from this man by convection in an environment at 17°C.

2-68   A 2-in-external-diameter, 30-ft-long hot-water pipe at 170°F is losing heat to the surrounding air at 40°F by natural convection with a heat transfer coefficient of 5 Btu/h·ft²·°F. Determine the rate of heat loss from the pipe by natural convection.

2-69   A 12-cm-diameter spherical ball whose surface is maintained at a temperature of 110°C is suspended in the middle of a room at 20°C. If the convection heat transfer coefficient is 15 W/m²·°C and the emissivity of the surface is 0.6, determine the total rate of heat transfer from the ball.

2-70   A 1000-W iron is left on the ironing board with its base exposed to the air at 23°C. The convection heat transfer coefficient between the base surface and the surrounding air is 35 W/m²·°C. If the base has an emissivity of 0.9 and a surface area of 0.02 m², determine the temperature of the base of the iron.

2-71   A thin metal plate is insulated on the back and exposed to solar radiation on the front surface. The exposed surface of the plate has an absorptivity of 0.8 for solar radiation. If solar radiation is incident on the plate at a rate of 350 W/m² and the surrounding air temperature is 25°C, determine the surface temperature of the plate when the heat loss by convection equals the solar energy absorbed by the plate. Assume the convection heat transfer coefficient to be 40 W/m²·°C, and disregard heat loss by radiation.

2-72   Which is not a mechanism of heat transfer?
(*a*) Convection   (*b*) Radiation   (*c*) Diffusion   (*d*) Conduction   (*e*) None of these

2-73   Which is not a mechanism of conduction in solids, liquids, and gases?
(*a*) Collisions of molecules   (*b*) Buoyancy motion   (*c*) Vibrations of molecules
(*d*) Transport of free electrons   (*e*) Diffusion of molecules

2-74   The *rate* of heat conduction through a medium does not depend on
(*a*) Surface roughness   (*b*) Geometry of the medium   (*c*) Material of the medium
(*d*) Temperature difference   (*e*) None of these

2-75   The *rate* of heat conduction through a plane layer is inversely proportional to
(*a*) Heat transfer area   (*b*) Thermal conductivity of the wall   (*c*) Wall thickness
(*d*) Temperature difference   (*e*) None of these

2-76   Which one doubles the rate of heat conduction through a plane layer?
(*a*) Decreasing area by half
(*b*) Decreasing thermal conductivity by half
(*c*) Decreasing temperature difference by half
(*d*) Decreasing the wall thickness by half
(*e*) None of these

2-77   The thermal conductivities of copper and rigid foam insulations are, respectively (in W/m·K)
(*a*) 1000, 1   (*b*) 400, 0.025   (*c*) 50, 10   (*d*) 10, 0.5   (*e*) 6, 0.005

2-78   The thermal conductivity of water is given to be 0.607 W/m·K. This is equivalent to
(*a*) 0.607 W/m·°C   (*b*) 273.6 W/m·°C   (*c*) 1.09 W/m·°F
(*d*) 0.632 Btu/h·ft·°F   (*e*) 1.99 W/ft·R

2-79   The thermal conductivities of iron and chromium are 83 and 95 W/m·K, respectively. What is the thermal conductivity of steel containing 1 percent of chrome?
(*a*) 97 W/m·K   (*b*) 95 W/m·K   (*c*) 88 W/m·K   (*d*) 83 W/m·K   (*e*) 62 W/m·K

2-80   The convection heat transfer coefficient does not depend on
(*a*) Surface geometry   (*b*) Surface conductivity   (*c*) Fluid velocity
(*d*) Fluid type   (*e*) Fluid viscosity

2-81   For which case does the convection heat transfer coefficient take the highest values?
(*a*) Forced convection of liquids   (*b*) Forced convection of gases   (*c*) Boiling and condensation
(*d*) Free convection of liquids   (*e*) Free convection of gases

2-82   The emissivity of white paint is
(*a*) 1   (*b*) 0.9   (*c*) 0.5   (*d*) 0.25   (*e*) 0

2-83   The emissivity and absorptivity of a blackbody are, respectively
(*a*) 0, 1   (*b*) 0.5, 1   (*c*) 0, 0   (*d*) 1, 0   (*e*) 1, 1

2-84   When a surface is *completely enclosed* by a much larger surface separated by air, the net rate of radiation heat transfer between these two surfaces does not depend on
(*a*) Surface area of small surface
(*b*) Surface temperature
(*c*) Temperature of surrounding surface
(*d*) Air temperature
(*e*) Stefan-Boltzmann constant

2-85   Which one is not a correct unit for combined heat transfer coefficient?
(*a*) Btu/ft²·°F   (*b*) W/m²·K   (*c*) W/m²·°C   (*d*) kJ/s·ft²·K   (*e*) kcal/h·ft²·°F

FLUID MECHANICS

2-86   Define fluid and fluid mechanics.

2-87   What is the difference between a liquid and a gas? Explain.

2-88   What is viscosity? What is it due to?

2-89   What is the difference between dynamic viscosity and kinematic viscosity? What are their units?

2-90   When the pressure drop in pipe is available, in kPa, how do you determine the pumping power requirement, in kW, to overcome this pressure drop?

2-91   The dynamic viscosity of air at 101 kPa and 10°C is 1.778×10^{-5} kg/m·s. What is the kinematic viscosity of air, in m²/s and stoke?

2-92   The kinematic viscosity of air at 1 atm and 100°F is 1.809×10^{-4} ft²/s. What is the dynamic viscosity of air, in lbm/ft·s?

2-93   The viscosity of a fluid is to be measured by a viscometer constructed of two 85-cm-long concentric cylinders. The outer diameter of the inner cylinder is 15 cm, and the gap between the two cylinders is 1 mm. The inner cylinder is rotated at 450 rpm, and the torque is measured to be 0.9 Nm. Determine the viscosity of the fluid.

2-94   The viscosity of a fluid is to be measured by a viscometer constructed of two 2-ft-long concentric cylinders. The inner diameter of the outer cylinder is 6 in, and the gap between the two cylinders is 0.05 in. The outer cylinder is rotated at 300 rpm, and the torque is measured to be 1.2 lbf·ft. Determine the viscosity of the fluid.

2-95   Water at 15°C ($\rho = 999.7$ kg/m³ and $\mu = 1.307 \times 10^{-3}$ kg/m·s) is flowing steadily in a 0.25-cm-diameter, 35-m-long pipe at an average velocity of 1.2 m/s. Determine (a) the pressure drop, (b) the head loss, and (c) the pumping power requirement to overcome this pressure drop.

2-96   Air enters a 12-m-long section of a rectangular duct of cross section 15 cm × 20 cm made of commercial steel at 1 atm and 23°C at an average velocity of 5 m/s. Disregarding the entrance effects, determine the fan power needed to overcome the pressure losses in this section of the duct. The properties of air at 1 atm and 35°C are $\rho = 1.145$ kg/m³, $\mu = 1.895 \times 10^{-5}$ kg/m·s, and $\nu = 1.655 \times 10^{-5}$ m²/s.

2-97   Water at 72°F passes through 0.75-in-internal-diameter copper tubes at a rate of 1.3 lbm/s. Determine the pumping power per foot of pipe length required to maintain this flow at the specified rate. The density and dynamic viscosity of water at 60°F are $\rho = 62.36$ lbm/ft³ and $\mu = 2.713$ lbm/ft·h = 7.536×10^{-4} lbm/ft·s.

2-98   Which science deals with liquid flows in pipes and open channels?
(a) Hydrology   (b) Hydraulics   (c) Fluid dynamics
(d) Aerodynamics   (e) Gas dynamics

2-99   Which one is not a correct unit for dynamic or kinematic viscosity?
(a) N·s/m²   (b) ft²/min   (c) kg/m·s   (d) m²/s   (e) Pa·s/m²

2-100   When the temperature increases
(a) Viscosity of a gas increases and viscosity of a liquid decreases.
(b) Viscosity of a gas and a liquid increase.
(c) Viscosity of a gas decreases and viscosity of a liquid increases.
(d) Viscosity of a gas and a liquid decrease.
(e) Viscosity of a gas and a liquid remain constant.

2-101   The pressure drop during laminar flow of a fluid in a pipe does not depend on
(a) Friction coefficient   (b) Pipe diameter   (c) Surface roughness
(d) Fluid density   (e) Fluid velocity

2-102   When the diameter of a pipe doubles, the pumping power requirement in a laminar-flow piping system decreases by a factor of
(a) 2   (b) 4   (c) 8   (d) 16   (e) 32

THERMOCHEMISTRY

2-103   What are the combustible elements in common fossil fuels?

2-104   What are LPG and LNG? Explain.

2-105   What is a fuel? What is combustion?

2-106   The total mass of reactants must be equal to that of products in a chemical reaction. Is this also true for mole numbers?

2-107   What is stoichiometric air?

2-108   Define enthalpy of reaction, enthalpy of combustion, and enthalpy of formation.

2-109   The enthalpy of formation of CO_2 at the standard reference state is −393,520 kJ/kmol. What does the negative sign represent?

2-110   What is the heating value of a fuel?

2-111   What is the difference between the higher heating value (HHV) and lower heating value (LHV)? How are they related to each other?

2-112   Propane fuel (C_3H_8) is burned in the presence of air. Assuming that the combustion is theoretical, determine (a) the mass fraction of carbon dioxide and (b) the mole and mass fractions of the water vapor in the products. (c) Determine how much carbon dioxide is produced when 5 kg of propane is burned.

2-113   Ethyl alcohol (C_2H_5OH) is burned with 50 percent excess air. (a) Calculate the mole fractions of the products formed and the reactants. (b) Calculate the mass of carbon dioxide, water, and oxygen contained in the products per unit mass of fuel burned. (c) Calculate the air-fuel ratio.

2-114   Propylene (C_3H_6) is burned with 100 percent excess air during a combustion process. Assuming complete combustion and a total pressure of 105 kPa, determine (a) the air-fuel ratio and (b) the temperature at which the water vapor in the products will start condensing.

2-115   Methane (CH_4) is burned with dry air. The volumetric analysis of the products on a dry basis is 5.20 percent CO_2, 0.33 percent CO, 11.24 percent O_2, and 83.23 percent N_2. Determine (a) the air-fuel ratio and (b) the percentage of theoretical air used.

2-116   The higher heating value of propane (C_3H_8) is 50,330 kJ/kg. Calculate its lower heating value.

2-117   The lower heating value of isopentane (C_5H_{12}) is 19,310 Btu/lbm. Calculate its higher heating value.

2-118   The enthalpy of combustion of butane (C_4H_{10}) at 25°C and 1 atm is calculated to be 2857 MJ/kmol when the water in the products is in the liquid form. Calculate the lower and higher heating values of butane.

2-119   Determine the enthalpy of combustion of methane (CH_4) at 25°C and 1 atm, using the enthalpy of formation data from Table A-6. Assume that the water in the products is in the liquid form.

2-120   Calculate the HHV and LHV of gaseous n-octane fuel (C_8H_{18}). Compare your results with the values in Table A-7.

2-121   Calculate the higher and lower heating values of a coal from Illinois, that has an ultimate analysis (by mass) of 67.40 percent C, 5.31 percent H_2, 15.11 percent O_2, 1.44 percent N_2, 2.36 percent S, and 8.38 percent ash (noncombustibles). The enthalpy of formation of SO_2 is −297,100 kJ/kmol.

2-122   Octane gas (C_8H_{18}) is burned with 100 percent excess air in a constant-pressure burner. The air and fuel enter this burner steadily at standard conditions, and the products of combustion leave at 257°C. Calculate the rate of heat transfer from this burner when fuel is burned at a rate of 2.5 kg/min. Enthalpy values are given in the following table.

Substance	$\bar{h}_f^\circ$ kJ/kmol	$\bar{h}_{298\,K}$ kJ/kmol	$\bar{h}_{530\,K}$ kJ/kmol
C_8H_{18} (g)	−208,450	—	—
O_2	0	8682	15,708
N_2	0	8669	15,469
H_2O (g)	−241,820	9904	17,889
CO_2	−393,520	9364	19,029

2-123   Diesel fuel ($C_{12}H_{26}$) at 77°F is burned in a steady-flow combustion chamber with 20 percent excess air that also enters at 77°F. The products leave the combustion chamber at 800 R. Assuming combustion is complete, determine the required mass flow rate of the diesel fuel to supply heat at a rate of 1250 Btu/s. Enthalpy values are given in the following table.

Substance	$\overline{h}_f^{\circ}$ Btu/lbmol	$\overline{h}_{537\,R}$ Btu/lbmol	$\overline{h}_{800\,R}$ Btu/lbmol
$C_{12}H_{26}$	−125,190	—	—
O_2	0	3725.1	5602.0
N_2	0	3729.5	5564.4
H_2O (g)	−104,040	4258.0	6396.9
CO_2	−169,300	4027.5	6552.9

2-124   A coal from Texas which has an ultimate analysis (by mass) of 39.25 percent C, 6.93 percent H_2, 41.11 percent O_2, 0.72 percent N_2, 0.79 percent S, and 11.20 percent ash (noncombustibles) is burned steadily with 40 percent excess air in a power plant boiler. The coal and air enter this boiler at standard conditions, and the products of combustion in the smokestack are at 127°C. Calculate the heat transfer, in kJ/kg fuel, in this boiler. Include the effect of the sulfur in the energy analysis by noting that sulfur dioxide has an enthalpy of formation of −297,100 kJ/kmol and an average specific heat at constant pressure of $\overline{c}_p = 41.7$ kJ/kmol·K. Enthalpy values are given in the following table.

Substance	$\overline{h}_f^{\circ}$ kJ/kmol	$\overline{h}_{298\,K}$ kJ/kmol	$\overline{h}_{400\,K}$ kJ/kmol
O_2	0	8682	11,711
N_2	0	8669	11,640
H_2O (g)	−241,820	9904	13,356
CO_2	−393,520	9364	13,372
SO_2	−297,100	—	—

2-125   Which is not a fuel?
(a) Hydrogen   (b) Water   (c) Carbon   (d) Sulfur   (e) Carbon monoxide

2-126   When a hydrocarbon fuel is burned completely, the mass of carbon dioxide formed is always _____ the mass of fuel.
(a) Less than   (b) Equal to   (c) Greater than

2-127   When a hydrocarbon fuel is burned completely, the mass of water formed is always _____ the mass of fuel.
(a) Less than   (b) Equal to   (c) Greater than

2-128   When a fuel is burned with 200 percent theoretical air, this is equivalent to
(a) 100% stoichiometric air   (b) 200% excess air   (c) 300% stoichiometric air
(d) 100% excess air   (e) 100% deficiency of air

2-129   When a fuel is burned with 75 percent theoretical air, this is equivalent to
(a) 175% stoichiometric air   (b) 75% excess air   (c) 25% stoichiometric air
(d) 75% deficiency of air   (e) 25% deficiency of air

2-130   When a fuel is burned completely, the products cannot contain
(a) H_2 (b) O_2 (c) CO_2 (d) N_2 (e) H_2O

2-131 When a fossil fuel is burned completely, the products can contain
(*a*) C   (*b*) H_2 (*c*) CO   (*d*) OH   (*e*) SO_2

2-132 When a fuel is burned completely with stoichiometric air, the products cannot contain
(*a*) SO_2 (*b*) O_2 (*c*) CO_2 (*d*) N_2 (*e*) H_2O

2-133 The enthalpy of a substance at a specified state due to its chemical composition is called
(*a*) Enthalpy of reaction   (*b*) Enthalpy of combustion   (*c*) Enthalpy of formation
(*d*) Lower heating value   (*e*) Higher heating value

2-134 The higher heating value of carbon monoxide (CO) is 10,100 kJ/kg, and the enthalpy of vaporization of water at 25°C is $h_{fg} = 2442$ kJ/kg. What is the lower heating value of CO?
(*a*) 7658 kJ/kg   (*b*) 10,100 kJ/kg   (*c*) 12,542 kJ/kg
(*d*) 5216 kJ/kg   (*e*) 8879 kJ/kg

HEAT ENGINES AND POWER PLANTS

2-135 What are the characteristics of all heat engines?

2-136 What is a working fluid?

2-137 A steam power plant receives heat from a furnace at a rate of 290 GJ/h. Heat losses to the surrounding air from the steam as it passes through the pipes and other components are estimated to be about 5 GJ/h. If the waste heat is transferred to the cooling water at a rate of 160 GJ/h, determine (*a*) net power output and (*b*) the thermal efficiency of this power plant.

2-138 A car engine with a power output of 140 hp has a thermal efficiency of 30 percent. Determine the rate of fuel consumption if the heating value of the fuel is 19,000 Btu/lbm.

2-139 A 900-MW steam power plant, which is cooled by a nearby river, has a thermal efficiency of 40 percent. Determine the rate of heat transfer to the river water. Also, determine the maximum thermal efficiency of this power plant if the temperatures of the furnace and river are 1000 and 20°C, respectively.

2-140 An automobile engine consumes fuel at a rate of 22 L/h and delivers 55 kW of power to the wheels. If the fuel has a heating value of 44,000 kJ/kg and a density of 0.75 g/cm³, determine the efficiency of this engine.

2-141 A coal power plant produces 5 million kWh of electricity per year consuming coal with a heating value of 23,500 kJ/kg. If the plant consumes 2400 tons of coal per year, determine the overall plant efficiency.

2-142 A natural gas-fired power plant produces 400 MW of power with an overall plant efficiency of 43 percent.
(*a*) Determine the rate of natural gas consumption, in therm/h. Note that 1 therm = 100,000 Btu = 105,500 kJ.
(*b*) Determine the amount of electricity produced per year if the plant operates 8200 h a year.

2-143 A natural gas-fired power plant generates 800 MW of power consuming natural gas at a rate of 33 kg/s. Determine the overall plant efficiency. The heating value of natural gas is 50,000 kJ/kg.

2-144 Which plant truly operates on a thermodynamic cycle?
(*a*) Gas turbine   (*b*) Steam power plant   (*c*) Gasoline automobile engine
(*d*) Diesel automobile engine   (*e*) Solar photovoltaic panel

2-145 Which is *not* a heat engine?
(*a*) Gas turbine   (*b*) Steam power plant   (*c*) Automobile engine
(*d*) Hydroelectric power plant   (*e*) Solar photovoltaic panel

2-146 Which equation is *not* correct for heat engines?
(*a*) $W_{net,out} = Q_{in} - Q_{out}$ (*b*) $W_{net,out} = W_{out} - W_{in}$ (*c*) $Q_{in} = W_{out} + W_{in}$

(*d*) $\eta_{th} = 1 - \dfrac{Q_{out}}{Q_{in}}$ (*e*) $\eta_{th} = \dfrac{W_{net,out}}{Q_{in}}$

REFRIGERATORS AND HEAT PUMPS

2-147   What is a refrigerator? What is an air conditioner?

2-148   The second law of thermodynamics requires that heat cannot flow from low-temperature regions to high-temperature ones. Does the operation of a refrigerator violate this principle? Explain.

2-149   What is a heat pump? What is the difference between a refrigerator and a heat pump? What is the difference between a heat pump and an air conditioner?

2-150   Define the coefficient of performance (COP) of a refrigerator and a heat pump in words. Can they be greater than 1?

2-151   Someone claims that the COP of a refrigerator cannot be greater than 1. She says this would be a violation of the first law of thermodynamics since a refrigerator cannot provide more cooling than the work it consumes. Can she be right? Explain.

2-152   How can the COP of a Carnot refrigerator and a heat pump be increased?

2-153   Why are we interested in the Carnot refrigerator and Carnot heat pump even though they cannot be approximated in practice?

2-154   Consider two identical freezers operating in the same environment. The temperature of freezer A is set to $-18°C$, while the temperature of freezer B is set to $-24°C$. Which freezer consumes more electricity for the same rate of heat removal? Why?

2-155   Consider two identical refrigerators, both maintaining the refrigerated compartment at the same temperature. Refrigerator A operates in a kitchen at $20°C$, while refrigerator B operates in a kitchen at $25°C$. Which refrigerator consumes more electricity for the same rate of heat removal? Why?

2-156   A food section is maintained at $-8°C$ by a refrigerator in an environment at $27°C$. The rate of heat gain to the food section is estimated to be 850 kJ/h and the heat rejection in the condenser is 1250 kJ/h. Determine (*a*) the power input to the compressor in kW and (*b*) the COP of the refrigerator.

2-157   A 12,000 Btu/h split air conditioner is used to maintain a small room at $25°C$ when the ambient temperature is $35°C$. The air conditioner is running at full load under these conditions. The power input to the compressor is 1.7 kW. Determine (*a*) the rate of heat rejected in the condenser in Btu/h, (*b*) the COP of the air conditioner, and (*c*) the rate of cooling in Btu/h if the air conditioner operated as a Carnot refrigerator for the same power input.

2-158   Heat is absorbed from a space at $44°F$ at a rate of 22 Btu/s by a cooling system. Heat is rejected to water in the condenser. Water enters the condenser at $60°F$ at a rate of 1.85 lbm/s. The COP of the system is estimated to be 1.85. Determine (*a*) the power input to the system in kW and (*b*) the temperature of the water at the exit of the condenser. The specific heat of water is 1.0 Btu/lbm·°F.

2-159   A room is maintained at $23°C$ by a heat pump by rejecting heat to an environment at $6°C$. The rate of heat loss from the room to the environment is estimated to be 36,000 kJ/h, and the power input to the compressor is 3.3 kW. Determine (*a*) the rate of heat absorbed from the environment in kJ/h, (*b*) the COP of the heat pump, and (*c*) the maximum rate of heat supply to the room for the given power input.

2-160   Which one is not a heating or cooling system?
(*a*) Compressor   (*b*) Heat pump   (*c*) Boiler
(*d*) Air conditioner   (*e*) Evaporative cooler

2-161   Which one cannot be smaller than 1?
(*a*) Efficiency of a boiler
(*b*) Thermal efficiency of a heat engine
(*c*) COP of an air conditioner
(*d*) COP of a refrigerator
(*e*) COP of a heat pump

2-162   The coefficient of performance of an electric resistance heater is
(*a*) 0   (*b*) 0.5   (*c*) 1.0   (*d*) 2.0   (*e*) 3.0

2-163   A heat pump supplies 5 kW of heat to a room while consuming 2.5 kW of electricity. If this heat pump is replaced by an electric resistance heater supplying the same rate of heat to the room, the rate of electricity consumption will be

(a) 0 kW   (b) 2.5 kW   (c) 5.0 kW   (d) 7.5 kW   (e) 10 kW

2-164   A household refrigerator consumes 1 kW of electricity and has a COP of 1.5. The net effect of this refrigerator on the kitchen air is

(a) 1 kW cooling   (b) 1 kW heating   (c) 1.5 kW cooling
(d) 1.5 kW heating   (e) 2.5 kW heating

2-165   A window air conditioner is placed in the middle of a room. The air conditioner consumes 2 kW of electricity and has a COP of 2. The net effect of this air conditioner on the room is

(a) 2 kW cooling   (b) 2 kW heating   (c) 4 kW cooling
(d) 4 kW heating   (e) 6 kW heating

CHAPTER 3

Fundamentals of Solar Energy

3-1 INTRODUCTION

The electromagnetic energy emitted by the sun is called *solar radiation* or solar energy (or solar heat). Tremendous amounts of energy are created within the sun, and only a fraction of this energy reaches earth. This keeps earth at a temperature suitable for life. The amount of solar energy reaching earth's surface can easily meet all energy needs of the world. However, this is not practical due to the low concentration of solar energy and its relatively high cost. Other renewable energies such as geothermal, wind, hydro, and biomass appear to be less costly than direct solar energy, but their potentials with the current technologies are also much less than direct solar energy.

Solar energy reaches earth by radiation. Below, we review the fundamentals of radiation and the characteristics of solar radiation that form a basis for a more complete understanding of solar energy and its applications (see Çengel and Ghajar, 2020, for more details).

3-2 RADIATION FUNDAMENTALS

The theoretical foundation of radiation was established in 1864 by physicist James Clerk Maxwell, who postulated that accelerated charges or changing electric currents give rise to electric and magnetic fields. These rapidly moving fields are called *electromagnetic waves* or *electromagnetic radiation*, and they represent the energy emitted by matter as a result of the changes in the electronic configurations of the atoms or molecules. In 1887, Heinrich Hertz experimentally demonstrated the existence of such waves. Electromagnetic waves transport energy just like other waves, and all electromagnetic waves travel at the *speed of light* in a vacuum, which is $c_0 = 2.9979 \times 10^8$ m/s. Electromagnetic waves are characterized by their *frequency* v or *wavelength* λ. These two properties in a medium are related by

$$\lambda = \frac{c}{v} \tag{3-1}$$

where c is the speed of propagation of a wave in that medium. The speed of propagation in a medium is related to the speed of light in a vacuum by $c = c_0/n$, where n is the *index of refraction* of that medium. The refractive index is essentially unity for air and most gases, about 1.5 for glass, and 1.33 for water. The commonly used unit of wavelength is the *micrometer* (μm) or micron, where 1 μm $= 10^{-6}$ m. Unlike the wavelength and the speed of propagation, the frequency of an electromagnetic wave depends only on the source and is independent of the medium through which the wave travels. The *frequency* (the number of oscillations per second) of an electromagnetic wave can range from less than a million Hz to a septillion Hz or higher, depending on the source. Note from Eq. (3-1) that the wavelength and the frequency of electromagnetic radiation are inversely proportional.

It has proven useful to view electromagnetic radiation as the propagation of a collection of discrete packets of energy called *photons* or *quanta*, as proposed by Max Planck in 1900 in conjunction with his *quantum theory*. In this view, each photon of frequency v is considered to have an energy of

$$e = hv = \frac{hc}{\lambda} \tag{3-2}$$

where $h = 6.626069 \times 10^{-34}$ J·s is *Planck's constant*. Note from the second part of Eq. (3-2) that the energy of a photon is inversely proportional to its wavelength. Therefore, shorter-wavelength radiation possesses larger photon energies. It is no wonder that we try to avoid very-short-wavelength radiation such as gamma rays and X-rays since they are highly destructive.

Although all electromagnetic waves have the same general features, waves of different wavelength differ significantly in their behavior. The electromagnetic radiation encountered in practice covers a wide range of wavelengths, varying from less than 10^{-10} μm for cosmic rays to more than 10^{10} μm for electrical power waves. The *electromagnetic spectrum* also includes gamma rays, X-rays, ultraviolet (UV) radiation, visible light, infrared (IR) radiation, thermal radiation, microwaves, and radio waves, as shown in Fig. 3-1.

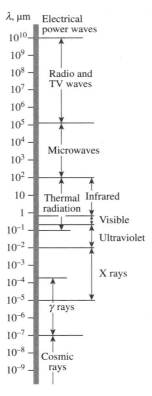

Figure 3-1 The electromagnetic wave spectrum.

Different types of electromagnetic radiation are produced through various mechanisms. For example, *gamma rays* are produced by nuclear reactions, X-rays by the bombardment of metals with high-energy electrons, *microwaves* by special types of electron tubes such as klystrons and magnetrons, and *radio waves* by the excitation of some crystals or by the flow of alternating current through electric conductors.

The short-wavelength gamma rays and X-rays are primarily of concern to nuclear engineers, while the long-wavelength microwaves and radio waves are of concern to electrical engineers. *Thermal radiation* is emitted as a result of energy transitions of molecules, atoms, and electrons of a substance. Temperature is a measure of the strength of these activities at the microscopic level, and the rate of thermal radiation emission increases with increasing temperature. Thermal radiation is continuously emitted by all matter whose temperature is above absolute zero. That is, everything around us, such as walls, furniture, and our friends, constantly emits (and absorbs) radiation (Fig. 3-2). Thermal radiation is also defined as the portion of the electromagnetic spectrum that extends from about 0.1 to 100 μm, since the radiation emitted by bodies due to their temperature falls almost entirely into this wavelength range. Thus, thermal radiation includes all visible and IR radiation as well as some UV radiation.

What we call *light* is simply the *visible* portion of the electromagnetic spectrum that lies between 0.40 and 0.76 μm. Light is characteristically no different than other electromagnetic radiation, except that it happens to trigger the sensation of seeing in the human eye. Light, or the visible spectrum, consists of narrow bands of color from violet (0.40 to 0.44 μm) to red (0.63 to 0.76 μm), as shown in Table 3-1.

Figure 3-2 Everything around us constantly emits thermal radiation.

TABLE 3-1   The Wavelength Ranges of Different Colors

Color	Wavelength Band (μm)
Violet	0.40–0.44
Blue	0.44–0.49
Green	0.49–0.54
Yellow	0.54–0.60
Orange	0.60–0.67
Red	0.63–0.76

A body that emits some radiation in the visible range is called a light source. The sun is obviously our primary light source. The electromagnetic radiation emitted by the sun is known as *solar radiation*, and nearly all of it falls into the wavelength band 0.3 to 3 μm. Almost *half* of solar radiation is light (i.e., it falls into the visible range), with the remaining being UV and IR.

The radiation emitted by bodies at room temperature falls into the IR region of the spectrum, which extends from 0.76 to 100 μm. Bodies start emitting noticeable visible radiation at temperatures above 800 K. The tungsten filament of a light bulb must be heated to temperatures above 2000 K before it can emit any significant amount of radiation in the visible range.

The UV radiation includes the low-wavelength end of the thermal radiation spectrum and lies between the wavelengths 0.01 and 0.40 μm. Ultraviolet rays are to be avoided since they can kill microorganisms and cause serious damage to humans and other living beings. About 12 percent of solar radiation is in the UV range, and it would be devastating if it were to reach the surface of the earth. Fortunately, the ozone (O_3) layer in the atmosphere acts as a protective blanket and absorbs most of this UV radiation. The UV rays that remain in sunlight are still sufficient to cause serious sunburns to sun worshippers, and prolonged exposure to direct sunlight is the leading cause of skin cancer, which can be lethal. Discoveries of "holes" in the ozone layer have prompted the international community to ban the use of ozone-destroying chemicals such as the refrigerant Freon-12 in order to save the earth. Ultraviolet radiation is also produced artificially in fluorescent lamps for use in medicine as a bacteria killer and in tanning parlors as an artificial tanner.

Blackbody Radiation

A body at a thermodynamic (or absolute) temperature above zero emits radiation in all directions over a wide range of wavelengths. The amount of radiation energy emitted from a surface at a given wavelength depends on the material of the body and the condition of its surface as well as the surface temperature. Therefore, different bodies may emit different amounts of radiation per unit surface area, even when they are at the same temperature. Thus, it is natural to be curious about the *maximum* amount of radiation that can be emitted by a surface at a given temperature. Satisfying this curiosity requires the definition of an idealized body, called a *blackbody,* to serve as a standard against which the radiative properties of real surfaces may be compared.

A *blackbody* is defined as *a perfect emitter and absorber of radiation*. At a specified temperature and wavelength, no surface can emit more energy than a blackbody. A blackbody absorbs *all* incident radiation, regardless of wavelength and direction. Also, a blackbody emits radiation energy uniformly in all directions per unit area normal to the direction of emission (Fig. 3-3). That is, a blackbody is a *diffuse* emitter. The term *diffuse* means "independent of direction."

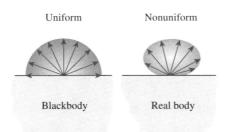

Figure 3-3   A blackbody is said to be a *diffuse* emitter since it emits radiation energy uniformly in all directions.

The radiation energy emitted by a blackbody per unit time and per unit surface area was determined experimentally by Joseph Stefan in 1879 and expressed as

$$E_b(T) = \sigma T^4 \qquad (3\text{-}3)$$

where $\sigma = 5.670 \times 10^{-8}$ W/m²·K⁴ is the *Stefan-Boltzmann constant* and T is the absolute temperature of the surface in K. This relation was theoretically verified in 1884 by Ludwig Boltzmann. Equation (3-3) is known as the *Stefan-Boltzmann law*, and E_b is called the *blackbody emissive power*. Note that the emission of thermal radiation is proportional to the *fourth power* of the absolute temperature.

Although a blackbody would appear *black* to the eye, a distinction should be made between the idealized blackbody and an ordinary black surface. Any surface that absorbs light (the visible portion of radiation) would appear black to the eye, and a surface that reflects it completely would appear white. Considering that visible radiation occupies a very narrow band of the spectrum from 0.4 to 0.76 μm, we cannot make any judgments about the blackness of a surface on the basis of visual observations. For example, snow and white paint reflect light and thus appear white. But they are essentially black for IR radiation since they strongly absorb long-wavelength radiation. Surfaces coated with lampblack paint approach idealized blackbody behavior.

The Stefan-Boltzmann law in Eq. (3-3) gives the *total* blackbody emissive power E_b, which is the sum of the radiation emitted over all wavelengths. Sometimes we need to know the *spectral blackbody emissive power*, which is *the amount of radiation energy emitted by a blackbody at a thermodynamic temperature T per unit time, per unit surface area, and per unit wavelength about the wavelength λ*. For example, we are more interested in the amount of radiation an incandescent light bulb emits in the visible wavelength spectrum than we are in the total amount emitted.

The relation for the spectral blackbody emissive power $E_{b\lambda}$ was developed by Max Planck in 1901 in conjunction with his famous quantum theory. This relation is known as *Planck's law* and is expressed as

$$E_{b\lambda}(\lambda, T) = \frac{C_1}{\lambda^5[\exp(C_2/\lambda T) - 1]} \qquad (\text{W/m}^2 \cdot \mu\text{m}) \qquad (3\text{-}4)$$

where

$$C_1 = 2\pi h c_0^2 = 3.74177 \times 10^8 \text{ W} \cdot \mu\text{m}^4/\text{m}^2$$

$$C_2 = h c_0/k = 1.43878 \times 10^4 \ \mu\text{m} \cdot \text{K}$$

Also, T is the absolute temperature of the surface, λ is the wavelength of the radiation emitted, and $k = 1.38065 \times 10^{-23}$ J/K is *Boltzmann's constant*. This relation is valid for a surface in a *vacuum* or a *gas*. For other mediums, it needs to be modified by replacing C_1 with C_1/n^2, where n is the index of refraction of the medium. Note that the term *spectral* indicates dependence on wavelength.

The variation of the spectral blackbody emissive power with wavelength is plotted in Fig. 3-4 for selected temperatures. Several observations can be made from this figure:

1. The emitted radiation is a continuous function of *wavelength*. At any specified temperature, it increases with wavelength, reaches a peak, and then decreases with increasing wavelength.

2. At any wavelength, the amount of emitted radiation *increases* with increasing temperature.

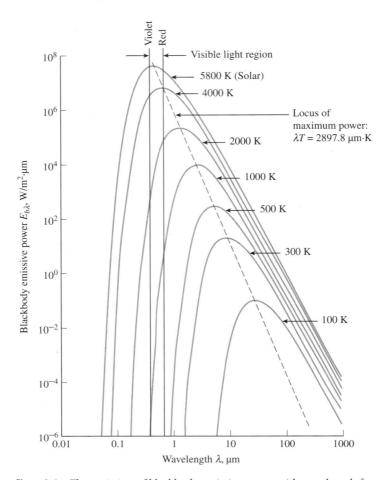

Figure 3-4 The variation of blackbody emissive power with wavelength for several temperatures.

3. As temperature increases, the curves shift to the left to the shorter wavelength region. Consequently, a larger fraction of the radiation is emitted at *shorter wavelengths* at higher temperatures.

4. The radiation emitted by the *sun,* which is considered to be a blackbody at 5780 K (or roughly at 5800 K), reaches its peak in the visible region of the spectrum. Therefore, the sun is in tune with our eyes. On the other hand, surfaces at $T < 800$ K emit almost entirely in the IR region and thus are not visible to the eye unless they reflect light coming from other sources.

As the temperature increases, the peak of the curve in Fig. 3-4 shifts toward shorter wavelengths. The wavelength at which the peak occurs for a specified temperature is given by *Wien's displacement law* as

$$(\lambda T)_{\text{max power}} = 2897.8 \ \mu\text{m·K} \tag{3-5}$$

A plot of Wien's displacement law, which is the locus of the peaks of the radiation emission curves, is also given in Fig. 3-4. The peak of solar radiation occurs at

$\lambda = 2897.8/5780 = 0.50$ μm, which is near the middle of the visible range. The peak of the radiation emitted by a surface at room temperature ($T = 298$ K) occurs at 9.72 μm, which is well into the IR region of the spectrum.

An electrical resistance heater starts radiating heat soon after it is plugged in, and we can feel the emitted radiation energy by holding our hands against the heater. But this radiation is entirely in the IR region and thus cannot be sensed by our eyes. The heater would appear dull red when its temperature reaches about 1000 K, since it starts emitting a detectable amount (about 1 W/m²·μm) of visible red radiation at that temperature. As the temperature rises even more, the heater appears bright red and is said to be *red hot*. When the temperature reaches about 1500 K, the heater emits enough radiation in the entire visible range of the spectrum to appear almost *white* to the eye, and it is called *white hot*.

Although it cannot be sensed directly by the human eye, IR radiation can be detected by IR cameras, which transmit the information to microprocessors to display visual images of objects at night. *Rattlesnakes* can sense the IR radiation or the "body heat" coming off warm-blooded animals, and thus they can see at night without using any instruments. Similarly, honeybees are sensitive to UV radiation. A surface that reflects all of the light appears *white,* while a surface that absorbs all of the light incident on it appears black. (Then how do we see a black surface?)

It should be clear from this discussion that the color of an object is not due to emission, which is primarily in the IR region, unless the surface temperature of the object exceeds about 1000 K. Instead, the color of a surface depends on the absorption and reflection characteristics of the surface and is due to selective absorption and reflection of the incident visible radiation coming from a light source such as the sun or an incandescent light bulb. A piece of clothing containing a pigment that reflects red while absorbing the remaining parts of the incident light appears "red" to the eye (Fig. 3-5). Leaves appear "green" because their cells contain the pigment chlorophyll, which strongly reflects green while absorbing other colors.

Integration of the *spectral* blackbody emissive power $E_{b\lambda}$ over the entire wavelength spectrum gives the *total* blackbody emissive power E_b:

$$E_b(T) = \int_0^\infty E_{b\lambda}(\lambda, T)\, d\lambda = \sigma T^4 \qquad (\text{W/m}^2) \qquad (3\text{-}6)$$

Thus, we obtained the Stefan-Boltzmann law [Eq. (3-3)] by integrating Planck's law [Eq. (3-4)] over all wavelengths. Note that on an $E_{b\lambda} - \lambda$ chart, $E_{b\lambda}$ corresponds to any value

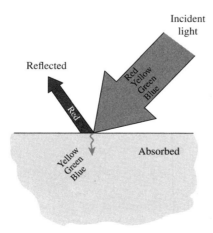

Figure 3-5 A surface that reflects red while absorbing the remaining parts of the incident light appears red to the eye.

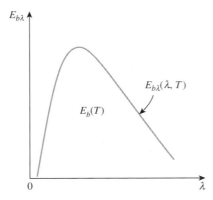

Figure 3-6 On an $E_{b\lambda} - \lambda$ chart, the area under a curve for a given temperature represents the total radiation energy emitted by a blackbody at that temperature.

on the curve, whereas E_b corresponds to the area under the entire curve for a specified temperature (Fig. 3-6). Also, the term *total* means "integrated over all wavelengths."

EXAMPLE 3-1
Radiation Emission
from a Blackbody

Consider a cylindrical body 30 cm in length and 15 cm in diameter at 327°C suspended in the air. Assuming the body closely approximates a blackbody, determine (*a*) the rate at which the cylinder emits radiation energy and (*b*) the spectral blackbody emissive power at a wavelength of 2.5 μm. (*c*) Also, determine the wavelength at which the emission of radiation from the filament peaks.

SOLUTION   (*a*) The surface area of the cylindrical body is

$$A_s = \pi DL + 2\frac{\pi D^2}{4} = \pi(0.15 \text{ m})(0.30 \text{ m}) + 2\frac{\pi(0.15 \text{ m})^2}{4} = 0.1767 \text{ m}^2$$

The total blackbody emissive power is determined from the Stefan-Boltzmann law to be

$$E_b(T) = \sigma T^4 A_s = (5.67 \times 10^{-8} \text{ W/m}^2 \cdot \text{K}^4)[(327 + 273) \text{ K}]^4 (0.1767 \text{ m}^2) = \mathbf{1298 \text{ W}}$$

(*b*) The spectral blackbody emissive power at a wavelength of 2.5 μm is determined from Planck's distribution law,

$$E_{b\lambda} = \frac{C_1}{\lambda^5 \left[\exp\left(\dfrac{C_2}{\lambda T}\right) - 1\right]} = \frac{3.74177 \times 10^8 \text{ W} \cdot \mu\text{m}^4/\text{m}^2}{(2.5 \text{ } \mu\text{m})^5 \left[\exp\left(\dfrac{1.43878 \times 10^4 \text{ } \mu\text{m} \cdot \text{K}}{(2.5 \text{ } \mu\text{m})(600 \text{ K})}\right) - 1\right]} = \mathbf{262 \text{ W}/\text{m}^2 \cdot \mu\text{m}}$$

(*c*) The wavelength at which the emission of radiation from the body is maximum is determined from Wien's displacement law to be

$$(\lambda T)_{\text{max power}} = 2897.8 \text{ } \mu\text{m} \cdot \text{K} \longrightarrow \lambda_{\text{max power}} = \frac{2897.8 \text{ } \mu\text{m} \cdot \text{K}}{600 \text{ K}} = \mathbf{4.83 \text{ } \mu\text{m}}$$

Note that the radiation emitted from the body peaks in the IR region.   ▲

3-3 RADIATIVE PROPERTIES

Most materials encountered in practice, such as metals, wood, and bricks, are *opaque* to thermal radiation, and radiation is considered to be a *surface phenomenon* for such materials. That is, thermal radiation is emitted or absorbed within the first few micrometers of the surface, and thus we speak of radiative properties of *surfaces* for opaque materials.

Some other materials, such as glass and water, allow visible radiation to penetrate to considerable depths before any significant absorption takes place. Radiation through such *semitransparent* materials obviously cannot be considered to be a surface phenomenon since the entire volume of the material interacts with radiation. On the other hand, both glass and water are practically opaque to IR radiation. Therefore, materials can exhibit different behavior at different wavelengths, and the dependence on wavelength is an important consideration in the study of radiative properties such as emissivity, absorptivity, reflectivity, and transmissivity of materials.

We defined a *blackbody* as a perfect emitter and absorber of radiation and said that nothing can emit more radiation than a blackbody at the same temperature. Therefore, a blackbody can serve as a convenient *reference* in describing the emission and absorption characteristics of real surfaces.

Emissivity

The *emissivity* of a surface represents *the ratio of the radiation emitted by the surface at a given temperature to the radiation emitted by a blackbody at the same temperature.* The emissivity of a surface is denoted by ε, and it varies between zero and one, $0 < \varepsilon < 1$. Emissivity is a measure of how closely a real surface approximates a blackbody, for which $\varepsilon = 1$.

Typical ranges of emissivity of various materials are given in Fig. 3-7. Note that metals generally have low emissivities, as low as 0.02 for polished surfaces, and nonmetals such as ceramics and organic materials have high ones. The emissivity of metals increases with temperature. Also, oxidation causes significant increases in the emissivity of metals. Heavily oxidized metals can have emissivities comparable to those of nonmetals.

Absorptivity, Reflectivity, and Transmissivity

Everything around us constantly emits radiation, and the emissivity represents the emission characteristics of those bodies. This means that every body, including our own, is constantly bombarded by radiation coming from all directions over a range of wavelengths.

The radiation flux incident on a surface from *all directions* is called *irradiation* or *incident radiation* and is denoted by G. It represents the rate at which radiation energy is incident on a surface per unit area of the surface. When radiation strikes a surface, part of it is absorbed, part of it is reflected, and the remaining part, if any, is transmitted, as illustrated in Fig. 3-8.

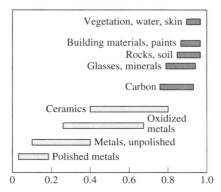

Figure 3-7 Typical ranges of emissivity for various materials.

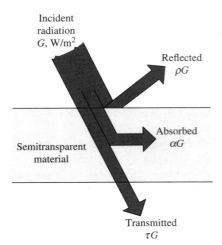

Incident
radiation
G, W/m^2

Reflected
ρG

Semitransparent
material

Absorbed
αG

Transmitted
τG

Figure 3-8   The absorption, reflection, and transmission of incident radiation by a semitransparent material.

The fraction of irradiation absorbed by the surface is called the *absorptivity* α, the fraction reflected by the surface is called the *reflectivity* ρ, and the fraction transmitted is called the *transmissivity* τ. That is,

Absorptivity: $\qquad \alpha = \dfrac{G_{abs}}{G} \qquad 0 < \alpha < 1$ (3-7)

Reflectivity: $\qquad \rho = \dfrac{G_{ref}}{G} \qquad 0 < \rho < 1$ (3-8)

Transmissivity: $\qquad \tau = \dfrac{G_{tr}}{G} \qquad 0 < \tau < 1$ (3-9)

where G is the radiation flux incident on the surface, and G_{abs}, G_{ref}, and G_{tr} are the absorbed, reflected, and transmitted portions of it, respectively. The first law of thermodynamics requires that the sum of the absorbed, reflected, and transmitted radiation be equal to the incident radiation. That is,

$$G_{abs} + G_{ref} + G_{tr} = G \qquad (3\text{-}10)$$

Dividing each term of this relation by G yields

$$\alpha + \rho + \tau = 1 \qquad (3\text{-}11)$$

For idealized blackbodies which are perfect absorbers, $\rho = 0$ and $\tau = 0$, and Eq. (3-11) reduces to $\alpha = 1$. For opaque surfaces such as most solids and liquids, $\tau = 0$, and thus

$$\alpha + \rho = 1 \qquad (3\text{-}12)$$

Unlike emissivity, the absorptivity of a material is practically independent of surface temperature. However, the absorptivity depends strongly on the temperature of the source from which the incident radiation is originating. This is also evident from Fig. 3-9, which shows the absorptivities of various materials at room temperature as functions of the temperature of the radiation source. For example, the absorptivity of the concrete roof of a house is about 0.6 for solar radiation (source temperature: 5780 K) and 0.9 for radiation

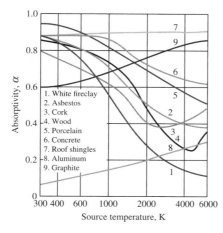

Figure 3-9 Variation of absorptivity with the temperature of the source of irradiation for various common materials at room temperature.

originating from the surrounding trees and buildings (source temperature: 300 K), as illustrated in Fig. 3-10.

Notice that the absorptivity of aluminum increases with the source temperature, a characteristic of metals, and the absorptivity of electric nonconductors, in general, decreases with temperature. This decrease is most pronounced for surfaces that appear white to the eye. For example, the absorptivity of a white painted surface is low for solar radiation, but it is rather high for IR radiation.

A relationship between the emissivity and absorptivity of a surface can be obtained as

$$\varepsilon(T) = \alpha(T) \tag{3-13}$$

That is, *the emissivity of a surface at temperature T is equal to its absorptivity for radiation coming from a blackbody at the same temperature.* This relation was first developed by Gustav Kirchhoff in 1860 and is now called *Kirchhoff's law*. Note that this relation is derived under the condition that the surface temperature is equal to the temperature of the source of irradiation, and the reader is cautioned against using it when a considerable difference (more than a few hundred degrees) exists between the surface temperature and the temperature of the source of irradiation.

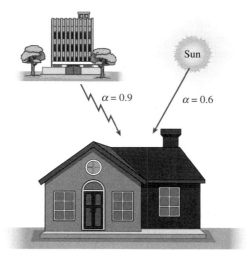

Figure 3-10 The absorptivity of a material may be quite different for radiation originating from sources at different temperatures.

EXAMPLE 3-2
Emissivity and Absorptivity
of a Plate

An opaque horizontal plate is well insulated on the edges and the lower surface (Fig. 3-11). The irradiation on the plate is 2500 W/m², of which 800 W/m² is reflected. The plate has a uniform temperature of 700 K and has an emissive power of 9000 W/m². Determine the total emissivity and absorptivity of the plate.

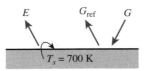

Figure 3-11   Schematic for Example 3-2.

SOLUTION   The total emissivity of the plate can be determined using

$$\varepsilon = \frac{E}{E_b} = \frac{E}{\sigma T_s^4} = \frac{9000 \text{ W/m}^2}{(5.67 \times 10^{-8} \text{ W/m}^2 \cdot \text{K}^4)(700 \text{ K})^4} = \mathbf{0.661}$$

The total absorptivity of the plate is determined using

$$\alpha + \rho + \tau = 1 \rightarrow \alpha = 1 - \rho \qquad (\text{for opaque surface, } \tau = 0)$$

The reflectivity of the plate is

$$\rho = \frac{G_{\text{ref}}}{G} = \frac{800 \text{ W/m}^2}{2500 \text{ W/m}^2} = 0.320$$

Hence, the total absorptivity of the plate is

$$\alpha = 1 - 0.320 = \mathbf{0.680} \quad \blacktriangle$$

The Greenhouse Effect

You have probably noticed that when you leave your car under direct sunlight on a sunny day, the interior of the car gets much warmer than the air outside, and you may have wondered why the car acts like a *heat trap*. The answer lies in the spectral transmissivity curve of the *glass,* which resembles an inverted U, as shown in Fig. 3-12. We observe from this figure that glass at thicknesses encountered in practice transmits over 90 percent of radiation in the visible range and is practically opaque (nontransparent) to radiation in the longer-wavelength IR regions of the electromagnetic spectrum (roughly $\lambda > 3$ μm). Therefore, glass has a transparent window in the wavelength range 0.3 μm $< \lambda <$ 3 μm in which over

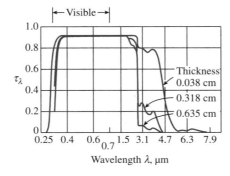

Figure 3-12   The spectral transmissivity of low-iron glass at room temperature for different thicknesses.

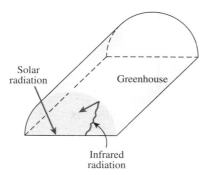

Figure 3-13 A greenhouse traps energy by allowing solar radiation to come in but not allowing IR radiation to go out.

90 percent of solar radiation is emitted. On the other hand, all radiation emitted by surfaces at room temperature falls in the IR region. Consequently, glass allows the solar radiation to enter but does not allow the IR radiation from the interior surfaces to escape. This causes a rise in the interior temperature as a result of the energy buildup in the car. This heating effect, which is due to the non-gray characteristic of glass (or clear plastics), is known as the *greenhouse effect*, since it is utilized extensively in greenhouses (Fig. 3-13).

The greenhouse effect is also experienced on a larger scale on earth. The surface of the earth, which warms up during the day as a result of the absorption of solar energy, cools down at night by radiating its energy into deep space as IR radiation. Greenhouse gases such as CO_2 and water vapor in the atmosphere transmit the bulk of the solar radiation but absorb the IR radiation emitted by the surface of the earth. Thus, there is concern that the energy trapped on earth will eventually cause global warming and thus drastic changes in weather patterns.

In *humid* places such as coastal areas, there is not a large change between the daytime and nighttime temperatures because the humidity acts as a barrier to the IR radiation coming from the earth, and thus it slows down the cooling process at night. In areas with clear skies such as deserts, there is a large swing between the daytime and nighttime temperatures because of the absence of such barriers for IR radiation.

3-4 SOLAR RADIATION

The sun is our primary source of energy. The energy coming off the sun, called *solar energy,* reaches us in the form of electromagnetic waves after multiple interactions with the atmosphere. The radiation energy emitted or reflected by the constituents of the atmosphere form the *atmospheric radiation.*

The *sun* is a nearly spherical body that has a diameter of $D = 1.393 \times 10^9$ m and a mass of $m \approx 2 \times 10^{30}$ kg and is located at a mean distance of $L = 1.496 \times 10^{11}$ m from the earth. It emits radiation energy continuously at a rate of $E_{sun} \approx 3.8 \times 10^{26}$ W. Less than a billionth of this energy (about 1.7×10^{17} W) strikes the earth, which is sufficient to keep the earth warm and to maintain life through photosynthesis. The sun generates energy by a continuous *fusion* reaction during which two hydrogen atoms fuse to form one atom of helium. Therefore, the sun is essentially a *nuclear reactor,* with temperatures as high as 40,000,000 K in its core region. The temperature drops to about 5800 K in the outer region of the sun, the convective zone, as a result of the dissipation of this energy by radiation. The solar energy reaching the earth's atmosphere is called the *total solar irradiance* G_s, whose value is

$$G_s = 1373 \text{ W/m}^2$$

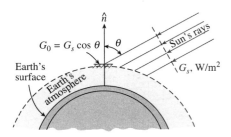

Figure 3-14   Solar radiation reaching the earth's atmosphere and the total solar irradiance.

The total solar irradiance (also called the *solar constant*) represents *the rate at which solar energy is incident on a surface normal to the sun's rays at the outer edge of the atmosphere when the earth is at its mean distance from the sun* (Fig. 3-14). The accepted value of the solar constant is 1373 W/m² (435.4 Btu/h·ft²), but its value changes by 3.5 percent from a maximum of 1418 W/m² on January 3, when the earth is closest to the sun, to a minimum of 1325 W/m² on July 4, when the earth is farthest away from the sun.

The value of the total solar irradiance can be used to estimate the effective surface temperature of the sun from the requirement that

$$(4\pi L^2)G_s = (4\pi r^2)\sigma T_{sun}^4 \tag{3-14}$$

where L is the mean distance between the sun's center and the earth and r is the radius of the sun. The left-hand side of this equation represents the total solar energy passing through a spherical surface whose radius is the mean earth-sun distance, and the right-hand side represents the total energy that leaves the sun's outer surface. The conservation of energy principle requires that these two quantities be equal to each other, since the solar energy experiences no attenuation (or enhancement) on its way through the vacuum (Fig. 3-15). The *effective surface temperature* of the sun is determined from Eq. (3-14) to be $T_{sun} = 5780$ K. That is, the sun can be treated as a blackbody at a temperature of 5780 K. This is also confirmed by the measurements of the spectral distribution of the solar radiation just outside the atmosphere, plotted in Fig. 3-16, which shows only small deviations from the idealized blackbody behavior.

The spectral distribution of solar radiation beyond the earth's atmosphere resembles the energy emitted by a *blackbody* (i.e., a perfect emitter and absorber of radiation) at 5780 K, with about 9 percent of the energy contained in the UV region (at wavelengths between 0.29 and 0.4 μm), 39 percent in the visible region (0.4 to 0.7 μm), and the remaining 52 percent

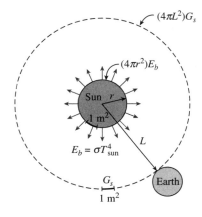

Figure 3-15   The total solar energy passing through concentric spheres remains constant, but the energy falling per unit area decreases with increasing radius.

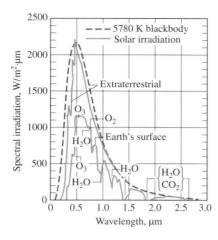

Figure 3-16 Spectral distribution of solar radiation just outside the atmosphere, at the surface of the earth on a typical day, and comparison with blackbody radiation at 5780 K.

in the near-IR region (0.7 to 3.5 μm). The peak radiation occurs at a wavelength of about 0.48 μm, which corresponds to the green portion of the visible spectrum.

The spectral distribution of solar radiation on the ground plotted in Fig. 3-16 shows that the solar radiation undergoes considerable *attenuation* as it passes through the atmosphere because of absorption and scattering. About 99 percent of the atmosphere is contained within a distance of 30 km from the earth's surface. The dips in the spectral distribution of radiation on the earth's surface are due to *absorption* by the gases O_2, O_3 (ozone), H_2O, and CO_2. Absorption by *oxygen* occurs in a narrow band about $\lambda = 0.76$ μm. The *ozone* absorbs UV radiation at wavelengths below 0.3 μm almost completely, and radiation in the range 0.3 to 0.4 μm considerably. Thus, the ozone layer in the upper regions of the atmosphere protects biological systems on earth from harmful UV radiation. In turn, we must protect the ozone layer from the destructive chemicals commonly used as refrigerants, cleaning agents, and propellants in aerosol cans. The use of these chemicals is now banned. The ozone gas also absorbs some radiation in the visible range. Absorption in the IR region is dominated by *water vapor* and *carbon dioxide*. The dust particles and other pollutants in the atmosphere also absorb radiation at various wavelengths.

As a result of these absorptions, the solar energy reaching the earth's surface is weakened considerably, to about 950 W/m² on a clear day and much less on cloudy or smoggy days. Also, practically all of the solar radiation reaching the earth's surface falls in the wavelength band from 0.3 to 2.5 μm.

Another mechanism that attenuates solar radiation as it passes through the atmosphere is *scattering* or *reflection* by air molecules and the many other kinds of particles such as dust, smog, and water droplets suspended in the atmosphere. Scattering is mainly governed by the size of the particle relative to the wavelength of radiation. The oxygen and nitrogen molecules primarily scatter radiation at very short wavelengths, comparable to the size of the molecules themselves. Therefore, radiation at wavelengths corresponding to violet and blue colors is scattered the most. This molecular scattering in all directions is what gives the sky its bluish color. The same phenomenon is responsible for red sunrises and sunsets. Early in the morning and late in the afternoon, the sun's rays pass through a greater thickness of the atmosphere than they do at midday, when the sun is directly overhead. Therefore, the violet and blue colors of the light encounter a greater number of molecules by the time they reach the earth's surface, and thus a greater fraction of them are scattered (Fig. 3-17). Consequently, the light that reaches the earth's surface consists primarily of colors corresponding to longer wavelengths such as red, orange, and yellow. The clouds appear in reddish-orange color during sunrise and sunset because the light they reflect is reddish-orange at those

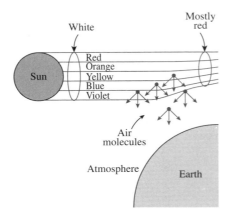

Figure 3-17 Air molecules scatter blue light much more than they do red light. At sunset, light travels through a thicker layer of atmosphere, which removes much of the blue from the natural light, allowing the red to dominate.

times. For the same reason, a red traffic light is visible from a longer distance than is a green light under the same circumstances.

The gas molecules and the suspended particles in the atmosphere emit radiation as well as absorb it. The atmospheric emission is primarily due to the CO_2 and H_2O molecules and is concentrated in the regions from 5 to 8 μm and above 13 μm. Although this emission is far from resembling the distribution of radiation from a blackbody, it is convenient in radiation calculations to treat the atmosphere as a blackbody at some lower fictitious temperature that emits an equivalent amount of radiation energy. This fictitious temperature is called the *effective sky temperature* T_{sky}. Then the radiation emission from the atmosphere to the earth's surface is expressed as

$$G_{sky} = \sigma T_{sky}^4 \qquad (\text{W/m}^2) \qquad (3\text{-}15)$$

The value of T_{sky} depends on the atmospheric conditions. It ranges from about 230 K for cold, clear-sky conditions to about 285 K for warm, cloudy-sky conditions. Note that the effective sky temperature does not deviate much from room temperature. Thus, in light of Kirchhoff's law, we can take the absorptivity of a surface to be equal to its emissivity at room temperature, $\alpha = \varepsilon$. Then the sky radiation absorbed by a surface can be expressed as

$$E_{sky,absorbed} = \alpha G_{sky} = \alpha \sigma T_{sky}^4 = \varepsilon \sigma T_{sky}^4 \qquad (\text{W/m}^2) \qquad (3\text{-}16)$$

The net rate of radiation heat transfer to a surface exposed to solar and atmospheric radiation is determined from an energy balance (Fig. 3-18):

$$\dot{q}_{net,rad} = E_{solar,absorbed} + E_{sky,absorbed} - E_{emitted}$$

$$= \alpha_s G_{solar} + \varepsilon \sigma T_{sky}^4 - \varepsilon \sigma T_s^4 \qquad (\text{W/m}^2) \qquad (3\text{-}17)$$

$$= \alpha_s G_{solar} + \varepsilon \sigma (T_{sky}^4 - T_s^4)$$

where T_s is the temperature of the surface in K and ε is its emissivity at room temperature. A positive result for $\dot{q}_{net,rad}$ indicates a radiation heat gain by the surface, and a negative result indicates a heat loss.

In solar energy applications, the spectral distribution of incident solar radiation is very different from the spectral distribution of emitted radiation by the surfaces, since the former

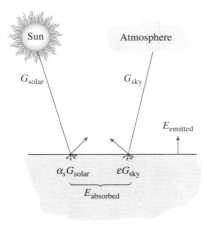

Figure 3-18   Radiation interactions of a surface exposed to solar and atmospheric radiation.

is concentrated in the short-wavelength region and the latter in the IR region. Therefore, the radiation properties of surfaces are quite different for the incident and emitted radiation, and the surfaces cannot be assumed to be gray. Instead, the surfaces are assumed to have two sets of properties: one for solar radiation and another for IR radiation at room temperature. Table 3-2 lists the *emissivity ε* and the *solar absorptivity α_s* of some common materials. Surfaces that are intended to *collect solar energy,* such as the absorber surfaces of

TABLE 3-2   Comparison of the Solar Absorptivity α_s of Some Surfaces with Their Emissivity ε at Room Temperature

Surface	α_s	ε
Aluminum		
Polished	0.09	0.03
Anodized	0.14	0.84
Foil	0.15	0.05
Copper		
Polished	0.18	0.03
Tarnished	0.65	0.75
Stainless steel		
Polished	0.37	0.60
Dull	0.50	0.21
Plated metals		
Black nickel oxide	0.92	0.08
Black chrome	0.87	0.09
Concrete	0.60	0.88
White marble	0.46	0.95
Red brick	0.63	0.93
Asphalt	0.90	0.90
Black paint	0.97	0.97
White paint	0.14	0.93
Snow	0.28	0.97
Human skin (Caucasian)	0.62	0.97

TABLE 3-3 Solar Radiative Properties of Materials

Description/Composition	Solar Absorptivity, α_s	Emissivity, ε, at 300 K	Ratio, α_s/ε	Solar Transmissivity, τ_s
Aluminum				
Polished	0.09	0.03	3.0	
Anodized	0.14	0.84	0.17	
Quartz-overcoated	0.11	0.37	0.30	
Foil	0.15	0.05	3.0	
Brick, red (Purdue)	0.63	0.93	0.68	
Concrete	0.60	0.88	0.68	
Galvanized sheet metal				
Clean, new	0.65	0.13	5.0	
Oxidized, weathered	0.80	0.28	2.9	
Glass, 3.2-mm thickness				
Float or tempered				0.79
Low iron oxide type				0.88
Marble, slightly off-white (nonreflective)	0.40	0.88	0.45	
Metal, plated				
Black sulfide	0.92	0.10	9.2	
Black cobalt oxide	0.93	0.30	3.1	
Black nickel oxide	0.92	0.08	11	
Black chrome	0.87	0.09	9.7	
Mylar, 0.13-mm thickness				0.87
Paints				
Black (Parsons)	0.98	0.98	1.0	
White, acrylic	0.26	0.90	0.29	
White, zinc oxide	0.16	0.93	0.17	
Paper, white	0.27	0.83	0.32	
Plexiglas, 3.2-mm thickness				0.90
Porcelain tiles, white (reflective glazed surface)	0.26	0.85	0.30	
Roofing tiles, bright red				
Dry surface	0.65	0.85	0.76	
Wet surface	0.88	0.91	0.96	
Sand, dry				
Off-white	0.52	0.82	0.63	
Dull red	0.73	0.86	0.82	
Snow				
Fine particles, fresh	0.13	0.82	0.16	
Ice granules	0.33	0.89	0.37	
Steel				
Mirror-finish	0.41	0.05	8.2	
Heavily rusted	0.89	0.92	0.96	
Stone (light pink)	0.65	0.87	0.74	
Tedlar, 0.10-mm thickness				0.92
Teflon, 0.13-mm thickness				0.92
Wood	0.59	0.90	0.66	

Source: V. C. Sharma and A. Sharma, "Solar Properties of Some Building Elements," *Energy* 14 (1989), pp. 805–810, and other sources.

solar collectors, should have high α_s but low ε values to maximize the absorption of solar radiation and to minimize the emission of radiation. Surfaces that are intended to *remain cool* under the sun, such as the outer surfaces of fuel tanks and refrigerator trucks, are desired to have just the opposite properties. Surfaces are often given the desired properties by coating them with thin layers of *selective* materials. A surface can be kept cool, for example, by simply painting it white. In practice, engineers pay close attention to the ratio α_s/ε when selecting appropriate materials for the purpose of heat collection or heat rejection. For heat collection, materials with large values of α_s/ε (such as clean galvanized sheet metal with $\alpha_s/\varepsilon = 5.0$) are required. For heat rejection, on the other hand, materials with small values of α_s/ε (such as anodized aluminum with $\alpha_s/\varepsilon = 0.17$) are desirable. Values of α_s/ε together with solar absorptivity for selected materials are listed in Table 3-3.

EXAMPLE 3-3
Calculation of Solar Constant

Determine the solar constant using Eq. (3-14). Take the effective surface temperature of the sun as $T_{sun} = 5778$ K.

SOLUTION   The sun is a nearly spherical body that has a diameter of $D = 1.393 \times 10^9$ m and is located at a mean distance of $L = 1.496 \times 10^{11}$ m from the earth. Also, the effective surface temperature of the sun is $T_{sun} = 5778$ K. Then, using Eq. (3-14), the solar constant is determined to be

$$(4\pi L^2)G_s = (4\pi r^2)\sigma T_{sun}^4$$

$$4\pi(1.496 \times 10^{11} \text{ m})^2 G_s = 4\pi(0.5 \times 1.393 \times 10^9 \text{ m})^2(5.67 \times 10^{-8} \text{ W/m}^2\cdot\text{K}^4)(5778 \text{ K})^4$$

$$G_s = \textbf{1370 W/m}^2 \quad \blacktriangle$$

EXAMPLE 3-4
Energy Balance on a Solar Collector

The absorber surface of a solar collector is made of aluminum coated with black chrome ($\alpha_s = 0.87$ and $\varepsilon = 0.09$). Solar radiation is incident on the surface at a rate of 720 W/m^2 (Fig. 3-19). The air and the effective sky temperatures are 25 and 15°C, respectively, and the convection heat transfer coefficient is 10 W/m$^2\cdot$K. For an absorber surface temperature of 70°C, determine the net rate of solar energy delivered by the absorber plate to the water circulating behind it.

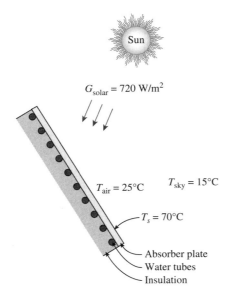

Figure 3-19   Schematic for Example 3-4.

SOLUTION   The net rate of solar energy delivered by the absorber plate to the water circulating behind it can be determined from an energy balance to be

$$\dot{q}_{\text{net}} = \dot{q}_{\text{gain}} - \dot{q}_{\text{loss}}$$

$$\dot{q}_{\text{net}} = \alpha_s G_{\text{solar}} - [\varepsilon\sigma(T_s^4 - T_{\text{sky}}^4) + h(T_s - T_{\text{air}})]$$

Then,

$$\dot{q}_{\text{net}} = 0.87(720 \text{ W/m}^2) - 0.09(5.67 \times 10^{-8} \text{ W/m}^2 \cdot \text{K}^4)[(70 + 273 \text{ K})^4 - (15 + 273 \text{ K})^4]$$

$$- (10 \text{ W/m}^2 \cdot \text{K})(70°\text{C} - 25°\text{C})$$

$$= \mathbf{141 \, W/m^2}$$

Therefore, heat is gained by the plate and transferred to water at a rate of 141 W per m^2 surface area.   ▲

3-5 ESTIMATION OF SOLAR RADIATION

The most accurate data on solar radiation are obtained for a large number of locations by long-term measurements. Solar tables are available in various literature, and they should be used whenever possible for the estimation of solar system performance. In the absence of such data, one can use one of the calculation methods proposed in the literature to estimate the amount of solar radiation in a particular location with specified time and surface orientation. In this section, we present a relatively simple method as outlined in Culp (1991), which is based on ASHRAE handbooks. More complicated and possibly more accurate methods are available in solar literature. Solar time and solar angles are essential parameters in the calculation of solar radiation, and they are presented first.

Solar Time

The *latitude* of a location is the angle between the equator plane and radial line to the location, and it is measured from the center of the earth. In another words, it is the angular distance of a location north or south of the equator. The lines of latitude are known as *parallels*. Constant latitude lines are circles parallel to the equator circle running in an east to west direction. The latitude is given in degrees. It is 0° at the equator, 90° at the north pole, and −90° at the south pole.

The *longitude* of a location is the angular distance east or west of the prime meridian, which runs from the north pole to the south pole through Greenwich, London. The lines of longitude are known as *meridians*. In the prime meridian, the longitude is 0°. Since one full revolution of the earth is 360° and takes 24 hours or 1440 minutes, one degree of longitude is equal to 4 minutes (1440 min/360° = 4 min/°). Each time zone has a width of 15° (equivalent to 1 hour). Note that in some countries and regions, a single time zone is used even though the width of longitude is greater than 15°.

A *standard meridian* is the longitude line running from the north pole to the south pole through the center of a time zone. The location of a standard meridian for a given local longitude is given by

$$\text{Longitude of standard meridian} = 15\left(\frac{\text{Local longitude}}{15}\right)_{\text{round to nearest integer}} \tag{3-18}$$

The standard meridians in the United States are:

60°W longitude (Atlantic ST)

75°W longitude (Eastern ST)

90°W longitude (Central ST)

105°W longitude (Mountain ST)

120°W longitude (Pacific ST)

135°W longitude (Yukon ST)

150°W longitude (Alaska-Hawaii ST)

The local standard time is equal to *mean sun time* (MST) at the location of the standard meridian in a given time zone. The mean sun time increases by 4 min to the east and decreases by 4 min to the west of the standard meridian for each degree of longitude. The mean sun time can be determined from

$$\text{Mean sun time} = \text{Local standard time} + (4 \text{ min}) \times (\text{Longitude of standard meridian} - \text{Local longitude}) \quad (3\text{-}19)$$

If the daylight savings time is used in a location during summer months, the local standard time should be taken as 1 hour less than the local daylight savings time.

The mean sun time is not used as the solar time in solar radiation calculations. Instead, the *apparent solar time* (or just *solar time*) is used. The mean sun time is not equal to the apparent solar time because the earth does not move around the sun at a constant velocity. This is due to the elliptical orbit of the earth around the sun. The difference between the actual solar time (apparent solar time) and the mean solar time is called *equation of time*. The equation of time changes between +16.3 min in November to −14.4 min in February depending on the time of the year. Equation of time values are given in Table 3-4.

TABLE 3-4   Some Solar Constants for Northern Latitudes (ASHRAE, 1974)

Date	Declination angle, δ	Equation of time, min	A_S, W/m²	B_S	C_S
January					
1	−23.0	−3.6	1231	0.142	0.057
11	−21.7	−8.0	1230	0.142	0.058
21	−19.6	−11.4	1229	0.142	0.058
31	−17.3	−13.5	1224	0.142	0.058
February					
1	−17.0	−13.6	1223	0.143	0.059
11	−13.9	−14.4	1218	0.143	0.050
21	−10.4	−13.8	1213	0.144	0.060
28	−7.6	−12.6	1206	0.147	0.063
March					
1	−7.4	−12.5	1205	0.148	0.063
11	−3.5	−10.2	1195	0.152	0.067
21	0.0	−7.4	1185	0.156	0.071
31	4.3	−4.3	1169	0.164	0.079
April					
1	4.7	−4.0	1167	0.165	0.080
11	8.5	−1.1	1151	0.172	0.088
21	12.0	1.2	1135	0.180	0.096
30	14.9	2.8	1126	0.185	0.104

(Continued)

TABLE 3-4 **Some Solar Constants for Northern Latitudes** (ASHRAE, 1974) (*Continued*)

Date	Declination angle, δ	Equation of time, min	A_s, W/m²	B_s	C_s
May					
1	15.2	2.9	1125	0.186	0.105
11	17.9	3.7	1114	0.191	0.113
21	20.3	3.6	1103	0.196	0.121
31	22.0	2.5	1098	0.199	0.125
June					
1	22.1	2.4	1097	0.199	0.126
11	23.1	0.6	1092	0.202	0.130
21	23.45	−1.5	1087	0.205	0.134
30	23.1	−3.4	1086	0.206	0.135
July					
1	23.1	−3.4	1086	0.206	0.135
11	22.1	−5.6	1085	0.206	0.135
21	20.6	−6.2	1084	0.207	0.136
31	18.1	−6.2	1091	0.205	0.132
August					
1	17.9	−6.2	1092	0.205	0.132
11	15.2	−5.1	1099	0.203	0.126
21	12.0	−3.1	1106	0.201	0.122
31	8.7	−1.5	1120	0.193	0.113
September					
1	8.2	0.0	1122	0.192	0.112
11	4.4	3.3	1136	0.185	0.097
21	0.0	6.8	1150	0.177	0.092
30	−2.9	9.9	1162	0.172	0.086
October					
1	−3.3	10.2	1164	0.171	0.086
11	−7.2	13.1	1177	0.166	0.079
21	−10.8	15.3	1191	0.160	0.073
31	−14.3	16.2	1200	0.157	0.070
November					
1	−14.6	16.3	1201	0.156	0.069
11	−17.5	15.9	1211	0.153	0.066
21	−20.0	15.1	1220	0.149	0.063
30	−21.7	11.3	1224	0.147	0.061
December					
1	−21.9	11.0	1224	0.147	0.061
11	−23.0	6.8	1228	0.144	0.059
21	−23.45	2.0	1232	0.142	0.057
31	−23.1	−3.1	1231	0.142	0.057

Once the value for equation of time is obtained from Table 3-4, the solar time can be calculated from

$$\text{Solar time} = \text{Mean sun time} + \text{Equation of time} \qquad (3\text{-}20)$$

At the solar time of 12:00, the sun's rays are perpendicular to a horizontal surface on the earth. Unless otherwise specified, the time is solar time in the estimation of solar radiation.

EXAMPLE 3-5
Calculation of Solar Time

Consider a location in the United States whose longitude is 80°W and the local time is 11:00 on October 21. What is the solar time?

SOLUTION   The longitude of standard meridian is calculated as

$$\text{Longitude of standard meridian} = 15\left(\frac{\text{Local longitude}}{15}\right)_{\text{round to nearest integer}}$$

$$= 15\left(\frac{80}{15}\right)_{\text{round to nearest integer}}$$

$$= (15)(5.33)_{\text{round to nearest integer}}$$

$$= (15)(5)$$

$$= 75°$$

The mean sun time of this location is

$$\text{Mean sun time} = \text{Local standard time} + (4\text{ min})(\text{Longitude of standard meridian} - \text{Local longitude})$$
$$= 11:00 + (4\text{ min})(75 - 80) = 10:40$$

From Table 3-4, the equation of time on October 21 is 15.3 min. Then the solar time becomes

$$\text{Solar time} = \text{Mean sun time} + \text{Equation of time}$$
$$= 10:40 + 15.3\text{ min} = \textbf{10:53}   ▲$$

Solar Angles

The imaginary plane on which the earth travels around the sun is called the *ecliptic plane* (Fig. 3-20). The axis of the earth's rotation (called *polar axis*) is inclined at a 23.45° angle from a normal to the ecliptic plane. This inclination is responsible for the production of seasons and much of the complications associated with solar radiation calculations.

The inclination angle of 23.45° is also responsible for a solar angle called the *declination angle* δ. It is the angle between sun's rays and the equatorial plane. It is also equal to

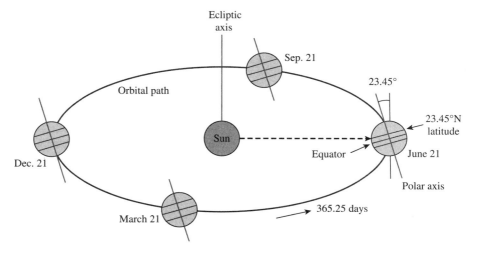

Figure 3-20   Earth motion around the sun.

that latitude on the earth at 12:00 solar time. The extreme values of the declination angle for northern latitudes are as follows:

0° on March 21 (vernal equinox)

0° on September 21 (autumnal equinox)

+23.45° on June 21 (summer solstice)

−23.45° on December 21 (winter solstice)

The declination angle depends on the time of the year, and its values are tabulated in Table 3-4. There are other important angles used in solar radiation calculations, as shown in Fig. 3-21.

The position of the sun can be expressed by two angles at a given latitude; solar altitude angle β_1 and solar azimuth angle α_1. The orientation of the collecting surface is expressed by the surface tilt angle β_2 and surface azimuth angle α_2. The most relevant angle for solar radiation calculation is incident angle θ. These angles as well as latitude angle and hour angle are defined as follows:

Altitude angle β_1: Altitude angle of the sun's rays. The angle between the sun's rays and the horizontal surface at the given location.

Solar azimuth angle α_1: Azimuth angle of the sun's rays. The angle between the horizontal projection of the sun's rays and the local longitude (north-south line).

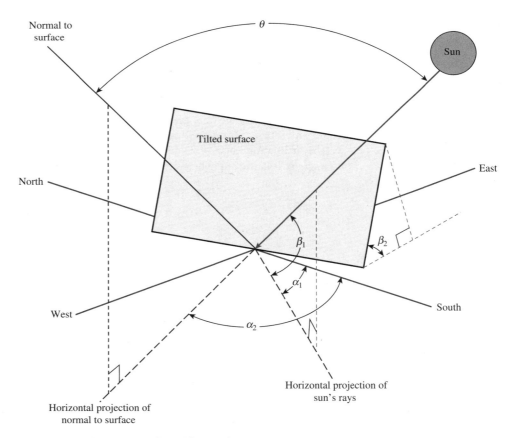

Figure 3-21   Solar angles. (*Adapted from Culp, 1991.*)

Tilt angle β_2: Tilt angle of the surface. The angle between the surface and the horizontal plane.

Surface azimuth angle α_2: Azimuth angle of the surface. The angle between the horizontal projection of the normal to the surface and due south line measured positive in the clockwise direction and negative in the counterclockwise direction. If the surface is horizontal, the tilt angle β_2 is zero and surface azimuth angle α_2 is not needed.

Incident angle θ: The angle that the sun's rays make with the normal of the surface.

Latitude angle L: The local latitude.

Hour angle H: The hour angle of the sun's rays is positive after noon and negative before noon.

The hour angle of the solar time (in degrees) is determined from

$$H = \frac{\text{Number of minutes past midnight} - 720}{4} \tag{3-21}$$

The altitude angle can be determined using certain solar angles as

$$\sin\beta_1 = (\cos L)(\cos\delta)(\cos H) + (\sin L)(\sin\delta) \tag{3-22}$$

If the calculated value of altitude angle is negative, the sun is below the horizon and there is no solar radiation incident on the surface. The solar azimuth angle for positive values of altitude angle can be calculated from

$$\sin\alpha_{1,c} = \frac{(\cos\delta)(\sin H)}{\cos\beta_1} \tag{3-23}$$

Here, the subscript c is a reminder that it is the calculated value.

The solar azimuth line is measured from the due south line in a positive clockwise direction or negative counterclockwise direction. The declination angle may exceed the local latitude during the summer season at latitudes smaller than 23.45°. When this happens, the value of solar azimuth angle α_1 may exceed 90°. Equation (3-23) for the solar azimuth angle gives values between −90° and 90° even though its absolute value can exceed 90°. This happens when the latitude is greater than the declination angle ($L > \delta$) at early morning or late afternoon hours. If $L \le \delta$, the sun is always north of a latitude line and the absolute value of the solar azimuth angle is greater than 90°. When $L > \delta$, we need to calculate the time that the sun is due east t_E, or the time that the sun is due west t_W, as follows:

$$t_E = \frac{180 - \cos^{-1}[(\tan\delta)/(\tan L)]}{15} \quad \text{(00:00 to 11:59, solar time)} \tag{3-24}$$

$$t_W = \frac{\cos^{-1}[(\tan\delta)/(\tan L)]}{15} \quad \text{(12:00 to 23:59, solar time)} \tag{3-25}$$

If the solar time is earlier than t_E or later than t_W, the sun is north of a constant-latitude line and the absolute value of solar azimuth angle is greater than 90°. Then the absolute value of solar azimuth angle can be calculated from

$$|\alpha_1| = 180° - |\alpha_{1,c}| \tag{3-26}$$

The incident angle can be calculated after evaluating other solar angles by the following relation:

$$\cos\theta = (\sin\beta_1)(\cos\beta_2) + (\cos\beta_1)(\sin\beta_2)[\cos(\alpha_1 - \alpha_2)] \tag{3-27}$$

For a horizontal surface ($\beta_2 = 0$):

$$\cos\theta = \sin\beta_1$$
$$\theta = 90° - \beta_1$$

For a vertical surface ($\beta_2 = 90°$ and $\alpha_2 = \alpha_1$):

$$\cos\theta = \cos\beta_1$$
$$\theta = \beta_1$$

If $\cos\theta$ is negative, the surface does not receive any direct solar radiation.

Components of Solar Radiation

The solar energy incident on a surface is usually called *solar irradiance, solar irradiation, solar insolation,* or just *insolation.* It is denoted by G with the unit W/m². Note that some sources make a distinction between solar irradiation and solar irradiance. They define solar irradiance as the rate of solar radiation incident on a surface (W/m²) and solar irradiation as the amount of solar radiation incident on a surface during a time period (J/m²). In this text, we make no such distinction and use these terms interchangeably. The context makes it clear if it is the rate or amount of solar radiation incident.

The solar radiation incident on the earth's surface is considered to consist of *direct, diffuse,* and *reflected* parts. The part of solar radiation that reaches the earth's surface without being scattered or absorbed by the atmosphere is called *direct solar radiation G_D* (also called *beam radiation*). The scattered radiation is assumed to reach the earth's surface uniformly from all directions and is called *diffuse solar radiation G_d.* When the direct component of solar radiation hits the surrounding ground-level surfaces, part of it reflects to the target surface, denoted by G_r. The *total* solar energy incident on the unit area of a *horizontal surface* on the ground is expressed as (Fig. 3-22)

$$G = G_D + G_d + G_r \tag{3-28}$$

where

$$G_D = G_{D,n}\cos\theta \tag{3-29}$$

In Eq. (3-29), $G_{D,n}$ is the direct solar radiation on a surface normal to the sun's rays and θ is the incident angle.

The diffuse radiation varies from about 10 percent of the total radiation on a clear day to nearly 100 percent on a totally cloudy day. The reflected component of solar radiation is difficult to estimate as it varies greatly depending on the application, orientation of surrounding surfaces, and their reflectivity. The reflected radiation should be evaluated

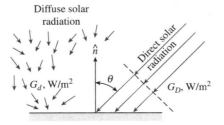

Figure 3-22 The direct and diffuse radiation incident on a horizontal surface on the earth's surface.

individually, and it is not considered here. Note that the reflected component of solar radiation plays a major role in some solar systems such as evacuated solar collectors, concentrating collectors, and solar-tower-power plant.

The direct solar radiation on a surface depends on the surface orientation, location, time of the day and year, the altitude, and atmospheric conditions. An estimate of normal component of direct solar radiation is given by

$$G_{D,n} = A_S \exp\frac{-B_S}{\sin\beta_1} \tag{3-30}$$

where A_S is the apparent extraterrestrial solar radiation (outside atmosphere) and B_S is the atmospheric extinction coefficient. The B_S value depends on time of the year and the amount of water vapor in the average atmosphere. A_S and B_S values are given for a number of days of the year in Table 3-4.

The behavior of atmosphere with respect to solar radiation may be expressed by the *clearness number*. It is defined as the ratio of the clear-day direct solar radiation at a given location and time to the value for the standard atmosphere at the same location and time. The values of clearness number range from about 0.85 to 1.15 in the United States.

For a surface receiving direct solar radiation, the diffuse component may be estimated using a relation that is a linear function of the direct solar radiation:

$$G_d = C_S F_{SS} G_{d,n} \tag{3-31}$$

Here, C_S is the ratio of the diffuse solar radiation to direct solar radiation falling on a horizontal surface. F_{SS} is the radiation view factor for radiation leaving the surface that hits the sky and is approximated by

$$F_{SS} = \frac{1+\cos\beta_2}{2} \tag{3-32}$$

C_S values are listed in Table 3-4.

EXAMPLE 3-6
Calculation of Direct and Diffuse Solar Radiation

Consider a location with a latitude of 40°N and the solar time is 11:00 on July 11. Estimate the direct and diffuse components of solar radiation on a surface at the southeast direction inclined at an angle of 30° from the horizontal.

SOLUTION   First, we obtain various parameters from Table 3-4:

Latitude $L = 40°$, Tilt angle $\beta_2 = 30°$, Surface azimuth angle $\alpha_2 = -45°$, Declination angle $\delta = 22.1°$, Equation of time $= -5.6$ min, $A_S = 1085$ W/m², $B_S = 0.206$, $C_S = 0.135$

Since the declination angle is greater than zero, we calculate the time the sun is due east as

$$t_E = \frac{180 - \cos^{-1}[(\tan\delta)/(\tan L)]}{15} = \frac{180 - \cos^{-1}[(\tan 22.1°)/(\tan 40°)]}{15} = 7.9295$$

or

$$t_E = 7:56$$

The solar time (11:00) is not earlier than t_E. The hour angle is

$$H = \frac{\text{Number of minutes past midnight} - 720}{4} = \frac{11 \times 60 - 720}{4} = -15°$$

The altitude angle is

$$\sin \beta_1 = (\cos L)(\cos \delta)(\cos H) + (\sin L)(\sin \delta)$$
$$= (\cos 40°)(\cos 22.1°)[\cos(-15°)] + (\sin 40°)(\sin 22.1°)$$
$$= 0.9274$$

or

$$\beta_1 = \sin^{-1} \beta_1 = \sin^{-1}(0.9274) = 68.03°$$

The solar azimuth angle is

$$\sin \alpha_{1,c} = \frac{(\cos \delta)(\sin H)}{\cos \beta_1} = \frac{(\cos 22.1°)[\sin(-15°)]}{\cos 68.03°} = 0.6410$$

or

$$\alpha_1 = \sin^{-1} \alpha_1 = \sin^{-1}(0.6410) = 39.86°$$

The incident angle is

$$\cos \theta = (\sin \beta_1)(\cos \beta_2) + (\cos \beta_1)(\sin \beta_2)[\cos(\alpha_1 - \alpha_2)]$$
$$= (\sin 68.03°)(\cos 30°) + (\cos 68.03°)(\sin 30°)[\cos(39.86° - (-45°))]$$
$$= 0.8199$$

or

$$\theta = \cos^{-1} \theta = \cos^{-1}(0.8199) = 34.93°$$

The radiation view factor is

$$F_{SS} = \frac{1 + \cos \beta_2}{2} = \frac{1 + \cos 30°}{2} = 0.9330$$

The direct solar radiation on this surface normal to the sun's rays is

$$G_{D,n} = A_S \exp \frac{-B_S}{\sin \beta_1} = (1085 \text{ W/m}^2) \exp \frac{-0.206}{\sin 68.03°} = 868.9 \text{ W/m}^2$$

The direct component of solar radiation on the surface is

$$G_D = G_{D,n} \cos \theta = (868.9 \text{ W/m}^2)(\cos 34.93°) = \textbf{712 W/m}^2$$

The diffuse component of solar radiation on the surface is

$$G_d = C_S F_{SS} G_{D,n} = (0.135)(0.9330)(868.9 \text{ W/m}^2) = \textbf{109 W/m}^2$$

The diffuse component is 13.3 percent of the combined direct and diffuse components of the solar radiation.   ▲

As already mentioned, the amount of solar radiation incident on a surface depends on the orientation of the surface, the latitude and elevation of the location, the humidity of air, the clearness of the sky, and the time of day. Table 3-5 gives hourly solar radiation incident on various surfaces at 40° latitude. Data are also available at other latitudes. Average daily solar radiation values on a horizontal surface in the United States are given for selected cities in Table 3-6. Extensive solar data at various latitudes and locations are available in

TABLE 3-5 Hourly Variation of Solar Radiation Incident on Various Surfaces and the Daily Totals throughout the Year at 40° Latitude (ASHRAE, 1993)

		Solar Radiation Incident on the Surface,* W/m²														
		Solar Time														
Date	Direction of Surface	5	6	7	8	9	10	11	12 noon	13	14	15	16	17	18	19	Daily Total
Jan.	N	0	0	0	20	43	66	68	71	68	66	43	20	0	0	0	446
	NE	0	0	0	63	47	66	68	71	68	59	43	20	0	0	0	489
	E	0	0	0	402	557	448	222	76	68	59	43	20	0	0	0	1863
	SE	0	0	0	483	811	875	803	647	428	185	48	20	0	0	0	4266
	S	0	0	0	271	579	771	884	922	884	771	579	271	0	0	0	5897
	SW	0	0	0	20	48	185	428	647	803	875	811	483	0	0	0	4266
	W	0	0	0	20	43	59	68	76	222	448	557	402	0	0	0	1863
	NW	0	0	0	20	43	59	68	71	68	66	47	63	0	0	0	489
	Horizontal	0	0	0	51	198	348	448	482	448	348	198	51	0	0	0	2568
	Direct	0	0	0	446	753	865	912	926	912	865	753	446	0	0	0	—
Apr.	N	0	41	57	79	97	110	120	122	120	110	97	79	57	41	0	1117
	NE	0	262	508	462	291	134	123	122	120	110	97	77	52	17	0	2347
	E	0	321	728	810	732	552	293	131	120	110	97	77	52	17	0	4006
	SE	0	189	518	682	736	699	582	392	187	116	97	77	52	17	0	4323
	S	0	18	59	149	333	437	528	559	528	437	333	149	59	18	0	3536
	SW	0	17	52	77	97	116	187	392	582	699	736	682	518	189	0	4323
	W	0	17	52	77	97	110	120	392	293	552	732	810	728	321	0	4006
	NW	0	17	52	77	97	110	120	122	123	134	291	462	508	262	0	2347
	Horizontal	0	39	222	447	640	786	880	911	880	786	640	447	222	39	0	6938
	Direct	0	282	651	794	864	901	919	925	919	901	864	794	651	282	0	—
July	N	3	133	109	103	117	126	134	138	134	126	117	103	109	133	3	1621
	NE	8	454	590	540	383	203	144	138	134	126	114	95	71	39	0	3068
	E	7	498	739	782	701	531	294	149	134	126	114	95	71	39	0	4313
	SE	2	248	460	580	617	576	460	291	155	131	114	95	71	39	0	3849
	S	0	39	76	108	190	292	369	395	369	292	190	108	76	39	0	2552
	SW	0	39	71	95	114	131	155	291	460	576	617	580	460	248	2	3849
	W	0	39	71	95	114	126	134	149	294	531	701	782	739	498	7	4313
	NW	0	39	71	95	114	126	134	138	144	203	383	540	590	454	8	3068
	Horizontal	1	115	320	528	702	838	922	949	922	838	702	528	320	115	1	3902
	Direct	7	434	656	762	818	850	866	871	866	850	818	762	656	434	7	—
Oct.	N	0	0	7	40	62	77	87	90	87	77	62	40	7	0	0	453
	NE	0	0	74	178	84	80	87	90	87	87	62	40	7	0	0	869
	E	0	0	163	626	652	505	256	97	87	87	62	40	7	0	0	2578
	SE	0	0	152	680	853	864	770	599	364	137	66	40	7	0	0	4543
	S	0	0	44	321	547	711	813	847	813	711	547	321	44	0	0	5731
	SW	0	0	7	40	66	137	364	599	770	864	853	680	152	0	0	4543
	W	0	0	7	40	62	87	87	97	256	505	652	626	163	0	0	2578
	NW	0	0	7	40	62	87	87	90	87	80	84	178	74	0	0	869
	Horizontal	0	0	14	156	351	509	608	640	608	509	351	156	14	0	0	3917
	Direct	0	0	152	643	811	884	917	927	917	884	811	643	152	0	0	—

*Multiply by 0.3171 to convert to Btu/h·ft².

Values given are for the 21st of the month for average days with no clouds. The values can be up to 15 percent higher at high elevations under very clear skies and up to 30 percent lower at very humid locations with very dusty industrial atmospheres. Daily totals are obtained using Simpson's rule for integration with 10-min time intervals. Solar reflectance of the ground is assumed to be 0.2, which is valid for old concrete, crushed rock, and bright green grass. For a specified location, use solar radiation data obtained for that location. The direction of a surface indicates the direction a vertical surface is facing. For example, W represents the solar radiation incident on a west-facing wall per unit area of the wall.

Solar time may deviate from the local time. Solar noon at a location is the time when the sun is at the highest location (and thus when the shadows are shortest). Solar radiation data are symmetric about the solar noon: the value on a west wall before the solar noon is equal to the value on an east wall two hours after the solar noon.

TABLE 3-6 Average Daily Solar Radiation on a Horizontal Surface in Selected Cities in the United States, in MJ/m²·day (NREL, 2018)

State and Location	Jan.	Feb.	Mar.	Apr.	May	June	July	Aug.	Sep.	Oct.	Nov.	Dec.	Average
Alabama, Birmingham	9.20	11.92	13.67	19.65	21.58	22.37	21.24	20.21	17.15	14.42	10.22	8.40	16.01
Alaska, Anchorage	1.02	3.41	8.18	13.06	15.90	17.72	16.69	12.72	8.06	3.97	1.48	0.56	8.63
Arizona, Tucson	12.38	15.90	20.21	25.44	28.39	29.30	25.44	24.08	21.58	17.94	13.63	11.24	20.44
Arkansas, Little Rock	9.09	11.81	15.56	19.19	21.80	23.51	23.17	21.35	17.26	14.08	9.77	8.06	16.24
California, San Francisco	7.72	10.68	15.22	20.44	24.08	25.78	26.46	23.39	19.31	13.97	8.97	7.04	16.92
Colorado, Boulder	7.84	10.45	15.64	17.94	17.94	20.47	20.28	17.12	16.07	12.09	8.66	7.10	14.31
Connecticut, Hartford	6.70	9.65	13.17	16.69	19.53	21.24	21.12	18.51	14.76	10.68	6.59	5.45	13.74
Delaware, Wilmington	7.27	10.22	13.97	17.60	20.33	22.49	21.80	19.65	15.79	11.81	7.84	6.25	14.65
Florida, Miami	12.72	15.22	18.51	21.58	21.46	20.10	21.10	20.10	17.60	15.67	13.17	11.81	17.38
Georgia, Atlanta	9.31	12.26	16.13	20.33	22.37	23.17	22.15	20.56	17.49	14.54	10.56	8.52	16.43
Hawaii, Honolulu	14.08	16.92	19.42	21.24	22.83	23.51	23.74	23.28	21.35	18.06	14.88	13.40	19.42
Idaho, Boise	5.79	8.97	13.63	18.97	23.51	26.01	27.37	23.62	18.40	12.26	6.70	5.11	15.90
Illinios, Chicago	6.47	9.31	12.49	16.47	20.44	22.60	22.03	19.31	15.10	10.79	6.47	5.22	13.85
Indiana, Indianapolis	7.04	9.99	13.17	17.49	21.24	23.28	22.60	20.33	16.35	11.92	7.38	5.79	14.76
Iowa, Waterloo	6.81	9.77	13.06	16.92	20.56	22.83	22.60	19.76	15.33	10.90	6.70	5.45	14.20
Kansas, Dodge City	9.65	12.83	16.69	21.01	23.28	25.78	25.67	22.60	18.40	14.42	10.11	8.40	17.49
Kentucky, Lousville	7.27	10.22	13.63	17.83	20.90	22.71	22.03	20.10	16.35	12.38	7.95	6.25	14.76
Lousiana, New Orleans	9.77	12.83	16.01	19.87	21.80	22.03	20.67	19.65	17.60	15.56	11.24	9.31	16.35
Maine, Portland	6.70	9.99	13.78	16.92	19.99	21.92	21.69	19.31	15.22	10.56	6.47	5.45	13.97
Maryland, Baltimore	7.38	10.33	13.97	17.60	20.21	22.15	21.69	19.19	15.79	11.92	8.06	6.36	14.54
Massachusetts, Boston	6.70	9.65	13.40	16.92	20.21	22.03	21.80	19.31	15.33	10.79	6.81	5.45	14.08
Michigan, Detroit	5.91	8.86	12.38	16.47	20.33	22.37	21.92	18.97	14.76	10.11	6.13	4.66	13.63
Minnesota, Minneapolis	6.36	9.77	13.51	16.92	20.56	22.49	22.83	19.42	14.65	9.99	6.13	4.88	13.97
Mississippi, Jackson	9.43	12.38	16.13	19.87	22.15	23.05	22.15	19.08	14.54	10.11	6.25	5.11	13.74
Missouri, Kansas City	7.95	10.68	14.08	18.28	21.24	23.28	23.62	20.78	16.58	12.72	8.40	6.70	15.44
Montana, Lewistown	5.22	8.40	12.72	17.15	20.33	23.05	24.53	20.78	15.10	10.22	5.91	4.32	13.97
Nebraska, Lincoln	7.33	10.10	13.65	16.22	19.26	21.21	22.15	18.87	15.44	11.54	7.76	6.20	14.16
Nevada. Las Vegas	10.79	14.42	19.42	24.87	28.16	30.09	28.28	25.89	22.15	17.03	12.15	9.88	20.33
New Mexico, AIbuquerque	11.47	14.99	19.31	24.53	27.60	29.07	27.03	24.76	21.12	17.03	12.49	10.33	19.99
New York, New York City	6.93	9.88	13.85	17.72	20.44	22.03	21.69	19.42	15.56	11.47	7.27	5.79	14.31
North Carolina, Charlotte	8.97	11.81	15.67	19.76	21.58	22.60	21.92	19.99	16.92	13.97	9.99	8.06	16.01
Ohio, Cleveland	5.79	8.63	12.04	16.58	20.10	22.15	21.92	18.97	14.76	10.22	6.02	4.66	13.51
Oklahoma, Oklahoma City	9.88	11.25	16.47	20.33	22.26	24.42	24.98	22.49	18.17	14.54	10.45	8.74	17.15
Oregon, Portland	4.20	6.70	10.68	15.10	18.97	21.24	22.60	19.53	14.88	9.20	4.88	3.52	12.61
Pennsylvania, Pittsburgh	6.25	8.97	12.61	16.47	19.65	21.80	21.35	18.85	15.10	10.90	6.59	5.00	13.63
South Carolina, Charleston	9.77	12.72	16.81	21.12	22.37	22.37	21.92	19.65	16.92	14.54	11.02	9.09	16.58
Tennessee, Memphis	8.86	11.58	15.22	19.42	22.03	23.85	23.39	21.46	17.38	14.20	9.65	7.84	16.24
Texas, Houston	9.54	12.26	15.22	18.06	20.21	21.69	21.35	20.21	17.49	15.10	11.02	8.97	15.90
Utah, Salt Lake City	6.93	10.45	14.76	19.42	23.39	26.46	26.35	23.39	18.85	13.29	8.06	6.02	16.47
Virginia, Norfolk	8.06	10.90	14.65	18.51	20.78	22.15	21.12	19.42	16.13	12.49	9.09	7.27	15.10
Washington, Seattle	3.52	5.91	10.11	14.65	19.08	20.78	21.80	18.51	13.51	7.95	4.20	2.84	11.92
West Virginia, Charleston	7.04	9.65	13.40	17.15	20.21	21.69	20.90	18.97	15.56	11.81	7.72	6.02	14.20
Wisconsin, Green Bay	6.25	9.31	13.17	16.81	20.56	22.49	22.03	18.85	14.20	9.65	5.79	4.88	13.74
Wyoming, Rock Springs	7.61	10.90	15.30	19.42	23.17	26.01	25.78	22.94	18.62	13.40	8.40	6.70	16.58

the literature. The hourly or instantaneous data such as that in Table 3-5 is useful for analyzing instantaneous performance of a solar system while daily, monthly, or yearly solar data for total amounts of solar radiation like that in Table 3-6 are useful to predict long-term performance of a solar system.

EXAMPLE 3-7
Comparison of Energies
from Solar and Coal

Estimate the amount of solar energy input per year to a horizontal collector whose surface area is 5.5 m² for Las Vegas, NV, and Cleveland, OH. Determine the equivalent amount of coal that would provide the same amount of energy. Take the higher heating value of coal to be 30,000 kJ/kg.

SOLUTION Using the average daily solar radiation values on a horizontal surface from Table 3-6, the amount of solar energy input to the collector surface is determined for each city as

$$G_{\text{Las Vegas}} = G_{\text{avg}} \times A \times \text{time} = (20,330 \text{ kJ/m}^2 \cdot \text{day})(5.5 \text{ m}^2)(365 \text{ days}) = \mathbf{4.081 \times 10^7 \text{ kJ}}$$

$$G_{\text{Cleveland}} = G_{\text{avg}} \times A \times \text{time} = (13,510 \text{ kJ/m}^2 \cdot \text{day})(5.5 \text{ m}^2)(365 \text{ days}) = \mathbf{2.712 \times 10^7 \text{ kJ}}$$

The equivalent amount of coal that would provide the same amount of energy for each city is

$$m_{\text{coal,Las Vegas}} = \frac{G_{\text{Las Vegas}}}{\text{HHV}_{\text{coal}}} = \frac{4.081 \times 10^7 \text{ kJ}}{30,000 \text{ kJ/kg}} = \mathbf{1360 \text{ kg}}$$

$$m_{\text{coal,Cleveland}} = \frac{G_{\text{Cleveland}}}{\text{HHV}_{\text{coal}}} = \frac{2.712 \times 10^7 \text{ kJ}}{30,000 \text{ kJ/kg}} = \mathbf{904 \text{ kg}}$$

REFERENCES

ASHRAE. 1993. American Society of Heating, Refrigeration, and Air Conditioning Engineers. *Handbook of Fundamentals*. Atlanta.

ASHRAE (American Society of Heating, Refrigeration, and Air Conditioning Engineers). 1974. *Handbook and Product Directory, Applications Volume*. New York: ASHRAE.

Çengel YA and Ghajar AJ. 2020. *Heat and Mass Transfer: Fundamentals and Applications*, 6th ed. New York: McGraw Hill.

Culp AW. 1991. *Principles of Energy Conversion*, 2nd ed. New-York: McGraw Hill.

NREL. 2018. National Renewable Energy Laboratory. Golden, CO. www.nrel.gov.

PROBLEMS

RADIATION FUNDAMENTALS

3-1 By what properties is an electromagnetic wave characterized? How are these properties related to each other?

3-2 What is visible light? How does it differ from the other forms of electromagnetic radiation?

3-3 Which regions of spectral distribution does solar radiation fall in? Which region is responsible for most solar radiation?

3-4 What is thermal radiation? How does it differ from the other forms of electromagnetic radiation?

3-5 What is the cause of color? Why do some objects appear blue to the eye while others appear red? Is the color of a surface at room temperature related to the radiation it emits?

3-6 How do ultraviolet and infrared radiation differ? Do you think your body emits any radiation in the ultraviolet range? Explain.

3-7 Why do skiers get sunburned so easily?

3-8 What is a blackbody? Does a blackbody actually exist?

3-9   Define the total and spectral blackbody emissive powers. How are they related to each other? How do they differ?

3-10   Consider two identical bodies, one at 1000 K and the other at 1500 K. Which body emits more radiation in the shorter-wavelength region? Which body emits more radiation at a wavelength of 20 μm?

3-11   Electricity is generated and transmitted in power lines at a frequency of 50 Hz (1 Hz = 1 cycle per second). Determine the wavelength of the electromagnetic waves generated by the passage of electricity in power lines.

3-12   Part of solar energy is emitted at a wavelength of 0.075 μm. Determine the frequency of these waves.

3-13   A microwave oven is designed to operate at a frequency of 5.8×10^9 Hz. Determine the wavelength of these microwaves and the energy of each microwave.

3-14   Determine the maximum rate of thermal radiation that can be emitted by a surface that is at a uniform temperature of (*a*) 300 K and (*b*) 600 K.

3-15   Consider a 15-cm × 15-cm × 15-cm cubical body at 800 K suspended in the air. Assuming the body closely approximates a blackbody, determine (*a*) the rate at which the cube emits radiation energy in W and (*b*) the spectral blackbody emissive power at a wavelength of 6 μm.

3-16   Daylight and incandescent light may be approximated as blackbodies at the effective surface temperatures of 5800 K and 2800 K, respectively. Determine the wavelength at maximum emission of radiation for each of the lighting sources.

3-17   A flame from a match may be approximated as a blackbody at the effective surface temperature of 1700 K, while moonlight may be approximated as a blackbody at the effective surface temperature of 4000 K. Determine the peak spectral blackbody emissive power for both lighting sources (match flame and moonlight). Also, determine the peak spectral blackbody emissive power for the sun, which may be approximated as a blackbody at the effective surface temperature of 5800 K.

3-18   Which has the shortest wavelength in the electromagnetic wave spectrum?
(*a*) Radio waves   (*b*) Microwaves   (*c*) Ultraviolet   (*d*) Infrared   (*e*) Visible

3-19   The *visible* portion of the electromagnetic spectrum that lies between 0.40 and 0.76 μm is called
(*a*) Solar radiation   (*b*) Infrared   (*c*) Ultraviolet   (*d*) X rays   (*e*) Light

3-20   Solar radiation falls almost entirely into the wavelength range of
(*a*) 0.3 to 3 μm   (*b*) 0.1 to 100 μm   (*c*) 1 to 100 μm   (*d*) 10^{-2} to 1 μm   (*e*) 10^2 to 10^5 μm

3-21   Which is not part of solar radiation in the electromagnetic wave spectrum?
(*a*) Thermal radiation   (*b*) Microwaves   (*c*) Ultraviolet   (*d*) Infrared   (*e*) Visible

3-22   About half of solar radiation falls into the _____ range in the electromagnetic spectrum.
(*a*) Radio wave   (*b*) Microwave   (*c*) Ultraviolet   (*d*) Infrared   (*e*) Visible

3-23   The radiation emitted by bodies at room temperature falls into the _____ range in the electromagnetic spectrum.
(*a*) Radio wave   (*b*) Microwave   (*c*) Ultraviolet   (*d*) Infrared   (*e*) Visible

3-24   The rate of radiation energy emitted by a blackbody is proportional to the _____ of the absolute temperature.
(*a*) First power   (*b*) Second power   (*c*) Third power
(*d*) Fourth power   (*e*) Fifth power

3-25   Which statement is not correct regarding the plot for the variation of the spectral blackbody emissive power with wavelength at several temperatures?
(*a*) At any wavelength, the amount of emitted radiation increases with increasing temperature.
(*b*) As temperature increases, the curves shift to the right to the longer wavelength region.
(*c*) The solar radiation reaches its peak in the visible region of the spectrum.
(*d*) The surfaces at $T < 800$ K emit almost entirely in the infrared region.
(*e*) The emitted radiation is a continuous function of wavelength.

3-26 The wavelength at which the peak for blackbody radiation occurs for a specified temperature is given by

(*a*) Stefan-Boltzmann law   (*b*) Planck's law   (*c*) Wien's displacement law
(*d*) Blackbody law   (*e*) Kirchhoff's law

3-27 The maximum rate of thermal radiation that can be emitted by a 5-m^2 surface that is at a uniform temperature of 200°C is

(*a*) 1000 kW   (*b*) 454 kW   (*c*) 14.2 kW   (*d*) 63.9 kW   (*e*) 23.7 kW

RADIATIVE PROPERTIES

3-28 Define the properties emissivity and absorptivity.

3-29 Define absorptivity, transmissivity, and reflectivity of a surface for solar radiation. What is the relationship between them for opaque surfaces?

3-30 Surfaces that are intended to collect solar energy, such as the absorber surfaces of solar collectors, should have (*a*) high absorptivity or (*b*) low absorptivity and (*a*) high emissivity or (*b*) low emissivity.

3-31 What is the greenhouse effect? Why is it a matter of great concern among atmospheric scientists?

3-32 A 7-in-diameter spherical ball is known to emit radiation at a rate of 340 Btu/h when its surface temperature is 880 R. Determine the average emissivity of the ball at this temperature.

3-33 Consider a 4-cm-diameter and 6-cm-long cylindrical rod at 800 K. If the emissivity of the rod surface is 0.75, determine the total amount of radiation emitted by all surfaces of the rod in 20 min.

3-34 Solar radiation is incident on a semitransparent body at a rate of 900 W/m^2. If 150 W/m^2 of this incident radiation is reflected back and 225 W/m^2 is transmitted across the body, determine the absorptivity of the body.

3-35 An opaque horizontal plate is well insulated on the edges and the lower surface. The irradiation on the plate is 3000 W/m^2, of which 500 W/m^2 is reflected. The plate has a uniform temperature of 550 K and has an emissive power of 5000 W/m^2. Determine the total emissivity and absorptivity of the plate.

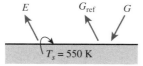

Figure P3-35

3-36 A horizontal opaque flat plate is well insulated on the edges and the lower surface. The top surface has an area of 8 m^2, and it experiences uniform irradiation at a rate of 6000 W. The plate absorbs 5000 W of the irradiation, and the surface is losing heat at a rate of 750 W by convection. If the plate maintains a uniform temperature of 350 K, determine the absorptivity, reflectivity, and emissivity of the plate.

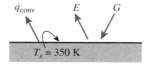

Figure P3-36

3-37 The emissivity of a blackbody is

(*a*) 0   (*b*) 0.25   (*c*) 0.5   (*d*) 1   (*e*) Infinity

3-38 The emissivity and absorptivity of a blackbody are, respectively
(a) 1, 1   (b) 0, 1   (c) 1, 0   (d) 0, 0   (e) 0, Infinity

3-39 Which surface is likely to have the lowest emissivity value?
(a) Water   (b) Metal   (c) Glass   (d) Soil   (e) Human skin

3-40 If the reflectivity of an opaque surface is 0.3, its absorptivity is
(a) 0   (b) 0.3   (c) 0.7   (d) 1   (e) Infinity

3-41 If the reflectivity of an opaque surface is 0.4, its transmissivity is
(a) 0   (b) 0.4   (c) 0.6   (d) 1   (e) Infinity

3-42 If the reflectivity of a surface is 0.3 and the absorptivity is 0.5, its transmissivity is
(a) 0   (b) 0.2   (c) 0.3   (d) 0.8   (e) 1

3-43 If the reflectivity of a surface is 0.3 and the absorptivity is 0.5, its emissivity is
(a) 0   (b) 1   (c) 0.2   (d) 0.3   (e) 0.5

3-44 Which statement is not correct regarding radiative properties?
(a) The absorptivity of a material is practically independent of surface temperature.
(b) The emissivity of a material depends on surface temperature.
(c) The absorptivity depends on the temperature of the source at which the incident radiation is originating.
(d) For blackbodies, reflectivity and transmissivity are both zero.
(e) The absorptivity of a surface for solar radiation is the same as that for radiation originating from the surrounding trees and buildings.

3-45 The rate of thermal radiation that can be emitted by a surface with an emissivity of 0.65 and a temperature of 500 K is
(a) 2303 W   (b) 3544 W   (c) 325 W   (d) 4039 W   (e) 76.9 W

SOLAR RADIATION

3-46 What is solar irradiance or solar constant? What is its accepted value in SI units and English units?

3-47 When the earth is closest to the sun, we have winter in the Northern Hemisphere. Explain why. Also explain why we have summer in the Northern Hemisphere when the earth is farthest away from the sun.

3-48 What is the effective sky temperature?

3-49 You have probably noticed warning signs on the highways stating that bridges may be icy even when the roads are not. Explain how this can happen.

3-50 Explain why surfaces usually have quite different absorptivities for solar radiation and for radiation originating from the surrounding bodies.

3-51 Why is the sky bluish at midday but reddish at sunset? Explain.

3-52 How is the solar constant used to determine the effective surface temperature of the sun? How would the value of the solar constant change if the distance between the earth and the sun were doubled?

3-53 What changes would you notice if the sun emitted radiation at an effective temperature of 2000 K instead of 5762 K?

3-54 If the distance between the sun and the earth were the half of what it is, $L = 0.5 \times 1.496 \times 10^{11}$ m, what would the solar constant be? The sun is a nearly spherical body that has a diameter of $D = 1.393 \times 10^9$ m, and the effective surface temperature of the sun is $T_{sun} = 5778$ K.

3-55 Solar radiation is incident on the outer surface of a spaceship at a rate of 400 Btu/h·ft². The surface has an absorptivity of $\alpha_s = 0.2$ for solar radiation and an emissivity of $\varepsilon = 0.7$ at room temperature. The outer surface radiates heat into space at 0 R. If there is no net heat transfer into the spaceship, determine the equilibrium temperature of the surface.

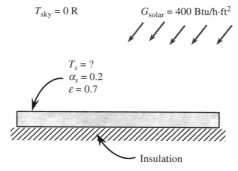

Figure P3-55

3-56 A surface has an absorptivity of $\alpha_s = 0.72$ for solar radiation and an emissivity of $\varepsilon = 0.6$ at room temperature. The surface temperature is observed to be 350 K when the direct and the diffuse components of solar radiation are $G_D = 350$ and $G_d = 400$ W/m^2, respectively, and the direct radiation makes a 30° angle with the normal of the surface. Taking the effective sky temperature to be 280 K, determine (*a*) the total solar energy incident on the surface, (*b*) the rate of solar radiation absorbed by the surface, and (*c*) the net rate of radiation heat transfer to the surface at that time, all in W/m^2.

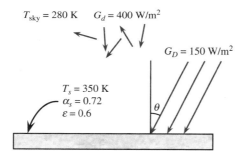

Figure P3-56

3-57 The absorber surface of a solar collector is made of aluminum coated with black nickel oxide ($\alpha_s = 0.92$ and $\varepsilon = 0.08$). Solar radiation is incident on the surface at a rate of 860 W/m^2. The air and the effective sky temperatures are 20 and 7°C, respectively, and the convection heat transfer coefficient is 15 W/m$^2\cdot$K. For an absorber surface temperature of 60°C, determine the net rate of solar energy delivered by the absorber plate to the water circulating behind it.

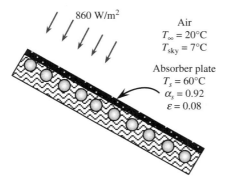

Figure P3-57

3-58   Determine the equilibrium temperature of the absorber surface in Prob. 3-57 if the back side of the absorber is insulated.

3-59   Which statement is not correct for the sun?
(*a*) The sun is a nearly spherical body.
(*b*) Solar energy reaches us in the form of electromagnetic waves.
(*c*) Less than a billionth of the energy created in the sun strikes the earth.
(*d*) The energy of the sun is due to a continuous fission reaction.
(*e*) The sun is essentially a nuclear reactor, with temperatures as high as 40,000,000 K in its core region.

3-60   The temperature in the outer region of the sun is about
(*a*) 40,000,000 K   (*b*) 5800 K   (*c*) 3000 K   (*d*) 1373 K   (*e*) 300 K

3-61   The rate of solar radiation incident at the outer edge of the atmosphere is called
(*a*) Solar constant   (*b*) Solar value   (*c*) Irradiation   (*d*) Radiosity   (*e*) Insolation

3-62   The rate of solar radiation incident at the outer edge of the atmosphere is equal to
(*a*) 5800 W/m²   (*b*) 2900 W/m²   (*c*) 1373 W/m²   (*d*) 1000 W/m²   (*e*) 950 W/m²

3-63   The rate of solar energy reaching the earth's surface on a clear day is about
(*a*) 5800 W/m²   (*b*) 2900 W/m²   (*c*) 1373 W/m²   (*d*) 1225 W/m²   (*e*) 950 W/m²

3-64   Which component in the atmosphere absorbs ultraviolet radiation at wavelengths below 0.3 μm almost completely?
(*a*) Oxygen   (*b*) Ozone   (*c*) Carbon dioxide   (*d*) Water vapor   (*e*) Argon

3-65   The diffuse solar radiation on a totally cloudy day is nearly
(*a*) 0%   (*b*) 10%   (*c*) 50%   (*d*) 70%   (*e*) 100%

3-66   The atmosphere can be treated as a blackbody at some lower fictitious temperature that emits an equivalent amount of radiation energy. This fictitious temperature is called
(*a*) Atmospheric temperature   (*b*) Effective sky temperature
(*c*) Solar temperature   (*d*) Radiative temperature
(*e*) Blackbody temperature

3-67   Surfaces that are intended to collect solar energy, such as the absorber surfaces of solar collectors, should have
(*a*) High emissivity, high absorptivity   (*b*) High emissivity, low absorptivity
(*c*) Low emissivity, high absorptivity   (*d*) High reflectivity, low transmissivity
(*e*) Low reflectivity, high transmissivity

ESTIMATION OF SOLAR RADIATION

3-68   Why is the mean sun time not equal to solar time?

3-69   How is the equation of time defined?

3-70   What is responsible for the production of seasons?

3-71   Which angles are used to express the position of the sun? Which angles are used to express the orientation of a surface?

3-72   What are the components of solar radiation on a surface?

3-73   Define the clearness number.

3-74   What does the amount of solar radiation incident on a surface depend on?

3-75   The longitude of Prague, Czech Republic, is 14.44°E. What is the solar time in Prague at 15:30 local daylight savings time on July 11?

3-76   The longitude of Reno, Nevada, is 120°W. What is the solar time in Reno at 8:15 local time on April 1?

3-77 The longitude of Dubai, UAE, is 55.3°E. What is the solar time in Dubai at 12:00 local time on Nov. 30?

3-78 Consider a location with a latitude of 40°N and the solar time is 13:00 on July 11. Estimate the direct and diffuse components of solar radiation on a surface at the southeast direction inclined at an angle of 30° from the horizontal.

3-79 Consider a location with a latitude of 40°N and the solar time is 13:00 on July 11. Estimate the direct and diffuse components of solar radiation on a horizontal surface at the southeast direction.

3-80 Consider a location with a latitude of 36°N and the solar time is 16:30 on October 31. Estimate the direct and diffuse components of solar radiation on a surface at the west direction inclined at an angle of 10° from the horizontal.

3-81 Consider a location with a latitude of 36°N and the solar time is 16:30 on October 31. Estimate the direct and diffuse components of solar radiation on a vertical surface at the west direction.

3-82 Consider a location with a latitude of 25°N and the solar time is 7:30 on April 1. Estimate the direct and diffuse components of solar radiation on a surface at the southwest direction inclined at an angle of 60° from the horizontal.

3-83 Consider a location with a latitude of 38°N and a longitude of 67°W. The local time is 9:00 (the daylight savings time) on July 11. Estimate the direct and diffuse components of solar radiation on a surface at the southeast direction inclined at an angle of 20° from the horizontal.

3-84 Estimate the amount of solar energy input per year to a horizontal collector whose surface area is 4 m² for Green Bay and Memphis. Determine the equivalent amount of natural gas that would provide the same amount of energy. Take the higher heating value of natural gas to be 35,000 kJ/m³.

3-85 Compare the amount of solar energy input to a horizontal surface for winter and summer months in Tucson and New York City. Take summer months to be May, June, July, August, and September and winter months to be December, January, and February for Tucson. Take summer months to be June, July, and August and winter months to be November, December, January, February, and March for New York. Also, determine the ratio of solar energy input in summer to that in winter for each city.

3-86 A house located in Boulder, Colorado (40°N latitude), has ordinary double-pane windows with 6-mm-thick glass, and the total window areas are 8, 6, 6, and 4 m² on the south, west, east, and north walls, respectively. Determine the total solar heat gain of the house through the windows at 9:00, 12:00, and 15:00 solar time in July. Also, determine the total amount of solar heat gain per day for an average day in January. The solar heat gain coefficient (SHGC) of the windows is 0.70. That is, 70 percent of solar energy input is transmitted through the window.

3-87 Consider a building in New York (40°N latitude) that has 90 m² of window area on its south wall. The windows are double-pane heat-absorbing type and equipped with light-colored venetian blinds. Determine the total solar heat gain of the building through the south windows at solar noon in April. What would your answer be if there were no blinds at the windows? The solar heat gain coefficient (SHGC) of the windows with no blinds is 0.50 and that with blinds is 0.30. The SHGC represents percentage of solar energy input transmitted through the window.

3-88 The difference between the solar time and the mean solar time is called
(*a*) Declination angle   (*b*) Equation of time   (*c*) Apparent sun time
(*d*) Daylight savings time   (*e*) Standard meridian

3-89 One degree of longitude is equal to _____ and each time zone has a width of _____.
(*a*) 15 min, 15°   (*b*) 30 min, 30°   (*c*) 2 min, 4°   (*d*) 10 min, 60°   (*e*) 4 min, 15°

3-90 A city is located at 30°E longitude. What is the solar time in this city at 10:00 local time on January 11?
(*a*) 10:00   (*b*) 10:04   (*c*) 09:56   (*d*) 09:52   (*e*) 10:08

3-91 A city is located at 60°W longitude. What is the mean sun time in this city at 13:00 local time on December 21?
(*a*) 13:00   (*b*) 13:02   (*c*) 13:04   (*d*) 12:58   (*e*) 12:56

3-92   Choose the incorrect statement regarding the declination angle.
(*a*) It is the angle between sun's rays and the equatorial plane.
(*b*) It depends on the time of the year.
(*c*) It is 23.45° on March 21.
(*d*) It is 23.45° on June 21.
(*e*) It is equal to latitude at 12:00 solar time.

3-93   The angle between the sun's rays and the equatorial plane is called
(*a*) Solar angle   (*b*) Declination angle   (*c*) Polar angle
(*d*) Altitude angle   (*e*) Incident angle

3-94   What is the most relevant angle for solar radiation calculation?
(*a*) Solar azimuth angle   (*b*) Declination angle   (*c*) Hour angle
(*d*) Altitude angle   (*e*) Incident angle

3-95   If the calculated value of _____ is negative, there is no solar radiation incident on the surface.
(*a*) Solar azimuth angle   (*b*) Declination angle   (*c*) Hour angle
(*d*) Altitude angle   (*e*) Incident angle

3-96   The components of solar radiation incident on a surface include
  I. Direct radiation
 II. Diffuse radiation
III. Reflected radiation
(*a*) Only I   (*b*) I and II   (*c*) I and III   (*d*) II and III   (*e*) I, II, and III

3-97   The amount of solar radiation incident on a surface does not depend on
(*a*) Orientation of the surface   (*b*) Latitude and elevation of the location
(*c*) Temperature of air   (*d*) Clearness of the sky
(*e*) Time of the day

CHAPTER 4

Solar Energy Applications

4-1 INTRODUCTION

There has been intense research for decades to develop solar technologies since solar energy is *free* and nonpolluting. However, the rate of solar radiation on a unit surface is quite low, and solar collectors with large surface areas must be installed. This is costly and requires a lot of space, which is not always available. Another disadvantage of solar energy is that it is available in reasonable quantities only in certain locations of the world and certain seasons of the year and times of the day. Note also that one of the most attractive applications of solar energy is heating of buildings, but this is not needed in summer when solar energy is readily available. Of course, solar heat can be used for cooling applications by absorption cooling systems, but they are complex devices involving high initial cost. Storage of solar energy for nighttime use is an option to tackle the noncontinuous feature of solar energy, but this adds to system cost, and it may not be effective in many applications. Nonetheless, we should try to get the best out of solar energy by utilizing the most current technologies, and we should continue to work on improving solar systems and making them more cost-effective. Solar energy has the most potential among all energy sources, and there is no limit to how much of it can be utilized for our energy needs. We should also note the cost of solar systems has been steadily decreasing.

The conversion of solar energy into other useful forms of energy can be accomplished by three conversion processes (Culp, 1991):

Heliochemical process: This is basically photosynthesis, and it is responsible for the production of fossil fuel and biomass.

Heliothermal process: Solar energy is collected and converted to thermal energy or heat. *Flat-plate* collectors, *concentrating* collectors, and *heliostats* are common devices that collect solar radiation for conversion to useful heat. Solar collectors are used for space heating and cooling and also for the production of hot water for buildings. Heliostats are mirrors that reflect solar radiation into a single receiver. The resulting high-temperature thermal energy is converted to electricity by a heat engine.

Helioelectrical process: The production of electricity by photovoltaic or solar cells is accomplished by a helioelectrical process. This process is different from heliostats in that solar energy is converted to electricity directly in solar cells, while it is first converted to thermal energy in heliostats.

Today, the single most common application of solar energy is flat-plate solar collectors used to meet hot water needs of residential and commercial buildings. This method of obtaining hot water is very cost-effective. Solar photovoltaic cells are becoming more common despite their high cost. However, solar cells are already cost-effective for off-grid electricity applications. Solar thermal power plants utilizing heliostats require large investments with large areas, and there have been only a few such installations worldwide.

Solar engineering processes can be divided into two main categories, *active* and *passive* solar applications. Active applications involve some mechanical/electrical operation and a device such as a pump in the system, and they include almost all solar collector, solar cell, and heliostat applications. Any application that aims to utilize solar energy by passive design qualifies as a passive solar application. The design of a house to collect solar energy in winter to reduce heating cost can be achieved by correct orientation of buildings, and selection of wall materials with the proper wall surface properties. The selection of windows with proper glazing can also help maximize solar heat gain in winter and minimize solar heat gain in summer. The use of a *trombe wall* on south walls can maximize the dissipation of solar heat into the house even after daytime hours.

Capturing solar energy and producing useful energy from it requires some special devices. In this chapter, we describe characteristics of common solar systems along with their thermal analyses.

4-2 FLAT-PLATE SOLAR COLLECTOR

The objective of a solar collector is to produce useful heat from solar energy. Most solar collectors in operation today are used to produce hot water. This hot water is normally used in residential and commercial buildings for kitchen use, bathrooms, showers, etc. Solar collectors can also be used for space heating in winter. Unfortunately, most solar heat is available in summer when space heating is not needed. Therefore, most solar collectors are used to produce hot water, and they are very common in southern Europe and Asia where solar energy is available for more than 200 days a year (Fig. 4-1). A complete unit that provides hot water needs of a family house costs as little as $1000 in some parts of the world. Solar-heated hot water is used for showers, everyday faucet uses, and washing machines.

Figure 4-1 Solar water collectors on the roof of residential buildings.

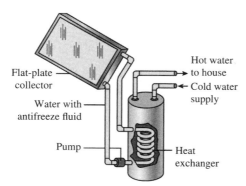

Figure 4-2 An active, closed loop solar water heater. (*Adapted from DOE/EERE.*)

The solar collector shown in Fig. 4-1 is a *thermosiphon solar water heater system*, which operates on natural circulation. Water flows through the system when warm water rises into the tank as cooler water sinks. An *active, closed loop solar water heater* uses a pump for the circulation of water containing antifreeze fluid (Fig. 4-2). The use of antifreeze fluid ensures that there is no freezing in subfreezing temperatures. Water containing antifreeze is heated in the collector and gives up its heat to water in a heat exchanger. The resulting hot water is used in the residence. This system may be equipped with an electric resistance heater to provide hot water when solar energy is not available.

Flat-plate solar collectors are much more common than parabolic solar collectors. Here, we provide a simplified analysis of a flat-plate solar collector. The thermal analysis below is based on those given in Hodge (2010), Goswami et al. (2000), Kreith and Kreider (2011), and Duffie and Beckman (2006), and they may be referred to for more details.

A flat-plate collector consists of a glazing, an absorber plate, flow tubes, insulation, glazing frame, and a box enclosure (Fig. 4-3). The absorber plate absorbs solar energy transmitted through the glazing, which is a type of glass. Flow tubes are attached to the absorber plate and water is heated as it flows in the tubes by absorbing heat from the absorber plate. The sides and back are insulated to minimize heat losses.

The rate of solar heat absorbed by the absorber plate is

$$\dot{Q}_{abs} = \tau \alpha A G \qquad (4\text{-}1)$$

where τ is the transmissivity of the glazing, α is the absorptivity of the absorber plate, A is the area of the collector surface, in m^2, and G is the *solar insolation* or *solar irradiation* (solar radiation incident per unit collector area), in W/m^2. This incident radiation consists of direct or beam radiation, diffuse radiation from the sky and the surrounding surfaces, and reflected radiation from the ground. There is heat loss from the collector by convection

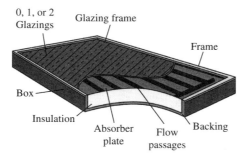

Figure 4-3 The cutaway view of a flat-plate solar collector. (*Adapted from DOE/EERE.*)

to the surrounding air and by radiation to surrounding surfaces and sky. The rate of heat loss can be expressed as

$$\dot{Q}_{loss} = UA(T_c - T_a) \tag{4-2}$$

where U is the overall heat transfer coefficient or heat loss coefficient, in W/m²·°C. It includes combined effects of conduction, convection, and radiation and obtained from the total thermal resistance between the absorber plate and the ambient air. T_c is the average collector temperature taken as the average temperature of the absorber plate and T_a is the ambient air temperature, both in °C. The useful heat transferred to the water is the difference between the heat absorbed and the heat lost:

$$\begin{aligned}\dot{Q}_u &= \dot{Q}_{abs} - \dot{Q}_{loss} \\ &= \tau\alpha AG - UA(T_c - T_a) \\ &= A[\tau\alpha G - U(T_c - T_a)]\end{aligned} \tag{4-3}$$

This relation indicates that the useful heat is maximized when the difference between the collector temperature and the air temperature is minimized. However, this also means that hot water is produced at a low temperature due to the lower temperature of the absorber plate. If the mass flow rate of water flowing through the collector $\dot{m}$ is known, the useful heat can also be determined from

$$\dot{Q}_u = \dot{m}c_p(T_{w,\text{out}} - T_{w,\text{in}}) \tag{4-4}$$

where c_p is the specific heat of water, in J/kg·°C, $T_{w,\text{in}}$ and $T_{w,\text{out}}$ are the inlet and outlet temperatures of water, respectively. For the same useful heat, a higher mass flow rate would yield a lower temperature rise for water in the collector.

The efficiency of a solar collector may be defined as the ratio of the useful heat delivered to the water to the radiation incident on the collector:

$$\eta_c = \frac{\dot{Q}_u}{\dot{Q}_{\text{incident}}} = \frac{\tau\alpha AG - UA(T_c - T_a)}{AG} = \tau\alpha - U\frac{T_c - T_a}{G} \tag{4-5}$$

Therefore, the collector efficiency is maximized for maximum values of transmissivity of the glazing and the absorptivity of the absorber plate. Also, the smaller the difference between the collector and air temperatures, the greater the collector efficiency. If the collector efficiency is plotted against the term $(T_c - T_a)/G$, we obtain a straight line, as shown in Fig. 4-4.

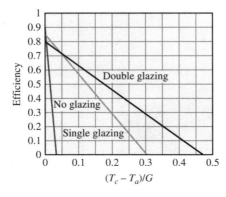

Figure 4-4   Collector efficiency for three different collectors. The data in Table 4-1 are used.

TABLE 4-1 **Typical Flat-Plate Solar Collector Properties** (Mitchell, 1983)

	$\tau\alpha$	U, W/m²·°C	U, Btu/h·ft²·°F
No glazing	0.90	28	5
Single glazing	0.85	2.8	0.5
Double glazing	0.80	1.7	0.3

Of course, the slope of this line is equal to $-U$. Typical values of transmissivity-absorptivity product $\tau\alpha$ and the overall heat transfer coefficient U are given in Table 4-1.

An unglazed collector allows more solar radiation input to the collector due to higher $\tau\alpha$ values but also involves higher heat transfer coefficients. However, even though the $\tau\alpha$ values go down slightly from no glazing to single and double glazing, the U value decreases significantly (see Table 4-1). As a result, a single-glazed collector is more efficient than an unglazed collector. Most flat-plate solar collectors have single glazing.

Equation (4-5) gives the collector efficiency as a function of average temperature of the absorber plate in the collector T_c. However, this temperature is usually not available. Instead, water temperature at the collector inlet is available. For this purpose, the *collector heat removal factor* F_R is defined to compare actual useful heat delivered to the working fluid (i.e., water) to the useful heat in an ideal case. The ideal case is recognized when the heat loss from the collector is a minimum. This would be achieved when the water remains at the inlet temperature $T_{w,in}$ as it flows in the collector tubes. This corresponds to the minimum possible temperature difference between the absorber plate and the ambient air and, thus, the minimum heat loss from the collector. Obviously, this also maximizes useful heat delivery. In practice, this condition is approached as the water flow rate increases. The collector heat removal factor is then defined as

$$F_R = \frac{\dot{Q}_u}{\dot{Q}_{u,\,ideal}} = \frac{\dot{Q}_u}{A[\tau\alpha G - U(T_{w,in} - T_a)]} \tag{4-6}$$

Solving for the rate of useful heat gives

$$\dot{Q}_u = F_R A[\tau\alpha G - U(T_{w,in} - T_a)] \tag{4-7}$$

This is a very useful relation because it relates the useful heat to the water inlet temperature, and it is applicable to all flat-plate collectors. Substituting this relation into the collector efficiency definition (Eq. 4-5), we obtain a more convenient form of collector efficiency:

$$\eta_c = F_R \tau\alpha - F_R U \frac{T_{w,in} - T_a}{G} \tag{4-8}$$

Equation (4-8) is known as the *Hottel-Whillier-Bliss equation*. This equation is also in a linear function format if the collector efficiency is plotted against the term $(T_{w,in} - T_a)/G$. Such a plot is given in Example 4-1. The slope of the resulting straight line is $-F_R U$. The maximum efficiency is obtained when the temperature difference and thus the term $F_R U(T_{w,in} - T_a)/G$ is zero. The maximum efficiency in this case is equal to intercept in the figure, and it is equal to $F_R \tau\alpha$.

The collector heat removal factor F_R increases as the water flow rate increases. At very high flow rates, the temperature rise of the water approaches zero. Even when this is the case, the temperature of the absorber plate is still higher than the water temperature. This temperature difference between the absorber plate and water is accounted for by another factor called the *collector efficiency factor F*.

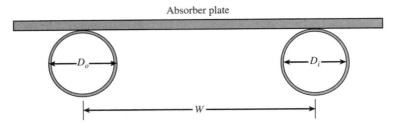

Figure 4-5 In the analysis of collector efficiency factor, the absorber plate is considered as a fin attached to the water tube.

A relation for the collector efficiency factor can be obtained by the heat transfer analysis of the absorber plate and water tubes combination. The absorber plate is considered as a fin attached to the tube (Fig. 4-5). A comprehensive heat transfer analysis yields the following relation for the collector efficiency factor:

$$F = \frac{1/U}{W\left[\dfrac{1}{U[D_o + (W - D_o)\eta_{\text{fin}}]} + \dfrac{1}{h_c} + \dfrac{1}{\pi D_i h_i}\right]} \qquad (4\text{-}9)$$

where W is the center-to-center distance between the tubes, D_o is the outer tube diameter, η_{fin} is the fin efficiency, h_c is contact conductance between the tube and absorber plate, D_i is the inner tube diameter, and h_i is the convection heat transfer coefficient on the inner surface of the tube. The numerator in this equation represents total thermal resistance between the absorber plate and the ambient air and the denominator represents total thermal resistance between the water in the tube and the ambient air. In another words, the collector efficiency factor is the ratio of actual useful heat gain to that when the absorber plate is at the water temperature.

The collector efficiency factor F is the functions of parameters included in Eq. (4-9). It is proportional to the convection heat transfer coefficient on the inner surface of the tube h_i and the fin efficiency η_{fin} but inversely proportional to the heat loss coefficient U. The fin efficiency increases as the distance between the tubes decreases and the thickness and thermal conductivity of the plate increase.

The collector efficiency factor is normally a design parameter for the collector. Higher values of collector efficiency factor are desirable for higher collector efficiency. The relation between the collector heat removal factor F_R and the collector efficiency factor F is given by

$$F_R = \frac{\dot{m}c_p}{UA}\left[1 - \exp\left(-\frac{UAF}{\dot{m}c_p}\right)\right] \qquad (4\text{-}10)$$

The collector efficiency factor F is the upper limit for the collector heat removal factor F_R. A complete analysis of the collector efficiency factor F can be found in Duffie and Beckman (2006).

The solar collector is normally fixed in position. As the angle of solar incident radiation changes throughout the day, the product $\tau\alpha$ also changes. This change can be accounted for by including an *incident angle modifier* $K_{\tau\alpha}$ in Eq. (4-7) as

$$\dot{Q}_u = F_R A[K_{\tau\alpha}\tau\alpha G - U(T_{w,\text{in}} - T_a)] \qquad (4\text{-}11)$$

Then the collector efficiency may be expressed as

$$\eta_c = F_R K_{\tau\alpha} \tau\alpha - F_R U \frac{T_{w,\text{in}} - T_a}{G}$$

(4-12)

The value of $K_{\tau\alpha}$ is a function of the incident angle. The standard collector test data are normally based on a value of 1 for $K_{\tau\alpha}$. Tests indicate that the incident angle modifier is in the following form (Souka and Safwat, 1966):

$$K_{\tau\alpha} = 1 - a\left(\frac{1}{\cos\theta} - 1\right)$$

(4-13)

where a is a constant and θ is the solar incident angle, which is the angle the direct solar radiation makes with the normal of the collector plane. The incident angle modifier $K_{\tau\alpha}$ is 1 when $\theta = 0$. For both a single-glass and a double-glass collector, $K_{\tau\alpha}$ remains close to 1 as θ increases until 30°. $K_{\tau\alpha}$ then decreases at higher incident angle values, and it reduces to 0.87 for single-glass and 0.78 for double-glass at $\theta = 60°$, as shown in Fig. 4-6 (ASHRAE Standard 93-77).

Flat-plate collectors provide hot water up to about 100°C. For higher temperatures, evacuated tube collectors and concentrating collectors may be used. Flat-plate collectors should be oriented such that solar radiation incident is maximized during the operating months. This is achieved by south orientation in the northern hemisphere and north orientation in the southern hemisphere. Total energy delivery is not affected significantly when the orientation is off from the south by up to 20° to the east or west. For better performance, the collectors should be tilted up from the horizon at about 15° greater than the local latitude (Kreith and Kreider, 2011).

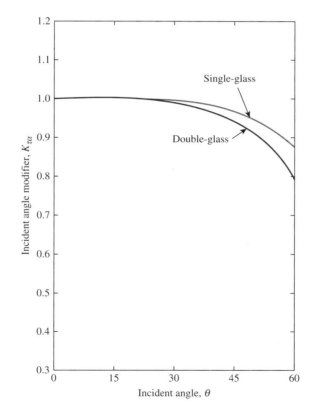

Figure 4-6 Incident angle modifier for single-glass and double-glass collectors as a function of incident angle. The absorptivity of absorber plate is 0.9. (*ASHRAE Standard 93-77.*)

EXAMPLE 4-1
Efficiency of a
Flat-Plate Solar
Collector

The specifications of two flat-plate collectors are given as follows:

Single glazing: $\tau = 0.96$, $\alpha = 0.96$, $U = 9$ W/m²·°C

Double glazing: $\tau = 0.93$, $\alpha = 0.93$, $U = 6.5$ W/m²·°C

The heat removal factor for both collectors is 0.95, the solar insolation is 550 W/m², and the ambient air temperature is 23°C. For each collector, determine (a) the collector efficiency when the water enters the collector at 45°C, (b) the temperature of water at which the collector efficiency is zero, and (c) the maximum collector efficiency. Take the incident angle modifier to be 1. (d) Also, plot the collector efficiency as a function of $(T_{w,in} - T_a)/G$ for each collector.

SOLUTION (a) The collector efficiency is determined from Eq. (4-12) for each collector to be

Single glazing:

$$\eta_c = F_R K_{\tau\alpha}\tau\alpha - F_R U \frac{T_{w,in} - T_a}{G}$$

$$= (0.95)(1)(0.96)(0.96) - (0.95)(9 \text{ W/m}^2\cdot°\text{C})\frac{45°\text{C} - 23°\text{C}}{550 \text{ W/m}^2}$$

$$= \mathbf{0.534}$$

Double glazing:

$$\eta_c = F_R K_{\tau\alpha}\tau\alpha - F_R U \frac{T_{w,in} - T_a}{G}$$

$$= (0.95)(1)(0.93)(0.93) - (0.95)(6.5 \text{ W/m}^2\cdot°\text{C})\frac{45°\text{C} - 23°\text{C}}{550 \text{ W/m}^2}$$

$$= \mathbf{0.575}$$

(b) Setting the collector efficiency to zero in Eq. (4-12) gives

Single glazing:

$$F_R K_{\tau\alpha}\tau\alpha - F_R U \frac{T_{w,in} - T_a}{G} = 0$$

$$F_R K_{\tau\alpha}\tau\alpha = F_R U \frac{T_{w,in} - T_a}{G}$$

$$(0.95)(1)(0.96)(0.96) = (0.95)(9 \text{ W/m}^2\cdot°\text{C})\frac{T_{w,in} - 23°\text{C}}{550 \text{ W/m}^2} = 0$$

$$T_{w,in} = \mathbf{79.3°C}$$

Double glazing:

$$F_R K_{\tau\alpha}\tau\alpha = F_R U \frac{T_{w,in} - T_a}{G}$$

$$(0.95)(1)(0.93)(0.93) = (0.95)(6.5 \text{ W/m}^2\cdot°\text{C})\frac{T_{w,in} - 23°\text{C}}{550 \text{ W/m}^2} = 0$$

$$T_{w,in} = \mathbf{96.2°C}$$

(c) The collector efficiency is maximum when the water temperature is equal to air temperature. Then,

Collector A: $\eta_{c,max} = F_R K_{\tau\alpha}\tau\alpha = (0.95)(1)(0.96)(0.96) = \mathbf{0.876}$

Collector B: $\eta_{c,max} = F_R K_{\tau\alpha}\tau\alpha = (0.95)(1)(0.93)(0.93) = \mathbf{0.822}$

(d) We plot the collector efficiency as a function of $(T_{w,in} - T_a)/G$ for each collector, as shown in Fig. 4-7.

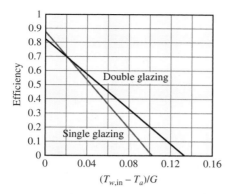

Figure 4-7 Collector efficiency for the two collectors considered in Example 4-1. ▲

EXAMPLE 4-2
Energy and Cost
Savings due to
Solar Collector

A solar collector provides the hot water needs of a family for a period of 7 months except for 5 months of the winter season. The collector supplies hot water at an average temperature of 55°C, and the average temperature of cold water is 18°C. An examination of water bills indicates that the family uses an average of 15 tons of hot water from the solar collector per month. An electrical resistance heater supplies hot water in the winter months. The cost of electricity is $0.13/kWh, and the efficiency of the electric heater can be taken to be 93 percent considering heat losses from the system. Determine the annual electricity and cost savings to this family due to the solar collector.

SOLUTION   The specific heat of water at room temperature is c_p = 4.18 kJ/kg·°C. The amount of solar water heating during a month is

$$Q = mc_p(T_1 - T_2) = (15,000 \text{ kg/month})(4.18 \text{ kJ/kg·°C})(55 - 18)°C = 2.320 \times 10^6 \text{ kJ/month}$$

The amount of gas that would be consumed (or saved) during the 7-month period is

$$\text{Electricity savings} = \frac{Q}{\eta_{\text{heater}}} = (7 \text{ months})\frac{(2.320 \times 10^6 \text{ kJ/month})}{0.93}\left(\frac{1 \text{ kWh}}{3600 \text{ kJ}}\right) = \textbf{4850 kWh}$$

The corresponding cost savings is

$$\text{Cost savings} = \text{Electricity savings} \times \text{Unit cost of electricity}$$
$$= (4850 \text{ kWh})(\$0.13/\text{kWh})$$
$$= \textbf{\$631}$$ ▲

EXAMPLE 4-3
Daily Analysis of a
Flat-Plate Collector

A residential unit uses a flat-plate collector with the following specifications:

Collector area A = 2.5 m²

Heat loss coefficient U = 4.0 W/m²·°C

Convection heat transfer coefficient at the inner surface of the tubes h_i = 200 W/m²·°C

Tube spacing W = 20 cm

Outer tube diameter D_o = 1.3 cm

Inner tube diameter D_i = 1.1 cm

Fin efficiency η_{fin} = 0.95

Contact resistance between the plate and the tube is negligible.

Flow rate of water in the tubes $\dot{m}_w$ = 0.025 kg/s

Water inlet temperature $T_{w,\text{in}}$ = 60°C

The product of transmissivity of glazing and absorptivity of the absorber $\tau\alpha$ = 0.82

Incident angle modifier $K_{\tau\alpha}$ = 1

Average hourly data for ambient air temperature and solar heat flux incident are given in the following table for a certain day.

Solar time	Ambient air temperature T_a, °C	Average rate of solar radiation incident G, W/m²
07:00	12	45
08:00	13	120
09:00	15	200
10:00	16	620
11:00	17	650
12:00	19	780
13:00	20	740
14:00	20	360
15:00	19	290
16:00	17	170
17:00	15	80
18:00	14	30

(a) Calculate the collector efficiency factor and the collector heat removal factor.

(b) For each hour, calculate the useful heat gain, the collector efficiency, and the temperature of water at the collector exit.

(c) Calculate the total useful heat gain and the overall collector efficiency for the entire day.

SOLUTION   (a) The collector efficiency factor is determined as

$$F = \frac{1/U}{W\left[\dfrac{1}{U[D_o + (W - D_o)\eta_{\text{fin}}]} + \dfrac{1}{h_c} + \dfrac{1}{\pi D_i h_i}\right]}$$

$$= \frac{1/(4.0\ \text{W/m}^2\cdot°\text{C})}{(0.2\ \text{m})\left[\dfrac{1}{(4\ \text{W/m}^2\cdot°\text{C})[(0.013\ \text{m}) + [(0.2\ \text{m}) - (0.013\ \text{m})](0.95)]} + 0 + \dfrac{1}{\pi(0.011\ \text{m})(200\ \text{Wm}^2\cdot°\text{C})}\right]}$$

$$= \mathbf{0.8585}$$

The collector heat removal factor is

$$F_R = \frac{\dot m c_p}{UA}\left[1 - \exp\left(-\frac{UAF}{\dot m c_p}\right)\right]$$

$$= \frac{(0.025\ \text{kg/s})(4180\ \text{J/kg}\cdot°\text{C})}{(4.0\ \text{W/m}^2\cdot°\text{C})(2.5\ \text{m}^2)}\left[1 - \exp\left(-\frac{(4\ \text{W/m}^2\cdot°\text{C})(2.5\ \text{m}^2)(0.8585)}{(0.025\ \text{kg/s})(4180\ \text{J/kg}\cdot°\text{C})}\right)\right]$$

$$= \mathbf{0.8242}$$

The specific of water is taken as $c_w = 4.18$ kJ/kg·°C (Table A-2).

(b) We show the calculation considering a representative hour of 12:00.

First, the hourly solar radiation incident is

$$G_h = G\Delta t = (780\ \text{J/s}\cdot\text{m}^2)(1\ \text{h})\left(\frac{3600\ \text{s}}{1\ \text{h}}\right) = 2.808 \times 10^6\ \text{J/m}^2$$

The amount of useful heat gain is

$$Q_u = F_R A[K_{\tau\alpha}\tau\alpha G_h - Ut(T_{w,in} - T_a)]$$
$$= (0.8242)(2.5 \text{ m}^2)\left[(1)(0.82)(2.808 \times 10^6 \text{ J/m}^2) - (4.0 \text{ W/m}^2 \cdot {}^\circ\text{C})(1 \text{ h})\left(\frac{3600 \text{ s}}{1 \text{ h}}\right)(60 - 19){}^\circ\text{C}\right]$$
$$= 3.528 \times 10^6 \text{ J}$$

The collector efficiency is

$$\eta_c = \frac{Q_u}{AG_h} = \frac{3.528 \times 10^6 \text{ J}}{(2.5 \text{ m}^2)(2.808 \times 10^6 \text{ J/m}^2)} = 0.503$$

The temperature of water at the collector exit is

$$Q_u = \dot{m}_w \Delta t c_w (T_{w,out} - T_{w,in})$$
$$3.528 \times 10^6 \text{ J} = (0.025 \text{ kg/s})(1 \text{ h})\left(\frac{3600 \text{ s}}{1 \text{ h}}\right)(4180 \text{ kJ/kg} \cdot {}^\circ\text{C})(T_{w,out} - 60){}^\circ\text{C}$$
$$T_{w,out} = 69.4{}^\circ\text{C}$$

Repeating the calculations at other hours, we obtain the following results:

Solar time	Ambient air temperature T_a, °C	Average rate of solar radiation incident G, W/m²	Hourly solar radiation incident G_h, kJ/m²	Useful heat gain Q_u, kJ	Collector efficiency η_c	Water temperature at the exit $T_{w,out}$, °C
07:00	12	45	162	−1151	—	56.9
08:00	13	120	432	−665	—	58.2
09:00	15	200	720	−119	—	59.7
10:00	16	620	2232	2466	0.442	66.6
11:00	17	650	2340	2678	0.458	67.1
12:00	19	780	2808	3528	0.503	69.4
13:00	20	740	2664	3314	0.498	68.8
14:00	20	360	1296	1002	0.310	62.7
15:00	19	290	1044	547	0.210	61.5
16:00	17	170	612	−242	—	59.4
17:00	15	80	288	−849	—	57.7
18:00	14	30	108	−1182	—	56.9
Total			12,384	13,535		

(*c*) During the hours 07:00-09:00 and 16:00-18:00, the useful heat gains are negative. That means the heat absorbed by the absorber plate is less than the heat loss. As a result, the temperature of water decreases through the collector. The collector should not be operated during these hours. We do not consider the results in these hours when calculating total useful heat gain:

$$Q_{u,total} = \mathbf{13{,}535 \text{ kJ}}$$

Using total values in the table without adding the values in the nonoperating hours, the collector efficiency for this day is determined to be

$$\eta_c = \frac{Q_{u,total}}{AG_{h,total}} = \frac{13{,}535 \text{ kJ}}{(2.5 \text{ m}^2)(12{,}384 \text{ kJ/m}^2)} = 0.437 = \mathbf{43.7\%} \quad \blacktriangle$$

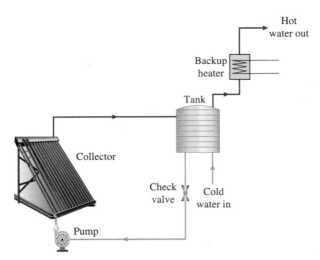

Figure 4-8 A common configuration of solar water collector with a pump and backup heater. (*Adapted from Duffie and Beckman, 2006.*)

Solar Water Heater Configurations

Flat-plate solar collectors can be arranged in various configurations. Figure 4-8 shows a relatively simple system with forced-circulation of water by a pump. Cold water is withdrawn from the bottom of tank; enters the collector at the bottom side; it is heated in the collector by solar energy; it leaves the collector as hot water; and it returns to the top portion of the tank. Hot water is discharged from the top portion of the tank. A backup heater system heats the water further when necessary. The most common backup heater is an electrical resistance heater because it is simple and has low initial cost. Alternatively, a heater burning natural gas, propane, or LPG could be installed.

The system configuration shown in Fig. 4-9 is better suited for climates with subfreezing temperatures in winter. An antifreeze fluid is added to water. Water-antifreeze fluid solution heated in the collector gives up its heat to the cold water as it circulates in the tank. A backup heater placed in the tank could be used in cloudy and rainy days or when the temperature of the water needs to be increased.

Ethylene glycol and propylene glycol are commonly used as the antifreeze fluid. If the mixture contains 10 percent ethylene glycol, the freezing temperature is reduced to −4°C. For ethylene glycol fractions of 20, 30, and 40 percent, the freezing temperatures are −9, −16, and −25°C, respectively. The corresponding temperatures are −2, −7, −13, and −21°C, respectively, if propylene glycol is used as the antifreeze fluid.

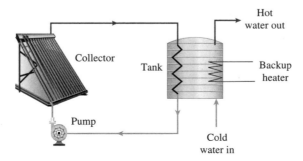

Figure 4-9 A common configuration of solar water collector with water-antifreeze fluid circulating through the collector. (*Adapted from Duffie and Beckman, 2006.*)

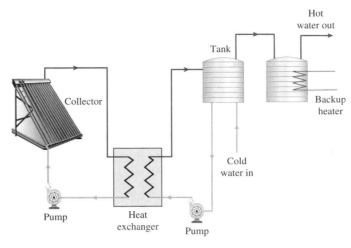

Figure 4-10   A common configuration of solar water collector with water-antifreeze fluid circulating through the collector and an external heat exchanger. (*Adapted from Duffie and Beckman, 2006.*)

A more complicated collector configuration involves an external heat exchanger between the collector and the tank, as shown in Fig. 4-10. There are two water loops; one on the collector side with water-antifreeze solution circulating, and another on the tank side with water circulating. A pump is required for each loop. The backup heater is placed in a separated tank.

Flat-plate solar water heaters are commonly manufactured and used in Mediterranean countries, Australia, and the United States (Florida and California) and elsewhere. Common sizes include 1.0 m × 2.0 m, 0.6 m × 1.2 m, and 1.2 m × 1.2 m. Any number of these collectors may be used together depending on the space and load requirements. In Mediterranean countries, two-tank configuration, one tank at the top of another, is common. The larger tank at the top is used as the cold-water storage. Cold water is used for lowering the temperature of the hot water for the desired use or as the backup storage when the city does not supply water for a period of time (Fig. 4-11).

Figure 4-11   Two-tank solar collectors placed on the roof of buildings.

Some flat-plate collectors do not have a pump. These are *natural circulation* or *thermo-siphon collectors*. The water tank must be located above the collector for natural circulation of water through the collector. As the water in the collector is heated by solar energy, a pressure difference caused by the density difference is established between the collector and the tank, which is the driving force for water flow. The larger the useful heat delivery in the collector, the larger the temperature difference, and the larger the density difference. Higher density difference causes higher flow rate through the collector.

In a natural circulation system, cold water leaves the tank at the bottom and enters the collector at the bottom. This water is heated by absorbing solar energy and flows through the higher levels of the collector. Hot water returns to the top portion of the tank where it is discharged for use (Fig. 4-12). An auxiliary heating system (usually electric heater) is placed in the top portion of the water tank to supply hot water on cloudy and rainy days.

Natural circulation solar heaters are preferred in mild climates to avoid the freezing of water. The majority of solar water heaters in Mediterranean countries and Australia are the natural circulation design. If these heaters are used in climates with the possibility of freezing in winter, water is drained in the coldest period of the season. In this period, a backup water heating system is used to provide hot water needs.

Results of various experimental studies suggest nearly constant temperature rise across the collector for properly designed and constructed natural circulation collectors. This temperature rise is usually between 8 and 11°C (Gupta and Garg, 1968; Close, 1962). Now, we wish to obtain a relation for the mass flow rate of water that will produce a constant temperature rise across the collector. The rate of useful heat delivery was expressed earlier as

$$\dot{Q}_u = F_R A\left[\tau\alpha G - U(T_{w,\text{in}} - T_a)\right]$$

and

$$\dot{Q}_u = \dot{m}_w c_w (T_{w,\text{out}} - T_{w,\text{in}})$$

Setting these equations equal and solving for the mass flow rate gives

$$\dot{m}_w = \frac{F_R A\left[\tau\alpha G - U(T_{w,\text{in}} - T_a)\right]}{c_w (T_{w,\text{out}} - T_{w,\text{in}})} \tag{4-14}$$

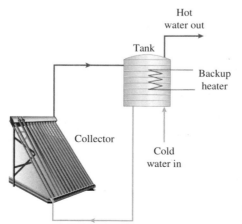

Figure 4-12   A natural circulation solar water collector. (*Adapted from Duffie and Beckman, 2006.*)

The relation between the collector heat removal factor F_R and the collector efficiency factor F was given by

$$F_R = \frac{\dot{m}c_p}{UA}\left[1 - \exp\left(-\frac{UAF}{\dot{m}c_p}\right)\right]$$

Substituting this relation into Eq. (4-14), we obtain a relation for the mass flow rate of water in terms of collector efficiency factor, water temperature rise, and other collector parameters:

$$\dot{m}_w = \frac{-UAF}{c_w \ln\left[1 - \dfrac{U(T_{w,\text{out}} - T_{w,\text{in}})}{\tau\alpha G - U(T_{w,\text{in}} - T_a)}\right]} \tag{4-15}$$

EXAMPLE 4-4
A Natural Circulation
Solar Water Heater

Consider a natural circulation solar collector with a constant temperature rise of 10°C. The collector size is 1.2 m × 1.2 m, the heat loss coefficient is 3.0 W/m²·°C, the product of transmissivity of glazing and absorptivity of the absorber plate is 0.79, the water inlet temperature is 40°C, the ambient air temperature is 20°C, the incident solar radiation is 880 W/m², and the collector efficiency factor is 0.89. Determine the mass flow rate of water through the collector, the rate of useful heat delivery, and the collector efficiency. Also, determine the rate of useful heat if the temperature rise of water is 20°C.

SOLUTION The mass flow rate of water is

$$\dot{m}_w = \frac{-UAF}{c_w \ln\left[1 - \dfrac{U(T_{w,\text{out}} - T_{w,\text{in}})}{\tau\alpha G - U(T_{w,\text{in}} - T_a)}\right]}$$

$$= \frac{-(3.0\ \text{W/m}^2\cdot°\text{C})(1.2\times1.2\ \text{m}^2)(0.89)}{(4180\ \text{J/kg}\cdot°\text{C})\ln\left[1 - \dfrac{(3.0\ \text{W/m}^2\cdot°\text{C})(10°\text{C})}{(0.79)(840\ \text{W/m}^2) - (3.0\ \text{W/m}^2\cdot°\text{C})(40-20)°\text{C}}\right]}$$

$$= \mathbf{0.0190\ kg/s}$$

The specific heat of water is taken as $c_w = 4180$ J/kg·°C (Table A-2). The rate of useful heat delivered is

$$\dot{Q}_u = \dot{m}_w c_w (T_{w,\text{out}} - T_{w,\text{in}}) = (0.0190\ \text{kg/s})(4180\ \text{J/kg}\cdot°\text{C})(10°\text{C}) = \mathbf{795\ W}$$

The collector efficiency is

$$\eta_c = \frac{\dot{Q}_u}{AG} = \frac{795\ \text{W}}{(1.2\times1.2\ \text{m}^2)(880\ \text{W/m}^2)} = 0.617 = \mathbf{61.7\%}$$

If the temperature rise of water is 20°C, the mass flow rate of water and the useful rate of heat delivery are

$$\dot{m}_w = \frac{-UAF}{c_w \ln\left[1 - \dfrac{U(T_{w,\text{out}} - T_{w,\text{in}})}{\tau\alpha G - U(T_{w,\text{in}} - T_a)}\right]}$$

$$= \frac{-(3.0\ \text{W/m}^2\cdot°\text{C})(1.2\times1.2\ \text{m}^2)(0.89)}{(4180\ \text{J/kg}\cdot°\text{C})\ln\left[1 - \dfrac{(3.0\ \text{W/m}^2\cdot°\text{C})(20°\text{C})}{(0.79)(840\ \text{W/m}^2) - (3.0\ \text{W/m}^2\cdot°\text{C})(40-20)°\text{C}}\right]}$$

$$= 0.00927\ \text{kg/s}$$

$$\dot{Q}_u = \dot{m}_w c_w (T_{w,\text{out}} - T_{w,\text{in}}) = (0.00927\ \text{kg/s})(4180\ \text{J/kg}\cdot°\text{C})(20°\text{C}) = \mathbf{775\ W}$$

This is less than 795 W calculated for a temperature rise of 10°C. This change is due to lower collector heat removal factor F_R. Remember that the higher the temperature rise of water through the collector, the lower the F_R. ▲

Solar Air Collector

We assumed the working fluid to be water in the analysis of flat-plate solar collector. This is normally the case when the collector is used to produce hot water. To avoid freezing of water in cold climates, an antifreeze fluid such as ethylene glycol can be added to water to decrease the freezing temperature. In this case, the properties of water-antifreeze solution should be used. Another solution to subfreezing temperatures is using air as the working fluid in the flat-plate collector. Air collectors have advantages over water collectors in that leakage of working fluid is less of a problem and an additional heat exchanger is not required.

Solar air collectors are usually preferred over water collectors for space heating and agricultural drying applications. In a common air collector design, water tubes are replaced by an air duct enclosed by the glass glazing at the top and an insulation layer at the bottom, as shown in Fig. 4-13. In various configurations, the absorber plate may be placed below

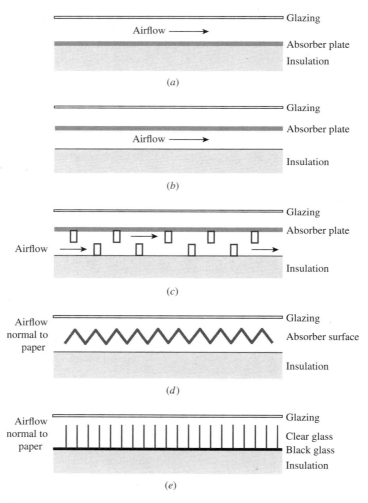

Figure 4-13   Various configurations for solar air collectors.

the air duct (Fig. 4-13*a*), above the air duct (Fig. 4-13*b*), or within the airflow with certain geometries to increase the contact area between the plate and the air. Any metallic plate or glass with a high absorptivity surface could be used as the absorber material.

Solar air collectors require a large heat transfer surface area and high flow rates for reasonable magnitudes of useful heat delivery. This is due to low specific heat of air and low convection heat transfer coefficient between the absorber plate and the air. The mass flow rate of air is relatively small compared to water flow; and as a result, air experiences a significant temperature rise in the collector.

Thermal analysis of a solar air collector requires information on collector heat removal factor F_R, and to determine F_R, we need to know the collector efficiency factor F. Relations for F as well as heat loss coefficient U are available in the literature for various air collector configurations.

EXAMPLE 4-5
Analysis of a Solar Air Collector

A house is heated in winter using an array of solar air collectors with a total area of 15 m². The solar irradiation is 775 W/m² and the product of transmissivity of glazing and absorptivity of the absorber plate is 0.81. The ambient air temperature is 10°C and the heat loss coefficient is 2.9 W/m²·°C. Air enters the collector at 50°C and leaves at 75°C. If the rate of heat supplied to the house by this collector is 6.0 kW, determine the collector heat removal factor, the mass flow rate of air through the collector, and the collector efficiency.

Also, determine the cost savings for a period of 10 h in a winter day due to using this air collector. Assume the conditions specified are applicable for this 10-h period. The backup heating system in the house is a natural gas furnace whose efficiency is 85 percent. The unit price of natural gas is $1.5/10⁵ kJ.

SOLUTION The collector heat removal factor can be determined from the relation for useful heat:

$$\dot{Q}_u = F_R A[\tau\alpha G - U(T_{a,\text{in}} - T_a)]$$

$$6000 \text{ W} = F_R(15 \text{ m}^2)[(0.81)(775 \text{ W/m}^2) - (2.9 \text{ W/m}^2\cdot°\text{C})(50 - 10)°\text{C}]$$

$$F_R = \mathbf{0.782}$$

The mass flow rate of air can be determined from

$$\dot{Q}_u = \dot{m}_a c_a (T_{a,\text{out}} - T_{a,\text{in}})$$

$$6000 \text{ W} = \dot{m}_a(1005 \text{ J/kg·°C})(75 - 50)°\text{C}$$

$$\dot{m}_a = \mathbf{0.239 \text{ kg/s}}$$

The specific heat of air is taken as $c_a = 1005$ J/kg·°C (Table A-1). The collector efficiency is

$$\eta_c = \frac{\dot{Q}_u}{AG} = \frac{6000 \text{ W}}{(15 \text{ m}^2)(775 \text{ W/m}^2)} = 0.516 = \mathbf{51.6\%}$$

The amount of natural gas saved in a 10-h period is

$$Q_{\text{gas}} = \frac{\dot{Q}_u \times \text{Operating time}}{\eta_{\text{gas heater}}} = \frac{(6 \text{ kJ/s})(10 \times 3600 \text{ s})}{0.85} = 254{,}120 \text{ kJ}$$

The corresponding cost savings is

$$\text{Cost savings} = Q_{\text{gas}} \times \text{Unit price} = (254{,}120 \text{ kJ})(\$1.5/10^5 \text{ kJ}) = \mathbf{\$3.81} \quad \blacktriangle$$

4-3 EVACUATED TUBE COLLECTORS

An effective method of improving collector performance is based on reducing heat losses from the collector. This is accomplished by placing a smaller tube inside a larger glass tube and remove the air in the annular space completely (Fig. 4-14). The fluid (we assume it is water) is heated as it flows inside the smaller tube whose walls act as a receiver (or absorber) of solar radiation. The incident solar radiation transmitted through the glass tube is absorbed by the surface of the receiver tube, and this heat is transferred to the water by conduction through the tube wall and by convection from the inner surface of the wall to the water. Since there is vacuum in the annular space between the tubes, heat loss from the outer surface of the receiver to the inner surface of the glass tube is by radiation only. Conduction and convection require a material medium to transfer heat. Heat is then lost through the glass wall by conduction and from the outer surface of the glass wall to the ambient by the combined effects of convection and radiation. This solar collector system is called the *evacuated tube collector*, also called the vacuum tube collector.

As a result of reduced heat losses in comparison to flat-plate collectors, evacuated tube collectors are more efficient, and water is heated to higher temperatures. The rate of heat loss from the collector can be expressed as

$$\dot{Q}_{\text{loss}} = UA_r(T_r - T_a) \tag{4-16}$$

where U is the overall heat loss coefficient, in W/m²·°C, A_r is the receiver area, T_r is receiver temperature, and T_a is ambient air temperature. The overall heat loss coefficient can be determined using a thermal resistance network between the outer surface of the receiver and the ambient air. It consists of a radiation resistance between the outer surface of the receiver tube and the inner surface of the glass tube, a conduction resistance through the glass wall, and a combined convection and radiation resistance between the outer surface of the glass tube and the ambient air. The result is

$$\frac{1}{U} = R_{\text{total}} = R_{\text{rad}} + R_{\text{cond}} + R_{\text{conv, rad}}$$

$$= \frac{1}{h_{\text{rad}}} + R_{\text{cond}} + \frac{1}{h_{\text{conv}} + h_{\text{rad}}} \tag{4-17}$$

$$= \frac{1}{\varepsilon_{\text{eff}}\sigma(T_r + T_g)(T_r^2 + T_g^2)} + \frac{\ln(D_{g,o}/D_{g,i})}{2\pi k_g L} + \frac{1}{h_{\text{conv}} + \varepsilon_g\sigma(T_g + T_a)(T_g^2 + T_a^2)}$$

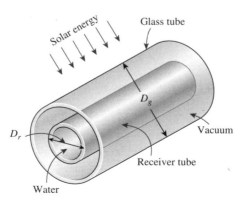

Figure 4-14 Schematic of a single evacuated tube.

where the effective emissivity over the evacuated space is

$$\frac{1}{\varepsilon_{\text{eff}}} = \frac{1}{\varepsilon_r} + \frac{1}{\varepsilon_g} - 1 \tag{4-18}$$

Also, $D_{g,o}$ and $D_{g,i}$ are the outer and inner diameters of the glass tube, respectively, k_g is thermal conductivity of the glass, L is tube length, h_{conv} is the convection heat transfer coefficient, h_{rad} is the radiation heat transfer coefficient, and σ is the Stefan-Boltzmann constant ($\sigma = 5.67 \times 10^{-8}$ W/m²·K⁴). Also, T is temperature, ε is emissivity, and the subscripts r and g stand for the receiver and glass, respectively.

The conduction resistance of the glass is small compared to other thermal resistances due to thin glass wall, and thus usually neglected. The first thermal resistance term is taken as the radiation resistance between two large parallel plates rather than a more complicated and accurate relation for a concentric cylinder. There is also some uncertainty in the estimation of h_{conv} value. Experiments indicate that U value for a single evacuated tube is between 0.5 and 1.0 W/m²·°C, which is much smaller than the values listed in Table 4-1 for flat-plate collectors. Therefore, more of the incident solar radiation is transferred to the water as useful heat and higher water temperatures are achieved as a result of reduced heat losses in evacuated tube collectors.

A common application of evacuated tube collector involves several tubes in parallel arrangement in a flat-plate stationary collector. In this arrangement, a plate with a highly reflective surface is placed at the bottom of the collector below the tubes. This allows solar radiation passing between the tubes to hit the back plate and reflect to the receiver tubes (Fig. 4-15).

Considering a collector with multiple tubes, the rate of useful heat may be determined from the water information (mass flow rate, specific heat, inlet and outlet temperatures):

$$\dot{Q}_u = \dot{m}_w c_p (T_{w,\text{out}} - T_{w,\text{in}}) \tag{4-19}$$

It may also be determined from

$$\dot{Q}_u = \tau_g \alpha_r G_{\text{eff}} A_{r,p} - U A_r (T_r - T_a) \tag{4-20}$$

where τ_g is the transmissivity of the glass, α_r is the absorptivity of the receiver, $A_{r,p}$ is the projected area of the receiver tube ($A_{r,p} = n_{\text{tube}} D_r L$) with n_{tube} is the number of evacuated

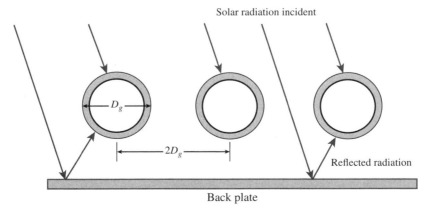

Figure 4-15 Schematic of a single evacuated tube.

tubes, and A_r is the receiver tube area ($A_r = n_{tube}\pi D_r L$). G_{eff} is the effective solar radiation on the tube, which includes the direct radiation from the sun, diffuse radiation from other surfaces, and the reflected radiation from the highly reflective back plate. More information on G_{eff} can be found in Kreith and Kreider (2011). The collector efficiency can be expressed as

$$\eta_c = \frac{\dot{Q}_u}{A_c G} = \frac{\dot{m}_w c_p (T_{w,out} - T_{w,in})}{A_c G} \tag{4-21}$$

where G is the incident solar radiation per unit collector aperture area and A_c is the collector aperture area. The collector efficiency can also be determined from

$$\eta_c = \frac{\dot{Q}_u}{A_c G} = \frac{\tau_g \alpha_r G_{eff} A_{r,p} - UA_r(T_r - T_a)}{A_c G} \tag{4-22}$$

The distance between the tubes is an important consideration in the design of evacuated tube collectors. Closely packed tubes cause shading losses as the angle of solar radiation incident changes throughout the day. Widely spaced tubes may cause unnecessary reduction of useful heat. The use of a reflector plate at the back of the tubes is a good solution to capture solar radiation between the tubes. It was demonstrated that the use of a back reflector increases the collector performance by 10 percent or more. Also, the tube spacing equal to the glass tube diameter (center-to-center distance = $2D_g$) maximizes daily solar energy gain (Beekley and Mather, 1975).

Evacuated tube collectors are commonly used in many locations in the world. They provide hot water for residential and commercial buildings at sufficiently high temperatures in locations and seasons with limited solar radiation and low ambient temperatures. They require little maintenance. If one tube is broken in the collector, it may be replaced easily by another one readily available in the market. High-temperature water output allows the use of these collectors for processing heat in industrial facilities. Solar air-conditioning and refrigeration by the means of absorption cooling is another viable application of an evacuated tube collector since these systems require high water temperatures for reasonable coefficient of performance (COP) values.

EXAMPLE 4-6
Analysis of an Evacuated Tube Collector

A solar collector with 15 evacuated tubes of 1.8-m length is used to provide hot water for a house. The tubes are placed in a rectangular frame with an aperture size of 1.8 m × 1.50 m. The specifications of the evacuated tubes and the operating conditions at a certain location and time are given as follows:

Glass tube: Diameter $D_g = 5$ cm, Emissivity $\varepsilon_g = 0.4$, Transmissivity $\tau_g = 0.8$, Temperature $T_g = 35°C$

Receiver tube: Diameter: $D_r = 4$ cm, Emissivity $\varepsilon_r = 0.1$, Absorptivity $\alpha_r = 0.8$, Temperature $T_r = 80°C$

Convection heat transfer coefficient at the outer surface of the glass tube $h_{conv} = 15$ W/m²·K

Ambient air temperature $T_a = 20°C$

Incident solar radiation $G = 600$ W/m²

Effective solar radiation $G_{eff} = 750$ W/m²

Water: Inlet and outlet temperatures $T_{w,in} = 75°C$, $T_{w,out} = 85°C$, Specific heat $c_w = 4.2$ kJ/kg·K

Calculate the overall heat loss coefficient, the rate of useful heat supplied to the water, the collector efficiency, and the mass flow rate of water.

SOLUTION   First, we calculate the total thermal resistance:

$$\frac{1}{\varepsilon_{\text{eff}}} = \frac{1}{\varepsilon_r} + \frac{1}{\varepsilon_g} - 1 \longrightarrow \frac{1}{\varepsilon_{\text{eff}}} = \frac{1}{0.1} + \frac{1}{0.4} - 1 \longrightarrow \varepsilon_{\text{eff}} = 0.08696$$

$$R_{\text{total}} = \frac{1}{\varepsilon_{\text{eff}}\sigma(T_r + T_g)(T_r^2 + T_g^2)} + \frac{1}{h_{\text{conv}} + \varepsilon_g\sigma(T_g + T_a)(T_g^2 + T_a^2)}$$

$$= \frac{1}{(0.08696)(5.67 \times 10^{-8} \text{ W/m}^2 \cdot \text{K}^4)[(353 \text{ K}) + (308 \text{ K})][(353 \text{ K})^2 + (308 \text{ K})^2]}$$

$$+ \frac{1}{15 \text{ W/m}^2 \cdot \text{K} + (0.4)(5.67 \times 10^{-8} \text{ W/m}^2 \cdot \text{K}^4)[(308 \text{ K}) + (293 \text{ K})][(308 \text{ K})^2 + (293 \text{ K})^2]}$$

$$= 1.455 \text{ m}^2 \cdot \text{K/W}$$

Then the overall heat loss coefficient is

$$U = \frac{1}{R_{\text{total}}} = \frac{1}{1.455 \text{ m}^2 \cdot \text{K/W}} = \mathbf{0.687 \text{ W/m}^2 \cdot \text{K}}$$

We neglected thermal resistance of the thin glass wall and used K for the temperature unit $[(T(\text{K}) = T(°\text{C}) + 273]$ as required in radiation calculations. The rate of useful heat is determined as

$$A_{r,p} = n_{\text{tube}}D_r L = (15)(0.04 \text{ m})(1.8 \text{ m}) = 1.08 \text{ m}^2$$

$$A_r = n_{\text{tube}}\pi D_r L = (15)\pi(0.04 \text{ m})(1.8 \text{ m}) = 3.393 \text{ m}^2$$

$$\dot{Q}_u = \tau_g\alpha_r G_{\text{eff}}A_{r,p} - UA_r(T_r - T_a)$$

$$= (0.80)(0.80)(750 \text{ W/m}^2)(1.08 \text{ m}^2) - (0.687 \text{ W/m}^2 \cdot \text{K})(3.393 \text{ m}^2)(353 - 293)\text{K}$$

$$= \mathbf{379 \text{ W}}$$

The collector efficiency is

$$\eta_c = \frac{\dot{Q}_u}{A_c G} = \frac{379 \text{ W}}{(1.8 \text{ m})(1.5 \text{ m})(600 \text{ W/m}^2)} = 0.234 = \mathbf{23.4\%}$$

The mass flow rate of water is

$$\dot{Q}_u = \dot{m}_w c_w(T_{w,\text{out}} - T_{w,\text{in}})$$

$$379 \text{ W} = \dot{m}_w(4200 \text{ J/kg} \cdot °\text{C})(85 - 75)°\text{C}$$

$$\dot{m}_w = 0.009012 \text{ kg/s} = \mathbf{32.4 \text{ kg/h}} \quad \blacktriangle$$

4-4 CONCENTRATING SOLAR COLLECTOR

The concentration of solar energy is low, and as a result, the temperature of hot water obtainable in a flat-plate collector is low (usually under 80°C). Hot fluid (water, steam, air, or another fluid) at much higher temperatures can be produced using concentrating collectors by concentrating solar radiation on a smaller area. The most common type of concentrating solar collector is the *parabolic trough collector* (Fig. 4-16).

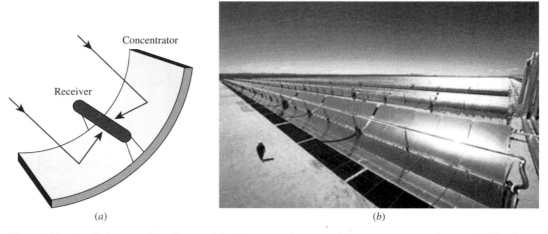

Figure 4-16   Parabolic trough collector. (*a*) Schematic diagram. (*U.S. Department of Energy.*) (*b*) Photo. (*NREL/Warren Gretz, staff photographer.*)

In a concentrating collector, solar radiation is incident on the collector surface, called aperture area A_a, and this radiation is reflected or redirected into a smaller receiver area A_r. The concentration factor CR is defined as

$$\text{CR} = \frac{A_a}{A_r} \tag{4-23}$$

The value of CR is greater than 1. The greater the value of CR, the greater is the hot fluid temperature. The effectiveness of the aperture-to-receiver process is a function of the orientation of surfaces and their radiative properties such as absorptivity and reflectivity. This effectiveness is expressed by an optical efficiency term η_{ar}. Then, the net rate of solar radiation supplied to the receiver is

$$\dot{Q}_r = \eta_{ar} A_a G \tag{4-24}$$

where G is the solar irradiation, in W/m². The rate of heat loss from the collector is expressed as

$$\dot{Q}_{\text{loss}} = U A_r (T_c - T_a) \tag{4-25}$$

The useful heat transferred to the fluid is

$$\dot{Q}_u = \dot{Q}_r - \dot{Q}_{\text{loss}} = \eta_{ar} A_a G - U A_r (T_c - T_a) \tag{4-26}$$

The efficiency of this solar collector is defined as the ratio of the useful heat delivered to the water to the radiation incident on the collector:

$$\eta_c = \frac{\dot{Q}_u}{\dot{Q}_{\text{incident}}} = \frac{\eta_{ar} A_a G - U A_r (T_c - T_a)}{A_a G}$$

$$= \eta_{ar} - \frac{U A_r (T_c - T_a)}{A_a G} = \eta_{ar} - \frac{U (T_c - T_a)}{\text{CR} \times G} \tag{4-27}$$

Therefore, the collector efficiency is maximized for maximum values of the optical efficiency of the aperture-to-receiver process η_{ar} and the concentration factor CR. The efficiency of concentrating collectors is greater than that of flat-plate collectors. For further details see Goswami et al. (2000), Kreith and Kreider (2011), and Duffie and Beckman (2006). If the collector efficiency is plotted against the term $(T_c - T_a)/(CR \times G)$, we obtain a straight line, similar to that for a flat-plate collector. The slope of this line is equal to $-U$.

Temperatures in the receiver of a concentrating collector can reach 400°C. The heated fluid is usually water, and it can be used for space and process heating and cooling or to drive a steam turbine for electricity production.

Linear concentrating solar power collectors are used to capture and reflect solar radiation onto a linear receiver tube. The fluid contained in the tube is heated. A common application is generating steam in the receiver tubes and running this steam through a turbine to generate electricity. In order to produce reasonable amounts of electrical power, a large number of collectors in parallel rows are used to collect solar heat. The most common linear concentrator is the *parabolic trough collector*. The receiver tube is positioned along the focal line of each parabolic reflector. Water coming out of a condenser is heated, boiled, and superheated by absorbing solar heat, and it is routed to a turbine, as shown in Fig. 4-17. Some existing parabolic trough systems produce 80 MW of electricity. In California, power plants with capacities of hundreds of megawatts were constructed using parabolic trough collectors combined with gas turbines.

If the parabolic trough collectors are oversized, excess heat can be stored, and this heat can be used during nighttime or cloudy days to produce electricity. These solar plants can be integrated with conventional power plants utilizing natural gas or coal. The system may be designed such that electricity is supplied by solar as much as possible and the conventional system is used as backup when solar heat is not available.

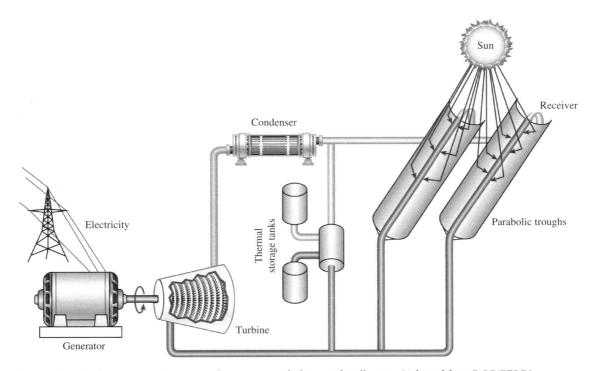

Figure 4-17   A solar concentrator power plant using parabolic trough collectors. (*Adapted from DOE/EERE.*)

The efficiency of a solar system used to produce electricity may be defined as the power produced divided by the total solar irradiation. That is,

$$\eta_{\text{th,solar}} = \frac{\dot{W}_{\text{out}}}{\dot{Q}_{\text{incident}}} = \frac{\dot{W}_{\text{out}}}{A_c G} \tag{4-28}$$

where A_c is the collector surface area receiving solar irradiation and G is the solar irradiation.

EXAMPLE 4-7
Comparison of Two
Concentrating Collectors

Two concentrating collectors (collector A and collector B) have the same concentration factor of CR = 9 and optical efficiency of $\eta_{ar} = 0.85$. The collector temperature for both collectors is 350°F, and the ambient air temperature is 85°F. The heat loss coefficient for collector A is 0.45 Btu/h·ft²·°F and that for collector B is 0.63 Btu/h·ft²·°F. The solar irradiation on collector A is 180 Btu/h·ft². Determine the solar irradiation rate of collector B so that both collectors have the same efficiency.

SOLUTION   The collector efficiency of collector A is determined from

$$\eta_{c,A} = \eta_{ar} - U_A \frac{T_c - T_a}{\text{CR} \times G_A} = 0.85 - (0.45 \text{ Btu/h·ft}^2 \cdot °\text{F})\frac{(350 - 85)\,°\text{F}}{(9)(180 \text{ Btu/h·ft}^2)} = 0.7764$$

Using the same relation for collector B at the same collector efficiency, we obtain the necessary solar irradiation rate:

$$\eta_{c,B} = \eta_{ar} - U_B \frac{T_c - T_a}{\text{CR} \times G_B}$$

$$0.7764 = 0.85 - (0.63 \text{ Btu/h·ft}^2 \cdot °\text{F})\frac{(350 - 85)\,°\text{F}}{9 G_B}$$

$$G_B = \textbf{252 Btu/h·ft}^2 \quad \blacktriangle$$

EXAMPLE 4-8
Electricity Production
from Parabolic Trough
Collectors in Two Cities

Oklahoma City and Portland are considered for the installation of a solar power plant utilizing parabolic trough collectors. The total area of the collectors is 6000 m², and the average efficiency of the plant is estimated to be 15 percent. Using the average daily solar radiation values on a horizontal surface in Table 3-6 in Chap. 3, determine the amount of electricity that can be produced in each city per year.

SOLUTION   The annual solar radiation value on a horizontal surface is obtained from Table 3-6 for each city as

$$G_{\text{Ok}} = (17.15 \text{ MJ/m}^2 \cdot \text{day})(365 \text{ days}) = 6260 \text{ MJ/m}^2$$

$$G_{\text{Po}} = (12.61 \text{ MJ/m}^2 \cdot \text{day})(365 \text{ days}) = 4603 \text{ MJ/m}^2$$

Noting that the average thermal efficiency is 15 percent and the total collector area is 6000 m², the amount of electricity that can be produced in each city per year would be

$$\text{Amount of electricity (Oklahoma City)} = \eta_{\text{th}} A G_{\text{Ok}}$$

$$= (0.15)(6000 \text{ m}^2)(6260 \text{ MJ/m}^2)\left(\frac{1000 \text{ kJ}}{1 \text{ MJ}}\right)\left(\frac{1 \text{ kWh}}{3600 \text{ kJ}}\right)$$

$$= \textbf{1.565} \times \textbf{10}^6 \text{ kWh}$$

Amount of electricity (Portland) $= \eta_{th} A G_{Po}$

$$= (0.15)(6000 \text{ m}^2)(4603 \text{ MJ/m}^2)\left(\frac{1000 \text{ kJ}}{1 \text{ MJ}}\right)\left(\frac{1 \text{ kWh}}{3600 \text{ kJ}}\right)$$

$$= 1.151 \times 10^6 \text{ kWh}$$

That is, 36 percent more electricity can be generated in Oklahoma City than in Portland. ▲

4-5 SOLAR-POWER-TOWER PLANT

Electricity can be produced from solar energy by solar cells by direct conversion of solar radiation into electricity or by using parabolic trough collectors. A *solar-power-tower plant* uses a large array of mirrors called *heliostats* that track the sun and reflect solar radiation into a receiver mounted on top of a tower (Fig. 4-18). Water is heated, boiled, and superheated by absorbing heat from the receiver system. The resulting steam is directed to a turbine to produce power. A generator is connected to the turbine to convert turbine shaft power into electricity.

The first large-scale solar-power-tower plant was Solar 1 located in Barstow, California. It has a capacity of 10 MW. The tower is 91 m high (300 ft.), and the receiver located at the top of the tower is water-cooled. There is an oil-sand storage unit that can help supply electricity for 3 to 4 h after sunset. The total cost of the Solar 1 plant was $14,000/kW, which is 5 to 10 times greater than the cost of electric power stations that run on fossil fuels and other renewables (Culp, 1991). The Solar 2 plant went into operation in 1996 and uses a molten nitrate salt thermal energy storage system.

Figure 4-18   A solar-power-tower plant uses large array of mirrors called *heliostats* that track the sun and reflect solar radiation into a receiver mounted on top of a tower. (*Kevin Burke/Corbis RF.*)

The Gemasolar power plant located in Seville, Spain, consists of 2650 heliostats that focus 95 percent of solar radiation onto a giant receiver. The plant started commercial operation in 2011, occupying a field of 185 hectares. Temperatures as high as 900°C are obtained at the receiver. Molten salt tanks are heated by concentrated solar heat reaching a temperature above 500°C. Water runs through the molten salt tanks, where it is boiled and superheated. The resulting steam is directed to turbines to produce power. Steam leaving the turbine is condensed and pumped back to the molten salt tanks to repeat the heat engine cycle. The plant can store solar heat and use it for a period of 15 h in the absence of daylight. The plant has an installed capacity of 19.9 MW and can produce 110 GWh of electricity per year. This is enough electricity for 25,000 homes. Electricity is produced for 270 days a year. The cost of the Gemasolar plant is $33,000/kW, which is even higher than that of Solar 1.

The Ivanpah solar power plant started commercial operation in 2013 after a 3-year construction period and consists of three separate units (Fig. 4-19). The electricity generated can serve 140,000 homes during the peak hours of the day. The plant is located in the Mojave desert in California and is the largest solar thermal power plant in the world with a capacity of 377 MW. It is located on 3500 acres (14.2 km^2) with 300,000 concentrated mirrors reflecting solar energy to receivers at the top of three towers in three plants. The towers are 140 m high. The solar energy is absorbed by water flowing in the pipes of the boiler. The water turns into superheated vapor, which is directed into a steam turbine located at the bottom of the tower (Fig. 4-20). Electrical output from the turbine is sent to transmission lines. This plant uses an air-cooled condenser, which uses 95 percent less water than wet-cooled solar thermal plants.

The Gemasolar plant in Spain, Solano plant in Arizona, and Crescent Dunes Solar Project in Nevada all store energy using molten salts. This ensures power generation during evening peak hours. The Solana and Crescent Dunes plants can produce power for up to 6 and 10 h after sunset, respectively. Ivanpah does not include any energy storage system.

Figure 4-19 Ivanpah solar thermal power plant. (*Courtesy of Brightsource Energy.*)

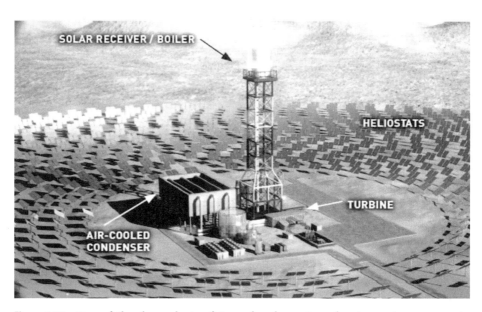

Figure 4-20 One of the three plants of Ivanpah solar system showing main components. (*Courtesy of Brightsource Energy.*)

EXAMPLE 4-9
Thermodynamic
Analysis of a Solar-
Power-Tower Plant

A solar-power-tower plant is considered for Tucson, Arizona. Heliostats with a total area of 80,000 m^2 are to be used to reflect solar radiation into a receiver. When the solar irradiation is 950 W/m^2, steam is produced at 2 MPa and 400°C at a rate of 20 kg/s. This steam is expanded in a turbine to 20 kPa pressure. The isentropic efficiency of the turbine is 85 percent.

(*a*) Determine the power output and the thermal efficiency of the plant when the solar radiation is 950 W/m^2.
(*b*) How much electricity can be produced per year if the average thermal efficiency is 15 percent and the generator efficiency is 96 percent?

SOLUTION (*a*) Using the turbine isentropic efficiency, the steam properties at the inlet and exit of the turbine are determined as follows (Tables A-3, A-4, and A-5):

$$\left. \begin{array}{l} P_1 = 2\,\text{MPa} \\ T_1 = 400°\text{C} \end{array} \right\} \quad \begin{array}{l} h_1 = 3248.4\ \text{kJ/kg} \\ s_1 = 7.1292\ \text{kJ/kg·K} \end{array}$$

$$\left. \begin{array}{l} P_2 = 20\,\text{kPa} \\ s_2 = s_1 \end{array} \right\} \quad h_{2s} = 2349.7\ \text{kJ/kg}$$

$$\eta_T = \frac{h_1 - h_2}{h_1 - h_{2s}} \longrightarrow h_2 = h_1 - \eta_T (h_1 - h_{2s})$$

$$= 3248.4 - (0.85)(3248.4 - 2349.7) = 2484.5\ \text{kJ/kg}$$

Then the power output is

$$\dot{W}_{\text{out}} = \dot{m}(h_1 - h_2) = (20\ \text{kg/s})(3248.4 - 2484.5)\ \text{kJ/kg} = \mathbf{15,280\ kW}$$

The thermal efficiency of this power plant is equal to power output divided by the total solar incident on the heliostats:

$$\eta_{\text{th}} = \frac{\dot{W}_{\text{out}}}{AG} = \frac{15,280\ \text{kW}}{(80,000\ \text{m}^2)(0.950\ \text{kW/m}^2)} = 0.201\ \text{or}\ \mathbf{20.1\%}$$

(*b*) The solar data for Tucson, Arizona, are given in Table 3-6 in Chap. 3. The daily average solar irradiation for an entire year on a horizontal surface is given to be 20.44 MJ/m²·day. Multiplying this value by 365 days of the year gives an estimate of solar irradiation on the heliostat surfaces. Using the definition of the thermal efficiency,

$$W_{out} = \eta_{th,avg} A G = (0.15)(80,000 \text{ m}^2)(20,440 \text{ kJ/m}^2\cdot\text{day})(365 \text{ days})\left(\frac{1 \text{ kWh}}{3600 \text{ kJ}}\right) = 2.487 \times 10^7 \text{ kWh}$$

This is total work output from the turbine. The electrical energy output from the generator is

$$W_{elect} = \eta_{gen} W_{out} = (0.96)(2.487 \times 10^7 \text{ kWh}) = \mathbf{2.387 \times 10^7 \text{ kWh}}$$

This solar power plant has a potential to generate 24 million kWh of electricity per year. If the electricity is sold at a price of $0.10/kWh, the potential revenue is $2.4 million per year. ▲

4-6 SOLAR POND

A promising method of power generation involves collecting and storing solar energy in large artificial lakes a few meters deep, called *solar ponds*. Solar energy is absorbed by all parts of the pond, and the water temperature rises everywhere. The top part of the pond, however, loses to the atmosphere much of the heat it absorbs, and as a result, its temperature drops. This cool water serves as insulation for the bottom part of the pond and helps trap the energy there. Usually, salt is planted at the bottom of the pond to prevent the rise of this hot water to the top. A power plant that uses an organic fluid, such as alcohol, as the working fluid can be operated between the top and the bottom portions of the pond, as shown in Fig. 4-21.

The main disadvantage of a solar pond power plant is the low thermal efficiency. For example, if the water temperature is 35°C near the surface and 75°C near the bottom of the pond, the maximum thermal efficiency can be determined to be

$$\eta_{th,max} = 1 - \frac{T_L}{T_H} = 1 - \frac{(35 + 273) \text{ K}}{(75 + 273) \text{ K}} = 0.115 \text{ or } 11.5\%$$

Actual thermal efficiency will be less than this value. Small experimental solar pond power plants have been installed. Several practical problems, such as mixing of pond water by the wind and fouling on heat exchanger surfaces, exist in addition to low thermal efficiency.

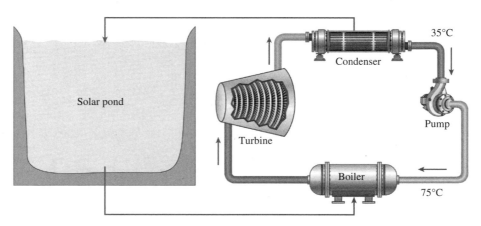

Figure 4-21   Operation of a solar pond power plant.

An *ocean thermal energy converter* (OTEC) system uses the same principle, but the water at the ocean surface is warmer as a result of solar energy absorption, and the water at a deeper location is cooler. Further discussion and analysis for the OTEC system is given in Chap. 10.

4-7 PASSIVE SOLAR APPLICATIONS

The use of solar collectors for water and space heating and electricity production and photovoltaic cells for electricity generation represent active solar energy applications. Less commonly, solar energy can be used for cooling by means of absorption refrigeration and desiccant cooling systems. The use of solar energy by means of engineering design without the involvement of mechanical equipment is called *passive* use of solar energy.

Significant energy savings can be accomplished if a house is designed and built to receive maximum solar heat in winter (to minimize heating energy consumption) and minimum solar heat gain in summer (to minimize cooling energy consumption). This may include selecting the correct orientation of walls and windows, window sizes and types, wall materials, and surface color and finishing of wall surfaces. Of course, particular preferences in the design and construction of buildings will be different in winter-dominated and summer-dominated climates. The solar heating of swimming pools, food drying, and solar cookers are some other examples of passive solar applications.

In this section, we describe a trombe wall and analyze solar heat gain through windows as common examples of passive solar applications.

Trombe Wall

Dark-painted thick masonry walls called *trombe walls* are commonly used on south sides of passive solar homes to absorb solar energy, store it during the day, and release it to the house during the night (Fig. 4-22). The idea was proposed by E. L. Morse of Massachusetts in 1881 and is named after Professor Felix Trombe of France, who used it extensively in his

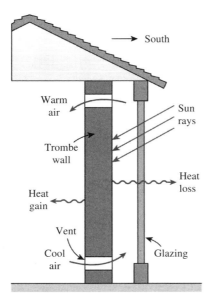

Figure 4-22 Schematic of a trombe wall.

designs in the 1970s. Usually a single or double layer of glazing is placed outside the wall and transmits most of the solar energy while blocking heat losses from the exposed surface of the wall to the outside. Also, air vents are commonly installed at the bottom and top of the trombe walls so that the house air enters the parallel flow channel between the trombe wall and the glazing, rises as it is heated, and enters the room through the top vent.

A trombe wall is normally built on the south of a building. It is particularly effective in reducing heating energy consumption for mild winter climates where solar energy is available during a significant period of time in winter. Southern and western United States and southern Europe are well suited for trombe wall applications.

Solar Heat Gain through Windows

The part of solar radiation that reaches the earth's surface without being scattered or absorbed is the *direct radiation*. Solar radiation that is scattered or reemitted by the constituents of the atmosphere is the *diffuse radiation*. Direct radiation comes directly from the sun following a straight path, whereas diffuse radiation comes from all directions in the sky. All radiation reaching the ground on an overcast day is diffuse radiation. The radiation reaching a surface, in general, consists of three components: direct radiation, diffuse radiation, and radiation reflected onto the surface from surrounding surfaces (Fig. 4-23). Common surfaces such as grass, trees, rocks, and concrete reflect about 20 percent of the radiation while absorbing the rest. Snow-covered surfaces, however, reflect 70 percent of the incident radiation. Radiation incident on a surface that does not have a direct view of the sun consists of diffuse and reflected radiation (Çengel and Ghajar, 2020).

A glazing material that transmits the visible part of the spectrum while absorbing the infrared portion is ideally suited for an application that calls for maximum daylight and minimum solar heat gain. Surprisingly, ordinary window glass approximates this behavior remarkably well. When solar radiation strikes a glass surface, part of it (about 8 percent for uncoated clear glass) is reflected back outdoors, part of it (5 to 50 percent, depending on composition and thickness) is absorbed within the glass, and the remainder is transmitted indoors, as shown in Fig. 4-24. The standard 3-mm-thick (⅛-in) single-pane double-strength

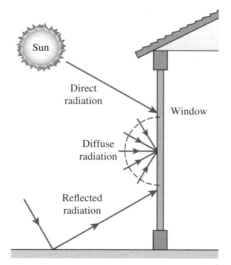

Figure 4-23 Direct, diffuse, and reflected components of solar radiation incident on a window.

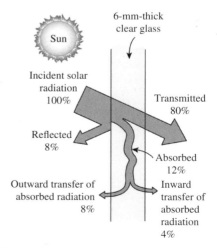

Figure 4-24 Distribution of solar radiation incident on a clear glass.

clear window glass transmits 86 percent, reflects 8 percent, and absorbs 6 percent of the solar energy incident on it.

The hourly variation of solar radiation incident on the walls and windows of a house is given in Table 3-5 in Chap. 3. Solar radiation that is transmitted indoors is partially absorbed and partially reflected each time it strikes a surface, but all of it is eventually absorbed as sensible heat by the furniture, walls, people, and so forth. Therefore, the solar energy transmitted inside a building represents a heat gain for the building. Also, the solar radiation absorbed by the glass is subsequently transferred to the indoors and outdoors by convection and radiation. The sum of the *transmitted* solar radiation and the portion of the *absorbed* radiation that flows indoors constitutes the *solar heat gain* of the building.

The fraction of incident solar radiation that enters through the glazing is called the *solar heat gain coefficient* (SHGC) and is expressed as

$$\text{SHGC} = \frac{\dot{q}_{\text{solar, gain}}}{G} = \tau_s + f_i \alpha_s \tag{4-29}$$

where α_s is the solar absorptivity of the glass and f_i is the inward flowing fraction of the solar radiation absorbed by the glass. Therefore, the dimensionless quantity SHGC is the sum of the fractions of the directly transmitted (τ_s) and the absorbed and reemitted ($f_i\alpha_s$) portions of solar radiation incident on the window. The value of SHGC ranges from 0 to 1, with 1 corresponding to an opening in the wall (or the ceiling) with no glazing. When the SHGC of a window is known, the total solar heat gain through that window is determined from

$$\dot{Q}_{\text{solar, gain}} = \text{SHGC} \times A_{\text{glazing}} \times G \tag{4-30}$$

where A_{glazing} is the glazing area of the window and G is the solar heat flux incident on the outer surface of the window, in W/m².

Another way of characterizing the solar transmission characteristics of different kinds of glazing and shading devices is to compare them to a well-known glazing material that can serve as a base case. This is done by taking the standard 3-mm-thick (⅛-in) double-strength clear window glass sheet whose SHGC is 0.87 as the *reference glazing* and defining a shading coefficient (SC) as

$$\text{SC} = \frac{\text{SHGC}}{\text{SHGC}_{\text{ref}}} = \frac{\text{SHGC}}{0.87} = 1.15 \times \text{SHGC} \tag{4-31}$$

Therefore, the shading coefficient of a single-pane clear glass window is SC = 1.0. The shading coefficients of other commonly used fenestration products are given in Table 4-2 for summer design conditions. The values for winter design conditions may be slightly lower because of the higher heat transfer coefficients on the outer surface due to high winds and thus higher rate of outward flow of solar heat absorbed by the glazing, but the difference is small.

Note that the larger the shading coefficient, the smaller the shading effect, and thus the larger the amount of solar heat gain. A glazing material with a large shading coefficient allows a large fraction of solar radiation to come in.

Solar heat entering a house through windows is preferable in winter since it reduces heating energy consumption, but it should be avoided as much as possible in summer since it increases cooling energy consumption.

Shading devices are used to control solar heat gain through windows. Shading devices are classified as *internal shading* and *external shading*, depending on whether the shading device is placed *inside* or *outside*. External shading devices are more effective in reducing the solar heat gain since they intercept the sun's rays before they reach the glazing. The solar heat gain through

TABLE 4-2 Shading Coefficient SC and Solar Transmissivity τ_{solar} for Some Common Glass Types for Summer Design Conditions (ASHRAE, 1993)

Types of Glazing	Nominal Thickness		τ_{solar}	SC*
	mm	in		
(a) Single glazing				
Clear	3	⅛	0.86	1.0
	6	¼	0.78	0.95
	10	⅜	0.72	0.92
	13	½	0.67	0.88
Heat absorbing	3	⅛	0.64	0.85
	6	¼	0.46	0.73
	10	⅜	0.33	0.64
	13	½	0.24	0.58
(b) Double glazing				
Clear in	3[†]	⅛	0.71[‡]	0.88
Clear out	6	¼	0.61	0.82
Clear in, heat absorbing out[§]	6	¼	0.36	0.58

*Multiply by 0.87 to obtain SHGC.
[†]The thickness of each pane of glass.
[‡]Combined transmittance for assembled unit.
[§]Refers to gray-, bronze-, and green-tinted heat-absorbing float glass.

a window can be reduced by as much as 80 percent by exterior shading. Roof overhangs have long been used for exterior shading of windows. The sun is high in the horizon in summer and low in winter. A properly sized roof overhang or a horizontal projection blocks off the sun's rays completely in summer while letting in most of them in winter, as shown in Fig. 4-25. Such shading structures can reduce the solar heat gain on the south, southeast, and

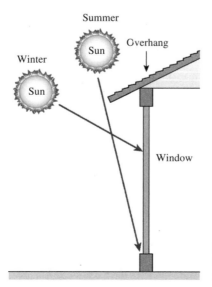

Figure 4-25 A properly sized overhang blocks off the sun's rays completely in summer while letting them through in winter.

southwest windows in the northern hemisphere considerably. A window can also be shaded from outside by vertical or horizontal or architectural projections, insect or shading screens, and sun screens. To be effective, air must be able to move freely around the exterior device to carry away the heat absorbed by the shading and the glazing materials.

Some type of internal shading is used in most windows to provide privacy and aesthetic effects as well as some control over solar heat gain. Internal shading devices reduce solar heat gain by reflecting transmitted solar radiation back through the glazing before it can be absorbed and converted into heat in the building.

Draperies reduce the annual heating and cooling loads of a building by 5 to 20 percent, depending on the type and the user habits. In summer, they reduce heat gain primarily by reflecting back direct solar radiation (Fig. 4-26). The semi-closed air space formed by the draperies serves as an additional barrier against heat transfer, resulting in a lower *U*-factor for the window and thus a lower rate of heat transfer in summer and winter. The solar optical properties of draperies can be measured accurately, or they can be obtained directly from the manufacturers. The shading coefficient of draperies depends on the openness factor, which is the ratio of the open area between the fibers that permits the sun's rays to pass freely, to the total area of the fabric. Tightly woven fabrics allow little direct radiation to pass through, and thus they have a small openness factor. The *reflectance* of the surface of the drapery facing the glazing has a major effect on the amount of solar heat gain. *Light-colored* draperies made of closed or tightly woven fabrics maximize the back reflection and thus minimize the solar gain. *Dark-colored* draperies made of open or semi-open woven fabrics, on the other hand, minimize the back reflection and thus maximize the solar gain.

The shading coefficients of drapes also depend on the way they are hung. Usually, the width of drapery used is twice the width of the draped area to allow folding of the drapes and to give them their characteristic "full" or "wavy" appearance. A flat drape behaves like an ordinary window shade. A flat drape has a higher reflectance and thus a lower shading coefficient than a full drape.

External shading devices such as overhangs and tinted glazing do not require operation, and they provide reliable service over a long time without significant degradation during their service life. Their operation does not depend on a person or an automated system, and these passive shading devices are considered fully effective when determining the peak

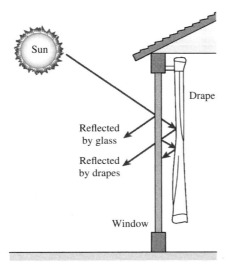

Figure 4-26 Draperies reduce heat gain in summer by reflecting back solar radiation, and they reduce heat loss in winter by forming an air space before the window.

cooling load and annual energy use. The effectiveness of manually operated shading devices, on the other hand, varies greatly depending on user habits, and this variation should be considered when evaluating performance.

The primary function of an indoor shading device is to provide *thermal comfort* for the occupants. An unshaded window glass allows most of the incident solar radiation in, and it also dissipates part of the solar energy it absorbs by emitting infrared radiation to the room. The emitted radiation and the transmitted direct sunlight may bother the occupants near the window. In winter, the temperature of the glass is lower than the room air temperature, causing excessive heat loss by radiation from the occupants. A shading device allows the control of direct solar and infrared radiation while providing various degrees of privacy and outward vision. The shading device is also at a higher temperature than the glass in winter, and thus reduces radiation loss from occupants. *Glare* from draperies can be minimized by using off-white colors. Indoor shading devices, especially draperies made of a closed-weave fabric, are effective in reducing *sounds* that originate in the room, but they are not as effective against the sounds coming from outside.

The type of climate in an area usually dictates the types of windows to be used in buildings. In *cold climates* where the heating load is much larger than the cooling load, the windows should have the highest transmissivity for the entire solar spectrum, and a high reflectivity (or low emissivity) for the far infrared radiation emitted by the walls and furnishings of the room. Low-e windows are well suited for such heating-dominated buildings. Properly designed and operated windows allow more heat into the building over a heating season than they lose, making them energy contributors rather than energy losers. In *warm climates* where the cooling load is much larger than the heating load, the windows should allow the visible solar radiation (light) in but should block off the infrared solar radiation. Such windows can reduce the solar heat gain by 60 percent with no appreciable loss in daylighting. This behavior is approximated by window glazings that are coated with a heat-absorbing film outside and a low-e film inside (Fig. 4-27). Properly selected windows can reduce the cooling load by 15 to 30 percent compared to windows with clear glass. Tinted glass and glass coated with reflective films reduce solar heat gain in summer and heat loss in winter.

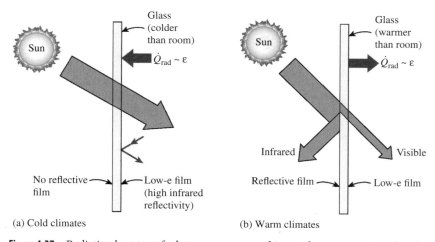

(a) Cold climates   (b) Warm climates

Figure 4-27 Radiation heat transfer between a room and its window is proportional to the emissivity of the glass surface, and low-e coatings on the inner surface of the windows reduce heat loss in winter and heat gain in summer.

Tinted glass and glass coated with reflective films reduce solar heat gain in summer and heat loss in winter. The conductive heat gains or losses can be minimized by using multiple-pane windows. Double-pane windows are usually called for in climates where the winter design temperature is less than 7°C (45°F). Double-pane windows with tinted or reflective films are commonly used in buildings with large window areas. Clear glass is preferred for showrooms since it affords maximum visibility from outside, but bronze-, gray-, and green-colored glass are preferred in office buildings since they provide considerable privacy while reducing glare.

EXAMPLE 4-10
Cooling Energy Consumption of a House for Two Window Options

Two window options are considered for a new house with a floor area of 250 m². Windows occupy 16 percent of the floor area and are equally distributed in all four sides. Window options are

Option 1: Double-glazed window, clear glass, air fill, U-factor = 4.18 W/m²·K, SHGC = 0.77

Option 2: Double-glazed window, low-e glass, argon gas fill, U-factor = 1.42 W/m²·K, SHGC = 0.39

The average daily solar heat fluxes incident on all four sides are given in kWh/m²·day during summer cooling months as follows:

June: 4.95   July: 4.80   August: 4.55   September: 3.95

(*a*) Determine the total amount of heat gain through the windows in summer. Take the average outside and indoor temperatures in summer to be 35 and 23°C, respectively.
(*b*) If the coefficient of performance (COP) of the cooling system is 2.3, determine the net cooling cost savings in summer due to using window option 2. Take the unit cost of electricity to be $0.11/kWh.

SOLUTION   (*a*) The total window area is 16 percent of the floor area:

$$A_{\text{window}} = (0.16)(250 \text{ m}^2) = 40 \text{ m}^2$$

The number of hours in summer months is

$$\text{Summer hours} = (30 + 31 + 31 + 30) \times 24 \text{ h} = 2928 \text{ h}$$

The total solar heat flux incident on the glazing during summer months is determined to be

$$\begin{aligned} q_{\text{solar}} &= (4.95 \text{ kWh/m}^2 \cdot \text{day} \times 30 \text{ days}) + (4.80 \text{ kWh/m}^2 \cdot \text{day} \times 31 \text{ days}) \\ &\quad + (4.55 \text{ kWh/m}^2 \cdot \text{day} \times 31 \text{ days}) + (3.95 \text{ kWh/m}^2 \cdot \text{day} \times 30 \text{ days}) \\ &= 556.9 \text{ kWh/m}^2 \end{aligned}$$

Calculations for window option 1:
The rate of heat transfer through the windows in winter is

$$\dot{Q}_{\text{transfer}} = U_{\text{overall}} A_{\text{window}} (T_o - T_i) = (4.18 \text{ W/m}^2 \cdot °C)(40 \text{ m}^2)(35 - 23)°C = 2006 \text{ W}$$

The amount of heat transfer through the windows is

$$Q_{\text{transfer}} = \dot{Q}_{\text{transfer}} \times \text{Summer hours} = (2.006 \text{ kW})(2928 \text{ h}) = 5875 \text{ kWh}$$

The amount of solar heat input is

$$Q_{\text{solar}} = \text{SHGC} \times A_{\text{window}} \times q_{\text{solar}} = (0.77)(40 \text{ m}^2)(556.9 \text{ kWh/m}^2) = 17,153 \text{ kWh}$$

The total amount of heat input through windows is

$$Q_{\text{total}} = Q_{\text{transfer}} + Q_{\text{solar}} = 5875 + 17,153 = \mathbf{23,028 \text{ kWh}}$$

Calculations for window option 2:
The rate and amount of heat transfers through the windows are

$$\dot{Q}_{transfer} = U_{overall} A_{window} (T_o - T_i) = (1.42 \text{ W/m}^2 \cdot °\text{C})(40 \text{ m}^2)(35 - 23)°\text{C} = 682 \text{ W}$$

$$Q_{transfer} = \dot{Q}_{transfer} \times \text{Summer hours} = (0.682 \text{ kW})(2928 \text{ h}) = 1996 \text{ kWh}$$

The amount of solar heat input in summer for this window is

$$Q_{solar} = \text{SHGC} \times A_{window} \times q_{solar} = (0.39)(40 \text{ m}^2)(556.9 \text{ kWh/m}^2) = 8688 \text{ kWh}$$

The total amount of heat input is

$$Q_{total} = Q_{transfer} + Q_{solar} = 1996 + 8688 = 10{,}684 \text{ kWh} \cong \mathbf{10{,}700 \text{ kWh}}$$

(*b*) The decrease in heat input in summer due to using window option 2 is

$$23{,}028 - 10{,}684 = 12{,}344 \text{ kWh}$$

This corresponds to a reduction of 53.6 percent.
The corresponding decrease in cooling cost is determined from

$$\text{Cooling cost savings} = \frac{\text{Cooling load decrease} \times \text{Unit cost of electricity}}{\text{COP}}$$

$$= \frac{(12{,}344 \text{ kWh})(\$0.11/\text{kWh})}{2.3} = \mathbf{\$590}$$

The window with lower *U*-factor and SHGC saves the house \$590 in summer cooling costs. It should be noted that the amount of heat loss through windows will also be lower in winter due to the lower *U*-factor of window option 2. However, there will be less solar heat gain in winter due to the lower SHGC.   ▲

REFERENCES

ASHRAE. 1993. American Society of Heating, Refrigeration, and Air Conditioning Engineers. *Handbook of Fundamentals*. Atlanta.

ASHRAE Standard 93-77. Methods of testing to determine the thermal performance of solar collectors. Atlanta.

Çengel YA and Ghajar AJ. 2020. *Heat and Mass Transfer: Fundamentals and Applications*, 6th ed. New York: McGraw Hill.

Beekley DC and Mather GR. 1975. *Analysis and Experimental Tests of a High-Performance Evacuated Tube Collector*. Toledo, OH: Owens-Illinois.

Close DJ, 1962. "The Performance of Solar Water Heaters with Natural Circulation." *Solar Energy*, 6(1): 33.

Culp AW. 1991. *Principles of Energy Conversion*, 2nd ed. New York: McGraw Hill.

DOE/EERE. Department of Energy, Energy Efficiency and Renewable Energy. www.eere.energy.gov

Duffie JA and Beckman W. 2006. *Solar Engineering of Thermal Processes*, 3rd ed. New York: Wiley.

Goswami Y, Kreith F, and Kreider JF. 2000. *Principles of Solar Engineering*, 2nd ed. New York: Taylor and Francis.

Gupta CL and Garg HP. 1968. "System Design in Solar Water Heaters with Natural Circulation." *Solar Energy*, 12: 163.

Hodge K. 2010. *Alternative Energy Systems and Applications*. New York: Wiley.

Kreith F and Kreider JF. 2011. *Principles of Sustainable Energy*. New York: Taylor and Francis.

Mitchell JW. 1983. *Energy Engineering*. New York: Wiley.

Souka AF and Safwat HH. 1966. "Optimum Orientations for the Double Exposure Flat-Plate Collector and its Reflectors." *Solar Energy*, 10: 170.

PROBLEMS

INTRODUCTION

4-1   What are the disadvantages of solar energy and its applications?

4-2   What are the three methods of conversion of solar energy into other useful forms of energy?

4-3   What is the heliothermal process? Which devices are used for this conversion process? Explain.

4-4   What is the helioelectrical process? Which devices are used for this conversion process?

4-5   How is a solar cell different from a heliostat?

4-6   Can solar energy be used for cooling applications? Explain.

4-7   Which is not a disadvantage of solar energy and its applications?
(*a*) Not suitable for cooling   (*b*) High system cost   (*c*) Intermittent
(*d*) Large space requirement   (*e*) Low concentration

4-8   Which conversion process is not used for the conversion of solar energy into other useful forms of energy?
(*a*) Heliochemical   (*b*) Helioelectrical   (*c*) Heliothermal
(*d*) Heliofluid   (*e*) None of these

4-9   The single most common application of solar energy is
(*a*) Heliostat   (*b*) Trombe wall   (*c*) Flat-plate solar collector
(*d*) Solar-power-tower plant   (*e*) Solar pond

4-10   Which system uses solar energy for a cooling application?
(*a*) Organic Rankine cycle   (*b*) Absorption refrigeration system
(*c*) Vapor-compression refrigeration system   (*d*) Brayton cycle
(*e*) Evaporative cooler

FLAT-PLATE SOLAR COLLECTOR

4-11   What is the most common use of flat-plate solar collectors?

4-12   What is the difference between the operation of a thermosiphon solar water heater system and an active, closed loop solar water heater?

4-13   Solar energy is not available during nighttime and cloudy days, and solar hot water collectors may not be able to provide hot water during these periods. How can this problem of solar collectors be dealt with?

4-14   Compare the efficiencies of unglazed, single-glazed and double-glazed flat-plate solar collectors. Which one is the most efficient? Why?

4-15   Under what conditions is the efficiency of a flat-plate solar collector maximum? Compare maximum efficiencies of unglazed, single-glazed, and double-glazed flat-plate solar collectors. Which one has the highest maximum efficiency? Why?

4-16   Define the collector heat removal factor. What is the upper limit for the collector heat removal factor?

4-17   Define the collector efficiency factor.

4-18   Freezing of water in solar water collectors can damage the collector in winter. What are the potential solutions to this danger of freezing?

4-19   Will the rate of useful heat remain constant when the temperature rise of water through the collector is doubled in a natural circulation solar water heater. Why? Explain.

4-20   What are the advantages and disadvantages of air-type solar collectors compared to water collectors?

4-21   Air-type collectors require a large heat transfer surface area and high flow rates for reasonable magnitudes of useful heat delivery. What is the reason for that?

4-22   Solar radiation is incident on a flat-plate collector at a rate of 930 W/m². The glazing has a transmissivity of 0.82, and the absorptivity of the absorber plate is 0.94. Determine the maximum efficiency of this collector.

4-23   Solar radiation is incident on a flat-plate collector at a rate of 750 W/m². The glazing has a transmissivity of 0.86, and the absorptivity of the absorber plate is 0.95. The heat loss coefficient of the collector is 3 W/m²·°C. The collector is at an average temperature of 45°C and the ambient air temperature is 23°C. Determine the efficiency of this collector.

4-24   Solar radiation is incident on a flat-plate collector at a rate of 260 Btu/h·ft². The product of the transmissivity of glazing and the absorptivity of the absorber plate is $\tau\alpha = 0.85$ for this single-glazed collector. The heat loss coefficient of the collector is 0.5 Btu/h·ft²·°F. The collector is at an average temperature of 120°F and the ambient air temperature is 67°F. (*a*) Determine the efficiency of this collector. (*b*) Determine the efficiency of a double-glazed collector whose $\tau\alpha$ value is 0.80 and the heat loss coefficient is 0.3 Btu/h·ft²·°F. Use the same collector and ambient air temperatures.

4-25   A solar collector provides the hot water needs of a family for a period of 8 months except for 4 months of winter season. The collector supplies hot water at an average temperature of 60°C, and the average temperature of cold water is 20°C. An examination of water bills indicates that the family uses an average of 6 tons of hot water from the solar collector per month. A natural gas water heater supplies hot water in winter months. The cost of natural gas is \$1.35/therm (1 therm = 100,000 Btu = 105,500 kJ), and the efficiency of the heater is 88 percent. Determine the annual cost of natural gas and cost savings to this family due to the solar collector.

4-26   Solar radiation is incident on a flat-plate collector at a rate of 880 W/m². The product of the transmissivity of the glazing and the absorptivity of the absorber plate is 0.82. The collector has a surface area of 33 m². This collector supplies hot water to a facility at a rate of 6.3 L/min. Cold water enters the collector at 18°C. If the efficiency of this collector is 70 percent, determine the temperature of hot water provided by the collector.

4-27   Solar radiation is incident on a flat-plate collector at a rate of 480 W/m². The glazing has a transmissivity of 0.85 and the absorptivity of the absorber plate is 0.92. The heat removal factor for the collector is 0.94 and the incident angle modifier is 1. Determine the maximum efficiency of this collector.

4-28   The specifications of a flat-plate collector are as follows: $\tau = 0.88$, $\alpha = 0.97$, $U = 1.3$ Btu/h·ft²·°F. The heat removal factor for the collector is 0.92, the solar insolation is 210 Btu/h·ft², and the ambient air temperature is 60°F. Determine (*a*) the collector efficiency if the water enters the collector at 115°F and (*b*) the temperature of water at which the collector efficiency is zero. Take the incident angle modifier to be 1.

4-29   Consider a flat-plate collector with the following specifications:
Collector area $A = 3.0$ m²
Incident radiation $G = 880$ W/m².
Heat loss coefficient $U = 3.3$ W/m²·°C
Convection heat transfer coefficient at the inner surface of the tubes $h_i = 250$ W/m²·°C
Tube spacing $W = 18$ cm
Outer tube diameter $D_o = 1.50$ cm
Inner tube diameter $D_i = 1.35$ cm
Fin efficiency $\eta_{fin} = 0.93$
Contact resistance between the plate and the tube is negligible.
Flow rate of water in the tubes $\dot{m}_w = 0.030$ kg/s
Water inlet temperature $T_{w,in} = 50$°C
Ambient air temperature $T_a = 15$°C
The product of transmissivity of glazing and absorptivity of the absorber $\tau\alpha = 0.80$

(*a*) Calculate the collector efficiency factor and the collector heat removal factor.
(*b*) Calculate the rate of useful heat gain, the temperature rise of water through the collector, and the collector efficiency.
(*c*) Repeat parts *a* and *b* for a water flow rate of 0.015 kg/s.

4-30   Consider a flat-plate collector with the following specifications:

Collector area $A = 2.2$ m^2

Heat loss coefficient $U = 2.8$ W/m$^2 \cdot$°C

Flow rate of water in the tubes $\dot{m}_w = 0.020$ kg/s

Water inlet temperature $T_{w,in} = 75$°C

The product of transmissivity of glazing and absorptivity of the absorber $\tau\alpha = 0.84$

The collector efficiency factor $F = 0.83$

The incident angle modifier is taken to be 1.

Average hourly data for ambient air temperature and solar heat flux incident are given in the following table for a July day at 40° latitude.

Time	T_a, °C	Average solar heat flux incident G, W/m^2
05:00	23	0
06:00	24	115
07:00	24	320
08:00	25	528
09:00	25	702
10:00	26	838
11:00	28	922
12:00	30	949
13:00	31	922
14:00	32	838
15:00	31	702
16:00	30	528
17:00	29	320
18:00	28	115
19:00	26	0

(*a*) Calculate the collector heat removal factor.

(*b*) For each hour, calculate the useful heat gain, the collector efficiency, and the temperature of water at the collector exit.

(*c*) Calculate the total useful heat gain and the overall collector efficiency for the entire day.

4-31   A solar flat-plate collector with a size of 1.5 m × 2.0 m is used to provide hot water for an entire day. The transmissivity of glazing is 0.88, and the absorptivity of absorber plate is 0.95. The water inlet temperature is 80°C, and the heat removal factor is 0.78. The average daily solar radiation on a horizontal surface in Tucson, Arizona, in June is 29.30 MJ/m^2. Assume that the collector is operated for a period of 12 hours in June 15. During this 12-h operating period, the average collector efficiency is 50 percent and the average ambient temperature is 30°C. Estimate the heat loss coefficient of the collector during this period.

4-32   Consider a single-glazing flat-plate collector with the following specifications:

Collector area $A = 2.0$ m^2

Heat loss coefficient $U = 3.5$ W/m$^2 \cdot$°C

Flow rate of water in the tubes $\dot{m}_w = 0.040$ kg/s

Water inlet temperature $T_{w,in} = 40$°C

The product of transmissivity of glazing and absorptivity of the absorber $\tau\alpha = 0.83$

The collector heat removal factor $F_R = 0.83$

Average hourly data for ambient air temperature and solar heat flux incident are given in the below table for a January day at 40° latitude.

Time	T_a, °C	Average solar heat flux incident G, W/m^2
07:00	24	0
08:00	25	220
09:00	25	560
10:00	26	720
11:00	28	830
12:00	30	860
13:00	31	820
14:00	32	730
15:00	31	540
16:00	30	240
17:00	29	0

(a) For each hour, calculate the useful heat gain, the collector efficiency, and the temperature of water at the collector exit.

(b) Calculate the total useful heat gain and the overall collector efficiency for the entire day.

(c) If 20 such collectors are used in an industrial facility, calculate the amount and cost of natural gas savings in the entire January if this solar collector is replacing a natural gas water heater with an efficiency is 80 percent. The unit price of natural gas is $1.2/therm.

4-33 Considering a flat-plate solar collector, develop a relation for the critical solar radiation for which the rate of useful heat is zero. Using this relation, determine the critical solar radiation value for a flat-plate collector with the following specifications:

Collector size = 1.0 m × 2.0 m
Transmissivity of glazing = 0.86
Absorptivity of the absorber plate = 0.93
Water inlet temperature = 55°C
Heat removal factor = 0.79
Ambient temperature = 10°C
Heat loss coefficient = 5.0 W/m^2·°C

4-34 Consider a natural circulation solar collector with a constant temperature rise of 18°F. The collector area is 15 ft^2, the heat loss coefficient is 0.53 Btu/h·ft^2·°F, the product of transmissivity of glazing and absorptivity of the absorber plate is 0.80, the water inlet temperature is 90°F, the ambient air temperature is 65°F, the incident solar radiation is 260 Btu/h·ft^2, and the collector efficiency factor is 0.91. Determine the mass flow rate of water through the collector and the rate of useful heat delivery.

4-35 Consider a natural circulation solar collector with a constant temperature rise of 10°C. The collector size is 1.0 m × 2.0 m, the heat loss coefficient is 2.5 W/m^2·°C, the product of transmissivity of glazing and absorptivity of the absorber plate is 0.82, the water inlet temperature is 30°C, the ambient air temperature is 12°C, the incident solar radiation is 630 W/m^2, and the collector efficiency factor is 0.92. Determine the mass flow rate of water through the collector and the rate of useful heat. Assuming this operation as the average conditions for a total period of 360 h in a given month, estimate the cost savings if the backup water heater runs on electricity. The unit price of electricity is $0.22/kWh.

4-36 An array of solar air collectors has a total area of 10 m^2 with a solar irradiation of 950 W/m^2. The product of transmissivity of glazing and absorptivity of the absorber plate is 0.80. Air enters the collector at 55°C with a flow rate of 0.14 kg/s. The ambient air temperature is 13°C, the heat loss coefficient is 3.5 W/m^2·°C, and the collector heat removal factor is 0.75. Determine the rate of useful heat, the air temperature at the collector outlet, and the collector efficiency.

4-37   The efficiency of a flat-plate solar collector is inversely proportional to
(a) Solar irradiation   (b) Transmissivity of the glazing   (c) Absorptivity of the absorber plate
(d) The difference between the collector and air temperatures   (e) None of these

4-38   The maximum efficiency of a flat-plate solar collector is equal to

(a) $\eta_c = \tau\alpha - U\dfrac{T_c - T_a}{G}$ (b) $\eta_c = U\dfrac{T_c - T_a}{G}$ (c) $\eta_c = \tau\alpha - \dfrac{U}{G}$ (d) $\eta_c = \tau\alpha - \dfrac{T_c - T_a}{G}$

(e) $\eta_c = \tau\alpha$

4-39   The efficiency of a solar collector is given by $\eta_c = \tau\alpha - U(T_c - T_a)/G$. If the collector efficiency is plotted against the term $(T_c - T_a)/G$, a straight line is obtained. The slope of this line is equal to
(a) U (b) $-U$ (c) $\tau\alpha$ (d) $-\tau\alpha$ (e) U/G

4-40   Which collectors have the highest and the lowest maximum efficiency values, respectively?
(a) Double glazing, single glazing   (b) Single glazing, double glazing
(c) Unglazed, single glazing   (d) Single glazing, unglazed   (e) Unglazed, double-glazing

4-41   Which collectors have the highest efficiencies under practical operating conditions?
(a) Single glazing   (b) Double glazing   (c) No glazing

4-42   What is the effect of increasing water flow rate through the collector on the collector heat removal factor, useful heat delivery, and the collector efficiency?
Collector heat removal factor:
(a) Increases   (b) Decreases   (c) Remains the same

Useful heat delivery:
(a) Increases   (b) Decreases   (c) Remains the same

Collector efficiency:
(a) Increases   (b) Decreases   (c) Remains the same

4-43   Choose the wrong statement regarding the collector efficiency factor and the collector heat removal factor.
(a) The collector heat removal factor cannot be greater than the collector efficiency factor.
(b) The collector heat removal factor increases as the water flow rate increases.
(c) The collector efficiency factor accounts for the temperature difference between the absorber plate and water.
(d) The collector efficiency factor is the ratio of actual useful heat gain to that when the absorber plate is at the water inlet temperature.
(e) The collector efficiency factor is proportional to heat loss coefficient.

4-44   In a flat-plate collector, the useful heat relation involves the temperature difference between the absorber plate and ambient air. A more convenient form of this relation can be obtained using the water inlet temperature instead of absorber plate temperature. This is made possible by the use of
(a) Collector efficiency factor   (b) Collector heat removal factor
(c) Collector efficiency   (d) Fin efficiency

4-45   The temperature rise of water through the collector is doubled in a natural circulation solar water heater. As a result, the rate of useful heat will
(a) Decrease a little   (b) Remain constant
(c) Increase a little   (d) Double
(e) More than double

4-46   Select the incorrect statement for air-type solar collectors.
(a) Air-type collectors are less common than water collectors.
(b) Air-type collectors require high-volume flow rates of air for reasonable magnitudes of useful heat.
(c) Air flows in a duct instead of tubes.
(d) They cannot be used for subfreezing temperatures.
(e) Temperature change of air across the collector is large compared to water collectors.

4-47 Solar radiation is incident on a flat-plate collector at a rate of 900 W/m^2. The product of transmissivity and absorptivity is $\tau\alpha = 0.88$, and the heat loss coefficient of the collector is 2.5 $W/m^2 \cdot °C$. The collector is at an average temperature of 60°C, and the ambient air temperature is 27°C. The efficiency of this collector is

(*a*) 70.0%   (*b*) 73.3%   (*c*) 76.4%   (*d*) 78.8%   (*e*) 81.2%

4-48 Solar radiation is incident on a flat-plate collector at a rate of 450 W/m^2. The product of transmissivity and absorptivity is $\tau\alpha = 0.85$, and the heat loss coefficient of the collector is 4.5 $W/m^2 \cdot °C$. The ambient air temperature is 10°C. The collector temperature at which the collector efficiency is zero is

(*a*) 95°C   (*b*) 104°C   (*c*) 112°C   (*d*) 87°C   (*e*) 73°C

4-49 Solar radiation is incident on a flat-plate collector at a rate of 600 W/m^2. The glazing has a transmissivity of 0.85, and the absorptivity of the absorber plate is 0.92. The heat loss coefficient of the collector is 3.0 $W/m^2 \cdot °C$. The maximum efficiency of this collector is

(*a*) 92%   (*b*) 85%   (*c*) 78%   (*d*) 73%   (*e*) 66%

EVACUATED TUBE COLLECTORS

4-50 In which type of collector is the water heated to higher temperatures?

4-51 What are the heat loss mechanisms in an evacuated tube collector?

4-52 What are the three components of effective solar radiation G_{eff} in an evacuated tube collector?

4-53 The hot water needs of a house is to be provided by a solar collector with 12 evacuated tubes of 6-ft length. The tubes are placed in a rectangular frame with an aperture size of 6 ft × 4 ft. The specifications of the evacuated tubes and the operating conditions at a certain location and time are given as follows:

Glass tube: Diameter $D_g = 2$ in, Emissivity $\varepsilon_g = 0.3$, Transmissivity $\tau_g = 0.85$, Temperature $T_g = 95°F$

Receiver tube: Diameter: $D_r = 1.6$ in, Emissivity $\varepsilon_r = 0.2$, Absorptivity $\alpha_r = 0.9$, Temperature $T_r = 180°F$

Convection heat transfer coefficient at the outer surface of the glass tube $h_{conv} = 2.5$ Btu/h·ft²·°F

Ambient air temperature $T_a = 70°F$

Incident solar radiation $G = 200$ Btu/h·ft²

Effective solar radiation $G_{eff} = 250$ Btu/h·ft²

Water: Inlet and outlet temperatures $T_{w,in} = 170°F$, $T_{w,out} = 190°F$, Specific heat $c_w = 1.0$ Btu/lbm·°F

Calculate the overall heat loss coefficient, the rate of useful heat supplied to the water, the collector efficiency, and the mass flow rate of water.

4-54 The process heating needs of a manufacturing facility is met by 25 evacuated tubes of 2.2-m length. The diameters of the glass tube and receiver tubes are 4.7 and 3.8 cm, respectively. The transmissivity of the glass tube is 0.86 and the absorptivity of the receiver tube is 0.93. The average temperature of the receiver tube is 70°C and the temperature of the ambient air is 15°C. The solar irradiation on the collector in a certain location and time is 580 W/m^2 and the effective solar radiation on the tube surface is 690 W/m^2. The overall heat loss coefficient is 0.6 $W/m^2 \cdot °C$. The tubes are placed in a rectangular frame with an aperture size of 2.2 m × 1.85 m. Determine the rate of useful heat delivery and the efficiency of the collector.

4-55 Space heating of a house is supplemented by 80 evacuated tubes of 2.2-m length. The diameters of the glass tube and receiver tubes are 4.0 and 3.2 cm, respectively. The transmissivity of the glass tube is 0.85 and the absorptivity of the receiver tube is 0.95. The average temperature of the receiver tube is 75°C and the temperature of the ambient air is 5°C. The effective solar radiation on the tube surface is estimated to be 550 W/m^2. The overall heat loss coefficient is 0.75 $W/m^2 \cdot °C$. Air flowing through the receiver tubes experience a temperature rise of 30°C. The main heating equipment for this house in winter is electric heaters.

(*a*) Determine the total flow rate of air in the tubes.

(*b*) Calculate the cost savings due to using evacuated tube collectors instead of electric heaters. Assume the average conditions specified in this problem can be used for a total period of 300 h in winter. The unit cost of electricity is \$0.15/kWh. The electricity used in the house come from a power plant burning natural gas. Also, determine the amount of natural gas saved due to using evacuated tube collectors in this house in winter. The heating value of natural gas is 55,000 kJ/kg. The fuel-to-electricity efficiency of the power plant is 35 percent.

4-56   A solar collector consists of 15 evacuated tubes of 1.8-m length and a receiver tube diameter of 4 cm. The transmissivity of the glass tube is 0.85 and the absorptivity of the receiver tube is 0.9. The average temperature of the receiver tube is 60°C, and the temperature of the ambient air is 25°C. The overall heat loss coefficient is 0.9 W/m²·°C. The working fluid is water, which enters the collector at 50°C at a rate of 20 kg/h and leaves at 70°C. Estimate the effective solar radiation on the tubes.

4-57   Which property should be high for the glass tube of an evacuated tube collector?
(*a*) Absorptivity   (*b*) Reflectivity   (*c*) Transmissivity   (*d*) Emissivity

4-58   A back plate is placed below the tubes in an evacuated tube collector. Which property should be high for this plate?
(*a*) Absorptivity   (*b*) Reflectivity   (*c*) Transmissivity   (*d*) Emissivity

4-59   What are the desired characteristics of receiver tubes?
(*a*) High absorptivity, high emissivity   (*b*) High absorptivity, low emissivity
(*c*) Low absorptivity, high emissivity   (*d*) Low absorptivity, low emissivity

CONCENTRATING SOLAR COLLECTOR

4-60   What are the advantages of a concentrating collector compared to a flat-plate solar collector?

4-61   How do we define the efficiency of a solar system used to produce electricity?

4-62   A concentrating collector has a concentration factor of CR = 15, and the optical efficiency of the aperture-to-receiver process $\eta_{ar} = 0.93$. The solar insolation is 520 W/m², and the ambient air temperature is 20°C. The heat loss coefficient is 4 W/m²·°C. If the collector temperature is 130°C, determine the collector efficiency.

4-63   Two concentrating collectors (collector A and collector B) have the same concentration factor of CR = 7 and the optical efficiency of $\eta_{ar} = 0.88$. The collector temperature for both collectors is 145°C and the ambient air temperature is 27°C. The heat transfer coefficient for collector A is 2.5 W/m²·°C and that for collector B is 3.5 W/m²·°C. The solar irradiation on collector A is 600 W/m². (*a*) At what solar irradiation rate does collector B have the same efficiency as collector A? (*b*) What is the efficiency change of collector A when the solar irradiation increases to 900 W/m²?

4-64   Electricity is generated by a solar system utilizing parabolic trough collectors. The temperature in the receiver of the collector is measured to be 300°C. If the ambient air temperature is 27°C, what is the maximum thermal efficiency of this solar power plant? If flat-plate solar collectors with a maximum fluid temperature of 95°C were used for electricity generation, what would be the maximum thermal efficiency?

4-65   A solar power plant utilizes parabolic trough collectors with a total collector area of 2500 m². The solar irradiation is 700 W/m². If the efficiency of this solar plant is 8 percent, how much power is generated?

4-66   Miami and Atlanta are considered for the installation of a solar power plant utilizing parabolic trough collectors. The total area of the collectors is 300,000 m², and the average efficiency of the plant is estimated to be 18 percent. Using the average daily solar radiation values on a horizontal surface in Table 3-6 in Chap. 3, determine the amount of electricity that can be produced in each city per year.

4-67   The efficiency of a concentrating solar collector is given by $\eta_c = \eta_{ar} - U(T_c - T_a)/(\text{CR} \times G)$. If the collector efficiency is plotted against the term $(T_c - T_a)/(\text{CR} \times G)$, a straight line is obtained. The slope of this line is equal to
(a) $-U$ (b) η_{ar} (c) $T_c - T_a$ (d) $(T_c - T_a)/\text{CR}$ (e) $(T_c - T_a)/G$

4-68   Which is not a component of a power plant using linear concentrating solar collectors?
(a) Receiver   (b) Turbine   (c) Mirror   (d) Parabolic troughs   (e) Condenser

4-69   Solar radiation is incident on a concentrating collector at a rate of 750 W/m². The optical efficiency of the aperture-to-receiver process is $\eta_{ar} = 0.90$, and the concentration factor is CR = 8. The heat loss coefficient of the collector is 3.9 W/m²·°C. The collector is at an average temperature of 120°C, and the ambient air temperature is 25°C. The efficiency of this collector is
(a) 75.1%   (b) 90.0%   (c) 78.2%   (d) 85.6%   (e) 83.8%

4-70   Solar radiation is incident on a concentrating collector at a rate of 640 W/m². The optical efficiency of the aperture-to-receiver process is $\eta_{ar} = 0.87$, and the concentration factor is CR = 6. The heat loss coefficient of the collector is 3.4 W/m²·°C. The collector is at an average temperature of 115°C, and the ambient air temperature is 30°C. The maximum efficiency of this collector is
(a) 75.1%   (b) 87.0%   (c) 79.5%   (d) 85.6%   (e) 83.8%

4-71   Consider a solar power plant with a total collector area of 5000 m². Solar energy is incident on collector surfaces at a rate of 800 W/m². The power output from the plant is 520 kW. What is the efficiency of this solar plant?
(a) 7.4 percent   (b) 13.0 percent   (c) 17.2 percent   (d) 4.9 percent   (e) 19.9 percent

4-72   Consider a solar power plant with a total collector area of 9000 m². Solar energy is incident on collector surfaces at a rate of 750 W/m². The efficiency of the plant is 9 percent. What is the power output from the plant?
(a) 6756 kW   (b) 3411 kW   (c) 1006 kW   (d) 608 kW   (e) 468 kW

SOLAR-POWER-TOWER PLANT

4-73   Describe the principle of a solar-power-tower plant.

4-74   What is a heliostat? What does it do?

4-75   The cost of the Solar 1 plant in California with a capacity of 10 MW was estimated to be $14,000/kW. If electricity generated is sold at a price of $0.09/kWh, how long will it take for this plant to pay for itself? Assume the plant operates 5000 h a year at the maximum capacity of 10 MW.

4-76   A solar-power-tower plant is considered for Houston, Texas. Heliostats with a total area of 400,000 ft² are to be used to reflect solar radiation into a receiver. When the solar irradiation is 250 Btu/h·ft², steam is produced at 160 psia and 600°F at a rate of 15 lbm/s. This steam is expanded in a turbine to 2 psia pressure. The isentropic efficiency of the turbine is 88 percent. (a) Determine the power output in kW and the thermal efficiency of the plant under these operating conditions. (b) How much electricity can be produced per year if the average thermal efficiency is 12 percent and the generator efficiency is 98 percent? Use the solar insolation value in Table 3-6 in Chap. 3. (c) The estimated cost of this plant is $17,000/kW, and the plant is expected to operate 4500 h a year at the power output determined in part (a). If electricity generated is to be sold at a price of $0.11/kWh, how long will it take for this plant to pay for itself?

4-77   In a solar-power-tower plant, the large array of mirrors is called
(a) Receiver   (b) Heliostat   (c) Turbine   (d) Parabolic troughs   (e) Condenser

4-78   In which solar system is the water heated to the highest temperature?
(a) Flat-plate collector   (b) Solar cell   (c) Parabolic trough collector
(d) Solar-power-tower plant   (e) Solar pond

4-79 A solar-power-tower plant produces 450 kW of power when solar radiation is incident at a rate of 1050 W/m². If the efficiency of this solar plant is 17 percent, the total collector area receiving solar radiation is
(*a*) 1750 m²   (*b*) 2090 m²   (*c*) 2520 m²   (*d*) 3230 m²   (*e*) 3660 m²

SOLAR POND

4-80 Describe the operation of a solar pond power plant. What is the main disadvantage of this system?

4-81 Describe the operation of an ocean thermal energy converter (OTEC). How is this system different from a solar pond system?

4-82 In a solar pond, the water temperature is 30°C near the surface and 75°C near the bottom of the pond. If the thermal efficiency of this solar pond power plant is 3.6 percent, what is the second-law efficiency of this power plant?

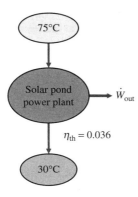

Figure P4-82

4-83 Solar energy is collected and stored in large artificial lakes a few meters deep. This system is called
(*a*) Flat-plate collector   (*b*) Parabolic trough collector   (*c*) Heliostat
(*d*) Ocean thermal energy conversion   (*e*) Solar pond

4-84 In a solar pond, the water temperature is 40°C near the surface and 90°C near the bottom of the pond. The maximum thermal efficiency a solar pond power plant can have is
(*a*) 13.8%   (*b*) 17.5%   (*c*) 25.7%   (*d*) 32.4%   (*e*) 55.5%

4-85 In an OTEC system, the water temperature is 40°C at the surface and 5°C at some depths. The maximum thermal efficiency an OTEC plant can have is
(*a*) 3.7%   (*b*) 19.1%   (*c*) 11.2%   (*d*) 34.4%   (*e*) 87.5%

PASSIVE SOLAR APPLICATIONS

4-86 What is the difference between active and passive solar applications? Give examples of active and passive solar applications.

4-87 How should a house be designed to receive solar heat for (*a*) winter-dominated and (*b*) summer-dominated climates? Why?

4-88 What is a trombe wall? Why is it used?

4-89 What is a shading device? Is an internal or external shading device more effective in reducing the solar heat gain through a window? How does the color of the surface of a shading device facing outside affect the solar heat gain?

4-90   Define the SHGC (solar heat gain coefficient), and explain how it differs from the SC (shading coefficient). What are the values of the SHGC and SC of a single-pane clear-glass window?

4-91   What does the SC (shading coefficient) of a device represent? How do the SCs of clear glass and heat-absorbing glass compare?

4-92   Describe the solar radiation properties of a window that is ideally suited for minimizing the air-conditioning load.

4-93   What is the effect of a low-e coating on the inner surface of a window glass on the (*a*) heat loss in winter and (*b*) heat gain in summer through the window?

4-94   Consider a building located near 40°N latitude that has equal window areas on all four sides. The building owner is considering coating the south-facing windows with reflective film to reduce the solar heat gain and thus the cooling load. But someone suggests that the owner will reduce the cooling load even more if she coats the west-facing windows instead. What do you think?

4-95   A manufacturing facility located at 32°N latitude has a glazing area of 60 m² facing west that consists of double-pane windows made of clear glass (SHGC = 0.766). To reduce the solar heat gain in summer, a reflective film that will reduce the SHGC to 0.35 is considered. The cooling season consists of June, July, August, and September, and the heating season, October through April. The average daily solar heat fluxes incident on the west side at this latitude are 2.35, 3.03, 3.62, 4.00, 4.20, 4.24, 4.16, 3.93, 3.48, 2.94, 2.33, and 2.07 kWh/day·m² for January through December, respectively. Also, the unit costs of electricity and natural gas are $0.15/kWh and $0.90/therm, respectively. If the coefficient of performance of the cooling system is 3.2 and the efficiency of the furnace is 0.90, determine the net annual cost savings due to installing reflective coating on the windows. Also, determine the simple payback period if the installation cost of reflective film is $15/m².

4-96   A house located in Boulder, Colorado (40°N latitude), has ordinary double-pane windows with 6-mm-thick glass and the total window areas are 8, 6, 6, and 4 m² on the south, west, east, and north walls, respectively. Determine the total solar heat gain of the house at 9:00, 12:00, and 15:00 solar time in July. Also, determine the total amount of solar heat gain per day for an average day in January.

4-97   Repeat Prob. 4-96 for double-pane windows that are gray-tinted.

4-98   Consider a building in New York (40°N latitude) that has 76 m² of window area on its south wall. The windows are double-pane heat-absorbing type, and are equipped with light-colored venetian blinds with a shading coefficient of SC = 0.30. Determine the total solar heat gain of the building through the south windows at solar noon in April. What would your answer be if there were no blinds at the windows?

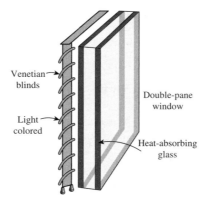

Figure P4-98

4-99   A typical winter day in Reno, Nevada (39°N latitude), is cold but sunny, and thus the solar heat gain through the windows can be more than the heat loss through them during daytime. Consider a house with double-door-type windows that are double paned with 3-mm-thick glass and 6.4 mm of air space and aluminum frames and spacers. The overall heat transfer coefficient for this window is 4.55 W/m²·°C. The house is maintained at 22°C at all times. Determine if the house is losing more or less heat than it is gaining from the sun through an east window on a typical day in January for a 24-h period if the average outdoor temperature is 10°C.

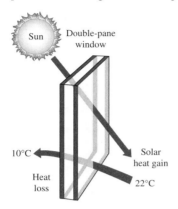

Figure P4-99

4-100   Repeat Prob. 4-99 for a south window.

4-101   Determine the rate of net heat gain (or loss) through a 9-ft-high, 15-ft-wide, fixed ⅛-in single-glass window with aluminum frames on the west wall at 3 p.m. solar time during a typical day in January at a location near 40°N latitude when the indoor and outdoor temperatures are 78 and 20°F, respectively. The overall heat transfer coefficient for this window is 1.17 Btu/h·ft²·°F.

4-102   Which is a passive solar application?
(*a*) Flat-plate solar collector   (*b*) Concentrating solar collector   (*c*) Solar pond
(*d*) Solar cell   (*e*) Trombe wall

4-103   The radiation reaching a surface consists of
  I. direct radiation
 II. diffuse radiation
III. radiation reflected onto the surface from surrounding surfaces
(*a*) Only I   (*b*) I and II   (*c*) I and III   (*d*) II and III   (*e*) I, II, and III

4-104   The solar heat gain of a building consists of
  I. transmitted radiation
 II. reflected radiation
III. outward transfer of absorbed radiation
 IV. inward transfer of absorbed radiation
(*a*) Only I   (*b*) I and II   (*c*) I and III   (*d*) I and IV   (*e*) I, III, and IV

4-105   The fraction of incident solar radiation that enters through the glazing is called
(*a*) Solar heat gain coefficient   (*b*) Shading coefficient
(*c*) Direct radiation coefficient   (*d*) Diffuse radiation coefficient
(*e*) Transmission coefficient

4-106   Comparing solar transmission characteristics of different types of glazing devices to a well-known glazing material can be done by defining a
(*a*) Solar heat gain coefficient   (*b*) Shading coefficient
(*c*) Direct radiation coefficient   (*d*) Diffuse radiation coefficient
(*e*) Transmission coefficient

4-107 The shading coefficient of a single-pane clear glass window is
(*a*) 0   (*b*) 1   (*c*) Between 0 and 1
(*d*) Greater than 1   (*e*) Less than 1

4-108 The standard 3-mm-thick (⅛-in) single-pane double-strength clear window glass transmits _____ of the solar energy incident on it.
(*a*) 100%   (*b*) 94%   (*c*) 86%   (*d*) 78%   (*e*) 72%

4-109 The shading coefficient of standard 3-mm-thick (⅛-in) single-pane double-strength clear window glass is
(*a*) 1   (*b*) 0.94   (*c*) 0.87   (*d*) 0.78   (*e*) 0

4-110 Which statement is not correct regarding windows?
(*a*) In cold climates, the windows should have the highest transmissivity for the entire solar spectrum.
(*b*) In cold climates, the windows should have a high reflectivity (or low emissivity) for the far infrared radiation emitted by the walls and furnishings of the room.
(*c*) Low-e windows are well suited for heating-dominated buildings.
(*d*) In warm climates, the windows should allow the infrared solar radiation in, but should block off the visible solar radiation.
(*e*) In warm climates, window glazings that are coated with a heat-absorbing film outside and a low-e film inside should be used.

4-111 A building has 35 m² of window area on its south wall. The windows have a solar heat gain coefficient of 0.696. The solar radiation incident at a south-facing surface at a particular time is 690 W/m². Determine the rate of solar heat gain through these windows at that time.
(*a*) 35.0 kW   (*b*) 29.1 kW   (*c*) 24.2 kW   (*d*) 19.3 kW   (*e*) 16.8 kW

4-112 A building has 35 m² of window area on its south wall. The windows have a shading coefficient of 0.8. The solar radiation incident at a south-facing surface at a particular time is 690 W/m². Determine the rate of solar heat gain through these windows at that time.
(*a*) 35.0 kW   (*b*) 29.1 kW   (*c*) 24.2 kW   (*d*) 19.3 kW   (*e*) 16.8 kW

CHAPTER 5

Solar Photovoltaic Systems

5-1 PHOTOVOLTAIC EFFECT

Solar cells or photovoltaic (PV) cells are devices that convert solar radiation into electricity directly. Solar cells have no moving parts or components, and they have a lightweight structure. A PV cell operates on the principle of *photovoltaic effect*. When photons from solar light are incident on a suitable material, electrons are released, which in turn generate voltage difference or electric current.

Becquerel noted the photoelectric effect for the first time in 1839. He observed this when light was incident on an electrode in a solution. The operating principle of a PV cell was discovered by Adams and Day in 1877 in solids with selenium as the material. Coblenz discovered the generation of voltage between the dark and illuminated regions of semiconducting crystals in 1919. Ohl discovered the photoelectric effect at a *p-n* junction of two semiconductors in 1941. Researchers at RCA and Bell Laboratories achieved a conversion efficiency of 6 percent in 1954 by working on *p*-type and *n*-type semiconductors.

PV cells are found in small devices such as calculators and watches, medium-size systems such as water pumps, traffic signs, satellites, space vehicles, residential units, and large systems such as power stations for utility grids (Fig. 5-1). Some advantages of PV systems include high reliability with no moving parts and lightweight structure, low operating and maintenance costs, flexibility in sizing, and no water consumption. Some disadvantages are relatively high initial cost, intermittent nature of solar energy, additional costs for energy storage, and dust collection on panel surfaces reducing system performance.

An understanding of the operation of solar cells requires physics of atomic theory and semiconductor theory. We learned from physics that an atom consists of a nucleus containing protons and neutrons and equal number of electrons orbiting about the nucleus. For the silicon element, the atomic number is 14, which is equal to the number of photons or electrons. Protons are contained in the nucleus while electrons are in orbitals or *bands* with respect to the nucleus. Inner bands are filled while outer bands may be partially filled (Hodge, 2010).

The energy of an electron depends on the band that it occupies. The most outside band an electron can be found is called the *valence band*. The chemical characteristics of an element depend on the number of electrons in the valence band. For a chemically inert element, the valence band is filled. In silicon, the valence band can accommodate a maximum of eight electrons, but it has four electrons.

When the electron bond in the valence band is strong, the neighboring atoms share electrons so that valence bands are filled. This is called the *covalent bond*. The electrons in a

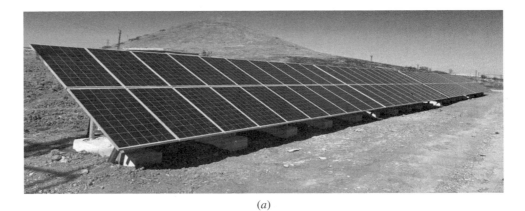

(a)

(b)

Figure 5-1 (*a*) A grid-connected PV system. (*b*) A cell phone charging station plus nighttime lighting powered by two 60-W solar panels.

valence band may attach themselves to a neighboring atom if their bond is not strong. As a result, the donor atom becomes a positively charged ion and the neighboring atom becomes a negatively charged ion. The two atoms form an *ionic bond*.

When the electrons in the valence band become too energetic, they jump into a band too far from the nucleus. This remote band is called the *conduction band*. The energy difference between an electron in the valence band and an electron in conduction band is called the *band gap energy*. A small amount of energy is sufficient to move electrons away from the atom when they are in the conduction band. This is the mechanism of heat and electrical conduction.

The common unit for band gap energy is electron-volt, eV. Note that 1 eV = 1.602 × 10⁻¹⁹ J. Insulators, conductors, and semiconductors are characterized by their band gap energies (Kreith and Kreider, 2011).

Conductors have very low band gap energies because their atoms have relatively empty valence bands with some electrons in the conduction band. Metals such as gold, copper, and iron are good conductors.

Insulators have high band gap energies because their atoms have full valence bands. Band gap energies for insulators are greater than 3 eV. Glass is an example of an insulator. Electrons in an insulator atom do not flow under the application of voltage or current.

Semiconductors have partially filled valence bands, and their band gap energies are less than 3 eV. Silicon is an example of semiconductor.

The band gap energy values for common semiconductor materials for PV cell applications are as follows (Hodge, 2010):

1.01 eV	copper indium diselenide
1.11 eV	silicon
1.27 eV	indium phosphide
1.40 eV	gallium arsenide
1.44 eV	cadmium telluride
2.24 eV	gallium phosphide
2.42 eV	cadmium sulfide

Silicon is a semiconductor material and commonly used in PV cells. There are four electrons in the valence band in silicon atoms. For PV cell applications, pure silicon, called an *intrinsic semiconductor*, is not used. Instead, silicon is doped with a small amount of another material (*dopant*) to yield an *extrinsic semiconductor*.

When the dopant has more electrons in the valence band compared to the number of electrons in silicon, the resulting material is an *n-type* (negative-type) *semiconductor*. Phosphorus has five valence electrons (one more than silicon atom), and it is used as a dopant to create a free electron, producing an *n*-type material. When the dopant has a smaller number of electrons in the valence band compared to pure silicon, the resulting material is a *p-type* (positive-type) *semiconductor*. A *p*-type semiconductor having fewer electrons in the valence band is also called having an excess of *holes*. Boron has three valence electrons (one less than silicon atom), and it is used as the dopant to create a shortage of electron (creating a hole), producing a *p*-type material. Despite having absence of electrons in the valence band, the semiconductor is electrically neutral. An *n*-type semiconductor is also electrically neutral despite having excess electrons in the valence band.

An *n*-type semiconductor acquires positive charge because it has tendency to lose electrons and gain holes. A *p*-type semiconductor acquires negative charge because it has tendency to lose holes and gain electrons. As a result of doping silicon with two different dopants, *n*-type and *p*-type semiconductors are obtained. Combining *n*-type and *p*-type semiconductors result in a *p-n* junction. Forming of the junction enhances the flow of electrons and holes. The negative charge on *p*-type semiconductor and positive charge on *n*-type semiconductor are responsible for preventing electrons and holes from crossing the junction.

The operation of a PV cell designated by a *p-n* junction is shown in Fig. 5-2. Photons from solar radiation penetrate the *p-n* junction and strike silicon atoms. If the energy level of the photon is sufficiently high, the electron in the valence band acquires more energy than the band gap energy, and the electron will jump to the conduction band. This initiates

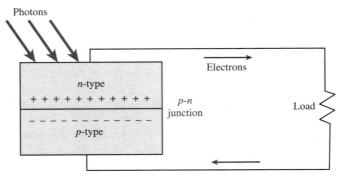

Figure 5-2   Photovoltaic cell operation. Electrons flow from the *n*-type semiconductor to the *p*-type semiconductor through the load.

an electron flow and an electric current flow. As a result, energy in photons of solar light is converted to electrical energy.

If the energy of photon is less than the band gap energy, the electron has insufficient energy to jump into the conduction band. This excess energy in the electron is converted to excess kinetic energy, which is converted to heat, manifesting itself as a rise in temperature of the PV system. This shows that only the photons of solar radiation having a minimum band gap energy contribute to the electricity production. The energy of a photon depends on the wavelength of solar light. The higher the wavelength, the lower the energy level of the photon. Therefore, in PV cells, only low-wavelength spectrum of solar radiation contributes to the PV energy conversion.

A single photon can only move a single valence electron to the conduction band. When the photon possesses more energy than the band gap energy, the part of energy that is equal to the gap energy is used to move the electron while the excess energy is converted to heat. Therefore, the excess energy of the photon does not contribute to the electricity production.

As discussed in Chap. 3, electromagnetic waves transport energy just like other waves, and all electromagnetic waves travel at the *speed of light* in a vacuum, which is $c_0 = 2.9979 \times 10^8$ m/s. Electromagnetic waves are characterized by their *frequency* v or *wavelength* λ. These two properties in a medium are related by

$$\lambda = \frac{c}{v}$$

(5-1)

where c is the speed of propagation of a wave in that medium. The speed of propagation in a medium is related to the speed of light in a vacuum by $c = c_0/n$, where n is the *index of refraction* of that medium. The refractive index is essentially unity for air and most gases, about 1.5 for glass, and 1.33 for water. The commonly used unit of wavelength is the *micrometer* (μm) or micron, where 1 μm = 10^{-6} m. It has proven useful to view electromagnetic radiation as the propagation of a collection of discrete packets of energy called photons or quanta, as proposed by Max Planck in 1900 in conjunction with his *quantum theory*. In this view, each photon of frequency v is considered to have an energy of

$$e = hv = \frac{hc}{\lambda}$$

(5-2)

where $h = 6.626069 \times 10^{-34}$ J·s is *Planck's constant*. Note from the second part of Eq. (5-2) that the energy of a photon is inversely proportional to its wavelength. Therefore, shorter-wavelength radiation possesses larger photon energies (Çengel and Ghajar, 2020).

EXAMPLE 5-1
Maximum Wavelength of
Solar Radiation for PV Cells

The band gap energy values are 1.11 eV for silicon and 1.40 eV for gallium arsenide. What are the maximum wavelengths of solar radiation for which solar radiation can be converted to electrical energy for silicon and gallium arsenide?

SOLUTION   The energy of each photon is $e = \dfrac{hc}{\lambda}$ where $h = 6.626069 \times 10^{-34}$ J·s and $c = 2.9979 \times 10^8$ m/s. For silicon, the band gap energy is 1.11 eV. Solving for the wavelength, we obtain

$$\lambda = \frac{hc}{e} = \frac{(6.626069 \times 10^{-34} \text{ J·s})(2.9979 \times 10^8 \text{ m/s})}{1.11 \text{ eV}} \left(\frac{1 \text{ eV}}{1.602 \times 10^{-19} \text{ J}} \right)$$

$$= 1.12 \times 10^{-6} \text{ m}$$

$$= \mathbf{1.12 \, \mu m}$$

Repeating the calculation for gallium arsenide, we find

$$\lambda = \frac{hc}{e} = \frac{(6.626069 \times 10^{-34} \text{ J·s})(2.9979 \times 10^8 \text{ m/s})}{1.40 \text{ eV}} \left(\frac{1 \text{ eV}}{1.602 \times 10^{-19} \text{ J}} \right)$$

$$= 0.885 \times 10^{-6} \text{ m}$$

$$= \mathbf{0.885 \, \mu m}$$

Therefore, the maximum wavelength of solar radiation for PV energy conversion is 1.12 μm for silicon and 0.885 μm for gallium arsenide.   ▲

This example shows that photons of solar radiation should have a maximum wavelength of 1.12 μm to convert solar light to electricity when silicon is used as the semiconductor material (Fig. 5-3). Remember that the energy of a photon is inversely proportional

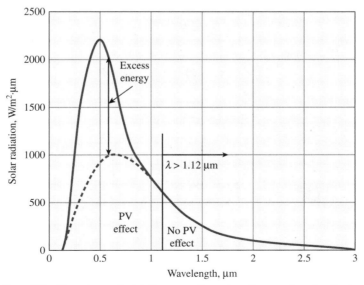

Figure 5-3   There is no electricity generation when the wavelength of solar radiation is greater than 1.12 μm when silicon is used as the semiconductor material. When the wavelength is less than 1.12 μm, the difference between the photon energy from the solar light and the band gap energy of the silicon is the excess energy. (*Adapted from Hodge, 2010.*)

to its wavelength. Solar radiation wavelengths greater than 1.12 μm do not have sufficient energy in photons to move a valence electron and initiate the current flow. When the wavelength is less than 1.12 μm, the difference between the photon energy from the solar light and the band gap energy of the silicon is the excess energy, which is also wasted. These characteristics explains why the conversion efficiency of solar radiation to electricity by PV effect is relatively low.

As discussed in Chap. 3, the electromagnetic spectrum of solar radiation falls between the wavelengths of 0.3 and 3.5 μm. About 40 percent of solar radiation is in visible range (between 0.4 and 0.7 μm) with peak radiation at a wavelength of 0.48 μm, which corresponds to the green portion of the visible spectrum. The visible spectrum of solar radiation contributes the most to the PV energy conversion. About 52 percent of solar radiation falls into near-infrared region with a wavelength range of 0.7 to 3.5 μm. The contribution of this infrared region to electricity generation by PV effect is very low. Ultraviolet region of solar radiation falls in wavelengths between 0.3 and 0.4 μm, and its contribution to PV energy conversion is small.

The conversion efficiency of a solar cell may be defined as the electrical power output (in W) divided by the incident solar radiation:

$$\eta_{cell} = \frac{\dot{W}_{out}}{AG_{solar}} \tag{5-3}$$

where A is the area of the solar cell in m² and G_{solar} is the solar irradiation in W/m².

EXAMPLE 5-2
Maximum Efficiency of a Silicon Solar Cell

Solar radiation is incident on a 1-m² silicon solar cell at a rate of 1000 W/m²·μm at a wavelength of 0.48 μm. Determine the maximum efficiency of this solar cell at this wavelength.

SOLUTION   The rate of solar radiation incident is

$$G = AG_{solar} = (1000 \text{ W/m}^2)(1.0 \text{ m}^2) = 1000 \text{ W}$$

Each photon with a wavelength λ has an energy of

$$e = \frac{hc}{\lambda} = \frac{(6.626069 \times 10^{-34} \text{ J·s})(2.9979 \times 10^8 \text{ m/s})}{0.48 \text{ μm}} \left(\frac{1 \text{ μm}}{1 \times 10^{-6} \text{ m}} \right) = 4.138 \times 10^{-19} \text{ J}$$

The rate of photons incident on the PV cell is

$$n_{photon} = \frac{G}{e} = \frac{1000 \text{ J/s}}{4.138 \times 10^{-19} \text{ J}} = 2.417 \times 10^{21} \text{ photons/s}$$

Each photon will move one electron and, therefore, the number of electrons are equal to the number of photons. The band gap energy of silicon is 1.11 eV. Then the maximum electrical power output is

$$\dot{W}_{out,max} = n_{photon}e = (2.417 \times 10^{21} \text{ photons/s})(1.11 \text{ eV}) \left(\frac{1.602 \times 10^{-19} \text{ J}}{1 \text{ eV}} \right) = 430 \text{ W}$$

The maximum efficiency at this wavelength is the maximum electrical power output divided by the incident solar radiation:

$$\eta_{\lambda,cell,max} = \frac{\dot{W}_{out,max}}{AG_{solar}} = \frac{430 \text{ W}}{(1 \text{ m}^2)(1000 \text{ W/m}^2)} = 0.430 = \textbf{43.0\%}$$

This is the maximum conversion efficiency of a silicon solar cell when solar radiation is incident at a wavelength of 0.48 μm. The maximum efficiency will be different at different wavelengths because photon energy depends on the wavelength. The value of maximum efficiency is 100 percent when the wavelength is 1.12 μm. The maximum efficiency is zero when the wavelength is greater than 1.12 μm. ▲

The overall maximum efficiency of a solar cell can be determined by integrating the cell efficiency at each wavelength over the entire solar irradiation spectrum:

$$\eta_{\text{cell,max}} = \frac{\int \eta_{\lambda,\text{cell,max}} G_\lambda \, d\lambda}{\int G_\lambda \, d\lambda} \tag{5-4}$$

where $\eta_{\lambda,\text{cell}}$ is the maximum cell efficiency at a given wavelength and G_λ is the solar irradiation at the given wavelength.

Actual efficiencies of solar cells are less than the maximum efficiencies based on Eq. (5-4). Some of the reasons are reflection of solar radiation from the cell surface, shading of the cell due to electrical contacts, internal electric resistance of the cell, and recombination of electrons and holes not contributing to electric flow. Using antireflective (AR) coatings on PV panels are now a standard practice to minimize reflection of solar light. This can reduce the reflection from 30 percent to 3 percent for a silicon cell. Undesired recombination of electrons and holes can be reduced by using hydrogen alloys in polycrystalline and amorphous cells (Kreith and Kreider, 2011).

Mismatch of the photon energy of the solar radiation and the band gap energy of the cell material results in relatively low maximum efficiencies for the solar cell. An effective solution to this issue is using multijunction solar cells for which two or more thin layers of solar cells are stacked on top of each other. This is shown in Fig. (5-4). Material a is used as the top layer, and the material b as the bottom layer. Material a will not produce PV effect when the wavelength is greater than λ_a. That is, only the area A1+A2 contributes to the solar radiation conversion to electricity. Material b has different characteristics

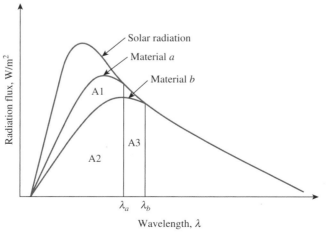

Figure 5-4 Photovoltaic energy conversion in multijunction solar cells. (*Adapted from Goswami et al., 2000.*)

such that its maximum wavelength for the PV effect is λ_b. By adding a second cell layer, a wider range of wavelength for solar radiation becomes available for energy conversion. The solar radiation with wavelength greater than λ_a will pass through layer a but absorbed by layer b for wavelengths up to λ_b. As a result, a greater area in the spectrum (A1+A2+A3) contributes to the solar radiation conversion to electricity. Multijunction solar cells allow higher cell efficiencies in comparison to the cell efficiencies using a single layer of material a or a single layer of material b by utilizing a wider spectrum of solar radiation in PV energy conversion. Using more layers would result in higher cell efficiencies. However, using very thin layers of crystalline and polycrystalline cells on top of each other has some serious challenges. This concept is mainly used for thin-film amorphous solar cells (Goswami et al., 2000). Different types of solar cells are described later in the chapter.

5-2 ANALYSIS OF SOLAR CELLS

Electrical analysis of PV cells is presented by following the simple model described in Hodge (2010) and Culp (1991). The cell involves a p-type semiconductor and an n-type semiconductor. Silicon is commonly used as a semiconductor material in solar cells. The silicon is doped with phosphorus to produce the n-type semiconductor, while it is doped with boron to produce the p-type semiconductor. There is a current density flow at the p-n junction of a solar cell (Fig. 5-5).

The *current density J* is defined as the current I over the cell surface area A. The current density flow from n-type semiconductor to p-type semiconductor is denoted by J_r and is called the *light-induced recombination current*, and that from p-type to n-type is denoted by J_o and is called the *dark current* or *reverse saturation current*. In an illuminated solar cell, J_r is proportional to J_o according to the relation

$$J_r = J_o \exp\left(\frac{e_o V}{kT}\right) \tag{5-5}$$

where $e_o = 1.6 \times 10^{-19}$ J/V is equal to the charge of one electron, $k = 1.381 \times 10^{-23}$ J/K is Boltzmann's constant, V is voltage, and T is the cell temperature. The junction current density J_j is equal to the algebraic sum of J_r and J_o:

$$J_j = J_r - J_o = J_o\left[\exp\left(\frac{e_o V}{kT}\right) - 1\right] \tag{5-6}$$

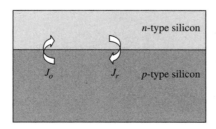

Figure 5-5   A simplified model for current density at p-n junction. (*Adapted from Hodge, 2010.*)

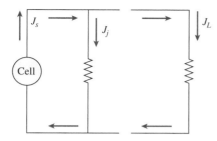

Figure 5-6 Equivalent circuit for solar cell. (*Adapted from Hodge, 2010.*)

The equivalent circuit for a solar cell is given in Fig. 5-6. The current output density J_s flows through the junction or load. The load current density J_L is given by

$$J_L = J_s - J_j = J_s - J_o \left[\exp\left(\frac{e_o V}{kT} \right) - 1 \right] \tag{5-7}$$

The voltage is zero, $V = 0$, when the cell is short-circuited, and thus $J_s = J_L$. The cell output is through the junction when the circuit is open and $J_L = 0$. The voltage in this case is called the open circuit voltage, V_{oc}. Equation (5-7) can be solved for V_{oc} to yield

$$V_{oc} = \frac{kT}{e_o} \ln\left(\frac{J_s}{J_o} + 1 \right) \tag{5-8}$$

An expression for the ratio of the load current density J_L to short circuit current density J_s may be obtained by dividing Eq. (5-7) by J_s:

$$\frac{J_L}{J_s} = 1 - \frac{J_o}{J_s} \left[\exp\left(\frac{e_o V}{kT} \right) - 1 \right] \tag{5-9}$$

The power output delivered to the load is

$$\dot{W} = J_L V A \tag{5-10}$$

where A is the cell area. Substituting J_L from Eq. (5-9) into Eq. (5-10) gives

$$\dot{W} = VAJ_s - VAJ_o \left[\exp\left(\frac{e_o V}{kT} \right) - 1 \right] \tag{5-11}$$

Differentiating Eq. (5-11) with respect to voltage V and setting the derivative equal to zero gives the maximum load voltage for the maximum power output:

$$\exp\left(\frac{e_o V_{max}}{kT} \right) = \frac{1 + J_s/J_o}{1 + \dfrac{e_o V_{max}}{kT}} \tag{5-12}$$

Note that the maximum voltage V_{max} is implicit in this equation. A trial-and-error approach or an equation solver is needed to solve for V_{max}. The maximum power output of the cell is

$$\dot{W}_{max} = \frac{A V_{max} (J_s + J_o)}{1 + \dfrac{kT}{e_o V_{max}}} \tag{5-13}$$

The conversion efficiency of a solar cell can be expressed as the power output divided by the incident solar radiation:

$$\eta_{cell} = \frac{\dot{W}}{AG} \tag{5-14}$$

where G is the solar irradiation. Using the maximum power output expression in Eq. (5-13), the maximum conversion efficiency of a solar cell can be written as

$$\eta_{cell,max} = \frac{\dot{W}_{max}}{AG} = \frac{AV_{max}(J_s + J_o)}{AG\left(1 + \frac{kT}{e_o V_{max}}\right)} = \frac{V_{max}(J_s + J_o)}{G\left(1 + \frac{kT}{e_o V_{max}}\right)} \tag{5-15}$$

Equation (5-9) along with Eq. (5-8) can be used to plot the current density ratio J_L/J_s as a function of load voltage for a specified value of open circuit voltage V_{oc}. Also, Eqs. (5-12) and (5-13) can be used to plot power output normalized with respect to maximum power against load voltage. The plots in Fig. 5-6 are obtained with $V_{oc} = 0.55$ V and $T = 300$ K. Note that a high-quality silicon solar cell can produce an open circuit voltage of about 0.6 V. For the short circuit case ($J_L = J_s$ or $J_L/J_s = 1$), the voltage is zero, and the power output is zero. For the open circuit voltage case ($J_L = 0$), the voltage is 0.55 V, and the power output is also zero. The maximum power occurs at a voltage close to open circuit voltage, which is 0.47 V in this case. The current density ratio remains close to one until the open circuit voltage is approached. Then it decreases rapidly before it becomes zero at the open circuit case. The trends and characteristics shown in Fig. 5-7 are typical of most solar cells.

The power output from a photovoltaic cell is proportional to solar radiation absorbed by the cell. It turns out that as the solar irradiation increases, the load current also increases, but the increase in open circuit voltage is small.

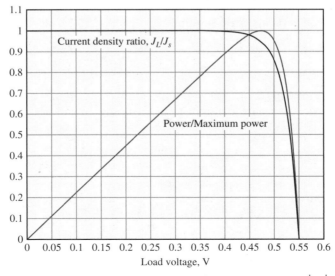

Figure 5-7 Current density ratio J_L/J_s and power output ratio $\dot{W}/\dot{W}_{max}$ in a solar cell as a function of load voltage.

EXAMPLE 5-3
Analysis of a Solar Cell

A solar cell has an open circuit voltage value of 0.62 V with a reverse saturation current density of 2.253×10^{-9} A/m².

(a) For a temperature of 20°C, determine the load voltage at which the power output is maximum.

(b) If the solar irradiation is 770 W/m², determine the efficiency of the solar cell at a load voltage of 0.5 V.

(c) Determine the cell area for a power output of 500 W at a load voltage of 0.5 V.

SOLUTION (a) The current output density is determined from Eq. (5-8) to be

$$V_{oc} = \frac{kT}{e_o} \ln\left(\frac{J_s}{J_o} + 1\right)$$

$$0.62 \text{ V} = \frac{(1.381 \times 10^{-23} \text{ J/K})(293 \text{ K})}{1.6 \times 10^{-19} \text{ J/V}} \ln\left(\frac{J_s}{2.253 \times 10^{-9} \text{ A/m}^2} + 1\right)$$

$$J_s = 100 \text{ A/m}^2$$

The load voltage at which the power output is maximum is determined from Eq. (5-12) to be

$$\exp\left(\frac{e_o V_{max}}{kT}\right) = \frac{1 + J_s/J_o}{1 + \frac{e_o V_{max}}{kT}}$$

$$\exp\left(\frac{(1.6 \times 10^{-19} \text{ J/V})V_{max}}{(1.381 \times 10^{-23} \text{ J/K})(293 \text{ K})}\right) = \frac{1 + (100 \text{ A/m}^2/2.253 \times 10^{-9} \text{ A/m}^2)}{1 + \frac{(1.6 \times 10^{-19} \text{ J/V})V_{max}}{(1.381 \times 10^{-23} \text{ J/K})(293 \text{ K})}}$$

$$V_{max} = \mathbf{0.5414 \text{ V}}$$

(b) The load current density is determined from Eq. (5-9):

$$\frac{J_L}{J_s} = 1 - \frac{J_o}{J_s}\left[\exp\left(\frac{e_o V}{kT}\right) - 1\right]$$

$$\frac{J_L}{100 \text{ A/m}^2} = 1 - \frac{2.253 \times 10^{-9} \text{ A/m}^2}{100 \text{ A/m}^2}\left[\exp\left(\frac{(1.6 \times 10^{-19} \text{ J/V})(0.5 \text{ V})}{(1.381 \times 10^{-23} \text{ J/K})(293 \text{ K})}\right) - 1\right]$$

$$J_L = 99.12 \text{ A/m}^2$$

The power output per unit area of the cell is

$$\dot{W}/A = J_L V = (99.12 \text{ A/m}^2)(0.5 \text{ V})\left(\frac{1 \text{ W}}{1 \text{ AV}}\right) = 49.56 \text{ W/m}^2$$

Then, the cell efficiency becomes

$$\eta_{cell} = \frac{\dot{W}/A}{G} = \frac{49.56 \text{ W/m}^2}{770 \text{ W/m}^2} = 0.0644 \text{ or } \mathbf{6.44\%}$$

(c) Finally, the cell area for a power output of 500 W is

$$A = \frac{\dot{W}}{\dot{W}/A} = \frac{500 \text{ W}}{49.56 \text{ W/m}^2} = \mathbf{10.1 \text{ m}^2} \quad \blacktriangle$$

5-3 PHOTOVOLTAIC TECHNOLOGIES AND SYSTEMS

Some well-known solar cell types and technologies include monocrystalline, polycrystalline, amorphous, thin-film, gallium arsenide, and multijunction (Gevorkian, 2007). The first three technologies involve silicon as the semiconductor materials.

Monocrystalline solar cell: The cells are produced from pure silicon by a process called the floating zone technique. Monocrystalline silicone grows on a seed out of a silicone melt. The resulting silicone rods are sliced into thin wafer disks (between 0.2 and 0.4 mm thickness). The wafers are further processed by grinding, polishing, cleaning, doping with impurities, and antireflective coating. Manufacturing of monocrystalline silicon cells is expensive, but they have the highest efficiency with respect to other cell technologies.

Polycrystalline solar cell: This is also called a multicrystalline solar cell. The silicon is produced in the form of an ingot as the silicon melt is cooled slowly. Grain boundaries are formed that separate the crystalline regions of the silicon ingot. The polycrystalline solar cell manufactured by this method has a lower efficiency than the monocrystalline solar cell due to the gaps in the grain boundaries. Polycrystalline solar cell is preferred by most cell manufacturers due to its lower manufacturing cost.

Amorphous solar cell: An amorphous silicon film is manufactured by depositing a thin layer of silicon on a carrier material and doping with suitable materials. The glass plates house the silicon film. The main advantage of the amorphous solar cell is that it is relatively simple and inexpensive to manufacture. The disadvantages with respect to other technologies include lower efficiency, larger installation surface, and continuous degradation over its lifetime.

Thin-film solar cell: This solar cell technology uses thin crystalline layers of cadmium telluride or copper indium diselenide deposited on the surface of a carrier base. The process is simple and inexpensive and has high cell efficiencies. However, these solar cells use a lot of space, which makes it unsuitable for a variety of applications. Their lifespan is short, and there are additional expenses due to cables and support structures.

Gallium arsenide solar cell: Multiple research institutions have active research programs in developing gallium arsenide solar cell technology. The interest is due to its high efficiency compared to other solar cell technologies. The process is very expensive, however, because gallium deposits are hard to find. They have some special uses such as space applications.

Multijunction cell: This technology involves the use of two cells, one on top of another. Silicon cell and gallium arsenide cell may be used. The result is higher efficiency compared to single cell systems.

Solar radiation incident on a solar energy conversion system originates from the sun. The upper limit for the efficiency of a solar thermal system for power generation may be determined from the Carnot efficiency by using the effective surface temperature of the sun (5780 K) and an ambient temperature of 298 K:

$$\eta_{\text{cell,max}} = 1 - \frac{T_L}{T_H} = 1 - \frac{298 \text{ K}}{5780 \text{ K}} = 0.948 = 94.8\%$$

The Carnot limit does not apply to solar cells because they are not heat engines. However, for the reasons explained in the previous section, the theoretical efficiency limit is about 34 percent for a single-junction solar cell. This limit can be exceeded by multijunction solar cells. If an infinite number of junctions is used with a high concentration of solar radiation, the limit becomes 86 percent (De Vos, 1980).

Silicon has been commonly used in solar cells, but commercial silicon solar cells have relatively low efficiencies (between 15 and 25 percent). Other materials have been tested extensively to increase solar cell efficiencies. They include cadmium telluride, cadmium sulfide, copper indium diselenide, gallium arsenide, gallium phosphide, and indium phosphide. Copper indium diselenide and gallium arsenide are among the most promising materials. An efficiency of 40 percent has been approached for gallium arsenide solar cells in a laboratory environment. Using the multijunction design with high solar irradiation has resulted in a research efficiency of 43 percent. However, the high cost of high-efficiency solar cells remains a concern.

Halide perovskites are a family of materials with the potential of high efficiency and low manufacturing costs in solar cells. Extensive research work underway in improving efficiencies of perovskite solar cells. Efficiencies as high as 25.5 percent were reported by combining a two-dimensional perovskite layer with a three-dimensional perovskite layer. The main concern for perovskite solar cells is their limited stability and corresponding short lifetime (NREL, 2022).

A single solar cell produces a small amount of voltage and current. Solar cells can be connected in series and parallel arrangement to increase voltage and current. When identical cells are connected in series, the current remains the same, but the voltages are added, just like DC circuits in series. For parallel arrangement of identical cells, the voltage remains the same while the currents are added (Fig. 5-8). Therefore, the series arrangement should be used to increase the voltage and the parallel arrangement should be used to increase the current. When two dissimilar cells are connected in series, the voltages are added ($V = V_1 + V_2$) while the current for the cell network is between the two current values ($I_1 < I < I_2$), but closer to I_1. When two dissimilar cells are connected in parallel, the currents are added ($I = I_1 + I_2$) while the voltage takes a value between the two voltage values ($V_1 < V < V_2$), but closer to V_1.

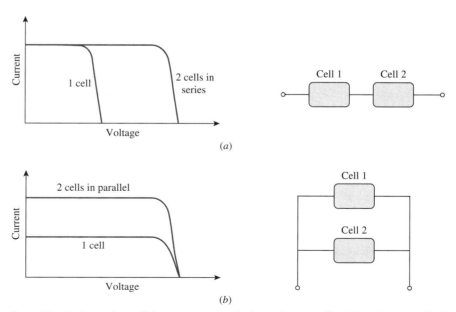

Figure 5-8 Series and parallel arrangements of identical solar cells. (*a*) Voltage is added in series arrangement. (*b*) Current is added in parallel arrangement.

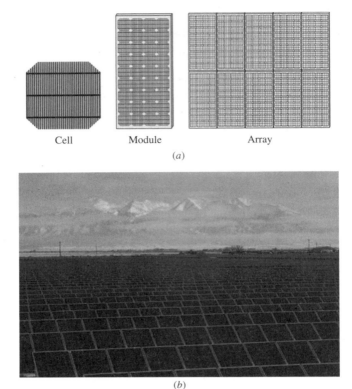

Cell   Module   Array

(a)

(b)

Figure 5-9 (a) A photovoltaic system typically consists of arrays, which are obtained by connecting modules, and modules consist of individual cells. (*DOE/EERE.*) (b) Solar arrays. (*Fotosearch/Photo Library RF.*)

A single solar cell produces only 1 to 3 W of power. Multiple cells should be connected to form modules, and modules should be connected into arrays so that reasonable amounts of power can be generated (Fig. 5-9). This way, both small and large PV systems can be installed, depending on the demand (Fig. 5-10).

EXAMPLE 5-4
Arrangement of Solar Cells

Consider a commercial solar cell with the following specifications:

Open circuit voltage = 0.60 V

Open circuit current = 5.1 A

Maximum power = 2.2 W

Voltage at maximum power = 0.48 V

Current at maximum power = 4.6 A

Efficiency = 18 percent

A module of these solar cells is to be constructed to provide a voltage output of 12 V and a power output of 275 W.

(a) How many cells should be used to satisfy the power output?

(b) How many cells should be arranged in series to satisfy the voltage output?

Figure 5-10 Solar array examples. (*Photos by Adem Atmaca.*)

(*c*) How many rows of the cells in series should be used?

(*d*) If the testing of this solar cell is made for a solar radiation value of 800 W/m², what is the required area of this module?

SOLUTION   (*a*) The total number of cells to provide a power output of 275 W is

$$n_{\text{cell, total}} = \frac{\dot{W}_{\text{total}}}{\dot{W}_{\text{one cell}}} = \frac{275 \text{ W}}{2.2 \text{ W}} = \textbf{125 cells}$$

(*b*) In series arrangement, the voltages are added. Then the number of cells in series should be

$$n_{\text{cell, series}} = \frac{V_{\text{total}}}{V_{\text{one cell}}} = \frac{12 \text{ V}}{0.48 \text{ V}} = \textbf{25 cells}$$

(c) Each line has 25 cells while the total number of cells is 125. Then, the number of the rows will be

$$n_{\text{row}} = \frac{n_{\text{cell, total}}}{n_{\text{cell, series}}} = \frac{125 \text{ cells}}{25 \text{ cells}} = 5 \text{ rows}$$

(d) The efficiency of this solar cell is specified to be 18 percent. Using the definition of the cell efficiency, we find the area of the module as

$$\eta_{\text{cell}} = \frac{\dot{W}_{\text{out}}}{A G_{\text{solar}}} \quad \rightarrow \quad 0.18 = \frac{275 \text{ W}}{A(800 \text{ W/m}^2)} \quad \rightarrow \quad A = 1.91 \text{ m}^2 \quad \blacktriangle$$

Characteristics of commercial PV modules are shown in Fig. 5-11, taken from the brochures of Akademi Energy (http://enerji.gantep.edu.tr). They are useful for technical specifications and practical information on PV systems. The PV module consists of 60 cells with 10 rows, each row with 6 cells in series.

There are three tables in Fig. 5-11. The first table gives mechanical characteristics of the solar module. The other two tables specify technical characteristics of monocrystalline and polycrystalline cells. The standard test conditions (STC) are based on a solar irradiance of 1000 W/m², a spectrum of AM 1.5, and a cell temperature of 25°C. Spectrum AM 1.5 refers to two standard terrestrial solar irradiance spectra. The nominal operating cell temperature (NOCT) conditions are as follows: solar irradiance = 800 W/m², ambient temperature = 20°C, wind speed = 1 m/s. The maximum power, voltage and current at maximum power, open circuit voltage, open circuit current, and the module efficiency are specified. As the power output increases, the voltage and current values increase, as expected, since power is equal to the product of voltage and current. The efficiency also increases at higher power outputs. The maximum power at NOCT conditions is lower than the that at STC because the power is based on an irradiance value of 1000 W/m² at STC while it is based on a value of 800 W/m² at NOCT.

The performance of a commercial PV module at various voltage values is given in Figs. 5-12a–c. Fig. 5-12a gives the effect of solar irradiance, and Fig. 5-12b gives the effect of cell temperature on the cell performance. The cell performance is given by the current-voltage curve. Fig. 5-12c is on the effect of solar irradiance on the power output. The higher the solar irradiance, the higher the power output. As the solar irradiance increases, the current and the power output increase as the open circuit voltage increases slightly. The performance of the solar module decreases at higher cell temperatures resulting in lower values of open circuit voltage and maximum power output. It is desirable to match the load to the best operating point in the current-voltage curve, which is the point for maximum power. The maximum power occurs at the knee of the current-voltage curve.

5-4 ENERGY PRODUCTION FROM PHOTOVOLTAIC SYSTEMS

The energy production from a PV system can be estimated using different methods, some detailed and more accurate, and some simple and less accurate. Free and commercial software with built-in solar data for various locations on the earth are also available. The characteristics of a PV system and solar radiation data in the location of interest affect such an estimate. A simplified approach for the estimation of energy production for an average day in a specified location is based on the following formula:

$$\text{Energy production} = n_{\text{days}} \eta_{\text{system}} \eta_{\text{cell}} A G_{\text{solar}} \qquad (5\text{-}16)$$

Mechanical Characteristics	
Solar Cell	Mono/Poly
No. of Cells	60 (6 × 10)
Dimensions	1658 × 996 × 35 mm
Weight	19.5 kgs
Front	Glass 3.2 mm tempered glass
Frame	Anodized aluminium alloy
Junction Box	IP 68 rated (3 by pass diodes)
Output Cables	4.0 mm^2, symmetrical length (−) 900 mm and (+) 900 mm
Connectors	MC4 compatible
Mechanical load test	5400 Pa

MONO

	GAP-310M		GAP-315M		GAP-320M		GAP-325M		GAP-330M	
	STC	NOCT	STC	NOCT	STC	NOCT	STC	NOCT	STC	NOCT
Maximum Power (Pmax)	310W	229W	315W	233W	320W	237W	325W	241W	330W	245W
Open Circuit Voltage (Voc)	39.80V	37.28V	40.12V	37.72V	40.43V	38.15V	40.74V	38.59V	41.05V	39.01V
Short Circuit Current (Isc)	10.05A	7.94A	10.12A	7.98A	10.19A	8.02A	10.26A	8.06A	10.33A	8.10A
Voltage at Maximum Power (Vmpp)	32.56V	30.57V	32.85V	30.94V	33.13V	31.31V	33.40V	31.67V	33.67V	32.03V
Current at Maximum Power (Impp)	9.52A	7.49A	9.59A	7.53A	9.66A	7.57A	9.73A	7.61A	9.80A	7.65A
Module Efficiency (%)	18.77%		19.08%		19.38%		19.68%		19.98%	

POLY

	GAP-275P		GAP-280P		GAP-285P		GAP-290P		GAP-295P	
	STC	NOCT	STC	NOCT	STC	NOCT	STC	NOCT	STC	NOCT
Maximum Power (Pmax)	275W	204W	280W	208W	285W	212W	290W	216W	295W	220W
Open Circuit Voltage (Voc)	38.27V	36.03V	38.57V	36.32V	38.84V	36.59V	39.14V	36.86V	39.41V	37.13V
Short Circuit Current (Isc)	9.37A	7.41A	9.45A	7.49A	9.54A	7.57A	9.62A	7.65A	9.71A	7.73A
Voltage at Maximum Power (Vmpp)	30.88V	29.39V	31.15V	29.63V	31.39V	29.86V	31.66V	30.08V	31.90V	30.30V
Current at Maximum Power (Impp)	8.91A	6.94A	8.99A	7.02A	9.08A	7.10A	9.16A	7.18A	9.25A	7.26A
Module Efficiency (%)	16.65%		16.96%		17.26%		17.56%		17.86%	

Figure 5-11   Mechanical and technical characteristics of commercial PV modules. (*Courtesy of Akademi Energy, http://enerji.gantep.edu.tr.*)

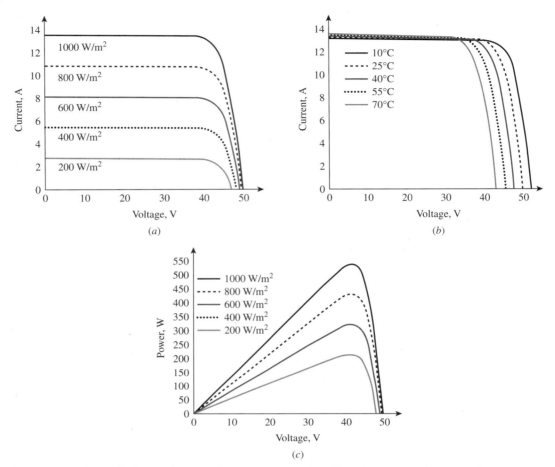

Figure 5-12 Effects of solar irradiance and temperature on the cell performance and power-voltage curve at different solar irradiance values for a commercial PV module (*Akademi Energy, http://enerji.gantep.edu.tr.*).

Here, n_{days} is the number of days, η_{cell} is the solar cell efficiency, A is area of the solar array, and G_{solar} is the solar irradiation per unit area per day (typically in MJ/m²·day or kWh/m²·day). Also, η_{system} is the combined efficiency of the PV system, which accounts for efficiencies of the components other than the PV cell, such as the inverter, battery, and distribution system.

EXAMPLE 5-5
Energy Production from a PV Array
A PV array uses 20 commercial PV modules each with an area of 1.85 m², a typical power output of 350 W, and a solar module efficiency of 19 percent. The system efficiency may be taken as 75 percent. This PV array is installed in Atlanta, Georgia. The average daily solar radiation on a horizontal surface in Atlanta is given in Table 3-6, in MJ/m²·day, as follows:

January	9.31	July	22.15
February	12.26	August	20.56
March	16.13	September	17.49
April	20.33	October	14.54
May	22.37	November	10.56
June	23.17	December	8.52
		Average	16.43 MJ/m²·day

(*a*) Estimate the amount of energy production from this solar array per year in Atlanta, in kWh.

(*b*) If the price of electricity is \$0.15/kWh, what is the annual potential revenue from this PV array?

(*c*) If the initial cost of this PV system is specified as \$2/W, what is the simple payback period?

SOLUTION (*a*) Each PV module has an area of 1.85 m². The total area for 20 such modules is

$$A_{total} = (20)(1.85 \text{ m}^2) = 37 \text{ m}^2$$

We now calculate amount of energy production for the month of January:

$$\text{Energy production (Jan)} = n_{days}\eta_{system}\eta_{cell}AG_{solar}$$

$$= (31 \text{ days})(0.75)(0.19)(37 \text{ m}^2)(9.31 \text{ MJ/m}^2\cdot\text{day})\left(\frac{1 \text{ kWh}}{3.6 \text{ MJ}}\right)$$

$$= 423 \text{ kWh}$$

We repeat the calculations at other months and obtain the energy production in each month and for an entire year, as shown in the following table.

January	423	July	1006
February	503	August	933
March	732	September	768
April	893	October	660
May	1016	November	464
June	1018	December	387
		Total	**8803 kWh**

Alternatively, we can calculate annual energy production by using the annual average solar radiation value for Atlanta (16.43 MJ/m²·day):

$$\text{Energy production (total)} = n_{days}\eta_{system}\eta_{cell}AG_{solar}$$

$$= (365 \text{ days})(0.75)(0.19)(37 \text{ m}^2)(16.43 \text{ MJ/m}^2\cdot\text{day})\left(\frac{1 \text{ kWh}}{3.6 \text{ MJ}}\right)$$

$$= 8783 \text{ kWh}$$

The results are sufficiently close, as expected.

(*b*) The price of electricity is \$0.15/kWh. Then, the annual potential revenue of this PV array is

$$\text{Potential revenue} = (\text{Annual energy production})(\text{Unit electricity price})$$

$$= (\$0.15/\text{kWh})(8803 \text{ kWh/yr})$$

$$= \textbf{\$1320/yr}$$

(*c*) The total power output from 20 PV modules is

$$\dot{W}_{total} = (20)(350 \text{ W}) = 7000 \text{ W}$$

The initial cost of this PV system is specified as \$2/W. Then the initial cost of this PV system is

$$\text{Initial cost} = (\text{Annual energy production})(\text{Unit initial cost})$$

$$= (7000 \text{ W})(\$2/\text{W})$$

$$= \$14,000$$

The simple payback period of this system is

$$\text{Payback period} = \frac{\text{Initial cost}}{\text{Potential revenue}} = \frac{\$14,000}{\$1320/\text{yr}} = \mathbf{10.6\ yr}$$

This PV system will pay for itself in about 10 years. ▲

The design of a PV system for a particular application involves identification of the system configuration and the component requirements. Whether the load is DC or AC, the need for battery storage, and the availability of a backup energy source are among the options and criteria to consider. Once the system configuration is known, the next task is usually the selection of the size of the system and estimation of the amount of energy use.

EXAMPLE 5-6
Peak Load and Energy Consumption of a PV System

A PV system consisting of solar modules of 200 W power will be installed on a remote cabin with the following loads. Only one device other than the refrigerator runs at any one time during the day. Lights and TV run at the same time during the night. The refrigerator is on all the time, but it operates one-third of the time.

Device	Power rating, W	Daily operating period
Lights	80	4 h, nighttime
Refrigerator	400	on 24 h, operates 8 h
Microwave	400	1 h, daytime
Computer	250	3 h, daytime
Dishwasher	350	1 h, daytime
Oven	450	2 h, daytime
Television	300	4 h, nighttime

(a) Find the required capacity of the PV system and the number of modules.

(b) Find the daily energy consumption from the PV system.

SOLUTION   (a) The required capacity of the system is equal to the total power requirement of the devices at the nighttime (peak load period) when the lights and TV are on, and the refrigerator is operating.

$$\text{Capacity} = 80 + 400 + 300 = \mathbf{780\ W}$$

Each solar module supplies 200 W, and thus, the required number of modules is

$$\text{Number of modules} = 780\ \text{W}/200\ \text{W} = 3.9 \approx \mathbf{4}$$

The result is rounded off to 4 modules.

(b) We need to consider the operation of all devices over a 24-h period to find the daily energy consumption.

$$\text{Energy consumption} = (80\ \text{W} \times 4\ \text{h}) + (400\ \text{W} \times 8\ \text{h}) + (400\ \text{W} \times 1\ \text{h}) + (250\ \text{W} \times 3\ \text{h}) + (350\ \text{W} \times 1\ \text{h})$$
$$+ (450\ \text{W} \times 2\ \text{h}) + (300\ \text{W} \times 4\ \text{h}) = \mathbf{7120\ Wh}\ ▲$$

Peak-Hours Approach to Estimate Energy Production

The energy yield performance of a solar array depends on characteristics of the PV array, power conditioning unit, and its components, as well as solar radiation data of the location. Here, we discuss an alternative method of the energy yield estimate using the peak-hours approach. The amount of electricity delivered from a PV array can be written as

$$\text{Energy production (kWh/day)} = \eta_{\text{avg}} AG \tag{5-17}$$

where A is the area of the PV array in m², G is the amount of solar radiation incident (i.e., insolation) per unit area per day in kWh/m²·day, and η_{avg} is the average PV system efficiency over the day. The AC power output from the PV system may be expressed as

$$\text{Power}_{AC} \text{ (kW)} = \eta_{1\text{-sun}} \times A \times (1\text{-sun}) \qquad (5\text{-}18)$$

where $\eta_{1\text{-sun}}$ is the PV system efficiency at 1-sun. 1-sun represents a power output of 1 kW per unit array area. That is, 1-sun = 1 kW/m². Remember that standard testing of PV modules is done at 1000 W/m². Solving Eq. (5-18) for the area A and substituting into Eq. (5-17) gives

$$\text{Energy production (kWh/day)} = \text{Power} \times \frac{G}{1\text{-sun}} \times \frac{\eta_{avg}}{\eta_{1\text{-sun}}} \qquad (5\text{-}19)$$

Assuming the same efficiency for η_{avg} and $\eta_{1\text{-sun}}$,

$$\text{Energy production (kWh/day)} = \text{Power}_{AC} \times \frac{G}{1\text{-sun}} \qquad (5\text{-}20)$$

PV manufacturers list STC power ratings of their PV modules, as shown in Fig. 5-11. These STC power ratings (DC power) may be converted into realistic AC power outputs by

$$\text{Power}_{AC} = \text{Derate factor} \times \text{Power}_{DC,STC} \qquad (5\text{-}21)$$

The derate factor can be taken to be 0.70 to 0.75. Substituting Eq. (5-21) into Eq. (5-20), we obtain

$$\text{Energy production (kWh/day)} = \text{Derate factor} \times \text{Power}_{DC,STC} \times \frac{G}{1\text{-sun}} \qquad (5\text{-}22)$$

The ratio $G/1\text{-sun}$ in Eq. (5-22) is called peak-hours. For example, if the annual average solar insolation in a location is specified as 5.5 kWh/m²·day, this ratio becomes

$$\frac{G}{1\text{-sun}} = \frac{5.5 \text{ kWh/m}^2 \cdot \text{day}}{1 \text{ kW/m}^2} = 5.5 \text{ h/day}$$

In other words, a solar insolation of 5.5 kWh/m²·day is equivalent to 5.5 hours of 1-kW solar insolation in a day or 5.5 peak hours. It is also called 5.5 h/day of peak sun. Using this terminology, Eq. (5-22) is sometimes given as

$$\text{Energy production (kWh/day)} = \text{Derate factor} \times \text{Power}_{DC,STC} \times \text{Peak-sun} \qquad (5\text{-}23)$$

Equation (5-22) assumes that the system efficiency remains the same during the day. Note that grid-connected PV systems operate near the maximum power point (knee of the current-voltage curve) throughout the day. The power at this maximum point is proportional to solar radiation incident. As a result, the efficiency of the system remains reasonably constant. The system efficiency is higher in the mornings due to lower temperatures, which causes slight underestimation of energy production (Masters, 2013).

The National Renewable Energy Laboratory (NREL) developed the PVWATTS calculator for PV performance (available on their website). This calculator allows the estimate of overall derate factor based on other factors considering various system and operating conditions such as temperature. The accuracy of energy production estimate from Eq. (5-22) can be improved by obtaining overall derate factor from the PVWATTS calculator.

EXAMPLE 5-7
Energy Production Estimate by the Peak-Hours Approach

A PV array is installed in Houston, Texas, with a power rating of 10 kW (DC, STC). The annual average solar insolation is given in Table 3-6 for Houston as 15.90 MJ/m²·day. Assuming a derate factor of 0.70, estimate the annual energy production using peak-hours approach.

SOLUTION   A solar insolation of 15.90 MJ/m²·day is equal to 4.42 kWh/m²·day since 1 kWh = 3600 kJ. The annual energy production by the peak-hours approach can be estimated from Eq. (5-22):

$$\text{Energy production} = \text{Derate factor} \times \text{Power}_{DC,STC} \times \frac{G}{1\text{-sun}} \times \text{Annual days}$$

$$= (0.70)(10\text{ kW})\frac{4.42\text{ kWh/m}^2\cdot\text{day}}{1\text{ kW/m}^2}(365\text{ day/yr})$$

$$= \mathbf{11{,}290\text{ kWh/yr}} \quad \blacktriangle$$

5-5 PHOTOVOLTAIC SYSTEM CONFIGURATIONS

The heart of any PV system is solar cells, arranged into modules and arrays. A PV system incorporates various other components depending on the application. The application areas can be divided into two main categories as *off-grid* (*stand-alone*) and *grid-connected* systems.

An off-grid PV system may be as simple as a PV array feeding direct current (DC) power to a load. This is called a *direct-coupled* PV system (Fig. 5-13). These systems are used for low power uses. They only produce power when there is solar irradiation (Hodge, 2010).

For remote applications such as residences not connected to the grid, the PV system must have a battery storage for meeting loads during nighttime hours and cloudy periods. An off-grid system with battery storage is shown in Fig. 5-14. The system has flexibility to supply both DC power and AC (alternating current) power. The charge controller takes DC power from the PV array and directs to an application working with DC power. When there is excess power output from the PV array, the charge controller directs the power to the battery for storage. When there is no solar irradiation, the charge controller absorbs power from the battery and directs it to the DC load. For the operation of devices using AC power, an inverter must be used. The inverter receives DC power from the battery and converts it to AC power.

Stand-alone PV systems can be used to power many different applications such as calculators, watches, street lighting, traffic signs, battery charging, telephones, radios, weather stations, and water purification (Fig. 5-15). These simple systems usually consist of a solar

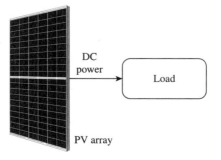

Figure 5-13   An off-grid direct-coupled PV system supplying DC power to a load.

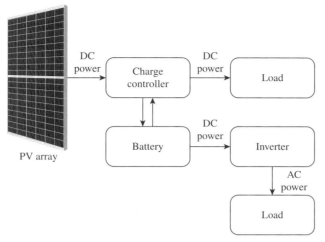

Figure 5-14   An off-grid PV system with energy storage. It is designed to supply both DC power and AC power.

module, a charge controller, and a battery. Low-power stand-alone PV systems are usually cost-competitive to conventional methods such as conventional batteries, grid power, and small diesel generators. They also have advantages of reliability, durability, and, of course, no fuel cost.

Figure 5-15   A stand-alone solar panel used in a small bus converted to a caravan.

Water pumping in remote homes, villages, and farms is a common application of off-grid systems. In this application, the system is not equipped with a battery. Another important application of an off-grid PV system is supplying electricity to remote homes and villages. The cost of extending grid power to remote homes is usually much higher than the cost of PV installation. A commonly used alternative to PV power for water pumping and remote homes is diesel generators. Diesel generators are less expensive to install but the fuel cost and operating and maintenance expenses are very high compared to PV systems.

A disadvantage of the PV system is its inability to supply power during winter with long periods of cloudy days. An effective solution is a hybrid system consisting of a diesel generator and a PV array. The PV system is used to provide power most of the time when solar radiation or battery storage are available, and the diesel generator provides power at other times.

Grid-connected PV systems supply AC power at grid voltage, phase, and frequency for compatibility with the grid. Such a system is shown in Fig. 5-16. The inverter does not only convert DC power to AC power but also condition the power (works as a power conditioner) to make it compatible with the grid specifications. In this configuration, there is no waste of solar power because all the power produced from the PV array are transferred to the grid, and the applications use only the required amount of power from the grid.

A main advantage of grid-connected PV systems is the absence of a battery for energy storage. This greatly simplifies the overall system. They provide an effective and green method of peak shaving during the periods of peak demand from the utility. Most of the PV systems in the world are grid connected. The grid connected systems can be as small as a few kilowatts for residential applications, hundreds of kilowatts for industrial uses, and thousands of kilowatts for supplying power to utilities (Fig. 5-17).

A battery can be added to a grid-connected PV system if the user demands power when the grid is down. This may be needed for industrial users to meet critical loads. The system becomes more complicated in this scenario with advanced controls to allow disconnecting and reconnecting to the grid.

Large installations of grid-connected PV systems are mostly owned by the utilities while the capacity is usually small for privately owned systems. There are significant variations in the specifications of the agreement between the private owner and the utility depending on the country, region, and the utility. In one case, the utility is required to pay for its levelized cost of producing 1 kWh of electricity. This is usually not attractive for the

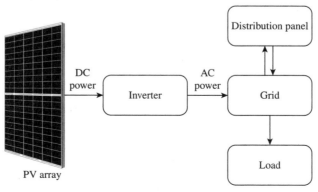

Figure 5-16   A grid-connected PV system.

Figure 5-17   A grid-connected PV array on top of a campus building supplying power to the grid.

owner of the PV system because this cost is usually much lower than the rate the consumers pay for the electricity. Another example of the agreement between the owner and the utility is called *net metering*. In this arrangement, the consumption of the owner increases when power from the grid is used; and the consumption decreases when solar power is supplied to the grid. The same pricing (in $/kWh) is used for both the purchased electricity and sold electricity. Net metering is financially attractive for the PV owner. Some governments provide significant incentives to allow a profitable agreement for the PV owner to promote greater production of solar electricity.

Photovoltaic arrays can be installed at a fixed angle, or they can involve one-axis tracking or two-axis tracking capabilities. For fixed-angle systems, the best performance in the northern hemisphere is obtained when the arrays are placed with the azimuth south, and the tilt angle is selected to be equal to the latitude of the location. Most PV installations have no option of tracking the sun. One-axis tracking involves an axis of rotation at the north-south or east-west line. Two-axis tracking has an additional control of changing the tilt angle to track the sun. Passive tracking systems involving no motors or gears are also available, but they are not effective in windy areas. PV systems with tracking capabilities can produce 25 to 45 percent more power in the summer and can pump up to 50 percent more water compared to fixed PV systems (Nelson, 2011).

The lifespan of a solar cell is about 20 to 35 years. The collection of dust on panel surfaces over time reduces the performance of solar panels. There is no easy solution to this problem, and this is one of the main obstacles preventing widespread use of solar panels in some solar-rich countries. The cost of solar panels has been decreasing steadily over the years while the efficiencies have been increasing. As a result, PV systems are likely to remain as a main actor in green energy production.

5-6 COMPONENTS OF PHOTOVOLTAIC POWER SYSTEMS[*]

A *PV generator* consists of a number of PV modules that are electrically connected to each other. The PV generator is the core component of a PV power system, which utilizes PV modules as a power source to generate electrical energy. However, the PV power systems require a variety of other components in order to provide usable electrical energy for the energy consumers. The PV industry refer to all these components, equipment, and services (including labor) other than PV modules as the *balance of system* (BOS).

PV power systems can be categorized according to where the PV modules are installed, including but not limited to, installations on the ground, building roofs or even water. In this section, we focus on the PV generators and most important components of BOS such as inverters and batteries.

Another common categorization of PV power systems is based on the form of the electrical energy and where it is supplied, both of which depend on the requirements of the energy consumers. A PV power system supplies either DC or AC as a form of electrical energy. PV power systems that are not connected to the electricity grid are defined as *stand-alone*, in other words autonomous or *off-grid systems*. A PV power system supplying solely DC energy operates mostly as a stand-alone system. Stand-alone PV systems can also provide AC energy. On the other hand, PV power systems that are connected to the electricity grid feed only AC energy into the grid. Such PV power systems are defined as *grid-connected* or *on-grid systems*.

PV Generator

The power of a PV generator is defined in *watt-peak* (maximum power rating in watt) similar to that of a PV module and varies from a few watts (e.g. stand-alone systems for LED lighting) to several gigawatts (e.g. large grid-connected systems). In a PV generator, a number of PV modules are firstly connected in series to form a *string*. Here the principles of electrical circuitry apply: The voltage across a string is the sum of the voltages of individual PV modules, while the lowest current value of a particular PV module defines the total current that flows through the string. Therefore, PV modules with similar short-circuit currents should be connected in a string. Several strings with the same number of PV modules can later be connected in parallel to form a *PV array*, so that the string currents are added to each other, while the voltages across each string are supposed to be the same. Figure 5-18 reveals how the voltages and the currents of PV modules are added up when they are connected in series and parallel, respectively.

PV modules of the same watt-peak value might have unidentical current and voltage values. In case such PV modules are combined in a PV array, the total power of the PV array is not equal to the sum of the individual PV module powers. This power loss in a PV generator is defined as a *mismatch loss*. Such losses also occur when a PV module in a PV array malfunctions or gets shaded, as well as when strings with different number of PV modules are connected in parallel to each other. The negative effects of mismatching and shading are so complex that it is better to calculate them by means of a simulation software, as shown in Fig. 5-19.

---

[*]This section is contributed by Muammer Kabaçam. He is an engineer and general manager of a PV company with years of experience in PV projects.

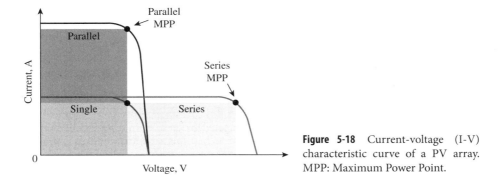

Figure 5-18 Current-voltage (I-V) characteristic curve of a PV array. MPP: Maximum Power Point.

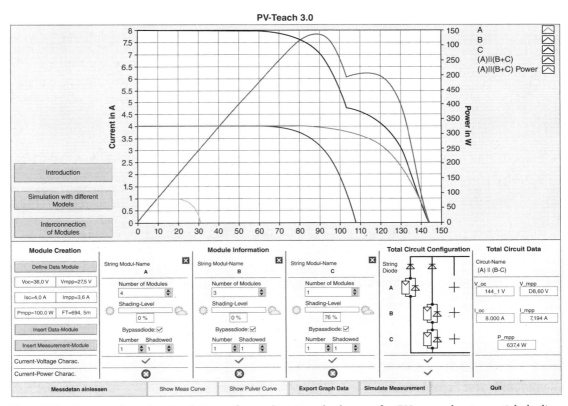

Figure 5-19 A screenshot of a simulation software depicting the losses of a PV array due to partial shading. (*The screenshot is from the free PV software PV-Teach accompanying the book by Mertens, 2018.*)

Inverter

A PV generator produces DC for starters, which cannot be fed into the electricity grid directly. The interconnected network for the distribution of electricity operates with either single- or three-phase AC depending on the energy consumers (industries, domestic customers, etc.). The voltage and frequency of the grid depend on the regions of the world.

The inverter transforms the DC energy from the PV generator into AC energy depending on the requirements of the electricity grid and energy consumers. Inverters deliver either single- or three-phase AC.

Various inverter types and power classes are available depending on the concept and design of the PV power system. The most widespread types are explained as follows.

Central inverters are used for large PV systems and deliver three-phase electric power ranging from several hundred kilowatts to megawatts. The outputs of individual strings from the PV generator are connected in parallel in a generator connection box, also referred to as *junction box* (not to be confused with the junction box of a PV module) and the DC energy is fed into the DC side of the central inverter from the junction box. The main advantage of central inverters compared to other inverter concepts is the lower investment cost both for the inverter itself and rest of the BOS. On the other hand, the operation and maintenance (O&M) costs for central inverters are higher because of their complex structure. Their O&M is also very crucial since they supply considerable amount of power, which would cause elevated power losses in case of an inverter defect. Another disadvantage of central inverters is the fact that the mismatch losses of the PV system due to shading, PV module defects, or ageing, for example, would be higher than other inverter concepts because of the parallel connection of strings.

String inverters can be either in single-phase or three-phase depending on their power classes. In small residential PV systems, mostly single-phase string inverters up to 5 kW are deployed. In case the residence is connected to a three-phase electricity grid, a small-scale (mostly up to 5 kW) single-phase inverter can be connected to one phase of the grid, as long as the grid regulation permits (there is no such problem if the residence is connected to a single-phase electricity grid). This solution cannot be used for higher powers, because such a phase unbalance can cause problems for the electricity grid. Three-phase string inverters range up to 150 kW. In case the PV generator's power is higher than a single inverter, multiple inverters can be deployed, each of which operates independently. Multiple strings can be connected to a string inverter and each string can be controlled by an individual *maximum power point tracker* (MPPT). Therefore, PV systems with string inverters suffer less mismatch losses than the ones with central inverters. In case of an inverter defect, the power loss of the PV system is limited. A defective inverter can easily be substituted by a backup product in the field and repaired afterward.

String inverters enable more flexibility for design, engineering, and operation of PV systems. As for central inverters, lower investment cost at the expense of this flexibility is a tradeoff for megascale projects that the investors must decide. The rest of this discussion focuses on string inverters because they are more widely deployed than central inverters.

Maximum Power Point Tracking

Maximum power point (MPP) is a fundamental term for PVs that characterize solar cells, PV modules, strings, arrays, and generators. The maximum power point tracker (MPPT) is a device that is programmed for controlling the DC-DC converter to make sure that the PV generator is operated at its MPP. The MPP of a PV generator varies with ambient conditions such as irradiance and temperature, as well as with mismatch losses such as shading effects. The MPPT's task is to find the MPP on the current-voltage (I-V) curve of a PV generator under the conditions above-mentioned.

There are MPPT algorithms that aim to find the I_{MMP}; however, the most common method for MPPT is to tune the voltage of the PV generator until its V_{MPP} is found. The MPPT changes the duty cycle D of the DC-DC converter while measuring the output voltage and

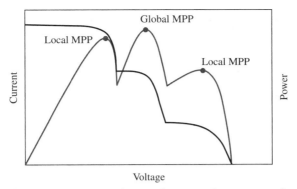

Figure 5-20   Current-voltage and power-voltage curves of a partially shaded PV generator with local and global MPPs.

current of the PV generator simultaneously until the PV generator is operated at its MPP. It is especially challenging to track the MPP in case the PV generator is partially shaded. The illustrative current-voltage (I-V) and power-voltage (P-V) curves of a partially shaded PV generator are depicted in Fig. 5-20, where the P-V curve has various local maxima. It is the task of the MPPT algorithm to find the global MPP. String inverters complete this task easier than the central inverters. In case multiple strings are connected to a string inverter, it is more effective to control each string by an individual MPPT.

Design of a Three-Phase String Inverter

Here, we consider a three-phase inverter design with a power class of 30 kW. In the first step, the DC-DC boost converter increases the DC voltage of the PV generator, if necessary. It is equipped with two MPP trackers, each of which having five input strings. Each MPPT operates its own PV array at its MPP independently, so that each PV array may:

- Consist of a different number of PV modules.
- Consist of different PV module types or power classes.
- Have PV modules with different installation conditions regarding orientation to the south and inclination from the horizontal plane.

However, each string within a PV array must:

- Consist of the same number of PV modules in series.
- Consist of the same PV module types and power classes.
- Have PV modules with the same installation conditions regarding orientation to the south and inclination from the horizontal plane.

The DC bus (also known as a link) capacitor provides a smooth DC bus voltage to the DC-AC inverter by eliminating the ripple currents and voltage transients. The DC-AC inverter converts the DC voltage into a sinusoidal AC voltage in compliance with the grid code. The low-pass filter attached to the DC-AC inverter suppresses the high-order harmonics of the AC voltage, so that the quality of the AC energy fed into the grid is increased.

The inverter has protective devices both on the DC and the AC side. The string fuses protect each string against return currents. In case of an earth fault or a short circuit in a string, the total current of all other parallel-connected strings that are not involved in the fault pass through the defective string. Such return currents may cause the risk of fire and damage to the PV generator. While a DC disconnect switch physically disconnects an individual PV array from the inverter, an AC disconnect switch physically disconnects the inverter from the electricity grid. An overvoltage protection system consisting of varistors, and surge protection devices (also known as surge arresters) are integrated in both DC and AC sides for protection against voltage peaks induced by lightning strikes, for example.

Batteries

Rechargeable (also known as secondary or storage) batteries can be used both in stand-alone and in grid-connected PV systems whenever the electrical energy from the PV generator needs to be stored. The term *rechargeable* is very vital, because the primary (single-use) batteries cannot be used in PV power systems, and thus are not relevant within the scope of this coverage. Rechargeable batteries, which are also known as accumulators, are electrochemical storages that can be charged, discharged into a load, and recharged many times.

An electric battery consists of electrochemical cells that can generate electrical energy from chemical reactions, as well as, initiate chemical reactions by using electrical energy. A battery converts chemical energy directly into electrical energy by means of a redox reaction. In a redox reaction, oxidation and reduction occur simultaneously. During this process, the reductant is transferring electrons to the oxidant and getting oxidized, while the oxidant is gaining electrons and getting reduced. This electrochemical reaction is reversible under electrical energy from an external source, so that the battery "accumulates" and stores energy. The amount of energy stored in a battery in comparison to its mass is denominated as energy density (or specific energy) and commonly expressed in watt-hours per kilogram (Wh/kg). A battery with a larger energy density than another one contains the same amount of energy although it has a smaller mass.

The operating principle of a battery is explained by the examples of lead-acid batteries and lithium-ion batteries. These two battery types are widely used for small-scale storages, which are the focus of this coverage. Megascale batteries are also available that can be used by utilities as a buffer between the intermittently produced electricity by the renewable energy sources and the difficult-to-predict electricity demands. Sodium-sulfur (NaS) batteries, redox flow batteries, and, to some extent, lithium-ion batteries can be used for this purpose.

Lead-Acid Battery

Invented during the second half of the 19th century, the lead-acid battery is the most established rechargeable battery technology. The electrochemical cells of a battery are electrically connected to each other. Each electrochemical cell consists of two half-cells connected in series and each half-cell consists of an electrode and an electrolyte. In a lead-acid battery, the negative electrode is made up of lead, while the positive electrode is made up of lead-dioxide. The half-cells are filled with sulfuric acid (H_2SO_4) solution as an electrolyte. The electrolyte is electrically conducting as it provides the flow of ions (anions and cations) between the electrodes. In case of a lead-acid battery, the electrolyte also participates in redox reactions inside the electrochemical cell. While the ions are conducted through the electrolyte, the electrons are conducted through an external conductor between the half-cells. A separator between the half-cells isolates the positive and negative electrodes but still allows the flow of ions between the electrodes.

Because lead sulfate at the surface of both electrodes decomposes during charging, sulfate ions are delivered back to the electrolyte from both electrodes. In case the battery is charged insufficiently or is not fully charged right after a deep discharge, the lead sulfate may partially remain at the electrodes, which may cause sulfation. Sulfation of the electrodes decreases their active masses and thus the capacity of the battery. On the other hand, a battery cell should not be overcharged. *Overcharging* occurs when the charging voltage is so high that excessive current is flown into the battery or when the battery is charged excessively even though it has become fully charged. In case of overcharging, the water inside the battery is electrolyzed so that oxygen gas is formed at the positive electrode and hydrogen gas is at the negative electrode. This process is known as *gassing* and may cause premature aging or even a risk of explosion of the battery.

There are various types of lead-acid batteries depending on the purpose of use. For small-scale stand-alone PV systems, *gel batteries* are usually preferred. In a gel battery, the sulfuric acid is thickened with mostly silica so that the electrolyte is gel-like. Compared to wet-cell (i.e., flooded-cell) batteries where the electrolyte is liquid as described above, gel batteries are completely sealed, have a much smaller risk of gas or electrolyte leakage, can be maintained infrequently, and have a longer lifespan. On the other hand, gassing should still be avoided or else the battery may dry out. Another disadvantage of gel batteries is that the mobility of the ions inside the electrolyte is limited because of thicker and less conductive electrolyte. Therefore, gel batteries cannot deliver high discharge currents like a starter battery in a car. However, they have a better deep-cycle characteristic than many other lead-acid battery types.

Compared to starter batteries, *deep-cycle batteries* can be deeply discharged on a regular basis and are less susceptible to degradation due to cycling. The latter prolongs the *battery lifespan*, which means the maximum number of charge-discharge cycles before the battery fails. This is also called the *cycle life*. Although possessing gel electrolytes and thicker plates (electrodes) compared to starter batteries extend the cycle life of deep-cycles batteries, it is also the reason why they cannot deliver discharge currents as high as starter batteries.

The capacity (C) of a lead-acid battery is measured in ampere hour (Ah), which is the amount of electric charge delivered under specific conditions. These conditions are explained by means of the characteristic curves of a commercial deep-cycle lead-acid gel battery with a nominal voltage of 12 V, as shown in Fig. 5-21.

The capacity of a battery is always given at a certain C-rate. C-rate defines a correlation between the discharge current and the discharge time. For example, 1C means that a fully charged battery with a capacity of 144 Ah at 1C can theoretically provide a discharge current of 144 A for 1 hour. The nominal capacity of lead-acid batteries is usually given at 0.05C (C_{20}) or 0.1C (C_{10}) corresponding to a 20-h discharge or a 10-h discharge, respectively.

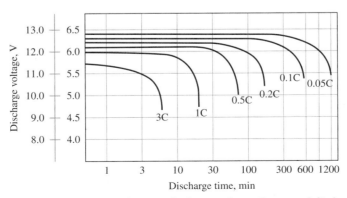

Figure 5-21 Correlation between discharge voltage, C-rate, and discharge time of a lead-acid battery at 25°C.

EXAMPLE 5-8   A lead-acid battery is operated at 0.2C with a capacity of 215 Ah. What is the discharge current I5 if
Discharge Current of a   the battery is discharged by a constant current for 5 hours at 0.2C?
Lead-Acid Battery

SOLUTION   Because 0.2C (C_5) corresponds to a 5-h discharge, the constant discharge current I_5 is
calculated as

$$I_5 = \frac{C_5}{\Delta t} = \frac{215\ \text{Ah}}{5\ \text{h}} = \textbf{43 A}$$

In the case of the battery shown in Fig. 5-21, only 0.05C corresponds to its theoretical
discharge time, which is 20 hours. All other C-rates fall short of their individual discharge
times. For example, 0.1C does not exactly correspond to 10 hours. As the C-rate increases,
the deviation gets bigger: the discharge time at 0.2C is approximately 3 hours, even though
it should have been 5 hours. Figure 5-21 also indicates that the operating voltage of a lead-
acid battery is usually below its nominal voltage (12 V). The end-of-discharge voltages at
various C-rates are also depicted in Fig. 5-21, meaning that in case of a longer discharge,
the operating voltage falls below this level and the depth of discharge is dangerously high.

Depth of discharge (DoD) defines to which extent a battery is discharged compared to
its fully charged state. The DoD is given in percentage and is the inverse of *state of charge*
(SoC) of a battery, which defines to which extent a battery is charged. That is,

$$\text{DoD} = 100\% - \text{SoC} \tag{5-24}$$

EXAMPLE 5-9   Calculate the DoD and SoC of a battery with a nominal capacity of 260 Ah, which is discharged for
Depth of Discharge of   30 minutes at a constant current of 100 A.
a Battery

SOLUTION   The depth of discharge and state of charge of this battery are determined as

$$\text{DoD} = \frac{I\Delta t}{C} = \frac{(100\ \text{A})(30\ \text{min})\left(\frac{1\ \text{h}}{60\ \text{min}}\right)}{260\ \text{Ah}} = 0.192 = \textbf{19.2\%}$$

$$\text{SoC} = 100\% - \text{DoD} = 100\% - 19.2\% = \textbf{80.8\%}$$

Therefore, 19.2 percent of the battery capacity is used and the remaining battery capacity is
80.8 percent.

The DoD has a negative effect on battery life. In case the battery shown in Fig. 5-22 is
fully discharged (DoD = 100%) on a regular basis, its capacity will reduce to 80 percent after
only approximately 500 cycles.

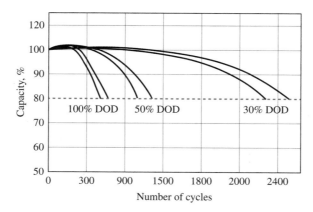

Figure 5-22   Correlation between capacity, depth of discharge, and cycle life of a lead-acid battery.

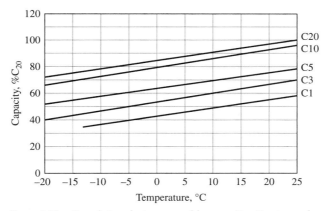

Figure 5-23   Correlation between usable capacity, C-rate, and ambient temperature of a lead-acid battery.

The usable capacity of a lead-acid battery strongly depends on its operating temperature. The battery capacity is reduced drastically when operating in subzero conditions. As depicted in Fig. 5-23, the capacity of most lead-acid batteries can be fully used at 25°C.

EXAMPLE 5-10
Effect of Temperature on the Capacity of a Battery

A lead-acid battery has a rated capacity of 250 Ah when operated at 0.1C and 25°C. It has a temperature behavior as shown in Fig. 5-23. How much of its rated capacity can be used when operated at 15°C?

SOLUTION   The lead-acid battery shown in Fig. 5-23 can deliver 90 percent of its rated capacity C_{10} (0.1C) when operated at 15°C. Therefore,

$$C_{10(15°C)} = (250 \text{ Ah})(0.9) = \textbf{225 Ah} \quad \blacktriangle$$

Lithium-Ion Battery

The lithium-ion (Li-ion) battery is a technology of the 20th century, with Li-ion batteries being commercially available from 1991. The original lithium battery consisted of lithium metals as electrodes; however, this technology was abandoned because of fire and explosion incidents due to highly reactive lithium. Nowadays, lithium is used in form of ions in order to avoid dangerous reactions.

In a Li-ion battery, a positive electrode is coated onto aluminum, whereas a negative electrode is coated on copper, both of which serve as current collecting metals. There is a variety of Li-ion batteries with different chemical compositions of the electrodes and the electrolyte. In most of the cases, the main difference is the material used for the positive electrode. In Li-ion batteries that are used in PV power systems, the following materials are mostly used as the positive electrode:

- Lithium nickel manganese cobalt oxide ($LiNiMnCoO_2$), also known as an NMC battery
- Lithium nickel cobalt aluminum oxide ($LiNiCoAlO_2$), also known as an NCA battery
- Lithium iron phosphate ($LiFePO_4$), also known as an LFP battery

All Li-ion battery types above-mentioned use graphite as the negative electrode. Because of its good electrical conductivity, graphite is used in many other Li-ion batteries too. Exceptions are the Li-ion batteries using lithium titanate (Li_2TiO_3) or lithium silicates (typically in form of Li_4SiO_4), the latter of which is still being a matter of intensive research (Su et al., 2022). The most common electrolyte type is lithium hexafluorophosphate ($LiPF_6$) salt solved in a mixture of organic carbonate solvents such as ethylene carbonate (EC), dimethyl carbonate (DMC), and diethyl carbonate (DEC). Most of the electrode types above-mentioned operate reasonably within the $LiPF_6$ solution. Nevertheless, solvent-free solid polymer electrolytes or gel polymer electrolytes have been investigated as an alternative to liquid-type electrolytes (Xu, 2014). Such batteries are denominated as lithium-ion polymer battery, abbreviated to LiPo battery or lithium polymer battery.

Lithium iron phosphate (LFP) batteries have been widely deployed as energy storages for PV power systems in recent years. Compared to other Li-ion battery types, LFP batteries have the following drawbacks:

- Lower nominal voltage of 3.2 volts per battery cell, whereas those of NMC and NCA batteries are 3.6 volts per cell.
- Lower energy density of approximately 120 Wh/kg, whereas those of NMC and NCA are approximately 220 Wh/kg and 260 Wh/kg, respectively.

Despite all these drawbacks, LFP batteries are preferred because of the following:

- They have a longer lifespan of over 2000 cycles, which is at least twice as much as that of an NMC battery and four times as much as that of an NCA battery.
- They can discharge at much higher C-rates, which also allows that the discharge voltage remains nearly constant at different discharge rates.
- They are very safe regarding explosion and fire.

During the discharge process of an LFP battery, lithium ions and electrons come off from the negative electrode (anode). The lithium ions are transported to the positive electrode (cathode) through the separator, while the electrons migrate over an external electrical circuit to the cathode. Both the lithium ions and the electrons integrate into iron phosphates at the cathode.

While the battery is getting charged, the reactions described above run the other way around: the positive electrode becomes the anode and both the lithium ions and the electrons are transported from the anode to the cathode (negative electrode). The electrolyte plays a key role for the conductivity of lithium ions, which is crucial for the power output of the battery. Similar to lead-acid batteries, a separator made of thin and porous polymer foils prevents a direct contact of the electrodes while allowing the transport of lithium ions from one electrode to the other. The overcharging of the battery should be avoided in order to prevent the destruction of the battery cell.

Li-ion batteries have much longer cycle lives, much better DoD rates, and higher discharge voltages, C-rates, and energy densities than lead-acid batteries. While the lead-acid battery technology is very mature, the technology for various types of Li-ion batteries is developing further. Despite their high costs, Li-ions batteries have been taking the place of lead-acid batteries in PV power systems.

EXAMPLE 5-11
Replacing a Lead-acid Battery with a Li-ion Battery

A 12-volt lead-acid battery with a nominal capacity of 100 Ah and a mass of 26.2 kg shall be replaced by a Li-ion battery. An LFP battery cell with a nominal discharge voltage of 3.2 volts and a nominal capacity of 102 Ah, which weighs 2.7 kg, shall be used as replacement.

(a) Determine the total number of LFP battery cells required for the replacement.

(b) Calculate the total mass of LFP battery and the energy densities of both battery types.

SOLUTION (*a*) If four LFP battery cells are connected in series, the system voltage of the new battery will be equivalent to the required system voltage of 12 volts, calculated as

$$V = (4)(3.2 \text{ V}) = 12.8 \text{ V}$$

Because of series connection, the capacity of the LFP battery remains 102 Ah. Therefore, **four** LFP battery cells are required to replace the existing lead-acid battery.

(*b*) Four LFP battery cells are used. Then, the mass of LFP battery is

$$m = (4)(2.7 \text{ kg}) = \textbf{10.8 kg}$$

The energy provided by the lead-acid battery is

$$E = VC = (12 \text{ V})(100 \text{ Ah}) = 1200 \text{ Wh}$$

Note that the unit VA is equivalent to W. The energy density of the lead-acid battery is

$$E_g = \frac{E}{m} = \frac{1200 \text{ Wh}}{26.2 \text{ kg}} = \textbf{45.8 Wh/kg}$$

The energy provided by the LFP battery is

$$E = VC = (12.8 \text{ V})(102 \text{ Ah}) = 1306 \text{ Wh}$$

Thus, the energy density of the LFP battery is calculated as

$$E_g = \frac{E}{m} = \frac{1305.6 \text{ Wh}}{10.8 \text{ kg}} = \textbf{120.9 Wh/kg}$$

The calculations clarify that the LFP battery, which is much lighter (and also smaller) than the lead-acid battery, meets the requirements regarding the energy output much better. ▲

REFERENCES

Çengel YA and Ghajar AJ. 2020. *Heat and Mass Transfer: Fundamentals and Applications*, 6th ed. New York: McGraw Hill.

Culp AW. 1991. *Principles of Energy Conversion,* 2nd ed. New York: McGraw Hill.

De Vos A. 1980. "Detailed Balance Limit of the Efficiency of Tandem Solar Cells." *Journal of Physics D: Applied Physics*, 13(5): 839–846.

Gevorkian P. 2007. *Sustainable Energy Systems Engineering*. New York: McGraw Hill.

Goswami Y, Kreith F, and Kreider JF. 2000. *Principles of Solar Engineering*, 2nd ed. New York: Taylor and Francis.

Hodge K. 2010. *Alternative Energy Systems and Applications*. New York: Wiley.

Kreith F and Kreider JF. 2011. *Principles of Sustainable Energy*. New York: Taylor and Francis.

Masters GM. 2013. *Renewable and Efficient Electric Power Systems*. 2nd ed. New Jersey: Wiley.

Mertens K. 2018. *Photovoltaics: Fundamentals, Technology and Practice*. 2nd ed. New York: Wiley.

Nelson V. 2011. *Introduction to Renewable Energy*. New York: CRC Press.

NREL (National Renewable Energy Laboratory). 2022. "Scientists Discover Way To Improve Perovskite Efficiency and Stability." https://www.nrel.gov/news/program/2021/scientists-discover-way-to-improve-perovskite-efficiency-and-stability.html. Access date: Sep. 16, 2022.

Su YS, Hsiao KC, Sireesha P, and Huang JY. 2022. "Lithium Silicates in Anode Materials for Li-Ion and Li-Metal Batteries." *Batteries*, 8(1): 2.

Xu K. 2014. "Electrolytes and Interphases in Li-Ion Batteries and Beyond." *Chemical Reviews*, 114(23): 11503–11618.

PROBLEMS

PHOTOVOLTAIC EFFECT

5-1 What is the photovoltaic effect?

5-2 Give at least one application for small-size, medium-size, and large-size PV systems.

5-3 What are the advantages and disadvantages of PV systems?

5-4 What is the band gap energy?

5-5 Describe semiconductor materials in terms of their valence bands and band gap energies.

5-6 Describe the operation of a PV cell using terminologies including *p-n* junction, silicon, photon energy, valence band, conduction band, and band gap energy.

5-7 Why is the conversion efficiency of a silicon solar cell low? Explain using concepts of wavelength of solar radiation, photon energy, and band gap energy.

5-8 How does using multijunction solar cells increase cell efficiencies?

5-9 What are the maximum wavelengths of solar radiation for which solar radiation can be converted to electrical energy for copper indium diselenide and cadmium telluride? The band gap energy values are 1.01 eV for copper indium diselenide and 1.44 eV for cadmium telluride.

5-10 Solar radiation is incident on a silicon solar cell. What is the excess energy (amount of photon energy not contributing to energy conversion) in a photon from solar radiation at a wavelength of 0.75 μm?

5-11 Solar radiation is incident on a silicon solar cell. What is the excess energy (amount of photon energy not contributing to energy conversion) in a photon from solar radiation at a wavelength of 1.20 μm?

5-12 Solar radiation is incident on a 1-m^2 gallium arsenide solar cell at a rate of 800 W/m^2·μm at a wavelength of 0.60 μm. Determine the maximum efficiency of this solar cell at this wavelength.

5-13 Solar radiation is incident on a 1-m^2 silicon solar cell at a rate of 1000 W/m^2·μm at a wavelength of 0.45 μm. Determine the rate of photons incident on the solar cell and the rate of excess solar energy at this wavelength.

5-14 Solar radiation is incident on a silicon solar cell at a rate of 650 W/m^2·μm at a wavelength of 1.12 μm. Determine the rate of photons incident on the solar cell and the maximum efficiency of this solar cell at this wavelength.

5-15 Solar radiation is incident on a silicon solar cell at a rate of 300 Btu/h·ft^2·μm at a wavelength of 0.70 μm. Determine the maximum efficiency of this solar cell at this wavelength.

5-16 Which of the following is not an advantage of PV systems?
(*a*) Lightweight (*b*) High efficiency (*c*) High reliability (*d*) No water consumption
(*e*) Flexibility in sizing

5-17 Which of the following is not a disadvantage of PV systems?
(*a*) High operating and maintenance cost (*b*) Low efficiency (*c*) Need for energy storage
(*d*) Intermittent energy source (*e*) Dust collection on panel surfaces

5-18 A material has high band gap energy, and their atoms have full valence bands. Electrons in the atom do not flow under the application of voltage or current. This material is most likely a(n)
(*a*) Conductor (*b*) Semiconductor (*c*) Insulator (*d*) Intrinsic semiconductor
(*e*) Extrinsic semiconductor

5-19 When the dopant has a smaller number of electrons in the valence band compared to pure silicon, the resulting material is a(n)
(*a*) Intrinsic semiconductor (*b*) Extrinsic semiconductor (*c*) *n*-type semiconductor
(*d*) *p*-type semiconductor

5-20 Select the *incorrect* statement regarding PV energy conversion.
(*a*) The higher the wavelength, the lower the energy level of the photon.
(*b*) Only the low-wavelength spectrum of solar radiation contributes to the PV energy conversion.
(*c*) When the photon possesses more energy than the band gap energy, the excess energy is converted to heat.
(*d*) If the energy of photon is less than the band gap energy, the electron in the valence band will not jump into the conduction band.
(*e*) A single photon can move multiple valence electrons to the conduction band.

5-21   Which of the following is not a reason for actual efficiencies of solar cells being less than the maximum efficiencies?
(*a*) Recombination of electrons and holes not contributing to electric flow.
(*b*) Reflection of solar radiation from the cell surface.
(*c*) Internal electric resistance of the cell.
(*d*) Low temperature operation in winter.
(*e*) Shading of the cell due to electrical contacts.

ANALYSIS OF SOLAR CELLS

5-22   How is current density defined? What are light-induced recombination current and dark current or reverse saturation current?

5-23   What is the approximate maximum voltage a high-quality silicon solar cell can produce? Under what conditions is the power output zero? At what voltage level is the power output maximum?

5-24   A solar cell has an open circuit voltage value of 0.55 V with a reverse saturation current density of $J_o = 1.9 \times 10^{-9}$ A/m^2. For a temperature of 25°C, determine (*a*) the current output density J_s, (*b*) the load voltage at which the power output is maximum, and (*c*) the maximum power output of the cell for a unit cell area.

5-25   A solar cell has an open circuit voltage value of 0.60 V with a reverse saturation current density of $J_o = 3.9 \times 10^{-9}$ A/m^2. The temperature of the cell is 27°C, the cell voltage is 0.52 V, and the cell area is 28 m^2. If the solar irradiation is 485 W/m^2, determine the power output and the efficiency of the solar cell.

5-26   A solar cell has an open circuit voltage value of 0.60 V with a reverse saturation current density of $J_o = 4.11 \times 10^{-10}$ A/ft^2.
(*a*) For a temperature of 75°F, determine the load voltage at which the power output is maximum.
(*b*) If the solar irradiation is 220 Btu/h · ft^2, determine the efficiency of the solar cell at a load voltage of 0.56 V.
(*c*) Determine the cell area, in ft^2, for a power output of 500 W at a load voltage of 0.56 V.

5-27   Reconsider Prob. 5-26. What is the maximum conversion efficiency of this solar cell?

5-28   The competition car developed by a group of engineering students uses solar cells with a total area of 8 m^2. The solar radiation is incident on the cells at a rate of 860 W/m^2. The shaft power output from the car is measured by a dynamometer to be 540 W. What is the thermal efficiency of this solar car?

5-29   The unit for current density is
(*a*) W/m^2 (*b*) W/m^3 (*c*) A/m^2 (*d*) A/m^3 (*e*) A/W

5-30   The theoretical efficiency limit for a single junction solar cell is considered to be about
(*a*) 34%   (*b*) 18%   (*c*) 50%   (*d*) 86%   (*e*) 95%

5-31   Which material is most commonly used in solar cells?
(*a*) Cadmium telluride   (*b*) Gallium arsenide   (*c*) Copper indium diselenide
(*d*) Silicon   (*e*) Cadmium sulfide

5-32   In a solar cell, the maximum power occurs at a voltage
(*a*) Equal to open circuit voltage   (*b*) Equal to zero   (*c*) Close to short circuit case
(*d*) Close to open circuit voltage   (*e*) Half of open circuit voltage

5-33   Which statement is not correct for solar cells?
(*a*) A high-quality silicon solar cell can produce an open circuit voltage of about 0.6 V.
(*b*) For the short circuit case, $J_L = J_s$.
(*c*) For the short circuit case, the voltage is zero.
(*d*) For the short circuit case, the power output is zero.
(*e*) For the open circuit voltage case, the power output is maximum.

5-34   In a solar cell, the load voltage is 0.5 V and the load current density is determined to be 80 A/m². If the solar irradiation is 650 W/m², the cell efficiency is
(*a*) 4.7%   (*b*) 6.2%   (*c*) 7.8%   (*d*) 9.1%   (*e*) 14.2%

PHOTOVOLTAIC TECHNOLOGIES AND SYSTEMS

5-35   List solar cell types and technologies. Which technologies have higher efficiencies with respect to others?

5-36   Can the Carnot efficiency be used as the upper limit for the efficiency of solar cells? Why?

5-37   What does a solar module consist of? What does a solar array consist of?

5-38   A PV system manufacturer lists the maximum power of a certain module as 273 W while the company lists the typical power as 365 W. Why is the typical power higher than the maximum power? Explain.

5-39   What is the effect of solar irradiation on the current, open circuit voltage, and maximum power?

5-40   What is the effect of cell temperature on the current, open circuit voltage, and maximum power?

5-41   Is it better to operate a solar cell in summer or in winter for the same solar irradiation? Why?

5-42   Consider a commercial solar cell with the following specifications:
Maximum power = 2.4 W
Voltage at maximum power = 0.53 V
Current at maximum power = 4.5 A
Efficiency = 20 percent
A module of these solar cells is to be constructed to provide a voltage output of 24 V and a power output of 150 W.
(*a*) How many cells should be used to satisfy the power output?
(*b*) How many cells should be arranged in series to satisfy the voltage output?
(*c*) How many rows of the cells in series should be used?
(*d*) What is the power rating of the solar module?
(*e*) If the testing of this solar cell is made for a solar radiation value of 1000 W/m², what is the required area of this module?

5-43   Consider a solar cell with the following specifications:
Typical power = 1.8 W
Voltage at typical power = 0.55 V
Current at maximum power = 3.3 A
Efficiency = 22 percent
A module of these solar cells is to be constructed to provide a voltage output of 8 V and a power output of 75 W.
(*a*) How many cells should be arranged in series to satisfy the voltage output?
(*b*) How many rows of the cells in series should be used?
(*c*) What is the power rating of the solar module?
(*d*) If the testing of this solar cell is made for a solar radiation value of 800 W/m², what is the required area of this module?

5-44   The silicon is produced in the form of an ingot as the silicon melt is cooled slowly. Grain boundaries are formed that separate the crystalline regions of the silicon ingot. Its efficiency is low due to the gaps in the grain boundaries. This type of solar cell is known as a(n) _____ solar cell.
(*a*) Amorphous   (*b*) Monocrystalline   (*c*) Polycrystalline   (*d*) Thin-film
(*e*) Multijunction

5-45   For monocrystalline fuel cells, the manufacturing cost is _____ and the efficiency is

_____.
(*a*) High, low   (*b*) High, high   (*c*) Low, high   (*d*) Low, low

5-46 Two identical solar cells each with a voltage of 0.75 V and current of 4 A are connected in series. What is the voltage and current of this cell network?
(*a*) 0.75 V, 4 A   (*b*) 0.75 V, 8 A   (*c*) 1.5 V, 4 A   (*d*) 1.5 V, 8 A

5-47 Two identical solar cells each with a voltage of 0.75 V and current of 4 A are connected in parallel. What is the voltage and current of this cell network?
(*a*) 0.75 V, 4 A   (*b*) 0.75 V, 8 A   (*c*) 1.5 V, 4 A   (*d*) 1.5 V, 8 A

5-48 Two dissimilar solar cells with $V_1 = 0.5$ V, $I_1 = 4$ A and $V_2 = 0.7$ V, $I_2 = 8$ A are connected in series. What are the most likely values of the voltage and current for this cell network?
(*a*) 0.5 V, 4 A   (*b*) 0.6 V, 12 A   (*c*) 1.2 V, 12 A   (*d*) 1.2 V, 5 A   (*e*) 0.6 V, 6 A

5-49 For higher power output from a solar cell, the solar radiation should be _____ and the cell temperature should be _____.
(*a*) Higher, higher   (*b*) Higher, lower   (*c*) Lower, higher   (*d*) Lower, lower

ENERGY PRODUCTION FROM PHOTOVOLTAIC SYSTEMS

5-50 A PV array uses 50 commercial PV modules each with an area of 1.7 m^2, a typical power output of 320 W, and a solar module efficiency of 16 percent. The system efficiency may be taken as 70 percent. This PV array is installed in Las Vegas, Nevada. The average daily solar radiation on a horizontal surface in Las Vegas is given in Table 3-6, in MJ/m^2·day, as follows:

January	10.79	July	28.28
February	14.42	August	25.89
March	19.42	September	22.15
April	24.87	October	17.03
May	28.16	November	12.15
June	30.09	December	9.88
		Average	20.33 MJ/m^2·day

(*a*) Estimate the amount of energy production from this solar array for each month and for the entire year in Las Vegas, in kWh.
(*b*) If the price of electricity is $0.22/kWh, what is the annual potential revenue from this PV array?
(*c*) If the initial cost of this PV system is specified as $2.5/W, what is the simple payback period?

5-51 A PV array has an area of 28 m^2 and a power capacity of 5 kW. The annual average daily solar radiation on a horizontal surface in a certain city is given as 4.95 kWh/m^2·day. The efficiency of the solar cell is 20 percent and the system efficiency may be taken as 65 percent. Estimate the amount of energy production from this PV system per year.

5-52 A PV array has an area of 15 m^2 and a power capacity of 2.4 kW. The annual average daily solar radiation on a horizontal surface in Miami, Florida is 17.38 MJ/m^2·day. The efficiency of the solar cell is 17 percent and the system efficiency may be taken as 70 percent. Estimate the amount of energy production from this PV system per year and calculate the capacity factor of this installation.

5-53 A PV array has an area of 650 ft^2 and a power capacity of 11 kW. The annual average daily solar radiation on a horizontal surface in a certain city is given as 1700 Btu/ft^2·day. The efficiency of the solar cell is 23 percent and the system efficiency may be taken as 60 percent. Estimate the amount of energy production from this PV system per year, in kWh and calculate the capacity factor of this installation.

5-54 A homeowner decides to install a PV cell system on the roof of his house to meet the electricity needs of the house. The capacity of the solar system is 6 kW and the cost of solar cells is $1.30/W. If the house owner currently pays an average of $125 for the electricity per month, determine how long it will take for the PV system to pay for itself. Assume the homeowner can meet approximately 80 percent of the electricity needs of the house with the solar system.

5-55 A PV array is installed in Hartford, Connecticut, with a power rating of 50 kW (DC, STC). The annual average solar insolation is given in Table 3-6 for Hartford as 13.74 MJ/m²·day. Assuming a derate factor of 0.75, estimate the annual energy production using peak-hours approach.

5-56 Observations over a year period shows that a PV array installed in Las Vegas, Nevada, provided 45,000 kWh of electricity. The total power rating of the PV array is 28 kW (DC, STC). What is the overall derate factor of this PV system operation? Use the peak-hours approach.

5-57 A PV array is installed in a location with a power rating of 20 kW (DC, STC). If the annual average peak-hours or peak-sun value is 4.75 h/day and overall derate factor is 0.72, estimate the annual energy production by this PV system. What is the annual average insolation value in this location in MJ/m²·day?

5-58 A PV array is installed in a location with a power rating of 5 kW (DC, STC). If the annual average peak-hours or peak-sun value is 3.5 h/day and overall derate factor is 0.7, estimate the annual energy production by this PV system.
(*a*) 12.3 kWh/yr   (*b*) 1230 kWh/yr   (*c*) 4470 kWh/yr   (*d*) 12,300 kWh/yr
(*e*) 36,800 kWh/yr

5-59 What is the annual average insolation value in a location with a peak-hour value of 4.3 h/day?
(*a*) 4.3 MJ/m²·day   (*b*) 5.5 MJ/m²·day   (*c*) 11.8 MJ/m²·day   (*d*) 15.5 MJ/m²·day
(*e*) 43.0 MJ/m²·day

PHOTOVOLTAIC SYSTEM CONFIGURATIONS

5-60 What are the two general categories for the application areas of PV systems?

5-61 How do you compare water pumping powered by a PV system or a diesel generator?

5-62 List two advantages of grid-connected PV systems compared to off-grid systems.

5-63 What is the best placement of PV arrays in the northern hemisphere in terms of direction and tilt angle?

5-64 Which component is used to convert DC power to AC power in PV applications?
(*a*) Charge controller   (*b*) Battery   (*c*) Distribution panel   (*d*) PV array
(*e*) Inverter

5-65 Select the *incorrect* statement regarding the grid-connected PV systems.
(*a*) These systems include an inverter.
(*b*) They provide an effective method of peak shaving.
(*c*) They supply DC power or AC power to the grid.
(*d*) Large installations of grid-connected PV systems are mostly owned by the utilities.
(*e*) They supply power at grid voltage, phase, and frequency.

COMPONENTS OF PHOTOVOLTAIC POWER SYSTEMS

5-66 What is a mismatch loss in a PV generator? Explain.

5-67 What is the function of an inverter in PV systems? Do the inverters deliver single-phase or three-phase AC? What are the common types of inverters?

5-68 What are the advantages and disadvantages of central inverters compared to string inverters?

5-69 What is the most common method for MPPT?

5-70 What is the function of rechargeable batteries in PV systems? Are they used in stand-alone or grid-connected PV systems?

5-71 How is energy density for a battery defined? What is its common unit?

5-72 What are the common types of batteries for PV systems? Which battery types can be used by utilities to meet demand during peak hours?

5-73 What is overcharging of a battery? What are the negative consequences of overcharging?

5-74   Give definitions for depth of discharge (DoD) and state of charge (SoC) for a battery. What is the relation between them?

5-75   What are the effects of higher values of depth of discharge (DoD) and lower values of temperature on the capacity of a battery?

5-76   List advantages and disadvantages of Li-ion batteries compared to lead-acid batteries.

5-77   A lead-acid battery is operated at 0.5C with a capacity of 144 Ah. What is the discharge current I_2 if the battery is discharged by a constant current for 2 hours at 0.5C?

5-78   Consider a lead-acid battery operating at 0.1C. If the constant discharge current is $I_{10} = 20$ A, what is the capacity of this battery?

5-79   Calculate the depth of discharge (DoD) and state of charge (SoC) of a battery with a nominal capacity of 180 Ah, which is discharged for 45 minutes at a constant current of 60 A.

5-80   The nominal capacity of a battery is 300 Ah and it is discharged at a constant current of 120 A. If this battery has an state of charge (SoC) value of 65 percent, what is the discharge time?

5-81   A lead-acid battery has a rated capacity of 175 Ah when operated at 0.05C and 25°C. It has a temperature behavior as shown in Fig. 5-23. How much of its rated capacity can be used when operated at 0°C?

5-82   A lead-acid battery has a rated capacity of 320 Ah when operated at 0.33C and 25°C. It has a temperature behavior as shown in Fig. 5-23. What are the usable capacities of this lead-acid battery when operated at 0.33C at a temperature of 10°C and 25°C?

5-83   A 24-volt lead-acid battery with a nominal capacity of 150 Ah and a mass of 40 kg shall be replaced by a Li-ion battery. An LFP battery cell with a nominal discharge voltage of 3.2 volts and a nominal capacity of 155 Ah, which weighs 4.2 kg, shall be used as replacement. Determine the number of LFP battery cells required for the replacement, the total mass of LFP battery, and the energy densities of both battery types.

5-84   A 12-volt lead-acid battery with a nominal capacity of 200 Ah and a mass of 25 kg shall be replaced by a Li-ion battery. An NMC battery cell with a nominal discharge voltage of 3.6 volts and a nominal capacity of 205 Ah, which weighs 3.3 kg, shall be used as replacement. Determine the number of NMC battery cells required for the replacement, the total mass of NMC battery, and the energy densities of both battery types.

5-85   Order the following components from the smaller power capacity to the larger ones:
I. Module   II. Cell   III. Array   IV. String
(*a*) I, II, III, IV   (*b*) I, II, IV, III   (*c*) II, I, III, IV   (*d*) II, I, IV, III   (*e*) IV, II, I, III

5-86   Which of the following is *not* an advantage of string inverters compared to central inverters?
(*a*) PV systems with string inverters suffer less mismatch losses.
(*b*) In case of an inverter defect, the power loss of the PV system is limited.
(*c*) String inverters enable more flexibility for design, engineering, and operation of PV systems.
(*d*) They can be either in single-phase or three-phase depending on their power classes.
(*e*) They involve lower investment cost both for the inverter itself and rest of the BOS.

5-87   The maximum power point (MPP) of a PV generator varies with which of the following:
I. Temperature   II. Irradiance   III. Mismatch losses
(*a*) I, II, and III   (*b*) I and II   (*c*) I and III   (*d*) II and III   (*e*) Only II

5-88   Which unit is commonly used to express the energy density of a battery?
(*a*) kW/kg   (*b*) Wh/kg   (*c*) kJ/kg   (*d*) kJ/m³   (*e*) W/m³

5-89   When a battery is charged extensively, the water inside the battery is electrolyzed so that oxygen gas and hydrogen gas are formed. This process is known as
(*a*) Undercharging   (*b*) Flooding   (*c*) Gassing   (*d*) Aging   (*e*) Overcharging

5-90   The battery capacity _____ with higher values of depth of discharge (DoD) and _____ with lower values of temperature.
(*a*) Increases, increases   (*b*) Increases, decreases   (*c*) Decreases, increases
(*d*) Decreases, decreases

5-91   Select the *incorrect* notation regarding the capacity of a battery.
(*a*) 0.1C (C_{10})   (*b*) 0.5C (C_2)   (*c*) 0.05C (C_{20})   (*d*) 0.2C (C_5)   (*e*) None of these

5-92   The nominal capacity of lead-acid batteries is usually given at which of the following:
I. 0.05C (C_{20})   II. 0.1C (C_{10})   III. 0.2C (C_5)
(*a*) I, II, and III   (*b*) I and II   (*c*) I and III   (*d*) II and III   (*e*) Only II

5-93   Which of the following is *not* an advantage of LFP batteries compared to NMC and NMA batteries?
(*a*) Longer lifespan   (*b*) Discharging at higher C-rates   (*c*) Higher energy density
(*d*) Safer regarding explosion and fire

CHAPTER 6

Wind Energy

6-1 INTRODUCTION

Wind energy has been used since 4000 BC to power sailboats, grind grain, pump water for farms, and, more recently, generate electricity. In the United States alone, more than 6 million small windmills, most of them under 5 hp, have been used since the 1850s to pump water. Small windmills have been used to generate electricity since 1887, but the development of modern wind turbines occurred only recently in response to the energy crises in the early 1970s. We note the distinction between the terms *windmill* used for *mechanical* power generation (grinding grain, pumping water, etc.) and *wind turbine* used for electrical power generation, although technically both devices are turbines since they extract energy from the fluid.

Areas with an average wind speed of 6 m/s (or 14 mph) or higher are potential sites for economical wind power generation. Most commercial wind turbines generate from 1 to 5 MW of electric power each at peak design conditions. The rotation speed of rotors of wind turbines is usually under 40 rpm (under 20 rpm for large turbines). The blade span (or rotor) diameter of the 14-MW wind turbine built by General Electric is 240 m and the tower height is 260 m.

Although the wind is "free" and renewable, modern wind turbines are expensive and suffer from one obvious disadvantage compared to most other power generation devices—they produce power only when the wind is blowing, and the power output of a wind turbine is thus inherently unsteady. Furthermore and equally obvious is the fact that wind turbines need to be located where the wind blows, which is often far from traditional power grids, requiring construction of new high-voltage power lines. Nevertheless, wind turbines are expected to play an ever increasing role in the global supply of energy for the foreseeable future (Fig. 6-1).

Much of the material in this chapter is taken from Çengel and Cimbala (2018), and some of this material is condensed from Manwell et al. (2010). The readers are referred to Manwell et al. (2010) for additional information and analysis.

6-2 WIND TURBINE TYPES AND POWER PERFORMANCE CURVE

Numerous innovative wind turbine designs have been proposed and tested over the centuries as sketched in Fig. 6-2. We generally categorize wind turbines by the orientation of their axis of rotation: *horizontal axis wind turbines* (HAWTs) and *vertical axis wind turbines* (VAWTs). An alternative way to categorize them is by the mechanism that provides torque to the rotating shaft: lift or drag. So far, none of the VAWT designs or drag-type designs has achieved the efficiency or success of the lift-type HAWT. This is why the vast majority of

Figure 6-1   Wind farms are popping up all over the world to generate green electricity.

wind turbines being built around the world are of this type, often in clusters affectionately called *wind farms*. For this reason, the lift-type HAWT is the only type of wind turbine discussed in any detail in this section. [See Manwell et al. (2010) for a detailed discussion as to why drag-type devices have inherently lower efficiency than lift-type devices.]

Every wind turbine has a characteristic power performance curve; a typical one is sketched in Fig. 6-3, in which electrical power output is plotted as a function of wind speed V at the height of the turbine's axis. We identify three key locations on the wind-speed scale:

Cut-in speed is the minimum wind speed at which useful power can be generated.

Rated speed is the wind speed that delivers the rated power, usually the maximum power.

Cut-out speed is the maximum wind speed at which the wind turbine is designed to produce power. At wind speeds greater than the cut-out speed, the turbine blades are stopped by some type of braking mechanism to avoid damage and for safety issues. The short section of dashed curve line indicates the power that *would* be produced if cut-out were not implemented.

The design of HAWT turbine blades includes tapering and twist to maximize performance. While the fluid mechanics of wind turbine design is critical, the power performance curve also is influenced by the electrical generator, the gearbox, and structural issues. Inefficiencies appear in every component, of course, as in all machines.

6-3 WIND TURBINE OPERATION AND AERODYNAMICS

A wind turbine converts the kinetic energy of the wind into electricity. The shape of the turbine blades is similar to an airplane wing. When air flows across the blade, the air pressure on one side of the blade becomes higher than the other side, and this pressure difference

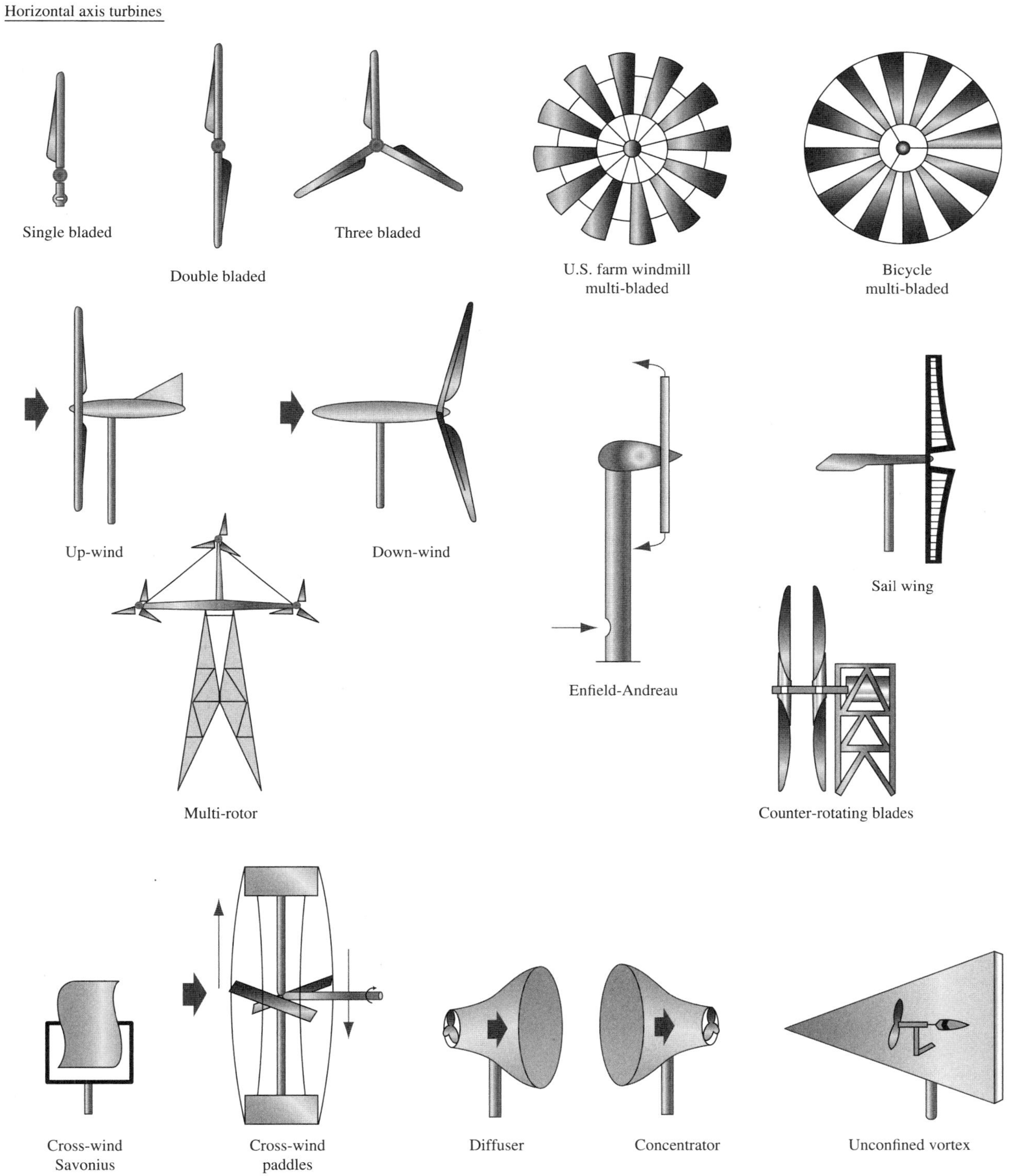

Horizontal axis turbines

Single bladed

Double bladed

Three bladed

U.S. farm windmill
multi-bladed

Bicycle
multi-bladed

Up-wind

Down-wind

Enfield-Andreau

Sail wing

Multi-rotor

Counter-rotating blades

Cross-wind
Savonius

Cross-wind
paddles

Diffuser

Concentrator

Unconfined vortex

Figure 6-2 Various wind turbine designs and their categorization. (*Adapted from Manwell et al., 2010.*)

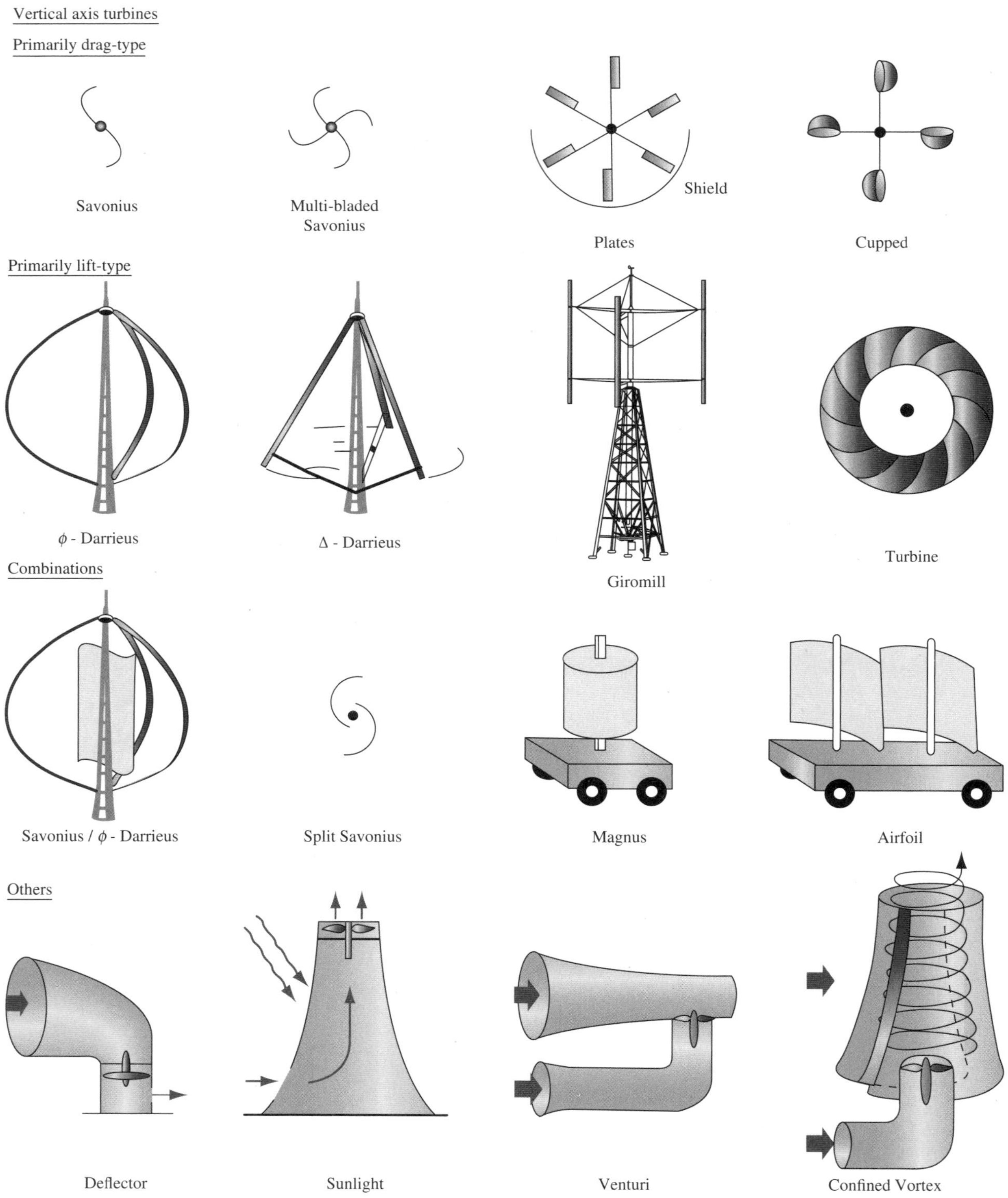

Vertical axis turbines

Primarily drag-type

Savonius

Multi-bladed
Savonius

Shield

Plates

Cupped

Primarily lift-type

ϕ - Darrieus

Δ - Darrieus

Giromill

Turbine

Combinations

Savonius / ϕ - Darrieus

Split Savonius

Magnus

Airfoil

Others

Deflector

Sunlight

Venturi

Confined Vortex

Figure 6-2   (*Continued*)

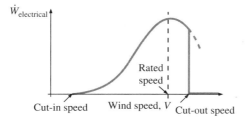

Figure 6-3 Typical qualitative wind-turbine power performance curve with definitions of cut-in, rated, and cut-out speeds.

creates a lifting force. There is also drag but the lift is stronger than drag. The result is the rotation of the rotor (blades and hub assembly). In a direct-drive turbine, the rotor is directly connected to a generator. This system is mostly used in offshore wind turbines. In most wind turbines, the rotor is connected to a shaft with a low rate of rotation. The gearbox increases the rotation from 20–100 rpm levels to 2000–3000 rpm levels. This allows the use of a smaller-sized generator and effective conversion of mechanical power into electricity.

Most commercial wind turbines are designed such that the blades are positioned to face the wind. This is called an *upwind turbine* (Fig. 6-4). The propeller of an airplane has a similar positioning. In a *downwind turbine*, the blades are positioned behind the unit. In downwind turbines, the blades are subjected to alternating higher and lower forces, which cause higher levels of noise and fatigue failure. An advantage of downwind turbines is that they may be designed without a yaw drive.

The electricity generated from the turbines is transferred to the grid over high-voltage transmission lines. *Transformers* in the wind farm are used to increase the voltage in order to reduce power losses during the transmission of electricity over long distances. Transformers are also used in a *substation* to reduce the voltage of electricity before it is safely delivered to the users (Fig. 6-5).

The heavy turbine structure is supported by a steel *tower*. The tower comes in multiple sections that are assembled on-site. Taller towers are preferred because the wind speed increases with height and more power is generated as a result. However, the cost of the tower, installation, and access to the turbine for maintenance limit the tower height. Early wind turbines had two blades, but modern wind turbines have three blades, which is considered optimum for power generation. Blades are usually made of fiberglass. They fit into the hub that is connected to the main shaft.

The powerhouse (also called drivetrain or nacelle) of a wind turbine is shown at the top of the tower in Figs. 6-6 and 6-7. It contains the main bearing, low-speed main shaft,

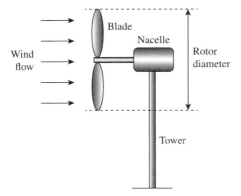

Figure 6-4 Horizontal axis upwind turbine.

Figure 6-5   A wind farm showing multiple turbines, a transformer, and a substation. (*DOE, 2022.*)

gearbox, generator, controller, brake, and yaw system. *Main shaft bearing* supports the low-speed shaft and reduces friction between moving parts. The high-speed shaft is connected to the gearbox and drives the *generator*. Electricity is generated as copper windings turn through a magnetic field.

Figure 6-6   The powerhouse (nacelle) of a wind turbine showing the rotor (blades and hub), main bearing, main shaft, gearbox, generator, controller, brake, and yaw system. A wind vane and an anemometer are also attached to the powerhouse. (*DOE, 2022.*)

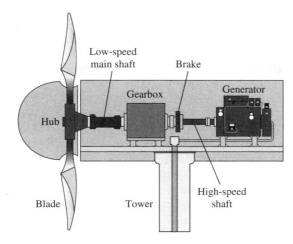

Figure 6-7 A closer look at the powerhouse of a horizontal axis wind turbine showing major components and their interaction.

There is a wind vane and an anemometer on the corner of the power box. The *wind vane* detects the wind direction in communication with the yaw drive and adjusts the turbine properly with respect to the wind. The *yaw drive* is only used in upwind turbines, and it rotates the nacelle when the wind direction changes. The *anemometer* measures the wind speed and sends this data to the controller. For example, if the wind speed is above a certain value for safe operation, the *controller* stops the rotation. The turbine control usually allows the operation at wind speeds of 3–5 m/s and usually stops it at about 25 m/s.

The angle of the blades with respect to wind is adjusted by a *pitch system* that controls the rotor speed. The pitch system can also adjust the blade angle to lower the lift and the resulting rotation of the rotor when the wind speed is too high for safe operation. When the pitch system stops the rotor rotation, the brake keeps the turbine blades from moving.

More details on the component and operation of both direct-drive turbines and the turbines with a gearbox are available online (NREL, 2022b).

Operation of a wind turbine is based on the rotation of blades when the wind blows toward the blades. The shape of the blades is designed to generate a lift force, which in turn causes rotation of the blades. Figure 6-8 shows the cross section of a blade section and the resulting lift and drag forces. The shape that produces lift is called airfoil. The blades are shaped such that the air velocity over the airfoil (curved surface) is higher than that below the airfoil (almost flat surface). This corresponds to a higher-pressure region below the airfoil and a low-pressure region above it. As a result, a lift force develops which is perpendicular to the direction of the drag force.

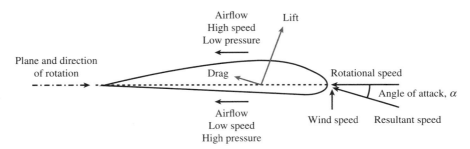

Figure 6-8   Generation of lift on a blade section near the blade tip. (*Adapted from Kreith and Kreider, 2011.*)

The drag force is parallel to the resultant speed, which is a vector addition of the wind speed and rotational speed at that location. The rotational speed is a function of the diameter of the blade at that cross section and the rotational speed. It is equal to $\pi D \dot{n}$ where $\dot{n}$ is the rotational speed in rpm, which is converted to rev/s to obtain the rotational speed in m/s. The straight dashed line connecting the front and rear edges of the airfoil is called the chord. The angle between the resultant speed and the chord line is called the angle of attack α. Lift force increases with angle of attack until stall occurs. The rotational speed and the chord line are not necessarily parallel.

In a wind turbine, the blades are designed to maximize the lift force and minimize the drag force. This is because the higher lift force means higher rotational force and higher power generation for a given wind speed. Therefore, maximizing the lift-to-drag ratio is a main design objective for turbine blades. Modern three-bladed horizontal axis upwind turbines are superior to other types of wind turbines such as American Multiblade, Darrieus Vertical Axis, and Dutch Four Arm mainly due to their higher lift-to-drag ratios and corresponding higher efficiency for converting wind energy to output power.

The blade geometry varies along the length of the blade. The position of blade cross section rotates in the clockwise direction with respect to the plane of rotation near the blade root. The rotational speed near the blade tip is much higher than that near the blade root due to lower local diameter. Therefore, near the blade root, the resultant speed is smaller, and the angle of attack is larger since the wind speed is the same. As a result, the lift force is smaller near the blade root than near the blade tip. The direction of the lift force near the root facilitates the application of torque to start blade rotation even at low wind speeds. Also, the surface area at the blade root is larger compared to the blade tip. Thus, the lift force is applied over a large area generating necessary torque for blade rotation (Kreith and Kreider, 2011).

6-4 WIND POWER POTENTIAL

The *mechanical energy* can be defined as *the form of energy that can be converted to mechanical work completely and directly by an ideal mechanical device such as an ideal turbine.* The mechanical energy of a flowing fluid can be expressed as

$$\dot{E}_{mech} = \dot{m}\left(\frac{P}{\rho} + \frac{V^2}{2} + gz\right) \tag{6-1}$$

where P/ρ is the flow energy, $V^2/2$ is the kinetic energy, and gz is the potential energy of the fluid, all per unit mass, and $\dot{m}$ is the mass flow rate of the fluid. The pressures at the inlet and exit of a wind turbine are both equal to the atmospheric pressure, and the elevation does not change across a wind turbine. Therefore, flow energy and potential energy do not change across a wind turbine. A wind turbine converts the kinetic energy of the fluid into power. If the wind is blowing at a location at a velocity of V, the available wind power is expressed as

$$\dot{W}_{available} = \frac{1}{2}\dot{m}V^2 \quad \text{(kW)} \tag{6-2}$$

This is the *maximum power* a wind turbine can generate for the given wind velocity V. The mass flow rate is given by

$$\dot{m} = \rho A V \quad \text{(kg/s)} \tag{6-3}$$

where ρ is the density and A is the disk area of a wind turbine (the circular area swept out by the turbine blades as they rotate). Substituting,

Wind power potential: $$\dot{W}_{available} = \frac{1}{2}\rho A V^3 \tag{6-4}$$

Equation (6-4) indicates that the power potential of a wind turbine is proportional to the cubic power of the wind velocity. For example, consider a location where the wind with a density of 1.2 kg/m³ is blowing at a velocity of 4 m/s. The maximum power a wind turbine with a rotor diameter of 1.03 m can generate is determined from Eq. (6-4) to be

$$\dot{W}_{available} = \frac{1}{2}\rho A V^3 = \frac{1}{2}(1.2 \text{ kg/m}^3)\frac{\pi(1.03 \text{ m})^3}{4}(4 \text{ m/s})^3 = 32 \text{ W}$$

If the wind velocity is doubled, the available power becomes 256 W (Fig. 6-9). That is, doubling the wind velocity will increase the power potential by a factor of 8. For this cubical relationship, a wind turbine investment is not justified if the location does not have a steady wind at a velocity of about 5 m/s or higher.

One way of increasing wind speed is using taller towers because the wind speed increases with height. In a taller tower, the blades, gearbox, generator, and other heavy components at the top of the tower are subjected to higher forces. This may require better strength values of the turbine components and, consequently, higher overall cost of the system.

The available power relation indicates that the power potential of a wind turbine is proportional to the density of air. As a result, cold air has a higher wind power potential than warm air. The density of air in Eq. (6-4) can be determined from the ideal gas relation $P = \rho R T$ when the pressure P and temperature T of air are known. Here, R is the gas constant. The disk area is equal to $A = \pi D^2/4$, where D is the blade diameter. Substituting these into Eq. (6-4), we obtain

$$\dot{W}_{available} = \frac{\pi}{8}\frac{PD^2 V^3}{RT} \tag{6-5}$$

Therefore, the power potential of a wind turbine is proportional to the square of the blade diameter. As a result, doubling blade diameter increases the power potential by a factor of 4.

Wind velocity, m/s	Available power, W
1	0.5
2	4
3	13.5
4	32
5	62.5
6	108
7	172
8	256
9	365
10	500

Figure 6-9 The power potential of a wind turbine is proportional to the cubic power of the wind velocity. Therefore, doubling the wind velocity will increase the power potential by a factor of 8.

Increasing the blade diameter requires nonconventional manufacturing techniques to make the blades strong enough to withstand extreme level of forces in high wind speeds. Also, cost effectiveness of using longer blades need to be evaluated. A major technology trend for wind turbines is designing higher power capacity turbines with large-diameter blades.

EXAMPLE 6-1
Wind Power
Potential in
a Location

A wind turbine with a blade diameter of 90 ft is to be installed in a location where average wind velocity is 20 ft/s (Fig. 6-10). The average temperature and pressure of ambient air in this location are 75°F and 14.5 psia, respectively. Determine the wind power potential.

SOLUTION The gas constant of air is $R = 0.3704$ psia·ft³/lbm·R. The density of air is determined from the ideal gas relation to be

$$\rho = \frac{P}{RT} = \frac{14.5 \text{ psia}}{(0.3704 \text{ psia·ft}^3/\text{lbm·R})[(75+460) \text{ R}]} = 0.07317 \text{ lbm/ft}^3$$

The blade span area is

$$A = \frac{\pi D^2}{4} = \frac{\pi(90 \text{ ft})^2}{4} = 6362 \text{ ft}^2$$

The mass flow rate is

$$\dot{m} = \rho AV = (0.07317 \text{ lbm/ft}^3)(6362 \text{ ft}^2)(20 \text{ ft/s}) = 9310 \text{ kg/s}$$

Figure 6-10 A wind turbine considered in Example 6-1.

Then the wind power potential is

$$\dot{W}_{\text{available}} = \frac{1}{2}\dot{m}V^2 = \frac{1}{2}(9310 \text{ lbm/s})(20 \text{ ft/s})^2\left(\frac{1 \text{ lbf}}{32.174 \text{ lbm·ft/s}^2}\right)\left(\frac{1 \text{ kW}}{737.56 \text{ lbf·ft/s}}\right) = \mathbf{78.5 \text{ kW}}$$

Alternatively, the wind power potential can be determined directly from Eq. (6-5) to be

$$\dot{W}_{\text{available}} = \frac{\pi}{8}\frac{PD^2V^3}{RT}$$

$$= \frac{\pi}{8}\frac{(14.5 \text{ psia})(90 \text{ ft})^2(20 \text{ ft/s})^3}{(0.3704 \text{ psia·ft}^3/\text{lbm·R})[(75+460) \text{ R}]}\left(\frac{1 \text{ lbf}}{32.174 \text{ lbm·ft/s}^2}\right)\left(\frac{1 \text{ kW}}{737.56 \text{ lbf·ft/s}}\right)$$

$$= 78.5 \text{ kW}$$

which is the same result, as expected. ▲

6-5 WIND POWER DENSITY

For comparison of various wind turbines and locations, it is more useful to think in terms of the available wind power *per unit area*, which we call the *wind power density* (WPD), typically in units of W/m²,

Wind power density: $$\qquad \text{WPD} = \frac{\dot{W}_{\text{available}}}{A} = \frac{1}{2}\rho V^3 \qquad\qquad (6\text{-}6)$$

Equation (6-6) is an instantaneous equation. As we all know, however, wind speed varies greatly throughout the day and throughout the year. For this reason, it is useful to define the *average wind power density* in terms of annual average wind speed $\overline{V}$, based on hourly averages as

Average wind power density: $$\quad \text{WPD}_{\text{avg}} = \frac{\overline{\dot{W}}_{\text{available}}}{A} = \frac{1}{2}\rho\overline{V}^3 \qquad\qquad (6\text{-}7)$$

The average WPD should be calculated based on hourly wind speed averages for the entire year. As a general rule of thumb, a location is considered poor for construction of wind turbines if the average WPD is less than about 100 W/m², good if it is around 400 W/m², and great if it is greater than about 700 W/m² (Fig. 6-11). Note that a WPD of 100 W/m² corresponds to a wind speed of 5.5 m/s for an air density of 1.2 kg/m³. Other factors affect the choice of a wind turbine site, such as atmospheric turbulence intensity, terrain, obstacles (buildings, trees, etc.), environmental impact, etc. See Manwell et al. (2010) for further details.

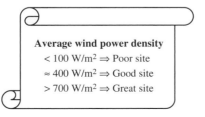

Average wind power density
< 100 W/m² ⇒ Poor site
≈ 400 W/m² ⇒ Good site
> 700 W/m² ⇒ Great site

Figure 6-11 Rule of thumb criteria for construction of wind turbines in a proposed site.

EXAMPLE 6-2
Comparison of Wind Power Potentials Based on Wind Power Densities

Consider two locations, location A and location B, with average wind power densities of 250 W/m² and 500 W/m², respectively. Determine the average wind speed in each location. Take the density of air to be 1.18 kg/m³. If a turbine with a diameter of 40 m is to be installed in location A and a turbine with a diameter of 20 m is to be installed in location B, what is the ratio of wind power potentials in location A to location B?

SOLUTION   The relation for WPD is

$$\text{WPD}_{avg} = \frac{1}{2}\rho \bar{V}^3$$

Solving for wind velocity at the two locations, we obtain

$$\text{WPD}_{avg,A} = \frac{1}{2}\rho \bar{V}_A^3 \longrightarrow 250 \text{ W/m}^2 = \frac{1}{2}(1.18 \text{ kg/m}^3)\bar{V}_A^3 \longrightarrow \bar{V}_A = 7.51 \text{ m/s}$$

$$\text{WPD}_{avg,B} = \frac{1}{2}\rho \bar{V}_B^3 \longrightarrow 500 \text{ W/m}^2 = \frac{1}{2}(1.18 \text{ kg/m}^3)\bar{V}_B^3 \longrightarrow \bar{V}_B = 9.46 \text{ m/s}$$

The WPD is given by

$$\text{WPD} = \frac{\dot{W}_{available}}{A}$$

and the blade span area is given by

$$A = \frac{\pi D^2}{4}$$

Therefore, wind power potential $\dot{W}_{available}$ is proportional to both the WPD and the blade span area A. Since the blade span area is proportional to the square of the diameter, the wind power potential is proportional to the product of WPD and the square of the blade diameter. Therefore,

$$\frac{\dot{W}_{available,A}}{\dot{W}_{available,B}} = \frac{\text{WPD}_A}{\text{WPD}_B}\frac{D_A^2}{D_B^2} = \left(\frac{250 \text{ W/m}^2}{500 \text{ W/m}^2}\right)\frac{(40 \text{ m})^2}{(20 \text{ m})^2} = 2$$

That is, the wind power potential in location A is twice the power potential in location B. ▲

6-6 WIND TURBINE EFFICIENCY

An actual wind turbine can produce only a percentage of available power potential into actual shaft power. This percentage is called the *wind turbine efficiency* and is determined from

$$\eta_{wt} = \frac{\dot{W}_{shaft}}{\dot{W}_{available}} = \frac{\dot{W}_{shaft}}{\frac{1}{2}\rho A V^3} \tag{6-8}$$

Here, $\dot{W}_{shaft}$ refers to rotor shaft power output. A gearbox/generator connected to the turbine converts shaft power into electrical power output $\dot{W}_{electric}$, and they are related to each other by

$$\dot{W}_{electric} = \eta_{gearbox/generator}\dot{W}_{shaft} \quad (kW) \tag{6-9}$$

where $\eta_{gearbox/generator}$ is the gearbox/generator efficiency and is typically above 80 percent.

We may also define an overall wind turbine efficiency as the electrical power output divided by the available wind power as

$$\eta_{wt,overall} = \frac{\dot{W}_{electric}}{\dot{W}_{available}} = \frac{\dot{W}_{electric}}{\frac{1}{2}\rho A V^3} \tag{6-10}$$

Note that a given wind turbine efficiency sometimes refers to overall wind turbine efficiency, and the context usually makes it clear. The overall wind turbine efficiency is related to wind turbine efficiency by

$$\eta_{\text{wt,overall}} = \eta_{\text{wt}}\eta_{\text{gearbox/generator}} = \left(\frac{\dot{W}_{\text{shaft}}}{\dot{W}_{\text{available}}}\right)\left(\frac{\dot{W}_{\text{electric}}}{\dot{W}_{\text{shaft}}}\right) = \frac{\dot{W}_{\text{electric}}}{\dot{W}_{\text{available}}} \qquad (6\text{-}11)$$

The efficiency of a wind turbine is usually referred to as power coefficient C_p. Using the wind turbine efficiency, the actual shaft power output from a wind turbine can be expressed as

$$\dot{W}_{\text{shaft}} = \frac{1}{2}\eta_{\text{wt}}\rho A V^3 \qquad (\text{kW}) \qquad (6\text{-}12)$$

The efficiency of wind turbines usually ranges between 30 and 40 percent.

If we neglect frictional effects in a wind turbine and take the wind velocity as the average velocity of air at the turbine inlet, we can state that the portion of incoming kinetic energy not converted to electric power leaves the wind turbine as outgoing kinetic energy (Fig. 6-12). That is,

$$\dot{m}\frac{V_1^2}{2} = \dot{W}_{\text{shaft}} + \dot{m}\frac{V_2^2}{2} \qquad (6\text{-}13)$$

The wind turbine efficiency is defined as

$$\eta_{\text{wt}} = \frac{\dot{W}_{\text{shaft}}}{\dot{m}\dfrac{V_1^2}{2}} \qquad (6\text{-}14)$$

Combining Eqs. (6-13) and (6-14) yields

$$\dot{m}\frac{V_2^2}{2} = \dot{m}\frac{V_1^2}{2}(1-\eta_{\text{wt}}) \qquad (6\text{-}15)$$

Solving for the exit velocity, we obtain

$$V_2 = V_1\sqrt{1-\eta_{\text{wt}}} \qquad (6\text{-}16)$$

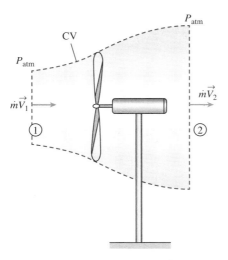

Figure 6-12 The flow of air across a wind turbine.

This relation enables us to determine the exit velocity when the turbine efficiency is known and the frictional effects are neglected.

Betz Limit for Wind Turbine Efficiency

A wind turbine converts the kinetic energy of air into work. This conversion is perfect (the wind turbine efficiency is 100 percent) under ideal conditions based on the second law of thermodynamics. Equation (6-13) shows that this would be the case only when the velocity of air at the turbine exit is zero. This is not possible for operational reasons because air must be taken away at the turbine exit to maintain the mass flow through the turbine. It turns out that there is a maximum possible efficiency for a wind turbine. This was first calculated by Albert Betz (1885–1968) in the mid-1920s.

We consider two control volumes surrounding the disk area—a large control volume and a small control volume—as sketched in Fig. 6-13, with upstream wind speed V taken as V_1.

The axisymmetric stream tube (enclosed by streamlines as drawn on the top and bottom of Fig. 6-13) can be thought of as forming an imaginary "duct" for the flow of air through the turbine. Since locations 1 and 2 are sufficiently far from the turbine, we take $P_1 = P_2 = P_{atm}$, yielding no net pressure force on the control volume. We approximate the velocities at the inlet (1) and outlet (2) to be uniform at V_1 and V_2, respectively, and the momentum flux correction factors to be unity. The momentum equation for this simplified case is written as

$$F_R = \dot{m}(V_2 - V_1) \tag{6-17}$$

where F_R is the reaction force on the turbine.

The smaller control volume in Fig. 6-13 encloses the turbine, but $A_3 = A_4 = A$, since this control volume is infinitesimally thin in the limit (we approximate the turbine as a disk). Since the air is considered to be incompressible, $V_3 = V_4$. However, the wind turbine extracts energy from the air, causing a pressure drop. Thus, $P_3 \neq P_4$. When we apply the streamwise component of the control volume momentum equation to the small control volume, we get

$$F_R + P_3 A - P_4 A = 0 \quad \rightarrow \quad F_R = (P_4 - P_3)A \tag{6-18}$$

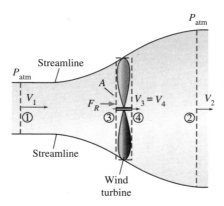

Figure 6-13   The large and small control volumes for analysis of ideal wind turbine performance bounded by an axisymmetric diverging stream tube.

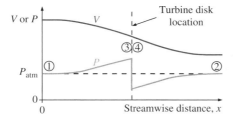

Figure 6-14 Qualitative sketch of average streamwise velocity and pressure profiles through a wind turbine.

The Bernoulli equation is certainly *not* applicable across the turbine, since it is extracting energy from the air. However, it is a reasonable approximation between locations 1 and 3 and between locations 4 and 2:

$$\frac{P_1}{\rho g} + \frac{V_1^2}{2g} + z_1 = \frac{P_3}{\rho g} + \frac{V_3^2}{2g} + z_3 \quad \text{and} \quad \frac{P_4}{\rho g} + \frac{V_4^2}{2g} + z_4 = \frac{P_2}{\rho g} + \frac{V_2^2}{2g} + z_2 \quad (6\text{-}19)$$

In this ideal analysis, the pressure starts at atmospheric pressure far upstream ($P_1 = P_{atm}$), rises smoothly from P_1 to P_3, drops suddenly from P_3 to P_4 across the turbine disk, and then rises smoothly from P_4 to P_2, ending at atmospheric pressure far downstream ($P_2 = P_{atm}$) (Fig. 6-14). We add Eqs. (6-17) and (6-18), setting $P_1 = P_2 = P_{atm}$ and $V_3 = V_4$. In addition, since the wind turbine is horizontally inclined, $z_1 = z_2 = z_3 = z_4$ (gravitational effects are negligible in air anyway). After some algebra, this yields

$$\frac{V_1^2 - V_2^2}{2} = \frac{P_3 - P_4}{\rho} \quad (6\text{-}20)$$

Substituting $\dot{m} = \rho V_3 A$ into Eq. (6-17) and then combining the result with Eqs. (6-18) and (6-20) yields

$$V_3 = \frac{V_1 + V_2}{2} \quad (6\text{-}21)$$

Thus, we conclude that *the average velocity of the air through an ideal wind turbine is the arithmetic average of the far upstream and far downstream velocities.* Of course, the validity of this result is limited by the applicability of the Bernoulli equation.

For convenience, we define a new variable a as the fractional loss of velocity from far upstream to the turbine disk as

$$a = \frac{V_1 - V_3}{V_1} \quad (6\text{-}22)$$

The velocity through the turbine thus becomes $V_3 = V_1(1-a)$, and the mass flow rate through the turbine becomes $\dot{m} = \rho V_3 A = \rho A V_1(1-a)$. Combining this expression for V_3 with Eq. (6-21) yields

$$V_2 = V_1(1-2a) \quad (6\text{-}23)$$

For an ideal wind turbine without irreversible losses such as friction, the power generated by the turbine is simply the difference between the incoming and outgoing kinetic energies. Performing some algebra, we get

$$\dot{W}_{ideal} = \dot{m}\frac{V_1^2 - V_2^2}{2} = \rho A V_1(1-a)\frac{V_1^2 - V_1^2(1-2a)^2}{2} = 2\rho A V_1^3 a(1-a)^2 \quad (6\text{-}24)$$

Again assuming no irreversible losses in transferring power from the turbine to the turbine shaft, the efficiency of the wind turbine is expressed as

$$\eta_{wt} = \frac{\dot{W}_{shaft}}{\frac{1}{2}\rho V_1^3 A} = \frac{\dot{W}_{ideal}}{\frac{1}{2}\rho V_1^3 A} = \frac{2\rho A V_1^3 a(1-a)^2}{\frac{1}{2}\rho V_1^3 A} = 4a(1-a)^2 \tag{6-25}$$

Finally, as any good engineer knows, we calculate the maximum possible value of η_{wt} by setting $d\eta_{wt}/da = 0$ and solving for a. This yields $a = 1$ or ⅓, and the details are left as an exercise. Since $a = 1$ is the trivial case (no power generated), we conclude that a must equal ⅓ for the maximum possible power coefficient. Substituting $a = ⅓$ into Eq. (6-25) gives

$$\eta_{wt,max} = 4a(1-a)^2 = 4\frac{1}{3}\left(1-\frac{1}{3}\right)^2 = \frac{16}{27} = 0.5926 \tag{6-26}$$

This value of $\eta_{wt,max}$ represents the *maximum possible efficiency of any wind turbine* and is known as the *Betz limit*. All real wind turbines have a maximum achievable efficiency less than this due to irreversible losses, which have been ignored in this ideal analysis.

Figure 6-15 shows wind turbine efficiency η_{wt} as a function of the ratio of turbine blade tip speed $V_{tip} = \omega R$ to wind speed V for several types of wind turbines, where ω is the angular velocity of the wind turbine blades and R is their radius. The angular velocity of rotating machinery is typically expressed in rpm (number of revolutions per minute) and denoted by $\dot{n}$. Noting that velocity is distance traveled per unit time and the angular distance traveled during each revolution is 2π, the angular velocity of a wind turbine is $\omega = 2\pi\dot{n}$ rad/min or $\omega = 2\pi\dot{n}/60$ rad/s.

The ratio of blade tip speed to wind speed is a design parameter for wind turbines. It varies between about 3 and 7 for a high-speed horizontal axis wind turbine. For safe operation, the blade tip speed should be below 90 m/s or 324 km/h.

From the wind turbine efficiency plot (Fig. 6-15), we see that an ideal propeller-type wind turbine approaches the Betz limit as $\omega R/V$ approaches infinity. However, the

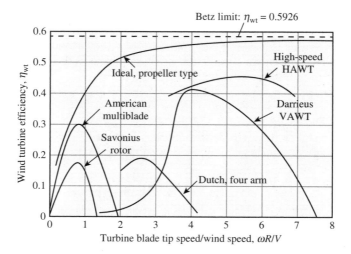

Figure 6-15 Wind turbine efficiency of various types of wind turbines as a function of the ratio of turbine blade tip speed to wind speed. So far, no design has achieved better performance than the horizontal axis wind turbine (HAWT). (*Adapted from Robinson, 1981.*)

efficiency of real wind turbines reaches a maximum at some *finite* value of $\omega R/V$ and then drops beyond that. In practice, three primary effects lead to a maximum achievable wind turbine efficiency that is lower than the Betz limit:

- Rotation of the wake behind the rotor (swirl)
- Finite number of rotor blades and their associated tip losses (tip vortices are generated in the wake of rotor blades for the same reason they are generated on finite airplane wings since both produce "lift")
- Nonzero aerodynamic drag on the rotor blades (frictional drag as well as induced drag)

See Manwell et al. (2010) for further discussion about how to account for these losses.

In addition, mechanical losses due to shaft friction lead to even lower maximum achievable power coefficients. Other mechanical and electrical losses in the gearbox, generator, etc., also reduce the overall wind turbine efficiency. As seen in Fig. 6-15, the "best" wind turbine is the high-speed HAWT, and that is why you see this type of wind turbine being installed throughout the world.

EXAMPLE 6-3
Efficiency of a
Wind Turbine

A wind turbine with a 50-m-diameter rotor is rotating at 25 rpm under steady winds at an average velocity of 10 m/s (Fig. 6-16). The electrical power output from the turbine is 375 kW. The combined efficiency of the gearbox/generator is 90 percent. Taking the density of air to be 1.20 kg/m³, determine (*a*) the wind turbine efficiency, (*b*) the tip speed of the blade, in km/h, and (*c*) the air velocity at the turbine exit if the turbine operated ideally at the Betz limit.

Figure 6-16   A wind turbine considered in Example 6-3.

SOLUTION   (*a*) Noting that the combined efficiency of the gearbox/generator is 90 percent, the shaft power output is determined as

$$\dot{W}_{\text{shaft}} = \frac{\dot{W}_{\text{electric}}}{\eta_{\text{gearbox/generator}}} = \frac{375 \text{ kW}}{0.90} = 416.7 \text{ kW}$$

The blade span area is

$$A = \pi D^2/4 = \pi(50 \text{ m})^2/4 = 1963 \text{ m}^2$$

The wind turbine efficiency is determined from its definition to be

$$\eta_{\text{wt}} = \frac{\dot{W}_{\text{shaft}}}{\frac{1}{2}\rho A V_1^3} = \frac{416.7 \text{ kW}}{\frac{1}{2}(1.20 \text{ kg/m}^3)(1963 \text{ m}^2)(10 \text{ m/s})^3 \left(\frac{1 \text{ kJ/kg}}{1000 \text{ m}^2/\text{s}^2}\right)} = 0.354 \text{ or } \mathbf{35.4\%}$$

(*b*) Noting that the tip of the blade travels a distance of πD per revolution, the tip velocity of the turbine blade for a 25 rpm speed becomes

$$V_{\text{tip}} = \pi D \dot{n} = \pi(50 \text{ m})(25/\text{min})\left(\frac{1 \text{ min}}{60 \text{ s}}\right) = 65.45 \text{ m/s} = \mathbf{236 \text{ km/h}}$$

(*c*) If the turbine is operated ideally at the Betz limit, its efficiency will be 0.5926. Then the air velocity at the turbine exit for this ideal operation could be determined from Eq. (6-16) to be

$$V_{2,\text{ideal}} = V_1\sqrt{1 - \eta_{\text{wt,max}}} = (10 \text{ m/s})\sqrt{1 - 0.5926} = \mathbf{6.38 \text{ m/s}}$$

The air velocity at the turbine exit would be 8.04 m/s if we used the actual wind turbine efficiency of 35.4 percent as calculated in part (*a*).   ▲

EXAMPLE 6-4
Electricity and Revenue
Generation by a
Wind Turbine

A wind turbine with a blade diameter of 55 m is to be installed in a location where average wind velocity is 6.5 m/s (Fig. 6-17). If the overall efficiency of the turbine is 38 percent, determine (*a*) the average electric power output, (*b*) the amount of electricity produced from this turbine for annual operating hours of 7500 h, and (*c*) the revenue generated if the electricity is sold at a price of $0.11/kWh. Take the density of air to be 1.25 kg/m³.

SOLUTION   (*a*) The blade span area is

$$A = \frac{\pi D^2}{4} = \frac{\pi(55 \text{ m})^2}{4} = 2376 \text{ m}^2$$

The wind power potential is

$$\dot{W}_{\text{available}} = \frac{1}{2}\rho A V^3 = \frac{1}{2}(1.25 \text{ kg/m}^3)(2376 \text{ m}^2)(6.5 \text{ m/s})^3\left(\frac{1 \text{ kJ/kg}}{1000 \text{ m}^2/\text{s}^2}\right) = 407.8 \text{ kW}$$

The electric power generated is

$$\dot{W}_{\text{electric}} = \eta_{\text{wt, overall}} \dot{W}_{\text{available}} = (0.38)(407.8 \text{ kW}) = \mathbf{155.0 \text{ kW}}$$

(*b*) The amount of electricity produced is determined from

$$W_{\text{electric}} = \dot{W}_{\text{electric}} \times \text{Operating hours} = (155.0 \text{ kW})(7500 \text{ h}) = \mathbf{1,162,500 \text{ kWh}}$$

(*c*) The revenue generated is

$$\text{Revenue} = W_{\text{electric}} \times \text{Unit price of electricity} = (1,162,500 \text{ kWh})(\$0.11/\text{kWh}) = \mathbf{\$127,900}$$ ▲

Figure 6-17 A wind turbine considered in Example 6-4. (©*Image Source/ Getty Images RF*)

6-7 ELECTRICITY PRODUCTION FROM WIND TURBINES

Wind resource maps have been developed for many locations in the world. A map at 100-m above the surface level for the United States is shown in Fig. 6-18. The maps are also available at other heights, which are useful because the wind speed and thus the power output increases by height. The highest values of wind speeds are available offshore along the east and west coasts. High-speed winds are also available in the middle parts of the country covering a very large area.

Wind maps are useful but long-term measurements of the wind distribution, characteristics, and speed should be obtained before selecting a site for a wind farm development. The wind speed should be sufficiently high for reasonable periods of time in an entire year. One can estimate the annual electricity generation when such data are available.

The probability of a given event may be estimated using a probability distribution function. The normal probability distribution has a bell-shaped curve, and it is successfully used to describe natural events. In this distribution, the vertical axis shows the frequency of the event, and the horizontal axis shows the magnitude of the event.

The probability of a given wind speed over a period may be estimated using the normal distribution. In the bell-shaped curve, the event may be taken as the wind speed and the frequency is the number of hours in a year. The probability of a certain wind speed V between wind speeds of V_1 and V_2 is equal to the area under the bell-shaped curve. That is,

$$P = \int_{V_1}^{V_2} f(V)\,dV \tag{6-27}$$

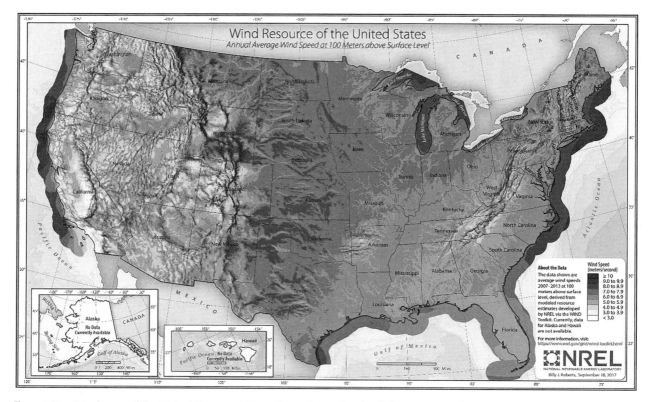

Figure 6-18   Wind map of the United States at 100-m above the surface level. (*Department of Energy, National Renewable Energy Laboratory, Washington, D.C.*)

The probability of all wind speeds is equal to 100 percent:

$$P = \int_0^\infty f(V)\,dV = 1 \tag{6-28}$$

The mean wind speed is

$$\overline{V} = \int_0^\infty V f(V)\,dV \tag{6-29}$$

or

$$\overline{V} = \frac{V_1 + V_2 + V_3 + \cdots\cdots + V_N}{N} = \frac{\sum_{i=1}^{N} V_i}{N} \tag{6-30}$$

where N is the number of wind speed data. The standard deviation is

$$\sigma = \sqrt{\int_0^\infty (V - \overline{V})^2\, f(V)\,dV} \tag{6-31}$$

or

$$\sigma = \sqrt{\frac{(V_1 - \overline{V})^2 + (V_2 - \overline{V})^2 + \cdots + (V_N - \overline{V})^2}{N}} = \sqrt{\frac{\sum_{i=1}^{N}(V_i - \overline{V})^2}{N}} \tag{6-32}$$

If a distribution is roughly bell shaped, then approximately 68 percent of the data will lie within one standard deviation of the mean wind speed (between $\bar{V} - 1\sigma$ and $\bar{V} + 1\sigma$), 95 percent of the data will lie within 2 standard deviations of the mean wind speed (between $\bar{V} - 2\sigma$ and $\bar{V} + 2\sigma$), and 99.7 percent of the data will lie within 3 standard deviations of the mean wind speed (between $\bar{V} - 3\sigma$ and $\bar{V} + 3\sigma$).

There are two common probability distributions used to describe wind speed distribution: Weibull and Rayleigh. The Weibull distribution function is given as

$$f(V) = \frac{k}{A}\left(\frac{V}{A}\right)^{k-1} \exp\left[-\left(\frac{V}{A}\right)^{k}\right] \tag{6-33}$$

where A is the Weibull scale parameter and k is the Weibull shape parameter. Both A and k are functions of the mean wind speed $\bar{V}$ and the standard deviation σ. The k value takes values between 1 and 3. A lower k value corresponds to highly variable winds while a higher value represents near constant winds. The Rayleigh distribution is a simplified version of the Weibull distribution for which $k = 2$ and $A = \bar{V}$. The Rayleigh distribution function and its cumulative distribution factor are expressed as

$$f(V) = \frac{\pi}{2}\left(\frac{V}{\bar{V}^2}\right)\exp\left[-\frac{\pi}{4}\left(\frac{V}{\bar{V}}\right)^2\right] \tag{6-34}$$

$$C(V) = 1 - \exp\left[-\frac{\pi}{4}\left(\frac{V}{\bar{V}}\right)^2\right] \tag{6-35}$$

The Rayleigh distribution is a function of the annual mean wind speed only. The Rayleigh wind speed distribution is given in Fig. 6-19 for three representative mean wind speeds. As the mean wind speed increases, there is a higher probability of higher wind speeds, and this corresponds to higher annual electricity production. For example, the probability of wind speed at 15 m/s at a mean speed of 6 m/s is 0.0048 or 0.48 percent. The probability

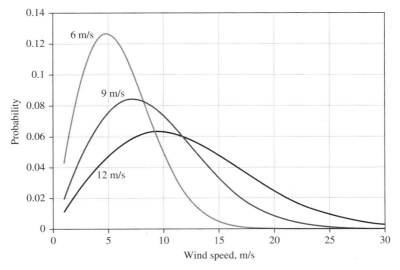

Figure 6-19 The Rayleigh wind speed distribution for three annual mean wind speeds of 6, 9, and 12 m/s.

of the same speed at a mean speed of 9 m/s is 3.3 percent and the probability at 12 m/s is 4.8 percent.

Suppose we need to find the probability of the wind speeds between V_1 and V_2 knowing that the wind data can reasonably be represented by the Rayleigh distribution. Using the above formulations, we find

$$P(V_1 \le V \le V_2) = \int_{V_1}^{V_2} f(V)\,dV = \int_0^{V_2} f(V)\,dV - \int_0^{V_1} f(V)\,dV = C(V_2) - C(V_1)$$

Now, suppose we need to find the probability of the wind speeds less than V_1. In this case, we find

$$P(V < V_1) = \int_0^{V_1} f(V)\,dV = C(V_1) - C(0) = C(V_1) - 0 = C(V_1)$$

Finally, suppose we need to find the probability of the wind speeds greater than V_2. In this case, we find

$$P(V > V_2) = \int_{V_2}^{\infty} f(V)\,dV = C(\infty) - C(V_2) = 1 - C(V_2)$$

We note that $C(0) = 0$ and $C(\infty) = 1$.

EXAMPLE 6-5
Rayleigh Wind Speed
Distribution

Wind speed measurements in a location are performed over a 1-year period. The mean speed is obtained to be 8 m/s. The Rayleigh wind speed distribution closely represents the measured data. Calculate the number of hours in a year with wind speeds less than 6 m/s, between 6 m/s and 9 m/s, and greater than 15 m/s.

SOLUTION   Using the Rayleigh wind speed distribution, first we determine the cumulative probability functions at the desired wind velocities:

$$C(V = 6 \text{ m/s}) = 1 - \exp\left[-\frac{\pi}{4}\left(\frac{V}{\overline{V}}\right)^2\right] = 1 - \exp\left[-\frac{\pi}{4}\left(\frac{6 \text{ m/s}}{8 \text{ m/s}}\right)^2\right] = 0.3571$$

$$C(V = 9 \text{ m/s}) = 1 - \exp\left[-\frac{\pi}{4}\left(\frac{V}{\overline{V}}\right)^2\right] = 1 - \exp\left[-\frac{\pi}{4}\left(\frac{9 \text{ m/s}}{8 \text{ m/s}}\right)^2\right] = 0.6299$$

$$C(V = 15 \text{ m/s}) = 1 - \exp\left[-\frac{\pi}{4}\left(\frac{15 \text{ m/s}}{8 \text{ m/s}}\right)^2\right] = 0.9368$$

The probability of wind speeds less than 6 m/s is determined to be

$$P(V < 6 \text{ m/s}) = C(6 \text{ m/s}) = 0.3571$$

The probability of wind speeds between 6 m/s and 9 m/s is

$$P(6 \text{ m/s} \le V \le 9 \text{ m/s}) = C(9 \text{ m/s}) - C(6 \text{ m/s}) = 0.6299 - 0.3571 = 0.2728$$

The probability of wind speeds above 15 m/s is

$$P(V > 25 \text{ m/s}) = 1 - C(25 \text{ m/s}) = 1 - 0.9368 = 0.0632$$

The total number of hours in a year is $365 \times 24 = 8760$ h. Then the number of hours in a year for these three cases are

$$\text{Number of hours}\,(V < 6\,\text{m/s}) = P(V < 6\,\text{m/s}) \times \text{Annual hours} = (0.3571)(8760\,\text{h}) = \mathbf{3128\,h}$$

$$\text{Number of hours}\,(6\,\text{m/s} \le V \le 9\,\text{m/s}) = P(6\,\text{m/s} \le V \le 9\,\text{m/s}) \times \text{Annual hours}$$
$$= (0.2728)(8760\,\text{h}) = \mathbf{2390\,h}$$

$$\text{Number of hours}\,(V > 15\,\text{m/s}) = P(V > 15\,\text{m/s}) \times \text{Annual hours} = (0.0632)(8760\,\text{h}) = \mathbf{554\,h} \quad \blacktriangle$$

EXAMPLE 6-6
Annual Electricity Production

Wind speed measurements in a location are performed over a 1-year period. The annual mean speed is obtained to be 10 m/s. The cut-in wind speed is 4 m/s, the cut-out wind speed is 22 m/s, and the rated wind speed is 14 m/s. The rotor diameter is 98 m. Take the air density to be 1.2 kg/m³. Use a constant overall wind turbine efficiency of 38 percent. Estimate the annual electricity production considering that the Rayleigh wind speed distribution closely represents the measured wind speed data. Assume that the power output remains constant between the rated speed and the cut-out speed. Also, determine the capacity factor of this wind turbine.

SOLUTION   We perform the calculations at a wind speed of 8 m/s. Using the Rayleigh wind speed distribution, we determine the probability of a wind speed of 8 m/s to be

$$f(8\,\text{m/s}) = \frac{\pi}{2}\left(\frac{V}{\overline{V}^2}\right)\exp\left[-\frac{\pi}{4}\left(\frac{V}{\overline{V}}\right)^2\right] = \frac{\pi}{2}\left[\frac{8\,\text{m/s}}{(10\,\text{m/s})^2}\right]\exp\left[-\frac{\pi}{4}\left(\frac{8\,\text{m/s}}{10\,\text{m/s}}\right)^2\right] = 0.07602$$

The total number of hours in a year is $365 \times 24 = 8760$ h. The operating hours at a wind speed of 8 m/s is

$$\text{Operating hours}\,(V = 8\,\text{m/s}) = (0.07602)(8760\,\text{h}) = 665.9\,\text{h}$$

The blade span area is

$$A = \frac{\pi D^2}{4} = \frac{\pi(98\,\text{m})^2}{4} = 7543\,\text{m}^2$$

The electrical power output is

$$\dot{W}_{\text{electric}} = \eta_{\text{wt,overall}}\dot{W}_{\text{available}}$$
$$= \eta_{\text{wt,overall}}\frac{1}{2}\rho A V^3 = (0.38)\frac{1}{2}(1.2\,\text{kg/m}^3)(7543\,\text{m}^2)(8\,\text{m/s})^3\left(\frac{1\,\text{kJ/kg}}{1000\,\text{m}^2/\text{s}^2}\right)$$
$$= 880.5\,\text{kW}$$

The amount of electricity produced is

$$W_{\text{electric}} = \dot{W}_{\text{electric}} \times \text{Operating hours} = (880.5\,\text{kW})(665.9\,\text{h}) = 586{,}355\,\text{kWh}$$

We repeat the calculations at other wind speeds and obtain the results listed in Table 6-1. The annual electricity production is calculated to be 17.1 million kWh.

$$W_{\text{electric,total}} = \mathbf{1.709 \times 10^7\,kWh}$$

TABLE 6-1   The Results Obtained in Example 6-6

Wind speed, m/s	Probability	Operating hours, h	Electrical power, kW	Electricity production, kWh
1	0.01559	136.5	1.72	—
2	0.03044	266.7	13.76	—
3	0.04391	384.6	46.43	—
4	0.05541	485.4	110.1	53,428
5	0.06454	565.4	215	121,537
6	0.07104	622.3	371.5	231,160
7	0.07483	655.5	589.9	386,685
8	0.07602	665.9	880.5	586,355
9	0.07483	655.5	1254	821,835
10	0.07162	627.4	1720	1,078,964
11	0.0668	585.2	2289	1,339,516
12	0.06083	532.9	2972	1,583,618
13	0.05415	474.4	3778	1,792,360
14	0.04717	413.2	4719	1,950,153
15	0.04025	352.6	4719	1,663,919
16	0.03365	294.8	4719	1,391,161
17	0.02759	241.7	4719	1,140,582
18	0.02219	194.4	4719	917,374
19	0.01752	153.5	4719	724,367
20	0.01358	118.9	4719	561,089
21	0.01033	90.49	4719	427,022
22	0.007721	67.63	4719	319,146
Total				17,090,270

The variations of electrical power and electricity production with wind speed are shown in Figs. 6-20 and 6-21, respectively. Note that there is no power generation below the cut-in speed (4 m/s). The power output is assumed to remain constant at the rated power of 4719 kW at wind speeds greater the rated speed of 14 m/s because the turbine is designed to operate at the loads corresponding to this maximum power. Higher power outputs may cause structural and operational problems. In an actual operation, the turbine power may be smaller than the rated power at wind speeds higher than 14 m/s.

The amount of electricity production decreases at wind speed values greater than 14 m/s while the power output remains constant. We note from Fig. 6-19 that the greatest amount of electricity production takes place at the rated speed even though the operating hours at this speed are not the highest. This shows that the design of this particular wind turbine for a rated speed of 14 m/s and the corresponding rated power of 4719 kW is correct based on the wind speed distribution and the power profile.

The *capacity factor* of a wind turbine is defined as the ratio of the actual annual power generation to the maximum power generation assuming that the turbine operates at the rated power throughout the year. If this wind turbine generated electricity at the rated power for an entire year of 8760 h, the amount of electricity production would be

$$W_{electric,max} = \dot{W}_{rated} \times \text{Annual hours} = (4719 \text{ kW})(8760 \text{ h}) = 4.134 \times 10^7 \text{ kWh}$$

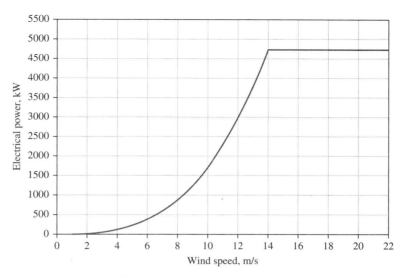

Figure 6-20 Variation of electrical power with wind speed in Example 6-6.

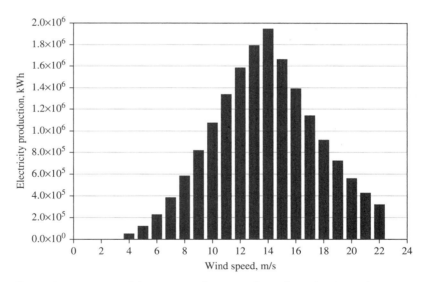

Figure 6-21 Variation of electricity production with wind speed in Example 6-6.

Then, the capacity factor of this wind turbine becomes

$$\text{Capacity factor} = \frac{W_{\text{electric,total}}}{W_{\text{electric,max}}} = \frac{1.709 \times 10^7 \ \text{kWh}}{4.1349 \times 10^7 \ \text{kWh}} = 0.413 = \mathbf{41.3\%}$$

The capacity factors of commercial wind turbines vary between 25 and 50 percent. Manufacturers of some very large wind turbines claim capacity factors up to 65 percent. ▲

6-8 WIND POWER COSTING

The cost estimate of a wind power project may be evaluated using the principles and methods of engineering economics presented in Chap. 12. Here, we present a simple method presented in Fingersh et al. (2006) for the estimation of the annual levelized cost of wind energy in dollars per kWh of electricity produced. The method considers a three-blade, upwind horizontal axis vertical wind turbine that is pitch controlled and has a variable rotor speed. It is applicable to both land-based and offshore wind projects.

The average levelized cost of energy (LCOE) is estimated from the following relation:

$$\text{LCOE} = \frac{\text{FCR} \times \text{ICC}}{\text{AEP}} + \text{AOE} \qquad (\$/\text{kWh}) \qquad (6\text{-}36)$$

Here, AOE stands for annual operating expenses in \$/kWh, and it is estimated from

$$\text{AOE} = \frac{\text{O\&M} + \text{LRC} + \text{LLC}}{\text{AEP}} \qquad (\$/\text{kWh}) \qquad (6\text{-}37)$$

Substituting into Eq. (6-36) gives

$$\text{LCOE} = \frac{(\text{FCR} \times \text{ICC}) + \text{O\&M} + \text{LRC} + \text{LLC}}{\text{AEP}} \qquad (\$/\text{kWh}) \qquad (6\text{-}38)$$

Various terms in this relation are described next.

ICC (initial capital cost): It includes turbine cost and various other costs associated with a wind farm project, named *balance of station*. The costs due to balance of station include foundation, support structure, transportation, roads, civil work, assembly, installation, electrical work (connections, cables, transformers), permits, and labor. The initial capital cost is expressed in \$. The turbine cost is usually expressed in dollars per kW of rated power (\$/kW). The balance of station is a single cost value for the given wind farm.

A breakdown of the capital cost of a wind turbine is given in Fig. 6-22. The largest cost is due to the gearbox, brake, and electrical system, with a contribution of 61 percent. Pitch, hub, and blade account for 27 percent of total capital cost. This is followed by the tower system with 12 percent (Yao et al., 2011).

For offshore turbines, there are additional initial cost items such as handling of marine environment, port, staging, and personal equipment, scour protection, and offshore insurance.

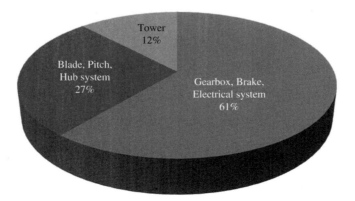

Figure 6-22 Capital cost breakdown of a wind turbine. (*Adapted from Yao et al., 2011.*)

FCR (fixed charge rate): This is used to convert the total ICC to annual payments considering the entire lifetime of the project. It includes the effects of interest rate, the return on dept and equity, and other fixed payments. It is expressed as percentage and enters Eq. (6-37) as a fraction.

AEP (annual energy production): This is the actual net annual electricity production expected from the installation. Its estimation is based on the capacity factor (actual amount of electricity produced/maximum possible electricity production) and the rated power. The annual energy production is calculated from

$$AEP = (Capacity\ factor)(Rated\ power)(Annual\ hours) \qquad (kWh/yr) \qquad (6\text{-}39)$$

The number of annual hours is taken as

$$Annual\ hours = 365 \times 24 = 8760\ h$$

O&M (operation and maintenance): This cost includes parts and supplies for turbine and facilities maintenance and equipment and labor for operation, administration, and support. It is expressed on an annual basis ($/yr).

LRC (levelized replacement cost): This cost may be due to scheduled or unscheduled replacements and overhaul. It is expressed on an annual basis over the entire life of the project ($/yr).

LLC (land lease cost): This is due to renting or leasing the land for the wind power installation. It is expressed in $/yr. If the land is purchased, the LLC is zero. In this case, the purchase price may be included in the initial capital cost.

In general, the estimated average levelized cost of energy is affected by many factors including variation in wind speed characteristics, regulatory and market structure, local financing parameters, technology status, market maturity, and estimation method (DOE, 2021).

EXAMPLE 6-7
Cost of Wind Energy

A wind farm consists of five wind turbines each with a power rating of 3.2 MW. The capacity factor is 45 percent. Various costs associated with this wind farm are as follows:

Cost of turbines = $1300/kW

Balance of station = $8,200,000

Fixed charge rate (FCR) = 12 percent

Levelized replacement costs (LRC) = $400,000/yr

Operation and Maintenance (O&M) = $340,000/yr

Land lease cost (LLC) = $63,000/yr

(*a*) Estimate the annual levelized cost of energy (ACOE).
(*b*) If the electricity is sold at a price of $0.11/kWh, calculate the annual profit.

SOLUTION   (*a*) The total power rating of the turbines and the total initial capital cost are

$$Power\ rating = (5)(3200\ kW) = 16{,}000\ kW$$

$$\begin{aligned} ICC &= (Turbine\ cost)(Power\ rating) + Balance\ of\ station \\ &= (\$1300/kW)(16{,}000\ kW) + \$8{,}200{,}000 \\ &= \$29{,}000{,}000 \end{aligned}$$

The net annual energy production is

$$AEP = (\text{Capacity factor})(\text{Rated power})(\text{Annual hours})$$
$$= (0.45)(16{,}000 \text{ kW})(8760 \text{ h/yr})$$
$$= 63{,}070{,}000 \text{ kWh/yr}$$

Then, the average levelized cost of energy is determined from

$$LCOE = \frac{(FCR \times ICC) + O\&M + LRC + LLC}{AEP}$$

$$= \frac{(0.12/\text{yr})(\$29{,}000{,}000) + \$340{,}000/\text{yr} + \$400{,}000/\text{yr} + \$63{,}000/\text{yr}}{63{,}070{,}000 \text{ kWh/yr}}$$

$$= \mathbf{\$0.0679/kWh}$$

That is, the cost of energy from this farm installation is 6.8 cents per kWh of electricity produced.

(*b*) The annual profit is determined from

$$\text{Annual profit} = (AEP)(\text{Unit price} - LCOE)$$
$$= (63{,}070{,}000 \text{ kWh/yr})[(0.11 - 0.0679)\$/\text{kWh}]$$
$$= \mathbf{\$2{,}655{,}000/yr} \quad \blacktriangle$$

6-9 CONSIDERATIONS IN WIND POWER APPLICATIONS

The first wind turbine for electricity generation was developed at the end of 19th century. Two important technological developments took place after 1940; the introduction of the three-blade structure and the replacement of DC generators with AC generators (Bansal et al., 2002). These were essential for stimulating the emergence of wind turbines. The technology for wind turbines has improved greatly over the years, but there is room for more improvements which will help promote more widespread use of this sustainable method of power generation.

Wind Farm

When a wind turbine project is underway on a windy site, many turbines are installed, and such a site is properly called a *wind farm* or a *wind park* (Fig. 6-23). The use of a wind farm is highly desirable due to reduced site development costs, simplified transmission lines, and centralized access for operation and maintenance. Single-use turbines are used for off-grid homes, offshore areas, and demonstration projects.

The wind farm usually covers an area of limited size. Turbines behave similarly in terms of their individual power output because the wind speed through each turbine is about the same. As a result, the total power output from a wind farm fluctuates as the wind speed changes. This fluctuation could be minimized if multiple wind farms are interconnected through the electric transmission grid. In this case, the group of wind farms behaves like a single wind farm due to near constant wind speed and, consequently, near constant power output.

There are several factors that need to be considered when a wind farm project is planned on a particular site. Ground conditions such as vegetation, topography, and ground roughness should be analyzed. Convenient access to the farm site, load capacity of the soil, and earthquake characteristics are some other considerations (Yao et al., 2011). Another important factor is to make sure that the planned wind farm site is not a flyway for birds.

Figure 6-23   A wind farm.

The number of wind turbines in a given site depends on the spacing between the turbines. If the turbines are spaced too close to each other, the flow through one turbine affects the flow through the next turbine, and this reduces turbine performance. If the turbines are too far from each other, the potential for the installation of additional turbines for greater power output is not realized. It turns out that there is an optimum spacing between the turbines, and it is estimated to be 3 to 5 blade diameters between the turbines in a row and 5 to 9 blade diameters between rows, as shown in Fig. 6-24 (Patel, 1999; Masters, 2004).

Offshore Wind Turbines

Installation of offshore wind turbines has been considered more expensive than land-based turbines. However, the cost has been reduced significantly due to technological improvements, increased competition, experience from large-scale installations, larger turbines, and supply chain optimization (DOE, 2021). The average levelized cost of energy for offshore wind power decreased by 28 to 51 percent between 2014 and 2020 (Wiser et al. 2021). As a result, the installations of offshore wind power have accelerated globally since 2015 (Fig. 6-25).

Offshore wind turbines have significant advantages including higher wind speed, better wind characteristics (less transient, less turbulent), and higher capacity factors (more annual power production). They also involve easier and less-costly access to electric transmission lines. This is because offshore wind installations are usually close to populated and industrial centers while land-based turbines are usually in relatively remote locations. Offshore wind turbines can be built with large diameter blades due to higher available wind speeds, and therefore the largest wind turbines in the world are offshore. Leading wind turbine manufacturers have been developing large offshore wind turbines with capacities between 12 and 15 MW.

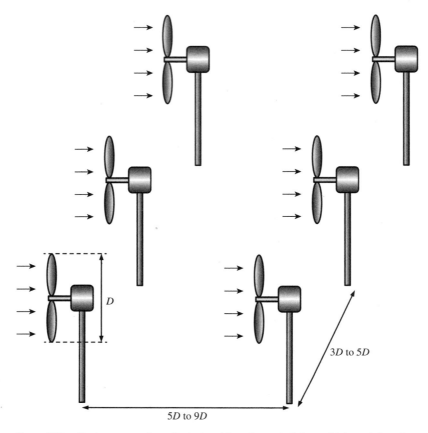

Figure 6-24 Optimum spacing of wind turbines in a wind farm. (*Adapted from Yao et al., 2011.*)

Figure 6-25 An offshore wind power installation. (*NREL, 2022a.*)

As shown in the wind map of the United States (Fig. 6-18), the highest values of wind speeds are available offshore along the east and west coasts. Over 400 GW offshore wind power potential is estimated in the United States considering water depths of less than 30 m. The estimate becomes nearly 1000 GW when water depths up to 60 m are considered. There has been a steady increase in offshore wind installations. For example, 5.5 GW capacity was added in 2020, mostly in China, the Netherlands, and other European countries. In Denmark, an offshore wind farm (Horns Rev 2 project) consists of 91 turbines with a total capacity of 209 MW (Wind Europe, 2018).

Most of the existing offshore wind turbines are installed in water depths of less than 30 m and within 20 km of the shoreline. These are fixed-bottom, foundation-type turbine installations. For water depths up to 60 m, the fixed-bottom turbines are the standard choice, and their contribution to overall offshore installations have been increasing. If the water depth is more than 60 m, floating offshore wind turbines are installed. It is estimated that about 60 percent of the offshore wind power potential in the United States is due to water depths greater than 60 m, while this percentage is 80 percent in Europe. Several floating offshore wind projects have been installed globally and many more are in construction or planning stages (DOE, 2021).

Tower Height

The wind tower should be strong enough to support the turbine's weight and withstand the forces of high wind speeds and the thrust on the wind turbine. Using tall wind towers minimizes the turbulence induced and allows flexibility in siting. Power output from the wind turbine changes with time of the day. For example, in early morning hours, the power output is 80 percent of the average power and in the afternoon, it is 120 percent of the average power. Such fluctuations can be reduced by using taller towers (Yao et al., 2011).

An important reason for using tall wind towers is that wind speed increases with tower height, and higher wind velocities translate into higher power outputs. The wind speed can be expressed as a function of tower height and the roughness of the earth's surface to be

$$\frac{V}{V_0} = \left(\frac{h}{h_0}\right)^\alpha \tag{6-40}$$

where V is the wind speed at height h, V_0 is the nominal wind speed at h_0, and α is the friction coefficient of the ground. Typical values of the friction coefficient are given in Table 6-2. Since the wind power is proportional to the third power of wind speed, a slight increase in wind speed translates into a significant increase in power output.

EXAMPLE 6-8
Wind Speed at the Top
of a Wind Tower

The wind speed at a height of 20 m in a location with smooth, hard ground is measured to be 4 m/s. Determine the wind speed at a tower height of 100 m. What will the wind speed be if the ground location is a small town with trees?

SOLUTION Using the friction coefficient value of 0.10 from Table 6-2, the wind speed at a wind tower height of 100 m in a location with smooth, hard ground is determined to be

$$\frac{V}{V_0} = \left(\frac{h}{h_0}\right)^\alpha \quad \longrightarrow \quad \frac{V}{4 \text{ m/s}} = \left(\frac{100 \text{ m}}{20 \text{ m}}\right)^{0.10} \quad \longrightarrow \quad V = \textbf{4.70 m/s}$$

TABLE 6-2 **Friction Coefficient for Various Ground Conditions** (Masters, 2004)

Ground Conditions	Friction Coefficient, α
Smooth hard ground, calm water	0.10
Tall grass on ground	0.15
High crops and hedges	0.20
Wooded countryside, many trees	0.30
Small town with trees	0.30
Large city with tall buildings	0.40

The wind speed increases by $4.70/4 = 1.175$ times from 20 to 100 m height. Since the power is proportional to third power of wind speed, the power potential increases by $(1.175)^3 = 1.62$ times.

If the location of the wind tower is a small town with trees, the friction coefficient becomes 0.30, and the wind speed would be

$$\frac{V}{V_0} = \left(\frac{h}{h_0}\right)^{\alpha} \longrightarrow \frac{V}{4 \text{ m/s}} = \left(\frac{100 \text{ m}}{20 \text{ m}}\right)^{0.30} \longrightarrow V = \textbf{6.48 m/s}$$

In this case, the wind speed increases by $6.48/4 = 1.62$ times, and the power potential increases by $(1.62)^3 = 4.25$ times. ▲

Blade Diameter and Generator Size

Selection of the turbine blade diameter and the power rating or the size of generator is important to maximize power output from the turbine. When the blade diameter is increased at the same generator rating, the power curve will move left, as shown in Fig. 6-26a. Then the rated power can be realized at a lower wind speed. When a larger generator is used, the rated power increases (Fig. 6-26b). This is particularly advantageous at high wind speeds. Therefore, the generator size can be increased to maximize turbine power output if the turbine mostly operates at high wind speeds (Masters, 2004).

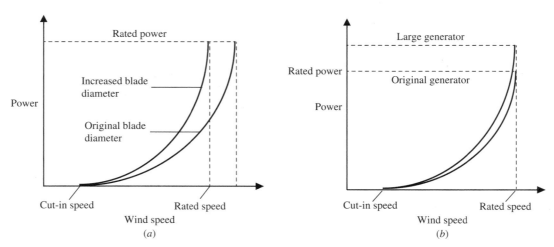

Figure 6-26 Effects of increasing blade diameter and a larger generator on power output of the wind turbine. (*Adapted from Masters, 2004.*)

For effective wind power generation, a reliable power grid/transmission network near the site and good grid stability are essential. The wind turbine generates power at 400 V, which is stepped up to 11 to 110 kV. Poor grid stability may cause 10 to 20 percent power loss (Bansal et al., 2002). In China, some wind turbines are not connected to the grid partly because of stability problems.

Control Schemes

Two major control schemes are used in wind turbines. In a simpler system, the blade pitch is varied while the rotor speed is held constant. Note that blade pitch refers to the angle of the blade chord to the plane of rotation. In a more complicated control system, both the rotor speed and the blade pitch are varied. This allows the wind turbine to operate at the highest efficiency level at all wind speeds. In the simpler control scheme, the turbine operates at the maximum efficiency at the rated speed only. As a result, more power could be generated.

Additional annual electricity generation due to the second type of control system should be weighed against the associated cost and complexities of the system. An analysis shows 15 percent more electricity generation when both rotor speed and blade pitch are varied with respect to varying only the blade pitch. The turbine reaches the rated power at a wind speed 13.5 m/s in the first more complicated control scheme but at 15 m/s in the second scheme (Kreith and Kreider, 2011).

Hybrid Power Systems

For remote places on the earth, there is sometimes no connection to the electrical grid. For such places, isolated power systems can be used. Diesel engine power plants are commonly used for such autonomous power systems since diesel systems can be put in operation in a relatively short time and the units are relatively small compared to other fossil fuel based–systems such as gas turbines and steam turbines. A *hybrid power system* consisting of a diesel system and a wind turbine system can be operated together in order to minimize diesel fuel consumption. In this operation, wind power should have priority in meeting the electrical needs of users, and the diesel engines should be operated to meet the demand when wind power is not sufficient. A hybrid system for an autonomous power system can integrate a diesel system with one or more renewable energy systems such as wind, photovoltaic, or a hydropower system.

REFERENCES

Bansal RC, Bhatti TS, and Kothari DP. 2002. "On Some of the Design Aspects of Wind Energy Conversion Systems." *Energy Conversion and Management*, 43: 2175–2187.

Çengel YA and Cimbala JM. 2018. *Fluid Mechanics: Fundamentals and Applications*, 4th ed. New York: McGraw Hill.

Department of Energy (DOE). 2021. *Offshore Wind Market Report: 2021 Edition*. Washington, D.C.

Department of Energy (DOE). 2022. "How a Wind Turbine Works." https://www.energy.gov/eere/wind/how-wind-turbine-works-text-version. Access date: Jan. 19, 2022.

Fingersh L, Hand M, and Laxson A. 2006. *Wind Turbine Design Cost and Scaling Model. Technical report*. NREL/TP-500-40566.

Kreith F and Kreider JF. 2011. *Principles of Sustainable Energy*. New York: Taylor & Francis.

Manwell JF, McGowan JG, and Rogers AL. 2010. *Wind Energy Explained—Theory, Design, and Application*, 2nd ed. West Sussex, England: John Wiley & Sons.

Masters GM. 2004. *Renewable and Efficient Electrical Power Systems*. New Jersey: John Wiley & Sons, Inc.

NREL. 2022a. "Offshore Wind Research." https://www.nrel.gov/wind/offshore-wind.html. Access date: Feb. 4, 2022.

NREL. 2022b. "Wind Resource Maps and Data." https://www.nrel.gov/gis/wind-resource-maps.html. Access date: Jan. 19, 2022.

Patel MR. 1999. *Wind and Solar Power Systems*. Boca Raton: CRC Press.

Robinson ML. 1981. "The Darrieus Wind Turbine for Electrical Power Generation." *Journal of Royal Aeronautical Society*, 85: 244–255.

Wind Europe. 2018. http://www.windeurope.org.

Wiser, R, J Rand, J Seel, P Beiter, E Baker, E Lantz, and P Gilman. 2021. "Expert Elicitation Survey Predicts 37% to 49% Declines in Wind Energy Costs by 2050." *Nature Energy*, 6: 555–565.

Yao F, Bansal RC, Dong ZY, Saket RK, and Shakya JS. 2011. "Wind Energy Resources: Theory, Design and Applications." In *Handbook of Renewable Energy Technology*, edited by AF Zobaa and RC Bansal, Singapore: World Scientific Publishing Co. Pte. Ltd.

PROBLEMS

WIND TURBINE TYPES AND POWER PERFORMANCE CURVE

6-1   What is the difference between a windmill and a wind turbine?

6-2   Can electricity be generated from wind turbines at all wind speeds? Is there a minimum wind speed for economical wind power generation? If so, what is it?

6-3   Classify wind turbines by the orientation of their axis of rotation and by the mechanism that provides torque to the rotating shaft. Which classes are normally used today and why?

6-4   What are the three key locations on the wind-speed scale? Explain each one briefly.

6-5   The minimum wind speed at which useful power can be generated is called
(*a*) Cut-in speed   (*b*) Cut-out speed   (*c*) Norm speed   (*d*) Rated speed   (*e*) Betz speed

6-6   The maximum wind speed at which power can be generated is called
(*a*) Cut-in speed   (*b*) Cut-out speed   (*c*) Norm speed   (*d*) Rated speed   (*e*) Betz speed

6-7   The most common design of wind turbine _____ type and _____ axis.
(*a*) Lift, horizontal   (*b*) Lift, vertical   (*c*) Drag, horizontal   (*d*) Drag, vertical

WIND TURBINE OPERATION AND AERODYNAMICS

6-8   What is a direct-drive turbine?

6-9   What is an upwind turbine and a downwind turbine? Which design is more common?

6-10   What is the purpose and advantage of the gearbox in a wind turbine?

6-11   Which components are included in the drivetrain of a wind turbine?

6-12   Describe how the blade rotation is stopped in a wind turbine when the wind speed is greater than the cut-out speed.

6-13   Describe how the lift force develops when air flows toward the blades.

6-14   Why are the three-bladed horizontal axis upwind turbines superior to other types of wind turbines?

6-15   The tower of a wind turbine is usually made of _____ and the blades are made of _____.
(*a*) Steel, copper   (*b*) Concrete, aluminum   (*c*) Fiberglass, aluminum
(*d*) Steel, fiberglass   (*e*) Steel, aluminum

6-16   In a wind farm, the _____ is used to increase the voltage.
(*a*) Generator   (*b*) Anemometer   (*c*) Transformer   (*d*) Drivetrain   (*e*) Nacelle

6-17   Which statement is incorrect regarding the components of wind turbines?
(*a*) The anemometer measures the wind speed and sends this data to the controller.
(*b*) Main shaft bearing supports the low-speed shaft and reduces friction between moving parts.
(*c*) Modern wind turbines have three blades.

(*d*) The angle of the blades with respect to wind is adjusted by a pitch system that controls the rotor speed.

(*e*) Upwind turbines may be designed without a yaw drive.

6-18   Lift force is perpendicular to the _____ speed.
(*a*) Wind   (*b*) Resultant   (*c*) Radial   (*d*) Rotational   (*e*) Tip

6-19   Which statement is incorrect regarding the aerodynamics of wind turbines?
(*a*) The angle of attack is smaller at the blade root compared to blade tip.
(*b*) Maximizing the lift-to-drag ratio is a main design objective for turbine blades
(*c*) The blade geometry varies along the length of the blade.
(*d*) The lift force is perpendicular to the direction of the drag force.
(*e*) The drag force is parallel to the resultant speed.

WIND POWER POTENTIAL

6-20   What does the mechanical energy of a flowing fluid consist of? Express each term in an equation and explain. Which part of mechanical energy is used in wind power conversion?

6-21   Consider a location where the wind is blowing at a speed of V. How do you calculate maximum power potential of a wind turbine at this location if the blade span area of the turbine is A and the density of air is ρ?

6-22   Consider two locations, both at the same altitude with the same wind speed, but the air temperature at one location is higher than the other. Which location has more wind power potential? Why?

6-23   Consider two locations with the same wind speed and ambient air temperature, but one location is at a higher altitude than the other. Which location has more wind power potential? Why?

6-24   A wind turbine with a blade diameter of 50 m is to be installed in a location where average wind velocity is 7.5 m/s. The average temperature and pressure of ambient air in this location are 23°C and 96 kPa, respectively. Determine the wind power potential.

6-25   Consider two locations (location A and B) with average wind velocity of 9 m/s for location A and 6 m/s for location B. If two wind turbines with the same characteristics are to be installed at both sites, what is the ratio of power generated from location A and location B? Assume the same air density at both locations.

6-26   Which statement is incorrect regarding wind turbines?
(*a*) Flow energy does not change across a wind turbine.
(*b*) Potential energy does not change across a wind turbine.
(*c*) A commonly used wind turbine type is lift-type HAWT.
(*d*) The power output of a wind turbine is constant.
(*e*) The rotation speed of rotors of wind turbines is usually under 40 rpm.

6-27   Which statement is incorrect regarding wind turbines?
(*a*) Cut-in speed for a wind turbine is zero.
(*b*) Electrical power output of a wind turbine at rated speed is greater than that at cut-out speed.
(*c*) The power performance curve of a wind turbine is influenced by the gearbox and generator.
(*d*) The design of HAWT turbine blades includes tapering and twist to maximize performance.
(*e*) A wind turbine produces maximum power at rated speed.

6-28   Which statement is incorrect regarding wind turbines?
(*a*) Power potential of a wind turbine is proportional to cube of velocity.
(*b*) Power potential of a wind turbine is proportional to air pressure and temperature.
(*c*) Power potential of a wind turbine is proportional to square of blade diameter.
(*d*) None of these

6-29   The power potential of a wind turbine is proportional to the _____ power of the wind velocity.

(*a*) First   (*b*) Square   (*c*) Cubic   (*d*) Fourth   (*e*) Fifth

6-30   The power potential of a wind turbine at a wind speed of 5 m/s is 50 kW. The power potential of the same turbine at a velocity of 8 m/s is

(*a*) 80 kW   (*b*) 128 kW   (*c*) 180 kW   (*d*) 205 kW   (*e*) 242 kW

6-31   Wind is blowing through a turbine at a velocity of 7 m/s. The turbine blade diameter is 25 m and the density of air is 1.15 kg/m³. The power potential of the turbine is

(*a*) 180 kW   (*b*) 153 kW   (*c*) 131 kW   (*d*) 116 kW   (*e*) 97 kW

6-32   The power potential of a wind turbine at a wind speed of 5 m/s is 100 kW. The blade diameter of this turbine is 40 m. The power potential of a similar turbine with a blade diameter of 60 m at the same velocity is

(*a*) 150 kW   (*b*) 225 kW   (*c*) 266 kW   (*d*) 338 kW   (*e*) 390 kW

6-33   The available power from wind turbine with a rotor area of 1300 m² is 568 kW. The density of air is 1.16 kg/m³. The wind speed is

(*a*) 9.1 m/s   (*b*) 7.8 m/s   (*c*) 7.3 m/s   (*d*) 6.9 m/s   (*e*) 5.7 m/s

6-34   The measurements over an entire week period indicates that a wind turbine with a blade diameter of 40 m has produced 11,000 kWh of electricity. The average rate of electricity production is

(*a*) 32.1 kW   (*b*) 50.7 kW   (*c*) 65.5 kW   (*d*) 81.2 kW   (*e*) 99.0 kW

WIND POWER DENSITY

6-35   Define WPD (wind power density) and average WPD. Which is more useful?

6-36   There are three sites with the following average wind power densities: site A: 75 W/m², site B: 500 W/m², and site C: 800 W/m². Do you recommend a wind turbine installation for all three sites? Explain.

6-37   Consider two locations (location A and B) with average WPD (wind power density) of 500 W/m² for location A and 250 W/m² for location B. If two wind turbines with the same characteristics are to be installed at both sites, what is the ratio of power generated from location A and location B? Also, determine the ratio of average wind velocities in location A and B.

6-38   Consider a location with an average WPD (wind power density) of 200 W/m². What is the corresponding average wind speed? Another location has a WPD of 400 W/m². What is the average wind speed for this location? Take the density of air to be 1.18 kg/m³.

6-39   The WPD (wind power density) in a location is estimated to be 600 W/m². The average wind speed for an air density of 1.2 kg/m³ is

(*a*) 6.3 m/s   (*b*) 7.8 m/s   (*c*) 10.0 m/s   (*d*) 11.5 m/s   (*e*) 12.4 m/s

WIND TURBINE EFFICIENCY

6-40   How is the wind turbine efficiency η_{wt} defined? How is it different from the power coefficient C_p? What are the typical values of wind turbine efficiency?

6-41   How is the overall wind turbine efficiency defined? How is it related to the wind turbine efficiency?

6-42   What is the theoretical limit for wind turbine efficiency based on the second law of thermodynamics? Is this limit the same as the Betz limit? Why? Explain.

6-43   Three different inventors come up with three wind turbine designs with these claimed efficiencies: turbine A: 41 percent, turbine B: 59 percent, turbine C: 67 percent. How do you evaluate these claimed efficiencies? Explain.

6-44   A wind turbine with a blade diameter of 25 m is to be installed in a location where average wind velocity is 6 m/s. If the overall efficiency of the turbine is 34 percent, determine (*a*) the average electric power output, (*b*) the amount of electricity produced from this turbine for annual operating hours of 8000 h, and (*c*) the revenue generated if the electricity is sold at a price of $0.09/kWh. Take the density of air to be 1.3 kg/m³.

6-45   The measurements over an entire week indicate that a wind turbine with a blade diameter of 40 m has produced 11,000 kWh of electricity. If the overall efficiency of the wind turbine is estimated to be 28 percent, determine the average wind velocity during this period. Take the density of air to be 1.16 kg/m³.

6-46   The electric power produced from a wind turbine is measured to be 50 kW when the wind velocity is 5.5 m/s. What is the wind turbine efficiency if the gearbox/generator efficiency is 90 percent? The ambient conditions during the time of measurement are 7°C and 100 kPa, and the blade span area is 1500 m².

6-47   An investor is to install a total of 40 identical wind turbines in a location with an average wind speed of 7.2 m/s. The blade diameter of each turbine is 18 m, and the average overall wind turbine efficiency is 33 percent. The turbines are expected to operate under these average conditions 6000 h per year, and the electricity is to be sold to a local utility at a price of $0.075/kWh. If the total cost of this installation is $1,200,000, determine how long it will take for these turbines to pay for themselves. Take the density of air to be 1.18 kg/m³.

Figure P6-47   (@ *Bear Dancer Studios/Mark DierkerRF*)

6-48   The wind velocity in a location varies between 16 and 24 ft/s. A wind turbine with a blade span diameter of 185 ft is to be installed. The wind turbine efficiency is 30 percent at a wind velocity of 16 m/s and 35 percent at 24 ft/s. It is estimated that wind blows at 16 ft/s for 3000 h per year and at 24 ft/s for 4000 h per year. The gearbox/generator efficiency is 93 percent. Determine the amount of electricity that can be produced by this turbine in kWh. Take the density of air to be 0.075 lbm/ft³. Also, determine the blade tip speed in miles per hour (mph) when the blades rotate at 15 rpm.

6-49   The efficiency of a wind turbine is 40 percent when the wind speed is 7 m/s. What is the air velocity at the turbine exit if the frictional effects are neglected?

6-50   The air velocity is 25 ft/s at the inlet of a wind turbine and 19.5 ft/s at the exit. What is the wind turbine efficiency if the frictional effects are neglected?

6-51   A school currently pays $23,000 per year for the electricity it uses at a unit price of $0.11/kWh. The school management decides to install a wind turbine with a blade diameter of 20 m and an average overall efficiency of 30 percent in order to meet its entire electricity need. What is the required average velocity of wind in this location? Take the density of air to be 1.2 kg/m³, and assume the turbine operates at the required average speed 7500 h per year.

6-52   Reconsider Prob. 6-51. If the school management goes for a larger turbine with a blade diameter of 30 m, what is the required average velocity of wind?

6-53   A wind turbine is designed to operate at a maximum blade rotation of 20 rpm. What is the maximum allowable blade span diameter for safe operation? What is the ratio of blade tip speed to wind speed at this rotor diameter and a cut-out speed of 18 m/s?

6-54   Which statement is not correct regarding wind turbines?
(*a*)   The efficiency of wind turbines usually ranges between 30 and 40 percent.
(*b*)   The overall efficiency of a wind turbine is smaller than wind turbine efficiency.
(*c*)   In a wind turbine, the shaft power output is smaller than the electric power output.
(*d*)   For a wind turbine, the gearbox/generator efficiency is typically above 80 percent.
(*e*)   The maximum efficiency of a wind turbine is given by the Betz limit.

6-55   What is the tip speed of the blades of a wind turbine when the blades rotate at 23 rpm and the blade span diameter is 75 m?
(*a*) 54.2 km/h   (*b*) 90.3 km/h   (*c*) 138 km/h   (*d*) 222 km/h   (*e*) 325 km/h

6-56   Wind is blowing through a turbine at a velocity of 9 m/s. The turbine blade diameter is 35 m. The air is at 95 kPa and 20°C. If the power output from the turbine is 115 kW, the efficiency of the turbine is
(*a*) 29%   (*b*) 32%   (*c*) 35%   (*d*) 38%   (*e*) 42%

ELECTRICITY PRODUCTION FROM WIND TURBINES

6-57   Why are the wind resource maps for the United States obtained at various heights above the surface level?

6-58   How are the wind speed distributions over a year in a particular location used?

6-59   How is the capacity factor of a wind turbine defined? How do you compare typical capacity factor of wind turbines to solar, hydro and geothermal power systems?

6-60   Wind speed measurements in a location are performed over a 1-year period. The mean speed is 10 m/s and the Rayleigh wind speed distribution represents the measured data. Calculate the number of hours in a year with wind speeds less than 5 m/s, between 9 m/s and 11 m/s, and greater than 20 m/s.

6-61   Wind speed measurements in a location are performed over a 1-year period. The mean speed is calculated as 11.5 m/s. Assume the Rayleigh wind speed distribution represents the measured data. The cut-in and cut-out wind speeds for the turbine project are projected to be 4.5 m/s and 19.5 m/s, respectively. Determine the percent time in a year when the power could not be generated from the turbine? Also, calculate the number of hours in a year when the power could be generated.

6-62   Wind speed measurements are performed in a location over a 1-year period. The mean speed is calculated as 9 m/s. Estimate cut-in and cut-out speeds if it is desired that the percent time in a year when power could not be generated is 10 percent at the lower end of the wind distribution and 5 percent above the higher end. Assume the Rayleigh wind speed distribution represents the measured data.

6-63   Wind speed measurements in a location are performed over a 1-year period. The annual mean speed is obtained to be 8.5 m/s. The cut-in wind speed is 4 m/s, the cut-out wind speed is 18 m/s, and the rated wind speed is 11 m/s. The blade span diameter is 84 m. Take the air density to be 1.2 kg/m³.

Use a constant overall wind turbine efficiency of 36 percent. Assume that the power output remains constant between the rated speed and the cut-out speed and the Rayleigh wind speed distribution closely represents the measured wind speed data.

(a) Estimate the annual electricity production.

(b) Determine the capacity factor of this wind turbine.

(c) Determine the potential revenue if the electricity is to be sold at a price of $0.22/kWh.

6-64   Wind speed measurements in a location are performed over a 1-year period. The annual mean speed is 11 m/s. The cut-in wind speed is 5 m/s, the cut-out wind speed is 24 m/s, and the rated wind speed is 13 m/s. The blade span diameter is 110 m. Take the air density to be 1.2 kg/m³ and use a constant overall wind turbine efficiency of 40 percent. Assume that the power output remains constant between the rated speed and the cut-out speed and the Rayleigh wind speed distribution represents the measured wind speed data.

(a) Estimate the annual electricity production.

(b) Determine the percentages of annual hours when the wind speed is less than cut-in speed and when the wind speed is greater than cut-out speed.

(c) Determine the capacity factor of this wind turbine.

6-65   Wind speed measurements in a location are performed over a 1-year period. The annual mean speed is 7 m/s. The cut-in wind speed is 3 m/s, the cut-out wind speed is 16 m/s, and the rated wind speed is 9 m/s. Take the air density to be 1.2 kg/m³ and use a constant overall wind turbine efficiency of 40 percent. Assume that the power output remains constant between the rated speed and the cut-out speed and the Rayleigh wind speed distribution represents the measured wind speed data. Calculate the annual electricity production for a rotor diameter of (a) 50 m and (b) 100 m. (c) What is the ratio of electricity production from part b to part a.

6-66   Wind speed measurements in a location are performed over a 1-year period. The annual mean speed is 10 m/s. The cut-in wind speed is 4 m/s and the cut-out wind speed is 21 m/s. Assume that the power output remains constant between the rated speed and the cut-out speed and the Rayleigh wind speed distribution represents the measured wind speed data. Take the air density to be 1.2 kg/m³ and use a constant overall wind turbine efficiency of 40 percent.

One of the biggest wind turbines as of 2022 was an offshore turbine with a rotor diameter of 236 m and a capacity of 16 MW. Estimate the rated wind speed for the operation of this turbine and annual power production using the wind speed data in this specified location.

6-67   Which statement is not correct regarding wind speed, normal, and the Rayleigh distributions?

(a) The normal probability distribution has a bell-shaped curve.

(b) The normal probability distribution is successfully used to describe natural events.

(c) The Rayleigh distribution is a simplified version of the Weibull distribution.

(d) The Rayleigh distribution is functions of the annual mean wind speed and standard deviation.

(e) As the mean wind speed increases there is a higher probability of higher wind speeds.

6-68   Consider the Rayleigh wind speed distribution for a wind speed data. The cut-in wind speed is 5 m/s and a cut-out wind speed is 15 m/s. The cumulative distribution factors are $C(5 \text{ m/s}) = 0.10$ and $C(15 \text{ m/s}) = 0.93$. What is the probability of the wind speeds less than 5 m/s?

(a) 3%   (b) 7%   (c) 10%   (d) 83%   (e) 93%

6-69   Consider the Rayleigh wind speed distribution for a wind speed data. The cut-in wind speed is 5 m/s and a cut-out wind speed is 15 m/s. The cumulative distribution factors are $C(5 \text{ m/s}) = 0.10$ and $C(15 \text{ m/s}) = 0.93$. What is the probability of the wind speeds greater than 15 m/s?

(a) 0.03   (b) 0.07   (c) 0.10   (d) 0.83   (e) 0.93

6-70   Consider the Rayleigh wind speed distribution for a wind speed data. The cut-in wind speed is 5 m/s and a cut-out wind speed is 15 m/s. The cumulative distribution factors are $C(5 \text{ m/s}) = 0.10$ and $C(15 \text{ m/s}) = 0.93$. What is number of hours in a year with the wind speeds between 5 m/s and 15 m/s?

(a) 263 h   (b) 8147 h   (c) 876 h   (d) 6130 h   (e) 7271 h

6-71 Annual electricity generation from a wind turbine for a rotor diameter of 50 m is estimated using the Rayleigh wind speed distribution to be 2×10^6 kWh. What is the estimate of annual electricity generation for a rotor diameter of 100 m?
(a) 2×10^6 kWh   (b) 4×10^6 kWh   (c) 6×10^6 kWh   (d) 8×10^6 kWh
(e) 16×10^6 kWh

6-72 The capacity factor of a wind turbine is 50 percent. If the capacity of the turbine is 2.5 MW, what is annual electricity generation from this wind turbine?
(a) 1.095×10^7 kWh   (b) 2.190×10^7 kWh   (c) 2.50×10^7 kWh   (d) 4.386×10^7 kWh
(e) 4.875×10^7 kWh

WIND POWER COSTING

6-73 Which cost items are included in the initial capital cost of a wind turbine project?

6-74 Which cost items are included in the annual operating expenses of a wind turbine project?

6-75 Estimate average levelized cost of energy (LCOE) of a wind turbine with a power rating of 7.5 MW. The capacity factor is 37 percent. Various costs associated with this wind turbine are as follows:
Turbine cost = $1250/kW
Balance of station = $4,600,000
Fixed charge rate (FCR) = 10 percent
Levelized replacement costs (LRC) = $220,000/yr
Operation and Maintenance (O&M) = $165,000/yr
Land lease cost (LLC) = $28,000/yr

6-76 An offshore wind turbine has a power rating of 15 MW and a capacity factor of 48 percent. Various costs associated with this wind turbine are as follows:
Cost of turbine = $1600/kW
Balance of station = $19,000,000
Fixed charge rate (FCR) = 13 percent
Levelized replacement costs (LRC) = $570,000/yr
Operation and Maintenance (O&M) = $390,000/yr
Offshore land lease cost (LLC) = $73,000/yr
(a) Estimate average levelized cost of energy (ACOE).
(b) If the owner expects to make an annual profit of $2 million, what is the minimum unit price of selling the electricity produced from this turbine?

6-77 A wind farm consists of 10 wind turbines each with a rating of 2.5-MW. The capacity factor is 40 percent. Various costs associated with this wind turbine are as follows:
Cost of turbines = $1400/kW
Balance of station = $13,000,000
Fixed charge rate (FCR) = 15 percent
Levelized replacement costs (LRC) = $600,000/yr
Operation and Maintenance (O&M) = $500,000/yr
Land lease cost (LLC) = $130,000/yr
(a) Estimate average levelized cost of energy (LCOE).
(b) Calculate the annual profit if the electricity is sold to the local utility company at a unit price of $0.125/kWh.

6-78 A wind turbine is rated at 2.5 MW. The capacity factor is 35 percent. Various costs associated with this wind turbine are as follows:
Turbine cost = $3,625,000
Balance of station = $1,700,000

Fixed charge rate (FCR) = 8.5 percent

Levelized replacement costs (LRC) = $80,000/yr

Operation and Maintenance (O&M) = $75,000/yr

Land lease cost (LLC) = $16,000/yr

Estimate the average levelized cost of energy (ACOE) of this wind turbine. What fraction of this cost is due to initial capital cost?

6-79   Which cost has the greatest share in the capital cost of a wind turbine?

(*a*) Gearbox, brake, electrical system   (*b*) Blade, pitch, hub system   (*c*) Tower system

(*d*) Control system   (*e*) Safety system

6-80   Which parameter does not affect the calculation of annual energy production from a wind turbine?

(*a*) Power rating   (*b*) Turbine cost   (*c*) Capacity factor   (*d*) Number of hours per year

6-81   Which of the following units are appropriate for expressing the cost of a wind turbine?

I. $ II. $/kW   III. $/kWh

(*a*) Only I   (*b*) Only II   (*c*) I and II   (*d*) I and III   (*e*) I, II, and III

6-82   What is the appropriate unit for the average levelized cost of energy (LCOE)?

(*a*) $ (*b*) $/kW·yr   (*c*) $/kW   (*d*) $/yr   (*e*) $/kWh

CONSIDERATIONS IN WIND POWER APPLICATIONS

6-83   Describe the advantages of wind farms.

6-84   What is the best strategy on the spacing between the turbines in a wind farm, closely spaced or widely spaced? Explain.

6-85   What factors should be considered when a wind farm project is planned on a particular site?

6-86   What are the advantages of offshore wind turbines compared to land-based turbines?

6-87   What are the two common control schemes used in wind turbines?

6-88   Under what conditions is it a good idea to use a larger generator to maximize turbine power output?

6-89   Describe the proper operation of a hybrid power system to minimize fuel consumption.

6-90   What are the two important technological developments that stimulated the emergence of wind turbines?

6-91   The wind speed at a height of 50 ft in a location with smooth, hard ground is measured to be 15 ft/s. Determine the wind speed at a tower height of 200 ft.

6-92   The wind speed at a height of 20 m in a location with high crops is measured to be 4.5 m/s. The wind speed at the top of a tower in the same location is measured to be 6 m/s. What is the height of the tower?

6-93   The optimum spacing between the turbines in a row with a blade diameter D is

(*a*) $1D - 2D$ (*b*) $2D - 3D$ (*c*) $3D - 5D$ (*d*) $5D - 9D$ (*e*) $6D - 9D$

6-94   The optimum spacing between the rows of the turbines with a blade diameter D is

(*a*) $1D - 2D$ (*b*) $2D - 3D$ (*c*) $3D - 5D$ (*d*) $5D - 9D$ (*e*) $6D - 12D$

6-95   Which is not an advantage of a wind farm?

(*a*) Simplified transmission lines

(*b*) Centralized access for operation and maintenance

(*c*) Reduced site development costs

(*d*) Increased power generation potential

(*e*) Close proximity to city center

6-96 Which is not a primary consideration in the selection of the site for wind power development?
(a) Tower height and blade diameter
(b) Reliable power grid/transmission network near the site
(c) Earthquake characteristics
(d) Load capacity of the soil
(e) Convenient access to the wind farm site

6-97 Which is not an advantage of offshore wind turbines compared to land-based turbines?
(a) Easy access to electric transmission lines
(b) Steadier winds with less turbulence
(c) Higher capacity factor
(d) Less expensive to install
(e) Larger turbines

6-98 The wind speed at a height of 30 m in a location with tall grass on ground is measured to be 5 m/s. Determine the wind speed at a tower height of 100 m. The friction factor of the ground is 0.15.
(a) 5 m/s   (b) 5.15 m/s   (c) 5.75 m/s   (d) 5.99 m/s   (e) 7.5 m/s

CHAPTER 7

Hydropower

7-1 INTRODUCTION

Turbines have been used for centuries to convert freely available mechanical energy from rivers and water bodies into useful mechanical work, usually through a rotating shaft. The rotating part of a hydro turbine is called the *runner*. When the working fluid is water, the turbomachines are called *hydraulic turbines* or *hydro turbines*. Large dams are built in the flow path of rivers to collect water. The water, having potential energy, is run through turbines to produce electricity. Such an installation is called a *hydroelectric power plant*. Some dams are also used for irrigation of farms and flood control. A large dam takes a long time and a large investment to build, but the cost of producing electricity by hydropower is much lower than the cost of electricity production by fossil fuels.

The first hydraulic turbine-generator set was developed in 1882. It used a 10-m-high water reservoir and provided 6 kW of power to light bulbs. This was followed by the first known hydroelectric power plant in 1883 in Northern Ireland. This plant had two turbines with a power rating of 39 kW. An electric train was powered by this plant. In the same year, two generators powered by waterwheels were installed in Diamantina, Brazil. The water source was 5 m high, and 12 kW of power was produced (Farret and Simoes, 2006). A good review of the history of hydropower can be found in Lewis et al. (2014).

Most large hydroelectric power plants have *several* turbines arranged in parallel. This offers the power company the opportunity to turn off some of the turbines during times of low power demand and for maintenance. Hoover Dam in Boulder City, Nevada, for example, has 17 parallel turbines, 15 of which are identical large Francis turbines that can produce approximately 130 MW of electricity each (Fig. 7-1). The maximum gross head is 590 ft (180 m). The total peak power production of the power plant exceeds 2000 MW, while about 4 billion kWh of electricity is produced every year.

Building a hydroelectric power plant with an accumulation reservoir is expensive and takes a long time to build. It also involves complex hydrological and topographical studies for a careful assessment of site and plant characteristics. The size of the plant, including the dam, is large, and environmental impacts of the plant need serious consideration and studies. Unlike run-of-river plants, a plant with a dam can use most of the collected water for power production. It should be noted that the water reservoir of a hydroelectric power plant is also used for other purposes such as irrigation, fish farming, and recreation. For some hydroelectric installations, irrigation is more important than or as important as power production.

The powerhouse of a hydropower plant has hydraulic turbines and generators. A regulator, a water admission valve, an electrical command board, and an inertial flywheel are

(a)

(b)

Figure 7-1 (a) An aerial view of Hoover Dam. (© *Corbis RF.*) (b) The top (visible) portion of several of the parallel electric generators driven by hydraulic turbines at Hoover Dam. (© *Brand X Pictures RF.*)

among the supplemental equipment. A hydroelectric power plant project involves considerable civil works. This may include the construction of a dam, a water intake system for diversion of the river from its normal course, a spillway to return water to its normal course, piping, a balance chimney, and a machine house. There is less construction involved in river plants. As a result, river plants can be built in a relatively short time; they occupy less space; they have fewer environmental impacts; and they involve lower initial cost compared to large hydropower plants.

In hydroelectric power plants, large dynamic turbines are used to produce electricity. There are two basic types of dynamic turbine—*impulse* and *reaction*. Comparing the two power-producing dynamic turbines, impulse turbines require a higher head but can operate with a smaller volume flow rate. Reaction turbines can operate with much less head but require a higher volume flow rate. Before discussing characteristics of impulse and reaction turbines, we provide a general analysis of a hydroelectric power plant.

7-2 ANALYSIS OF A HYDROELECTRIC POWER PLANT

Mechanical energy can be defined as *the form of energy that can be converted to mechanical work completely and directly by an ideal mechanical device such as an ideal turbine.* The mechanical energy of a flowing fluid can be expressed on a unit mass basis as

$$e_{\text{mech}} = \frac{P}{\rho} + \frac{V^2}{2} + gz \tag{7-1}$$

where P/ρ is the flow energy, $V^2/2$ is the kinetic energy, and gz is the potential energy of the fluid, all per unit mass. Then the mechanical energy change of a fluid during incompressible flow becomes

$$\Delta e_{\text{mech}} = \frac{P_2 - P_1}{\rho} + \frac{V_2^2 - V_1^2}{2} + g(z_2 - z_1) \tag{7-2}$$

Therefore, the mechanical energy of a fluid does not change during flow if its pressure, density, velocity, and elevation remain constant. In the absence of any irreversible losses, the mechanical energy change represents the mechanical work supplied to the fluid (if $\Delta e_{\text{mech}} > 0$) or extracted from the fluid (if $\Delta e_{\text{mech}} < 0$). The maximum (ideal) power generated by a turbine, for example, is

$$\dot{W}_{\text{max}} = \dot{m}\Delta e_{\text{mech}}$$

where $\dot{m}$ is the mass flow rate of the fluid, as shown in Fig. 7-2. If we take point 1 as the surface of the water reservoir and point 4 as the surface of discharge water from the turbine (Fig. 7-2a), the maximum power that can be generated by the turbine is

$$\dot{W}_{\text{max}} = \dot{m}\Delta e_{\text{mech}} = \dot{m}g(z_1 - z_4) = \dot{m}gh \tag{7-3}$$

since
$$P_1 \approx P_4 = P_{\text{atm}} \quad \text{and} \quad V_1 = V_4 \approx 0$$

If we take point 2 as the turbine inlet and point 3 as the turbine outlet (Fig. 7-2a), the maximum power that can be generated by the turbine is

$$\dot{W}_{\text{max}} = \dot{m}\Delta e_{\text{mech}} = \dot{m}\frac{P_2 - P_3}{\rho} = \dot{m}\frac{\Delta P}{\rho} \tag{7-4}$$

since
$$V_2 \approx V_3 \quad \text{and} \quad z_2 \approx z_3$$

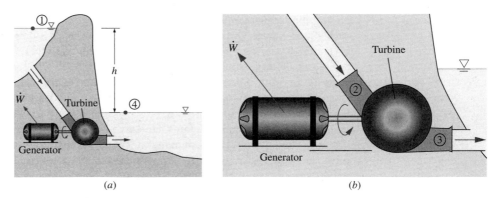

Figure 7-2   Mechanical energy is illustrated by an ideal hydraulic turbine coupled with an ideal generator. In the absence of irreversible losses, the maximum produced power is proportional to (*a*) the change in water surface elevation from the upstream to the downstream reservoir or (*b*) (close-up view) the drop in water pressure from just upstream to just downstream of the turbine.

In fluid systems, we are usually interested in increasing the pressure, velocity, and/or elevation of a fluid. This is done by *supplying mechanical energy* to the fluid by a pump, a fan, or a compressor. Or we are interested in the reverse process of *extracting mechanical energy* from a fluid by a turbine and producing mechanical power in the form of a rotating shaft that can drive a generator or any other rotary device. The degree of perfection of the conversion process between the mechanical work extracted and the mechanical energy change of the fluid is expressed by the *turbine efficiency*. In rate form, it is defined as

$$\eta_{\text{turbine}} = \frac{\dot{W}_{\text{shaft}}}{\Delta\dot{E}_{\text{mech,fluid}}} = \frac{\dot{W}_{\text{shaft}}}{\dot{m}\Delta e_{\text{mech}}} = \frac{\dot{W}_{\text{shaft}}}{\dot{m}gh} = \frac{\dot{W}_{\text{shaft}}}{\dot{W}_{\text{max}}} \tag{7-5}$$

where $\dot{W}_{\text{shaft}}$ is the shaft power output from the turbine and $\dot{m}\Delta e_{\text{mech}}$ is rate of decrease in the mechanical energy of the fluid, which is equal to maximum power $\dot{W}_{\text{max}} = \dot{m}gh$, based on the notation in Fig. 7-2*a*. We use the positive value for the mechanical energy change to avoid negative values for efficiencies. A turbine efficiency of 100 percent indicates perfect conversion between the shaft work and the mechanical energy of the fluid, and this value can be approached (but never attained) as the frictional effects are minimized.

The mechanical efficiency of a turbine should not be confused with the *generator efficiency*, which is defined as

$$\eta_{\text{generator}} = \frac{\dot{W}_{\text{electric}}}{\dot{W}_{\text{shaft}}} \tag{7-6}$$

where $\dot{W}_{\text{electric}}$ is the electrical power output from the generator. A turbine is usually packaged together with its generator. Therefore, we are usually interested in the *combined* or *overall efficiency* of a turbine-generator combination (Fig. 7-3), which is defined as

$$\eta_{\text{turbine-generator}} = \eta_{\text{turbine}}\eta_{\text{generator}} = \left(\frac{\dot{W}_{\text{shaft}}}{\dot{W}_{\text{max}}}\right)\left(\frac{\dot{W}_{\text{electric}}}{\dot{W}_{\text{shaft}}}\right) = \frac{\dot{W}_{\text{electric}}}{\dot{W}_{\text{max}}} \tag{7-7}$$

Most turbines have efficiencies approaching 90 percent. Large hydro turbines achieve overall efficiencies above 95 percent.

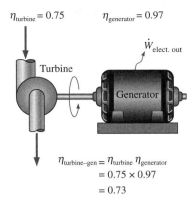

$\eta_{\text{turbine}} = 0.75$ $\eta_{\text{generator}} = 0.97$

$\dot{W}_{\text{elect. out}}$

Turbine

Generator

$\eta_{\text{turbine–gen}} = \eta_{\text{turbine}}\,\eta_{\text{generator}}$
$= 0.75 \times 0.97$
$= 0.73$

Figure 7-3 The overall efficiency of a turbine-generator is the product of the efficiency of the turbine and the efficiency of the generator, and represents the fraction of the mechanical power of the fluid converted to electrical power.

The analysis of a hydroelectric power plant involves that of the turbine and the *penstock*, which is the piping system between the upper and lower water levels, as shown in Fig. 7-4. The steady-flow energy equation on a unit-mass basis can be written for this penstock/turbine combination as

$$e_{\text{mech, in}} = e_{\text{mech, out}} + e_{\text{mech, loss}} \tag{7-8}$$

or

$$\frac{P_1}{\rho_1} + \frac{V_1^2}{2} + gz_1 = \frac{P_2}{\rho_2} + \frac{V_2^2}{2} + gz_2 + w_{\text{turbine}} + e_{\text{mech, loss}} \tag{7-9}$$

where w_{turbine} is the mechanical work output (due to a turbine). When the flow is incompressible, either absolute or gage pressure can be used for P since P_{atm}/ρ would appear on both sides and would cancel out. Multiplying Eq. (7-9) by the mass flow rate $\dot{m}$ gives

$$\dot{m}\left(\frac{P_1}{\rho_1} + \frac{V_1^2}{2} + gz_1\right) = \dot{m}\left(\frac{P_2}{\rho_2} + \frac{V_2^2}{2} + gz_2\right) + \dot{W}_{\text{turbine}} + \dot{E}_{\text{mech, loss, total}} \tag{7-10}$$

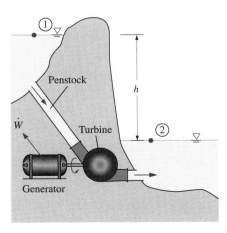

① ▽

Penstock

h

$\dot{W}$

Turbine

② ▽

Generator

Figure 7-4 The analysis of a hydroelectric power plant involves that of the turbine and the penstock.

where $\dot{W}_{\text{turbine}}$ is the shaft power output through the turbine's shaft, and $\dot{E}_{\text{mech, loss, total}}$ is the *total* mechanical power loss, which consists of turbine losses as well as the frictional losses in the piping network. That is,

$$\dot{E}_{\text{mech, loss, total}} = \dot{E}_{\text{mech, loss, turbine}} + \dot{E}_{\text{mech, loss, piping}} \tag{7-11}$$

By convention, irreversible turbine losses are treated separately from irreversible losses due to other components of the piping system. Thus, the energy equation is expressed in its most common form in terms of *heads* by dividing each term in Eq. (7-10) by $\dot{m}g$. The result is

$$\frac{P_1}{\rho_1 g} + \frac{V_1^2}{2g} + z_1 = \frac{P_2}{\rho_2 g} + \frac{V_2^2}{2g} + z_2 + h_{\text{turbine},e} + h_L \tag{7-12}$$

where

$$h_{\text{turbine},e} = \frac{w_{\text{turbine},e}}{g} = \frac{\dot{W}_{\text{turbine},e}}{\dot{m}g} = \frac{\dot{W}_{\text{turbine}}}{\eta_{\text{turbine}}\dot{m}g} \text{ is the \textit{extracted head removed from the fluid}}$$

by the turbine. Because of irreversible losses in the turbine, $h_{\text{turbine},e}$ is *greater* than $\dot{W}_{\text{turbine}}/\dot{m}g$ by the factor η_{turbine}.

$$h_L = \frac{e_{\text{mech, loss, piping}}}{g} = \frac{\dot{E}_{\text{mech, loss, piping}}}{\dot{m}g} \text{ is the \textit{irreversible head loss} between 1 and 2 due to}$$

all components of the piping system other than the turbine.

Note that the head loss h_L represents the frictional losses associated with fluid flow in the penstock, and it does not include the losses that occur within the turbine due to the inefficiencies of this device—these losses are taken into account by η_{turbine}. The total head loss in the penstock is determined from

$$h_{L,\text{total}} = h_{L,\text{major}} + h_{L,\text{minor}} = f\frac{L}{D}\frac{V^2}{2g} + \sum K_L \frac{V^2}{2g} = \left(f\frac{L}{D} + \sum K_L\right)\frac{V^2}{2g} \tag{7-13}$$

where

f = Darcy friction factor. It can be determined from the Moody chart (or Colebrook equation) for turbulent flow. For laminar flow, $f = 64/\text{Re}$, where Re is the Reynolds number.

L = length of the penstock

D = diameter of the penstock

V = velocity of water in the penstock

K_L = loss coefficient for minor losses in the piping system

The minor loss coefficients for various pipe components are available in Çengel and Cimbala (2018).

We sketch in Fig. 7-5 a typical hydroelectric dam that utilizes Francis reaction turbines to generate electricity. The overall or *gross head* H_{gross} is defined as the elevation difference between the reservoir surface upstream of the dam and the surface of the water exiting the dam,

$$H_{\text{gross}} = z_A - z_E \tag{7-14}$$

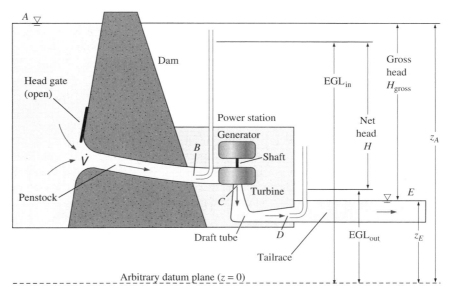

Figure 7-5 Typical setup and terminology for a hydroelectric plant that utilizes a Francis turbine to generate electricity (drawing not to scale). The pitot probes are shown for illustrative purposes only.

If there were no irreversible losses *anywhere* in the system, the maximum amount of power that could be generated per turbine would be

$$\dot{W}_{\text{max}} = \rho g \dot{V} H_{\text{gross}} \qquad (7\text{-}15)$$

where ρ is density, g is gravitational acceleration, and $\dot{V}$ is volume flow rate. Of course, there are irreversible losses throughout the system, so the power actually produced is lower than the maximum power given by Eq. (7-15).

We follow the flow of water through the whole system of Fig. 7-5, defining terms and discussing losses along the way. We start at point A upstream of the dam where the water is still, at atmospheric pressure, and at its highest elevation, z_A. Water flows at volume flow rate $\dot{V}$ through a large tube through the dam called the *penstock*. Flow to the penstock can be cut off by closing a large gate valve called a *head gate* at the penstock inlet. If we were to insert a pitot probe at point B at the end of the penstock just before the turbine, as illustrated in Fig. 7-5, the water in the tube would rise to a column height equal to the energy grade line EGL_{in} at the inlet of the turbine. Note that the energy grade line (EGL) represents the total head of the fluid

$$\text{EGL} = \text{Pressure head} + \text{Velocity head} + \text{Elevation head} = P/\rho g + V^2/2g + z \qquad (7\text{-}16)$$

This column height is lower than the water level at point A due to irreversible losses in the penstock and its inlet. The flow then passes through the turbine, which is connected by a shaft to the electric generator. Note that the electric generator itself has irreversible losses. From a fluid mechanics perspective, however, we are interested only in the losses through the turbine and downstream of the turbine.

After passing through the turbine runner, the exiting fluid (point C) still has appreciable kinetic energy (velocity head), and perhaps swirl. To recover some of this kinetic energy (which would otherwise be wasted), the flow enters an expanding area diffuser called a *draft tube*, which turns the flow horizontally and slows down the flow speed, while increasing the pressure prior to discharge into the downstream water, called the *tailrace*. If we were to imagine another pitot probe at point D (the exit of the draft tube), the water in the tube would rise to a column height equal to the energy grade line labeled EGL_{out} in Fig. 7-5. Since the draft tube is considered to be an integral part of the turbine assembly, the net head across the turbine is specified as the difference between EGL_{in} and EGL_{out},

$$H_{net} = EGL_{in} - EGL_{out} \qquad (7\text{-}17)$$

That is, the net head of a turbine is defined as the difference between the energy grade line just upstream of the turbine and the energy grade line at the exit of the draft tube.

At the draft tube exit (point D), the flow speed is significantly slower than that at point C upstream of the draft tube; however, it is *finite*. All the kinetic energy leaving the draft tube is dissipated in the tailrace. This represents an irreversible head loss and is the reason why EGL_{out} is higher than the elevation of the tailrace surface, z_E. Nevertheless, significant pressure recovery occurs in a well-designed draft tube. The draft tube causes the pressure at the outlet of the runner (point C) to decrease *below* atmospheric pressure, thereby enabling the turbine to utilize the available head most efficiently. In other words, the draft tube causes the pressure at the runner outlet to be lower than it would have been without the draft tube—increasing the change in pressure from the inlet to the outlet of the turbine. Designers must be careful, however, because subatmospheric pressures may lead to cavitation, which is undesirable for many reasons, as discussed previously.

If we were interested in the overall efficiency of the entire hydroelectric plant including penstock flow, we would define this efficiency as the ratio of actual electric power produced to maximum power, based on gross head. That is,

$$\eta_{plant} = \frac{\dot{W}_{electric}}{\dot{W}_{max}} = \frac{\dot{W}_{electric}}{\rho g \dot{V} H_{gross}} \qquad (7\text{-}18)$$

This is equivalent to turbine-generator efficiency [Eq. (7-7)], defined earlier. Usually, we are more interested in the efficiency of the turbine itself. The turbine efficiency given in Eq. (7-5) is based on the gross head. However, by convention, *turbine efficiency* $\eta_{turbine}$ should be based on net head H_{net} rather than gross head H_{gross}. Specifically, $\eta_{turbine}$ is defined as the ratio of actual turbine output shaft power to energy power extracted from the water flowing through the turbine (mechanical power difference between the end of the penstock and the exit of the draft tube),

$$\eta_{turbine} = \frac{\dot{W}_{shaft}}{\rho g \dot{V} H_{net}} \qquad (7\text{-}19)$$

This is a more realistic form of turbine efficiency as the turbine is not held responsible for irreversible losses in the penstock and its inlet and those after the exit of the draft tube. When the total irreversible head loss h_L in the piping is known, the corresponding power loss is determined from

$$\dot{E}_{mech,loss,piping} = \rho g \dot{V} h_L \qquad (7\text{-}20)$$

The turbine efficiency is then expressed as

$$\eta_{turbine} = \frac{\dot{W}_{shaft}}{\dot{W}_{max} - \dot{E}_{mech,loss,piping}} = \frac{\dot{W}_{shaft}}{\rho g \dot{V} H_{gross} - \rho g \dot{V} h_L} = \frac{\dot{W}_{shaft}}{\rho g \dot{V}(H_{gross} - h_L)} \qquad (7\text{-}21)$$

The effect of irreversible head losses in the piping system can be accounted for using an efficiency term η_{piping} as

$$\eta_{\text{piping}} = 1 - \frac{\dot{E}_{\text{mech, loss, piping}}}{\dot{W}_{\text{max}}} \tag{7-22}$$

Remembering that the generator efficiency is $\eta_{\text{generator}} = \dot{W}_{\text{electric}}/\dot{W}_{\text{shaft}}$, the overall efficiency of a hydroelectric power plant can be expressed as

$$\eta_{\text{plant}} = \eta_{\text{generator}} \eta_{\text{turbine}} \eta_{\text{piping}}$$

$$= \left(\frac{\dot{W}_{\text{electric}}}{\dot{W}_{\text{shaft}}} \right) \left(\frac{\dot{W}_{\text{shaft}}}{\dot{W}_{\text{max}} - \dot{E}_{\text{mech,loss,piping}}} \right) \left(1 - \frac{\dot{E}_{\text{mech,loss,piping}}}{\dot{W}_{\text{max}}} \right)$$

$$= \left(\frac{\dot{W}_{\text{electric}}}{\dot{W}_{\text{shaft}}} \right) \left[\frac{\dot{W}_{\text{shaft}}}{\dot{W}_{\text{max}} (1 - \dot{E}_{\text{mech,loss,piping}} / \dot{W}_{\text{max}})} \right] \left(1 - \frac{\dot{E}_{\text{mech,loss,piping}}}{\dot{W}_{\text{max}}} \right) \tag{7-23}$$

$$= \frac{\dot{W}_{\text{electric}}}{\dot{W}_{\text{max}}}$$

Therefore, the overall efficiency of a hydroelectric power plant is defined as the electrical power output divided by the maximum power potential, and it can be expressed as the product of generator, turbine, and piping efficiencies.

EXAMPLE 7-1
Power Potential of a Dam

River water is collected into a large dam whose water height is 240 ft. How much power can be produced by a set of ideal hydraulic turbines if water is run through the turbine at a total rate of 9000 gpm?

SOLUTION We take the density of water to be $\rho = 62.4$ lbm/ft³. The total mechanical energy the water in a dam possesses is equivalent to the potential energy of water at the free surface of the dam (relative to free surface of discharge water), and it can be converted to work entirely for an ideal operation. Therefore, the maximum power that can be generated is equal to the potential energy of the water. First, the total mass flow rate of the water is determined as

$$\dot{m} = \rho\dot{V} = (62.4 \text{ lbm/ft}^3)(9000 \text{ gal/min})\left(\frac{1 \text{ ft}^3}{7.4804 \text{ gal}} \right)\left(\frac{1 \text{ min}}{60 \text{ s}} \right) = 1251.3 \text{ lbm/s}$$

Then the maximum power is determined from

$$\dot{W}_{\text{max}} = \dot{m}gh = (1251.3 \text{ lbm/s})(32.2 \text{ ft/s}^2)(240 \text{ ft})\left(\frac{1 \text{ Btu/lbm}}{25{,}037 \text{ ft}^2/\text{s}^2} \right)\left(\frac{1 \text{ kJ}}{0.94782 \text{ Btu}} \right)\left(\frac{1 \text{ kW}}{1 \text{ kJ/s}} \right)$$

$$= \textbf{408 kW}$$

This is the maximum power that can be generated. Actual turbines generate less power. ▲

EXAMPLE 7-2
Analysis of a Hydraulic Turbine

The pressures just upstream and downstream of a hydraulic turbine are measured to be 1400 and 100 kPa, respectively. The turbine accepts water at the bottom level of a large water reservoir. (*a*) What is the shaft work, in kJ/kg, that can be produced by this turbine? (*b*) If the efficiency of the turbine is estimated to be 88 percent, what is the height of this reservoir?

SOLUTION (*a*) The density of water is taken to be $\rho = 1000$ kg/m³. The maximum work per unit mass of water flow is determined from

$$w_{max} = \frac{P_1 - P_2}{\rho} = \frac{(1400 - 100)\text{ kPa}}{1000\text{ kg/m}^3}\left(\frac{1\text{ kJ}}{1\text{ kPa}\cdot\text{m}^3}\right) = 1.30\text{ kJ/kg}$$

The actual shaft work per unit mass of water flow is determined using turbine efficiency to be

$$w_{shaft} = \eta_{turbine}w_{max} = (0.88)(1.30\text{ kJ/kg}) = \textbf{1.144 kJ/kg}$$

(*b*) The height of the reservoir is determined from

$$w_{shaft} = gh \quad\longrightarrow\quad h = \frac{w_{shaft}}{g} = \frac{1.144\text{ kJ/kg}}{9.81\text{ m/s}^2}\left(\frac{1000\text{ m}^2/\text{s}^2}{1\text{ kJ/kg}}\right) = \textbf{117 m} \quad \blacktriangle$$

EXAMPLE 7-3
Analysis of a
Hydroelectric
Power Plant

The water in a large water dam is to be used to generate electricity by the installation of a hydraulic turbine. The elevation difference between the free surfaces upstream and downstream of the dam is 320 m (Fig. 7-6). Water is to be supplied to the turbine at a rate of 8000 L/s. The turbine efficiency is 93 percent based on the net head, and the generator efficiency is 96 percent. The total irreversible head loss (major losses + minor losses) in the piping system including the penstock is estimated to be 7.5 m. Determine the overall efficiency of this hydroelectric plant, the electric power produced, and the turbine shaft power.

SOLUTION The density of water is taken to be $\rho = 1000$ kg/m³ = 1 kg/L. The total mechanical energy the water in a dam possesses is equivalent to the potential energy of water at the free surface of the dam (relative to the free surface of discharge water), and it can be converted to work entirely for an ideal operation. Therefore, the maximum power that can be generated is equal to the potential energy of the water. Noting that the mass flow rate is $\dot{m} = \rho\dot{V}$, the maximum power is determined from

$$\dot{W}_{max} = \dot{m}gH_{gross} = \rho\dot{V}gH_{gross}$$

$$= (1\text{ kg/L})(8000\text{ L/s})(9.81\text{ m/s}^2)(320\text{ m})\left(\frac{1\text{ kJ/kg}}{1000\text{ m}^2/\text{s}^2}\right)$$

$$= 25,115\text{ kW}$$

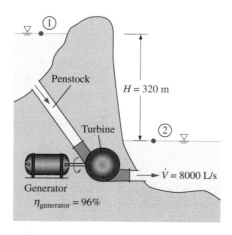

Figure 7-6 Schematic for Example 7-3.

The mechanical energy loss due to frictional and minor losses in the piping is

$$\dot{E}_{\text{mech,loss,piping}} = \dot{m}gh_L = \rho\dot{V}gh_L$$

$$= (1 \text{ kg/L})(8000 \text{ L/s})(9.81 \text{ m/s}^2)(7.5 \text{ m})\left(\frac{1 \text{ kJ/kg}}{1000 \text{ m}^2/\text{s}^2}\right)$$

$$= 589 \text{ kW}$$

Therefore, $589/25{,}115 = 0.0234$ or 2.3 percent of power potential is lost due to piping losses. This is expressed by the efficiency term η_{piping} as

$$\eta_{\text{piping}} = \left(1 - \frac{\dot{E}_{\text{mech, loss, piping}}}{\dot{W}_{\text{max}}}\right) = \left(1 - \frac{589 \text{ kW}}{25{,}115 \text{ kW}}\right) = 0.977$$

The overall plant efficiency is the product of turbine, generator, and piping efficiencies:

$$\eta_{\text{plant}} = \eta_{\text{generator}}\eta_{\text{turbine}}\eta_{\text{piping}} = (0.96)(0.93)(0.977) = 0.872 \text{ or } \mathbf{87.2\%}$$

The electric power produced is determined from the definition of plant efficiency to be

$$\eta_{\text{plant}} = \frac{\dot{W}_{\text{electric}}}{\dot{W}_{\text{max}}} \longrightarrow \dot{W}_{\text{electric}} = \eta_{\text{plant}}\dot{W}_{\text{max}} = (0.872)(25{,}115 \text{ kW}) = \mathbf{21{,}900 \text{ kW}}$$

The definition of turbine efficiency based on the net head is

$$\eta_{\text{turbine}} = \frac{\dot{W}_{\text{shaft}}}{\dot{W}_{\text{max}} - \dot{E}_{\text{mech, loss, piping}}}$$

Solving for the shaft power output, we obtain

$$\dot{W}_{\text{shaft}} = \eta_{\text{turbine}}(\dot{W}_{\text{max}} - \dot{E}_{\text{mech, loss, piping}})$$

$$= (0.93)(25{,}115 \text{ kW} - 589 \text{ kW})$$

$$= \mathbf{22{,}800 \text{ kW}}$$

This plant has an overall efficiency of 87.2 percent, which is considerably higher than the efficiency of steam and gas-turbine power plants (typically less than 50 percent). ▲

7-3 IMPULSE TURBINES

In an *impulse turbine*, the fluid is sent through a nozzle so that most of its available mechanical energy is converted into kinetic energy. The high-speed jet then impinges on bucket-shaped vanes that transfer energy to the turbine shaft, as sketched in Fig. 7-7. The modern and most efficient type of impulse turbine was invented by Lester A. Pelton (1829–1908) in 1878, and the rotating wheel is now called a *Pelton wheel* in his honor. The buckets of a Pelton wheel are designed to split the flow in half and turn the flow nearly 180° around (with respect to a frame of reference moving with the bucket), as illustrated in Fig. 7-7b. According to legend, Pelton modeled the splitter ridge shape after the nostrils of a cow's nose. A portion of the outermost part of each bucket is cut out so that the majority of the jet can pass through the bucket that is not aligned with the jet (bucket $n + 1$ in Fig. 7-7a)

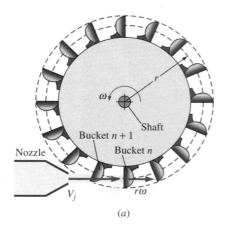

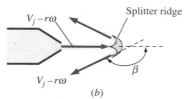

(a)

(b)

Figure 7-7 Schematic diagram of a Pelton-type *impulse turbine*; the turbine shaft is turned when high-speed fluid from one or more jets impinges on buckets mounted to the turbine shaft. (a) Side view, absolute reference frame, and (b) bottom view of a cross section of bucket n, rotating reference frame.

to reach the most aligned bucket (bucket n in Fig. 7-7a). In this way, the maximum amount of momentum from the jet is utilized. These details are seen in a photograph of a Pelton wheel (Fig. 7-8). Figure 7-9 shows a Pelton wheel in operation; the splitting and turning of the water jet is clearly seen.

Figure 7-8 A close-up view of a Pelton wheel showing the detailed design of the buckets; the electrical generator is on the right. This Pelton wheel is on display at the Waddamana Power Station Museum near Bothwell, Tasmania. (*Courtesy of Hydro Tasmania, www.hydro.com.au. Used with permission.*)

Figure 7-9 A view from the bottom of an operating Pelton wheel illustrating the splitting and turning of the water jet in the bucket. The water jet enters from the top, and the Pelton wheel is turning to the right. (*Courtesy of Andritz Hydro. Used with permission.*)

We analyze the power output of a Pelton wheel turbine by using the Euler turbomachine equation. The power output of the shaft is equal to

$$\dot{W}_{shaft} = \omega T_{shaft} = \rho \omega \dot{V}(r_2 V_{2,t} - r_1 V_{1,t}) \tag{7-24}$$

We must be careful of negative signs since this is an energy-*producing* rather than an energy-*absorbing* device. For turbines, it is conventional to define point 2 as the inlet and point 1 as the outlet. The center of the bucket moves at tangential velocity $r\omega$, as illustrated in Fig. 7-7. We simplify the analysis by assuming that since there is an opening in the outermost part of each bucket, the entire jet strikes the bucket that happens to be at the direct bottom of the wheel at the instant of time under consideration (bucket n in Fig. 7-7a). Furthermore, since both the size of the bucket and the diameter of the water jet are small compared to the wheel radius, we approximate r_1 and r_2 as equal to r. Finally, we make the approximation that the water is turned through angle β without losing any speed; in the relative frame of reference moving with the bucket, the relative exit speed is thus $V_j - r\omega$ (the same as the relative inlet speed) as sketched in Fig. 7-7b. Returning to the absolute reference frame, which is necessary for the application of Eq. (7-24), the tangential component of velocity at the inlet, $V_{2,t}$, is simply the jet speed itself, V_j. We construct a velocity diagram in Fig. 7-10 as an aid in calculating the tangential component of absolute velocity at the outlet, $V_{1,t}$. After some trigonometry, which you can verify after noting that $\sin(\beta - 90°) = -\cos \beta$,

$$V_{1,t} = r\omega + (V_j - r\omega)\cos \beta$$

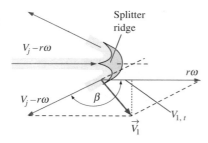

Figure 7-10 Velocity diagram of flow into and out of a Pelton wheel bucket. We translate outflow velocity from the moving reference frame to the absolute reference frame by adding the speed of the bucket ($r\omega$) to the right.

Upon substitution of this equation, Eq. (7-24) yields

$$\dot{W}_{\text{shaft}} = \rho r\omega \dot{V}\{V_j - [r\omega + (V_j - r\omega)\cos\beta]\}$$

which simplifies to

Output shaft power: $\dot{W}_{\text{shaft}} = \rho r\omega \dot{V}(V_j - r\omega)(1 - \cos\beta)$ (7-25)

Obviously, the maximum power is achieved theoretically if $\beta = 180°$. However, if that were the case, the water exiting one bucket would strike the back side of its neighbor coming along behind it, reducing the generated torque and power. It turns out that in practice, the maximum power is achieved by reducing β to around 160° to 165°. The efficiency factor due to β being less than 180° is

Efficiency factor due to β : $\eta_\beta = \dfrac{\dot{W}_{\text{shaft, actual}}}{\dot{W}_{\text{shaft, ideal}}} = \dfrac{1 - \cos\beta}{1 - \cos(180°)}$ (7-26)

When $\beta = 160°$, for example, $\eta_\beta = 0.97$—a loss of only about 3 percent. Finally, we see from Eq. (7-25) that the shaft power output $\dot{W}_{\text{shaft}}$ is zero if $r\omega = 0$ (wheel not turning at all). $\dot{W}_{\text{shaft}}$ is also zero if $r\omega = V_j$ (bucket moving at the jet speed). Somewhere in between these two extremes lies the optimum wheel speed. By setting the derivative of Eq. (7-25) with respect to $r\omega$ to zero, we find that this occurs when $r\omega = V_j/2$ (bucket moving at half the jet speed, as shown in Fig. 7-11).

For an actual Pelton wheel turbine, there are other losses besides that reflected in Eq. (7-26): mechanical friction, aerodynamic drag on the buckets, friction along the inside walls of the buckets, nonalignment of the jet and bucket as the bucket turns, back splashing,

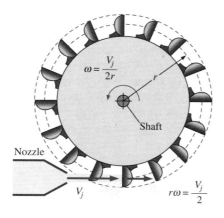

Figure 7-11 The theoretical maximum power achievable by a Pelton turbine occurs when the wheel rotates at $\omega = V_j/(2r)$, that is, when the bucket moves at half the speed of the water jet.

and nozzle losses. Even so, the efficiency of a well-designed Pelton wheel turbine can approach 90 percent. In other words, up to 90 percent of the available mechanical energy of the water is converted to rotating shaft energy.

EXAMPLE 7-4
Flow Rate and
Shaft Power Output
of a Pelton Turbine

A Pelton wheel turbine has an efficiency of 84 percent. The average radius of the wheel is 1.22 m, and the jet velocity is 87 m/s from a nozzle of exit diameter equal to 6.7 cm. The turning angle of the buckets is $\beta = 160°$. Determine (a) the volume flow rate through the turbine and (b) the output shaft power.

SOLUTION (a) We assume that frictional losses are negligible, so that the Euler turbomachine equation applies, and the relative exit speed of the jet is the same as its relative inlet speed. We take the density of water as 1000 kg/m³. The volume flow rate of the jet is equal to jet area times jet velocity:

$$\dot{V} = V_j \frac{\pi D_j^2}{4} = (87 \text{ m/s}) \frac{\pi (0.067 \text{ m})^2}{4} = \mathbf{0.3067 \ m^3/s}$$

(b) The maximum output shaft power occurs when the bucket moves at half the jet speed ($\omega r = V_j/2$). Thus, the rotational speed of the wheel for maximum power is

$$\omega = \frac{V_j}{2r} = \frac{87 \text{ m/s}}{2(1.22 \text{ m})} = 35.66 \text{ rad/s}$$

The maximum shaft power is determined from

$$\dot{W}_{\text{shaft,max}} = \rho r \omega \dot{V}(V_j - r\omega)(1 - \cos\beta)$$

$$= (1000 \text{ kg/m}^3)(1.22 \text{ m})(35.66 \text{ rad/s})(0.3067 \text{ m}^3/\text{s})$$

$$\times (87 \text{ m/s} - (1.22 \text{ m})(35.66 \text{ rad/s}))(1 - \cos 160°)$$

$$\times \left(\frac{1 \text{ N}}{1 \text{ kg} \cdot \text{m/s}^2} \right) \left(\frac{1 \text{ J/s}}{1 \text{ N} \cdot \text{m/s}} \right) \left(\frac{1 \text{ kW}}{1000 \text{ J/s}} \right)$$

$$= 1126 \text{ kW}$$

The actual shaft power is determined using the turbine efficiency as

$$\dot{W}_{\text{shaft,actual}} = \eta_{\text{turbine}} \dot{W}_{\text{shaft,max}} = (0.84)(1126 \text{ kW}) = \mathbf{946 \ kW} \quad \blacktriangle$$

7-4 REACTION TURBINES

The other main type of energy-producing hydro turbine is the *reaction turbine*, which consists of fixed guide vanes called *stay vanes*, adjustable guide vanes called *wicket gates*, and rotating blades called *runner blades* (Fig. 7-12). Flow enters tangentially at high pressure, is turned toward the runner by the stay vanes as it moves along the spiral casing or *volute*, and then passes through the wicket gates with a large tangential velocity component. Momentum is exchanged between the fluid and the runner as the runner rotates, and there is a large pressure drop. Unlike the impulse turbine, the water completely fills the casing of a reaction turbine. For this reason, a reaction turbine generally produces more power than an impulse turbine of the same diameter, net head, and volume flow rate. The angle of the wicket gates is adjustable so as to control the volume flow rate through the runner.

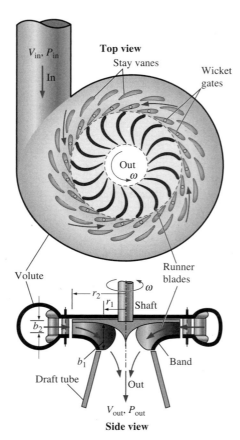

Top view

V_{in}, P_{in}

In

Stay vanes

Wicket gates

Out
ω

Runner blades

Volute

r_2 r_1 Shaft

ω

b_2

b_1 Band

Draft tube

Out

V_{out}, P_{out}

Side view

Figure 7-12   A *reaction turbine* differs significantly from an impulse turbine; instead of using water jets, a *volute* is filled with swirling water that drives the runner. For hydro turbine applications, the axis is typically vertical. Top and side views are shown, including the fixed *stay vanes* and adjustable *wicket gates*.

(In most designs the wicket gates can close on each other, cutting off the flow of water into the runner.) At design conditions the flow leaving the wicket gates impinges parallel to the runner blade leading edge (from a rotating frame of reference) to avoid shock losses. Note that in a good design, the number of wicket gates does not share a common denominator with the number of runner blades. Otherwise there would be severe vibration caused by simultaneous impingement of two or more wicket gate wakes onto the leading edges of the runner blades. For example, in Fig. 7-12 there are 17 runner blades and 20 wicket gates. These are typical numbers for many large reaction hydro turbines, as shown in Figs. 7-14 and 7-15. The number of stay vanes and wicket gates is usually the same (there are 20 stay vanes in Fig. 7-12). This is not a problem since none of them rotate, and unsteady wake interaction is not an issue.

There are two main types of reaction turbine—*Francis* and *Kaplan*. The *Francis turbine* is somewhat similar in geometry to a centrifugal or mixed-flow pump, but with the flow in the opposite direction. Note, however, that a typical pump running backward would *not* be a very efficient turbine. The Francis turbine is named in honor of James B. Francis (1815–1892), who developed the design in the 1840s. In contrast, the *Kaplan turbine* is somewhat like an *axial-flow* fan running backward. If you have ever seen a window fan start spinning in the wrong direction when a gust of wind blows through the window, you can visualize the basic operating principle of a Kaplan turbine. The Kaplan turbine is named in honor of its inventor, Viktor Kaplan (1876–1934). There are actually several subcategories of both Francis and Kaplan turbines, and the terminology used in the hydro turbine field is not always standard.

We classify reaction turbines according to the angle at which the flow *enters* the runner (Fig. 7-13). If the flow enters the runner radially as in Fig. 7-13*a*, the turbine is called a *Francis radial-flow turbine* (see also Fig. 7-12). If the flow enters the runner at some angle between radial and axial (Fig. 7-13*b*), the turbine is called a *Francis mixed-flow turbine*. The latter design is more common. Some hydro turbine engineers use the term "Francis turbine" only when there is a *band* on the runner as in Fig. 7-13*b*. Francis turbines are most suited for heads that lie between the high heads of Pelton wheel turbines and the low heads of Kaplan turbines. A typical large Francis turbine may have 16 or more runner blades and can achieve a turbine efficiency of 90 to 95 percent. If the runner has no band, and flow enters the runner partially turned, it is called a *propeller mixed-flow turbine* or simply a *mixed-flow turbine* (Fig. 7-13*c*). Finally, if the flow is turned completely axially *before* entering the runner (Fig. 7-13*d*), the turbine is called an *axial-flow turbine*. The runners of an axial-flow turbine typically have only three to eight blades, a lot fewer than Francis turbines. Of these there are two types: Kaplan turbines and propeller turbines. Kaplan turbines are called *double regulated* because the flow rate is controlled in two ways—by turning the wicket gates and by adjusting the pitch on the runner blades. *Propeller turbines* are nearly identical to Kaplan turbines except that the blades are fixed (pitch is not adjustable), and the flow rate is regulated only by the wicket gates (*single regulated*). Compared to the Pelton and Francis turbines, Kaplan turbines and propeller turbines are most suited for low head, high volume flow rate conditions. Their efficiencies rival those of Francis turbines and may be as high as 94 percent.

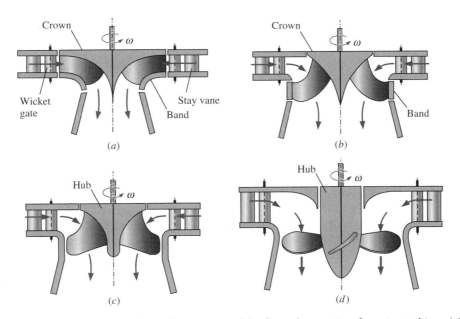

Figure 7-13 The distinguishing characteristics of the four subcategories of reaction turbines: (*a*) Francis radial flow, (*b*) Francis mixed flow, (*c*) propeller mixed flow, and (*d*) propeller axial flow. The main difference between (*b*) and (*c*) is that Francis mixed-flow runners have a *band* that rotates with the runner, while propeller mixed-flow runners do not. There are two types of propeller mixed-flow turbines: *Kaplan turbines* have adjustable pitch blades, while *propeller turbines* do not. Note that the terminology used here is neither universal among turbomachinery textbooks nor among hydro turbine manufacturers.

Figure 7-14   The runner of a Francis radial-flow turbine used at the Boundary hydroelectric power station on the Pend Oreille River north of Spokane, WA. There are 17 runner blades of outer diameter 18.5 ft (5.6 m). The turbine rotates at 128.57 rpm and produces 230 MW of power at a volume flow rate of 335 m³/s from a net head of 78 m. (© *American Hydro Corporation. Used with permission.*)

Figure 7-14 is a photograph of the radial-flow runner of a Francis radial-flow turbine. The workers are shown to give you an idea of how large the runners are in a hydroelectric power plant. Figure 7-15 is a photograph of the mixed-flow runner of a Francis turbine, and Fig. 7-16 is a photograph of an axial-flow propeller turbine. The view is from the inlet (top).

We perform preliminary design and analysis of turbines using the Euler turbomachine equation and velocity diagrams. For a turbine, the flow direction is opposite to that of a pump, so the inlet is at radius r_2 and the outlet is at radius r_1. For a first-order analysis we approximate the blades as being infinitesimally thin. We also assume that the blades are aligned such that the flow is always tangent to the blade surface, and we ignore viscous effects (boundary layers) at the surfaces.

Consider for example the top view of the Francis turbine of Fig. 7-12. Velocity vectors are drawn in Fig. 7-17 for both the absolute reference frame and the relative reference frame rotating with the runner. Beginning with the stationary guide vane (thick black line in Fig. 7-17), the flow is turned so that it strikes the runner blade (thick gray line) at absolute velocity $\vec{V}_2$. But the runner blade is rotating counterclockwise, and at radius r_2 it moves tangentially to the lower left at speed ωr_2. To translate into the rotating reference frame, we form the vector sum of $\vec{V}_2$ and the *negative* of ωr_2, as shown in the sketch. The resultant is vector $\vec{V}_{2,\text{ relative}}$, which is parallel to the runner blade leading edge (angle β_2 from the tangent

Figure 7-15 The runner of a Francis mixed-flow turbine used at the Smith Mountain hydroelectric power station in Roanoke, VA. There are 17 runner blades of outer diameter 20.3 ft (6.19 m). The turbine rotates at 100 rpm and produces 194 MW of power at a volume flow rate of 375 m^3/s from a net head of 54.9 m. (*Photo courtesy of American Hydro Corporation, York, PA. Used with permission.*)

Figure 7-16 The five-bladed propeller turbine used at the Warwick hydroelectric power station in Cordele, GA. There are five runner blades of outer diameter 12.7 ft (3.87 m). The turbine rotates at 100 rpm and produces 5.37 MW of power at a volume flow rate of 63.7 m^3/s from a net head of 9.75 m. (*Photo courtesy of American Hydro Corporation, York, PA. Used with permission.*)

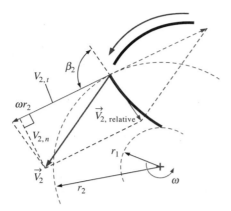

Figure 7-17 Relative and absolute velocity vectors and geometry for the outer radius of the runner of a Francis turbine. Absolute velocity vectors are bold.

line of circle r_2). The tangential component $V_{2,t}$ of the absolute velocity vector $\vec{V}_2$ is required for the Euler turbomachine equation [Eq. (7-24)]. After some trigonometry,

$$\text{Runner leading edge:} \qquad V_{2,t} = \omega r_2 - \frac{V_{2,n}}{\tan \beta_2} \qquad (7\text{-}27)$$

Following the flow along the runner blade in the relative (rotating) reference frame, we see that the flow is turned such that it exits parallel to the trailing edge of the runner blade (angle β_1 from the tangent line of circle r_1). Finally, to translate back to the absolute reference frame we vectorially add $\vec{V}_{1,\text{relative}}$ and blade speed ωr_1, which acts to the left as sketched in Fig. 7-18. The resultant is absolute vector $\vec{V}_1$. Since mass must be conserved, the normal components of the absolute velocity vectors $V_{1,n}$ and $V_{2,n}$ are related through

$$\dot{V} = 2\pi r_1 b_1 V_{1,n} = 2\pi r_2 b_2 V_{2,n}$$

where axial blade widths b_1 and b_2 are defined in Fig. 7-12. After some trigonometry (which turns out to be identical to that at the leading edge), we generate an expression for the tangential component $V_{1,t}$ of absolute velocity vector $\vec{V}_1$ for use in the Euler turbomachine equation,

$$\text{Runner trailing edge:} \qquad V_{1,t} = \omega r_1 - \frac{V_{1,n}}{\tan \beta_1} \qquad (7\text{-}28)$$

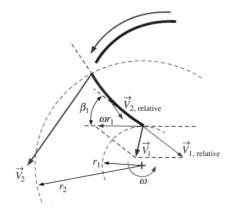

Figure 7-18 Relative and absolute velocity vectors and geometry for the inner radius of the runner of a Francis turbine. Absolute velocity vectors are bold.

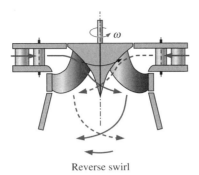

Figure 7-19   In some Francis mixed-flow turbines, high-power, high-volume flow rate conditions sometimes lead to *reverse swirl*, in which the flow exiting the runner swirls in the direction opposite to that of the runner itself, as sketched here.

Reverse swirl

For some hydro turbine runner applications, high power/high flow operation can result in $V_{1,t} < 0$. Here the runner blade turns the flow so much that the flow at the runner outlet rotates in the direction opposite to runner rotation, a situation called *reverse swirl* (Fig. 7-19). The Euler turbomachine equation predicts that maximum power is obtained when $V_{1,t} < 0$, so we suspect that reverse swirl should be part of a good turbine design. In practice, however, it has been found that the best efficiency operation of most hydro turbines occurs when the runner imparts a small amount of *with-rotation swirl* to the flow exiting the runner (swirl in the same direction as runner rotation). This improves draft tube performance. A large amount of swirl (either reverse or with-rotation) is not desirable, because it leads to much higher losses in the draft tube. (High swirl velocities result in "wasted" kinetic energy.) Obviously, much fine tuning needs to be done in order to design the most efficient hydro turbine system (including the draft tube as an integral component) within imposed design constraints. Also keep in mind that the flow is three-dimensional; there is an *axial* component of the velocity as the flow is turned into the draft tube, and there are differences in velocity in the *circumferential* direction as well. It doesn't take long before you realize that computer simulation tools are enormously useful to turbine designers. In fact, with the help of modern CFD codes, the efficiency of hydro turbines has increased to the point where retrofits of old turbines in hydroelectric plants are economically wise and common.

EXAMPLE 7-5
Analysis of a
Francis Radial-Flow
Hydro Turbine

A Francis radial-flow hydro turbine has the following dimensions, where location 2 is the inlet and location 1 is the outlet: $r_2 = 1.90$ m, $r_1 = 1.35$ m, $b_2 = 0.78$ m, and $b_1 = 2.15$ m. The runner rotates at $\dot{n} = 165$ rpm. The wicket gates turn the flow by angle $\alpha_2 = 55°$ from radial at the runner inlet, and the flow at the runner outlet is at angle $\alpha_1 = 25°$ from radial (Fig. 7-20). The volume flow rate at design conditions is 200 m³/s. Neglecting irreversible losses, calculate the inlet and outlet runner blade angles β_2 and β_1, respectively, and predict the power output (MW) and required net head (m).

SOLUTION   We assume that the flow is everywhere tangent to the runner blades and we neglect irreversible losses through the turbine. We solve for the normal component of velocity at the inlet,

$$V_{2,n} = \frac{\dot{V}}{2\pi r_2 b_2} = \frac{200 \text{ m}^3/\text{s}}{2\pi(1.90 \text{ m})(0.78 \text{ m})} = 21.48 \text{ m/s}$$

Using the figure provided with this problem as a guide, the tangential velocity component at the inlet is

$$V_{2,t} = V_{2,n}\tan\alpha_2 = (21.48 \text{ m/s})\tan 55° = 30.67 \text{ m/s}$$

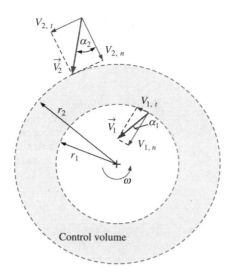

Figure 7-20   Schematic for Example 7-5.

The angular velocity is

$$\omega = \frac{2\pi \dot{n}}{60} = 2\pi(165 \text{ rad/min})\left(\frac{1 \text{ min}}{60 \text{ s}}\right) = 17.28 \text{ rad/s}$$

The tangential velocity component of the absolute velocity at the inlet is obtained from trigonometry to be

$$V_{2,t} = \omega r_2 - \frac{V_{2,n}}{\tan \beta_2}$$

From the above relationship, we solve for the runner leading edge angle β_2,

$$\beta_2 = \arctan\left(\frac{V_{2,n}}{\omega r_2 - V_{2,t}}\right) = \arctan\left[\frac{21.48 \text{ m/s}}{(17.28 \text{ rad/s})(1.90 \text{ m}) - 30.67 \text{ m/s}}\right] = \mathbf{84.3°}$$

We repeat the calculations for the runner outlet:

$$V_{1,n} = \frac{\dot{V}}{2\pi r_1 b_1} = \frac{200 \text{ m}^3/\text{s}}{2\pi(1.35 \text{ m})(2.15 \text{ m})} = 10.97 \text{ m/s}$$

$$V_{1,t} = V_{1,n}\tan\alpha_1 = (10.97 \text{ m/s})\tan 25° = 5.114 \text{ m/s}$$

$$\beta_1 = \arctan\left(\frac{V_{1,n}}{\omega r_1 - V_{1,t}}\right) = \arctan\left[\frac{10.97 \text{ m/s}}{(17.28 \text{ rad/s})(1.35 \text{ m}) - 5.114 \text{ m/s}}\right] = \mathbf{31.1°}$$

The shaft output power is estimated from the Euler turbomachine equation as

$$\dot{W}_{\text{shaft}} = \rho\omega\dot{V}(r_2 V_{2,t} - r_1 V_{1,t})$$

$$= (1000 \text{ kg/m}^3)(17.28 \text{ rad/s})(200 \text{ m}^3/\text{s})[(1.90 \text{ m})(30.67 \text{ m/s}) - (1.35 \text{ m})(5.114 \text{ m/s})]$$

$$= 1.775\times10^8 \; \frac{1 \text{ N}\cdot\text{s}^2}{1 \text{ kg}\cdot\text{m}}\left(\frac{1 \text{ MW}\cdot\text{s}^3}{10^6 \text{ kg}\cdot\text{m}^2}\right)$$

$$= \mathbf{177.5 \text{ MW}}$$

Finally, we calculate the required net head assuming that $\eta_{\text{turbine}} = 100$ percent since we are ignoring irreversibilities:

$$H = \frac{\dot{W}_{\text{shaft}}}{\rho \dot{V} g} = \frac{177{,}500 \text{ kJ/s}}{(1000 \text{ kg/m}^3)(200 \text{ m}^3/\text{s})(9.81 \text{ m/s}^2)\left(\dfrac{1 \text{ kJ/kg}}{1000 \text{ m}^2/\text{s}^2}\right)} = 90.5 \text{ m}$$

This is a preliminary design in which we are neglecting irreversibilities. Actual output power will be lower, and actual required net head will be higher than the values predicted here. ▲

7-5 TURBINE SPECIFIC SPEED

A useful parameter in the preliminary selection of hydraulic turbines is the *turbine specific speed* N_{St}. It is defined in terms of two independent dimensionless parameters for turbines, namely power coefficient C_P and head coefficient C_H, as

$$N_{\text{St}} = \frac{C_P^{1/2}}{C_H^{5/4}} = \frac{\left(\dfrac{\text{bhp}}{\rho \omega^3 D^5}\right)^{1/2}}{\left(\dfrac{gH}{\omega^2 D^2}\right)^{5/2}} = \frac{\omega(\text{bhp})^{1/2}}{\rho^{1/2}(gH)^{5/4}} \tag{7-29}$$

where ω is the rotational speed, bhp is brake horsepower (shaft power $\dot{W}_{\text{shaft}}$), ρ is the fluid density, and H is the net head.

Although N_{St} is by definition a dimensionless parameter, practicing engineers have grown accustomed to using inconsistent units that transform N_{St} into a cumbersome dimensional quantity. In the United States, most turbine engineers write the rotational speed in units of rotations per minute (rpm), bhp in units of horsepower, and H in units of feet. Furthermore, they ignore gravitational constant g and density ρ in the definition of N_{St}. (The turbine is assumed to operate on earth and the working fluid is assumed to be water.) We define

$$N_{\text{St, US}} = \frac{(\dot{n}, \text{rpm})(\text{bhp}, \text{hp})^{1/2}}{(H, \text{ft})^{5/4}} \tag{7-30}$$

There is some discrepancy in the turbomachinery literature over the conversions between the two forms of turbine specific speed. To convert $N_{\text{St,US}}$ to N_{St}, we divide by $g^{5/4}$ and $\rho^{1/2}$, and then use conversion ratios to cancel all units. We set $g = 32.174$ ft/s^2 and assume water at density $\rho = 62.40$ lbm/ft^3. When done properly by converting ω to rad/s, the conversion is $N_{\text{St,US}} = 0.02301 \, N_{\text{St}}$ or $N_{\text{St}} = 43.46 \, N_{\text{St,US}}$.

There is also a metric or SI version of turbine specific speed that is becoming more popular these days and is preferred by many hydro turbine designers. It is defined in the same way as the customary U.S. pump specific speed, except that SI units are used (m^3/s instead of gpm and m instead of ft),

$$N_{\text{St, SI}} = \frac{(\dot{n}, \text{rpm})(\dot{V}, \text{m}^3/\text{s})^{1/2}}{(H, \text{m})^{3/4}} \tag{7-31}$$

We may call this *capacity specific speed* to distinguish it from power specific speed [Eq. (7-29)].

Technically, turbine specific speed could be applied at any operating condition and would just be another function of C_P. That is not how it is typically used, however. Instead, it is common to define turbine specific speed only at the best efficiency point (BEP) of the turbine. The result is a single number that characterizes the turbine. Turbine specific speed is used to characterize the operation of a turbine at its optimum conditions (the BEP) and is useful for preliminary turbine selection.

As plotted in Fig. 7-21, impulse turbines perform optimally for N_{St} near 0.15, while Francis turbines and Kaplan or propeller turbines perform best at N_{St} near 1 and 2.5,

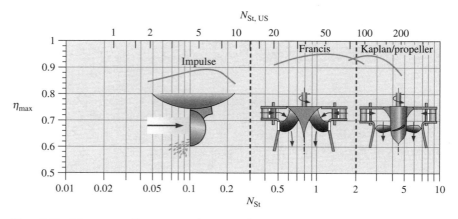

Figure 7-21   Maximum efficiency as a function of turbine specific speed for the three main types of dynamic turbine. Horizontal scales show nondimensional turbine specific speed (N_{St}) and turbine specific speed in customary U.S. units ($N_{St,US}$). Sketches of the blade types are also provided on the plot for reference.

respectively. It turns out that if N_{St} is less than about 0.3, an impulse turbine is the best choice. If N_{St} is between about 0.3 and 2, a Francis turbine is a better choice. When N_{St} is greater than about 2, a Kaplan or propeller turbine should be used. These ranges are indicated in Fig. 7-21 in terms of N_{St} and $N_{St,US}$.

EXAMPLE 7-6
Turbine Specific
Speed of a
Francis Turbine

Calculate the turbine specific speed of the Francis turbine in Example 7-5. Provide answers in both dimensionless form and in customary U.S. units. The values for the turbine of Example 7-5 are $\omega = 17.28$ rad/s, bhp $= 1.775 \times 10^8$ W, $H = 90.5$ m, $\dot{n} = 165$ rpm.

SOLUTION   We first calculate the nondimensional form of N_{St},

$$N_{St} = \frac{\omega(\text{bhp})^{1/2}}{\rho^{1/2}(gH)^{5/4}} = \frac{(17.28 \text{ rad/s})(1.775 \times 10^8 \text{ W})^{1/2}}{(1000 \text{ kg/m}^3)^{1/2}\left[(9.81 \text{ m/s}^2)(90.5 \text{ m})\right]^{5/4}}\left(\frac{1 \text{ kg·m}^2}{1 \text{ W·s}^3}\right)^{1/2} = \mathbf{1.50}$$

From Fig. 7-21, this is in the range for a Francis turbine. From the conversion given in the text,

$$N_{St,US} = 1.50 \times 43.46 = \mathbf{65.2}$$

Alternatively, we can use the original dimensional data to calculate $N_{St,US}$,

$$N_{St,US} = \frac{(\dot{n}, \text{rpm})(\text{bhp, hp})^{1/2}}{(H, \text{ft})^{5/4}} = \frac{(165 \text{ rpm})\left[(1.775 \times 10^5 \text{ kW})\left(\dfrac{1 \text{ hp}}{0.7457 \text{ kW}}\right)\right]^{1/2}}{\left[(90.5 \text{ m})\left(\dfrac{3.2808 \text{ ft}}{1 \text{ m}}\right)\right]^{5/4}} = \mathbf{65.3} \quad \blacktriangle$$

7-6 RUN-OF-RIVER PLANTS AND WATERWHEELS

The majority of the electricity from hydroelectric power plants is produced using accumulation reservoirs, also called dams. *Run-of-river plants* (also called river plants or small hydroelectric power plants) do not use water reservoirs. Instead, they are built along a water stream such as a river. In this system, a proper portion of the water stream is diverted to a turbine-generator unit, and the used water is returned to the river, as shown in Fig. 7-22 (Farret and Simoes, 2006).

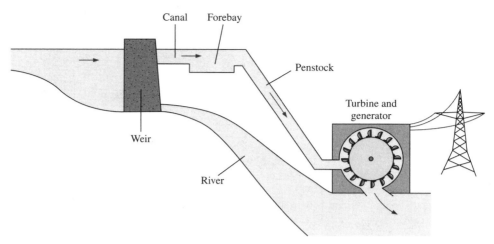

Figure 7-22   Working principle of a run-of-river plant. *Weir* is the barrier placed in the river to control water discharge. *Forebay* is the part of the canal from which water is taken to run the turbine.

The flow rate of a river is not constant as it varies with season, month, and day. As a consequence, this scheme of hydropower generation does not make full use of the water stream most of the time. This is because the system capacity is normally selected for an average water flow rate (i.e., rated flow rate). When the flow rate is more than the rated flow, excess water is diverted from the turbines and not used for power generation. Installing a large turbine-generator system to accommodate the highest possible water flow rate increases the amount of power generation, but this also increases the initial cost, and the system operates at part load (and possibly lower efficiency) most of the time. Therefore, the selection of system capacity is a trade-off between the initial cost, power output, and efficiency.

A run-of-river plant is installed when there is sufficient velocity and flow rate for a river. In order to estimate the power potential from a river, we consider a river water flow at a mass flow rate of $\dot{m}$, a velocity of V, and negligible elevation change. A hydraulic turbine converts the kinetic energy of this water into power. Then the power potential of this water stream can be determined from

$$\dot{W}_{\text{available}} = \dot{m}\frac{V^2}{2} = \rho A V \frac{V^2}{2} = \rho A \frac{V^3}{2} \tag{7-32}$$

For a water density of $\rho = 1000 \text{ kg/m}^3$, a water velocity of $V = 1$ m/s, and per unit cross-section area A, the rate of kinetic energy or power potential is calculated to be 500 W/m² of flow cross section. Unfortunately, only about 60 percent of this potential is theoretically available. In addition, the efficiency of turbine-generator units in river plants is lower than that of large hydropower plants. Equation (7-32) shows that the power potential is proportional to the cube of velocity. For a water velocity of 2 m/s, the power potential would be 4000 W/m² of flow section. Unfortunately, water speeds in rivers are not high in most cases, and the power potential is limited.

A *waterwheel* is different from a run-of-river plant in that a waterwheel is installed directly into falling or free-flowing water to convert the energy of the water into power. This is particularly illustrated for a lower bucket type of a waterwheel in which water flows through the lower blades (called buckets), causing the wheel to rotate. A more common type of waterwheel involves upper buckets for which water flows through the upper buckets, fills them, and forces the wheel to rotate due to the weight acting on the buckets (Fig. 7-23).

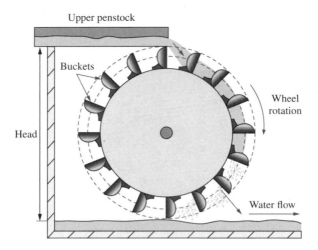

Upper penstock

Buckets

Wheel rotation

Head

Water flow

Figure 7-23 Operation of an upper bucket waterwheel.

The power potential from a waterwheel can be determined from

$$\dot{W}_{available} = \dot{m}gH = \rho\dot{V}gH \tag{7-33}$$

where $\dot{m}$ is the mass flow rate, $\dot{V}$ is the volume flow rate, ρ is the water density, and H is the head of the water that is the height difference between water streams at the input and output of the water channel. The actual power output can be estimated by multiplying Eq. (7-33) by the efficiency of the waterwheel. The efficiencies of waterwheels can be taken to be about 60 percent. For a water density of $\rho = 1000$ kg/m³, a water flow rate of $\dot{V} = 1$ m³/s, and a head of $H = 10$ m, the power potential is calculated to be 98 kW. For an efficiency of 60 percent, the actual power output is 59 kW.

Waterwheels were used since ancient times for milling flour in gristmills, and they are still used in some parts of the world for this application. Previously, they were also used for water-lifting for irrigation, grinding wood into pulp for papermaking, machining, hammering wrought iron, and powering mine hoists. They are usually made from wood or steel, with shovels of blades fixed regularly around their circumference. Water pushes the shovels (or buckets) tangentially around the wheel, causing a torque to develop on the shaft. The rotational speeds are typically low, and these machines have low efficiencies due to losses such as friction, turbulence, and incomplete filling of the buckets (Farret and Simoes, 2006).

Waterwheels are simple devices, and operation is not affected by dirty water. Since waterwheels have very low rotational speeds, a generator requires high rates of speed multipliers to produce electricity. This causes additional losses in the system. The efficiency of a waterwheel for the production of electricity is very low.

EXAMPLE 7-7
Efficiency of a Run-of-River Plant

A run-of-river plant operates on a water stream whose cross-sectional area is 18 m². Water is flowing at a speed of 2.5 m/s. If the electrical power output from the plant is 61 kW, what is the overall efficiency of this plant?

SOLUTION   The power potential of this water stream is

$$\dot{W}_{available} = \rho A \frac{V^3}{2} = (1000 \text{ kg/m}^3)(18 \text{ m}^2)\frac{(2.5 \text{ m/s})^3}{2} = 140{,}625 \text{ kg·m}^2/\text{s}^3$$

The unit of power potential does not mean much to us. Working with proper unit conversions, we obtain

$$\dot{W}_{available} = (140{,}625 \text{ kg} \cdot \text{m}^2/\text{s}^3)\left(\frac{1 \text{ W}}{1 \text{ J/s}}\right)\left(\frac{1 \text{ J}}{1 \text{ N} \cdot \text{m}}\right)\left(\frac{1 \text{ N}}{1 \text{ kg} \cdot \text{m/s}^2}\right) = 140{,}625 \text{ W}$$

Knowing the actual power output, the overall efficiency of the plant is determined to be

$$\eta_{overall} = \frac{\dot{W}_{actual}}{\dot{W}_{available}} = \frac{61{,}000 \text{ W}}{140{,}625 \text{ W}} = 0.434 = \textbf{43.4\%}$$

The overall efficiency of this plant is much lower than that available in large hydropower plants.   ▲

REFERENCES

Çengel YA and Cimbala JM. 2018. *Fluid Mechanics: Fundamentals and Applications*, 4th ed. New York: McGraw Hill.

Farret F and Simoes MG. 2006. *Integration of Alternative Sources of Energy*. Hoboken, New Jersey: John Wiley & Sons, Inc.

Lewis BJ, Cimbala HM, and Wouden AM. 2014. Major historical developments in the design of water wheels and Francis hydroturbines. IOP Conf. Series: Earth and Environmental Science 22(2014) 012020.

PROBLEMS

INTRODUCTION

7-1   What is a hydro turbine? What is the working fluid in a hydraulic turbine? What is the rotating part of a hydro turbine called?

7-2   How does a hydro turbine differ from a water pump?

7-3   What are the basic types of dynamic turbines? How do you compare them in terms of head and volume flow rate?

7-4   The rotating part of a turbine is called
(*a*) Propeller   (*b*) Scroll   (*c*) Blade row   (*d*) Impeller   (*e*) Runner

7-5   Which one is correct for the comparison of the operation of impulse and reaction turbines?
(*a*) Impulse: Higher flow rate   (*b*) Impulse: Higher head   (*c*) Reaction: Higher head
(*d*) Reaction: Smaller flow rate   (*e*) None of these

ANALYSIS OF HYDROELECTRIC POWER PLANT

7-6   How do you determine the absolute pressure of water at the inlet of a hydro turbine if the turbine is placed at the bottom of a dam whose height is *h*?

7-7   River water is collected into a large dam whose height is *h*. How much work, in kJ/kg, can be produced by an ideal hydraulic turbine?

7-8   The pressures just upstream and downstream of a hydraulic turbine are measured to be P_1 and P_2, respectively. How do you determine the maximum work, in kJ/kg, that can be produced by this turbine?

7-9   How is the efficiency of a hydraulic turbine defined? How is it related to the overall turbine efficiency?

7-10   Why are we usually more interested in the overall efficiency of a turbine-generator combination than in the turbine efficiency?

7-11   It is argued that hydraulic turbines cannot have an efficiency of 100 percent even in the absence of irreversible losses due to limitations created by the second law of thermodynamics. Do you agree? Explain.

7-12   It is argued that hydraulic turbines cannot have a maximum efficiency of 100 percent and that there must be a lower upper limit for the turbine efficiency similar to the Betz limit of wind turbines. Do you agree? Explain.

7-13   The total irreversible head loss h_L in the piping of a hydroelectric power plant (the penstock and its inlet and those after the exit of the draft tube), the gross head H_{gross}, and the volume flow rate of water through the turbine $\dot{V}$ are specified. How do you express the power loss and an efficiency term to account for irreversible head losses in the piping system? Simplify the efficiency relation as much as you can.

7-14   How do you express the efficiency of a hydroelectric power plant when turbine, generator, and piping efficiencies are given? Show that this equation reduces to electric power output divided by the maximum power potential.

7-15   The water height in a dam is 80 m. What is the absolute pressure of water at the inlet of a hydro turbine if the turbine is placed at the bottom of a dam? The atmospheric pressure is 101 kPa.

7-16   River water is collected into a large dam whose height is 65 m. How much power can be produced by an ideal hydraulic turbine if water is run through the turbine at a rate of 1500 L/s?

7-17   The pressures just upstream and downstream of a hydraulic turbine are measured to be 1325 and 100 kPa, respectively. What is the maximum work, in kJ/kg, that can be produced by this turbine? If this turbine is to generate a maximum power of 100 kW, what should be the flow rate of water through the turbine, in L/min?

7-18   Two hydraulic turbines (turbine A and turbine B) are considered. Turbine A uses a water body with a height of 200 m while turbine B uses one with 100 m height. The flow rate of water through turbine A is 150 kg/s and that through turbine B is 300 kg/s. Which turbine has more power producing potential?

7-19   Water is run through a hydraulic turbine at a rate of 11,500 L/min from a 160-m-high reservoir and produces 250 kW of shaft power. What is the efficiency of this turbine?

7-20   A hydraulic turbine-generator unit placed at the bottom of a 75-m-high dam accepts water at a rate of 1020 L/s and produces 630 kW of electricity. Determine (*a*) the overall efficiency of the turbine-generator unit and (*b*) the turbine efficiency if the generator efficiency is 96 percent, and (*c*) the power losses due to inefficiencies in the turbine and the generator.

7-21   The pressures just upstream and downstream of a hydraulic turbine are measured to be 95 and 15 psia, respectively, and the water flow through the turbine is 280 lbm/s. The turbine accepts water at the bottom level of a large water reservoir. If the efficiency of the turbine is estimated to be 86 percent, determine (*a*) the shaft power output from the turbine and (*b*) the height of this reservoir.

7-22   The irreversible losses in the penstock and its inlet and those after the exit of the draft tube are estimated to be 7 m. The elevation difference between the reservoir surface upstream of the dam and the surface of the water exiting the dam is 140 m. If the flow rate through the turbine is 4000 L/min, determine (*a*) the power loss due to irreversible head loss, (*b*) the efficiency of the piping, and (*c*) the electric power output if the turbine-generator efficiency is 84 percent.

7-23   The piping efficiency of a hydroelectric power plant is estimated to be 98 percent while the turbine efficiency based on the net head is 87 percent and the generator efficiency is 97 percent. If the elevation difference between the reservoir surface upstream of the dam and the surface of the water exiting the dam is 220 m, determine the overall efficiency of the hydroelectric plant and the electric power output. The flow rate through the turbine is 600 L/s.

7-24   A hydroelectric power plant consists of 18 identical turbine-generator units with an overall plant efficiency of 90 percent. The gross head of the dam is 150 m and the flow rate through each turbine is 3300 L/min. The plant operates 80 percent of the time throughout the year, and the electricity generated is sold to the utility company at a rate of $0.095/kWh. How much revenue can this plant generate in a year?

7-25   A village with 55 households currently spends $4600 for electricity per month at a rate of $0.12/kWh. To meet the electricity needs of the village, an engineer proposes to convert mechanical energy of a waterfall in the village into electricity by a hydraulic turbine. The waterfall extends to a height of 80 m from the ground level, and the flow rate of the waterfall is 350 L/min. Assuming a turbine-generator efficiency of 82 percent, determine if the proposed turbine can meet the electricity needs of the village. Assume the operating and maintenance expenses are negligible and the turbine can operate nonstop throughout the year.

7-26 An investor is to build a hydroelectric power plant if the plant can pay for itself in 5 years. The total cost of the plant including the dam is $24 million. The plant will consist of 10 identical turbines with an overall plant efficiency of 88 percent. The gross head of the dam is 90 m, and the flow rate through each turbine is 565 L/s. The plant will operate an average of 8200 h per year, and the electricity generated is sold to the utility company at a rate of $0.105/kWh. If the operating and maintenance expenses of the plant are $750,000/year, determine if the investor should go ahead with this project.

7-27 The water in a large dam is to be used to generate electricity by the installation of a hydraulic turbine. The elevation difference between the free surfaces upstream and downstream of the dam is 480 ft. Water is to be supplied to the turbine at a rate of 1075 gal/s. The turbine efficiency is 89 percent based on the net head, and the generator efficiency is 98 percent. The total irreversible head loss (major losses + minor losses) in the piping system including the penstock is estimated to be 13 ft. Determine (a) the overall efficiency of this hydroelectric plant, (b) the electric power produced, and (c) the turbine shaft power.

7-28 The demand for electric power is usually much higher during the day than it is at night, and utility companies often sell power at night at much lower prices to encourage consumers to use the available power generation capacity and to avoid building new, expensive power plants that will be used only a short time during peak periods. Utilities are also willing to purchase power produced during the day from private parties at a high price.

Suppose a utility company is selling electric power for $0.05/kWh at night and is willing to pay $0.12/kWh for power produced during the day. To take advantage of this opportunity, an entrepreneur is considering building a large reservoir 40 m above the lake level, pumping water from the lake to the reservoir at night using cheap power, and letting the water flow from the reservoir back to the lake during the day, producing power as the pump motor operates as a turbine-generator during reverse flow. Preliminary analysis shows that a water flow rate of 2 m³/s can be used in either direction. The combined pump-motor and turbine-generator efficiencies are expected to be 75 percent each. Disregarding the frictional losses in piping and assuming the system operates for 10 h each in the pump and turbine modes during a typical day, determine the potential revenue this pump-turbine system can generate per year.

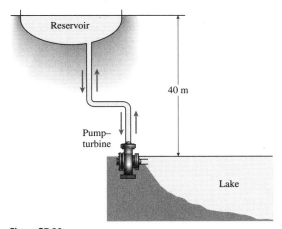

Figure P7-28

7-29 Which is not a component of mechanical energy of a flowing fluid?
(a) Kinetic energy   (b) Flow energy   (c) Potential energy   (d) Dynamic energy

7-30 Which statement is not correct regarding hydropower?
(a) Hydraulic turbine efficiency is greater than overall efficiency.
(b) Generator efficiency is the ratio of electric power to the shaft power.
(c) Overall efficiency of a hydroelectric plant is based on the net head.
(d) Piping efficiency is inversely proportional to head losses.
(e) Piping efficiency is greater than overall efficiency.

7-31 In a hydroelectric power plant, water flows through a large tube through the dam, which is called a
(a) Tailrace   (b) Draft tube   (c) Runner   (d) Penstock   (e) Propeller

7-32 A turbine is placed at the bottom of a 70-m-high water body. Water flows through the turbine at a rate of 15 m³/s. The power potential of the turbine is
(*a*) 10.3 MW   (*b*) 8.8 MW   (*c*) 7.6 MW   (*d*) 7.1 MW   (*e*) 5.9 MW

7-33 The efficiency of a hydraulic turbine-generator unit is specified to be 85 percent. If the generator efficiency is 96 percent, the turbine efficiency is
(*a*) 0.816   (*b*) 0.850   (*c*) 0.862   (*d*) 0.885   (*e*) 0.960

7-34 The efficiency of a hydraulic turbine is 82 percent. If the generator efficiency is 94 percent, the turbine-generator unit efficiency is
(*a*) 0.77   (*b*) 0.82   (*c*) 0.87   (*d*) 0.89   (*e*) 0.94

7-35 A hydraulic turbine is used to generate power by using the water in a dam. The elevation difference between the free surfaces upstream and downstream of the dam is 120 m. The water is supplied to the turbine at a rate of 150 kg/s. If the shaft power output from the turbine is 155 kW, the efficiency of the turbine is
(*a*) 0.77   (*b*) 0.80   (*c*) 0.82   (*d*) 0.85   (*e*) 0.88

7-36 A hydroelectric power plant is to be built in a dam with a gross head of 200 m. The head losses in the head gate and penstock are estimated to be 6 m. The flow rate through the turbine is 180,000 L/min. The efficiencies of the turbine and the generator are 88 and 96 percent, respectively. The electricity production from this turbine is
(*a*) 6.21 MW   (*b*) 5.89 MW   (*c*) 5.43 MW   (*d*) 5.17 MW   (*e*) 4.82 MW

IMPULSE TURBINES AND REACTION TURBINES

7-37 Consider a Pelton wheel turbine with an efficiency of 81 percent. The average radius of the wheel is 4.03 ft, and the jet velocity is 295 ft/s from a nozzle of exit diameter equal to 2.7 in. The turning angle of the buckets is $\beta = 170°$. Determine (*a*) the volume flow rate through the turbine and (*b*) the output shaft power.

7-38 The Pelton wheel of a hydroelectric power plant has the following specifications: The average radius of the wheel is 1.70 m; the exit diameter of the nozzle is 9.97 cm; the jet velocity is 90 m/s; and the turning angle of the buckets is 165°. Determine the volume flow rate through the turbine and the rotational speed of the wheel (in rpm) for maximum power. Also, determine the efficiency of the turbine if the shaft power output is 2100 kW.

7-39 A Francis radial-flow hydro turbine is to be designed with the following specifications: $r_2 = 2.00$ m, $r_1 = 1.42$ m, $b_2 = 0.731$ m, and $b_1 = 2.20$ m. The runner speed is 195 rpm. The wicket gates turn the flow by angle $\alpha_2 = 35°$ from radial at the runner inlet, and the flow at the runner outlet is at angle $\alpha_1 = 15°$ from radial (Fig. P7-39). The gross head of the dam is 95 m and the volume flow rate at design conditions is 330 m³/s. Calculate the inlet and outlet runner blade angles β_2 and β_1, respectively, and predict the power output and required net head. Neglect irreversible losses.

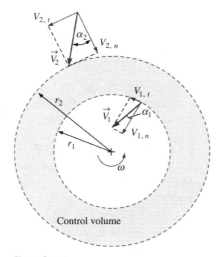

Figure P7-39

7-40   A Francis radial-flow hydro turbine has the following dimensions, where location 2 is the inlet and location 1 is the outlet: $r_2 = 2$ m, $r_1 = 1.30$ m, $b_2 = 0.85$ m, and $b_1 = 2.10$ m. The runner blade angles are $\beta_2 = 71.4°$ and $\beta_1 = 15.3°$ at the turbine inlet and outlet, respectively. The runner rotates at 150 rpm. The volume flow rate at design conditions is 80 m³/s. Irreversible losses are neglected in this preliminary analysis. Calculate the angle α_2 through which the wicket gates should turn the flow, where α_2 is measured from the radial direction at the runner inlet (Fig. P7-39). Calculate the swirl angle α_1, where α_1 is measured from the radial direction at the runner outlet (Fig. P7-39). Predict the power output and required net head.

7-41   A Francis radial-flow hydro turbine has the following dimensions, where location 2 is the inlet and location 1 is the outlet: $r_2 = 6.60$ ft, $r_1 = 4.40$ ft, $b_2 = 2.60$ ft, and $b_1 = 7.20$ ft. The runner blade angles are $\beta_2 = 80°$ and $\beta_1 = 45°$ at the turbine inlet and outlet, respectively. The runner rotates at 120 rpm. The volume flow rate at design conditions is 4.70×10^6 gpm. Irreversible losses are neglected in this preliminary analysis. Calculate the angle α_2 through which the wicket gates should turn the flow, where α_2 is measured from the radial direction at the runner inlet (Fig. P7-39). Calculate the swirl angle α_1, where α_1 is measured from the radial direction at the runner outlet (Fig. P7-39). Predict the power output in hp and required net head in feet.

TURBINE SPECIFIC SPEED

7-42   Calculate the turbine specific speed of the turbine in Prob. 7-38. Provide answers in both dimensionless form and in customary U.S. units. Is it in the normal range for an impulse turbine? Use the following values for this turbine: $\omega = 26.47$ rad/s, bhp = 2100 kW, $H = 322$ m, $\dot{n} = 253$ rpm.

7-43   Calculate the turbine specific speed of the turbine in Prob. 7-39. Provide answers in both dimensionless form and in customary U.S. units. Is it in the normal range for a Francis turbine? If not, what type of turbine would be more appropriate? The values for the turbine are $\omega = 20.42$ rad/s, bhp $= 2.958 \times 10^8$ W, $H = 91.4$ m, $\dot{n} = 195$ rpm.

7-44   Calculate the turbine specific speed of the turbine in Prob. 7-40. Provide answers in both dimensionless form and in customary U.S. units. Is it in the normal range for a Francis turbine? If not, what type of turbine would be more appropriate? The values for the turbine are $\omega = 15.71$ rad/s, bhp $= 6.713 \times 10^7$ W, $H = 85.5$ m, $\dot{n} = 150$ rpm.

7-45   Calculate the turbine specific speed of the turbine in Prob. 7-41 using customary U.S. units. Is it in the normal range for a Francis turbine? If not, what type of turbine would be more appropriate? The values for the turbine are $\omega = 12.57$ rad/s, bhp $= 1.96 \times 10^5$ hp, $H = 165$ ft, $\dot{n} = 120$ rpm.

7-46   Which turbine type is an impulse turbine?
(*a*) Kaplan   (*b*) Francis   (*c*) Pelton   (*d*) Propeller   (*e*) Centrifugal

7-47   Which turbine(s) are reaction turbines?
I. Kaplan   II. Pelton
III. Propeller   IV. Francis

(*a*) I   (*b*) I and II   (*c*) II and III   (*d*) I, III, and IV   (*e*) I, II, III, and IV

7-48   What is the specific speed of a turbine with the following specifications: $\omega = 15$ rad/s, bhp $= 5 \times 10^7$ W, $H = 80$ m.
(*a*) 0.052   (*b*) 0.81   (*c*) 1.74   (*d*) 2.20   (*e*) 4.78

RUN-OF-RIVER PLANTS AND WATERWHEELS

7-49   What is the difference between a hydroelectric power plant and a run-of-river plant?

7-50   What is the difference between a run-of-river plant and a waterwheel?

7-51   What is the power potential from a river per unit cross-sectional area (in W/m²) if the water velocity is 3 m/s? Show the unit conversions.

7-52   A run-of-river plant operates on a water stream whose cross-sectional area is 30 m². Water is flowing at a speed of 4 m/s. If the overall efficiency of this plant is 40 percent, what is the electrical power output from the plant?

7-53   Which power producing system has the highest efficiency?
(*a*) Waterwheel   (*b*) River plant   (*c*) Hydroelectric power plant   (*d*) Wind turbine

7-54   Most of electricity from hydroelectric power plants is produced using accumulation reservoirs, also called
(*a*) Waterwheels   (*b*) Dams   (*c*) River plants   (*d*) Pools

7-55   Which statement is not correct regarding hydropower?
(*a*) Turbomachines producing power when the working fluid is water are called hydraulic turbines.
(*b*) There is a larger amount of construction involved in river plants.
(*c*) River plants do not have an accumulation reservoir.
(*d*) River plants occupy less space compared to hydroelectric plants with a dam.

7-56   What is the power potential from a river per unit cross-sectional area if the water velocity is 2 m/s?
(*a*) 2 W/m²   (*b*) 4 W/m²   (*c*) 2000 W/m²   (*d*) 4000 W/m²   (*e*) 8000 W/m²

CHAPTER 8

Geothermal Energy

8-1 INTRODUCTION

Geothermal energy is the thermal energy within the earth's interior. It is a renewable energy source because heat is continuously transferred from within the earth to the water recycled by rainfall. The origin of geothermal energy is the earth's core (Fig. 8-1), and it is about 6500 km deep. The core is made up of an inner core (iron center) and an outer core made up of very hot magma. The temperature in the *magma* remains very high due to decay of radioactive particles. The outer core is surrounded by the *mantle*, whose thickness is about 3000 km. The mantle is made of magma and rock. The layer of the earth housing continents and ocean floors is called the *crust*. The thickness of the crust is 25 to 55 km on the continents and 5 to 8 km under the oceans. The crust is made up of tectonic plates. Volcanoes occur near the edges of these plates due to magma getting close to them. At some reasonable depths, the rocks and water absorb heat from this magma. These sites are characterized as geothermal resources. By digging wells and pumping the hot water to the surface, we make use of geothermal energy.

Geothermal resources can be classified based on their thermal and compositional characteristics.

Hydrothermal These are known geothermal fields containing high-temperature water in steam, mixture, or liquid phases.

Geopressurized These resources contain hot liquid water at 150 to 180°C at very high pressures (up to 600 bar). The fluid in these deposit-filled reservoirs also contains methane and high levels of dissolved solids. The fluid is highly corrosive and thus very difficult to harvest and handle.

Magma This is also called *molten rock*, typically contained under active volcanoes at temperatures above 650°C.

Enhanced They are also called hot, dry rock geothermal systems. These are not natural geothermal resources. The idea is that water is injected into the hot rock formation at high pressure, and then the resulting hot steam is brought back to the surface (Fig. 8-2). The system involves drilling of injection and production wells to a depth of 3 to 5 km. The temperature of the hot rock at this depth can be around 250°C.

Among these four geothermal resource categories, only hydrothermal resources are currently being exploited. The other three are estimated to have enormous energy potentials, but current technologies do not allow feasible energy production from these resources. The quality and life of a hydrothermal resource can be prolonged by reinjecting the waste

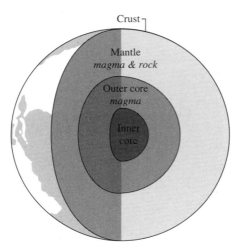

Figure 8-1 The interior of the earth. (*The NEED Project.*)

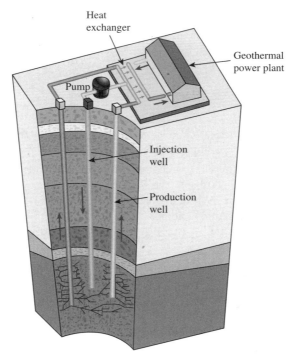

Figure 8-2 Operation of enhanced geothermal systems. (*Adapted from Geothermal Worldwide, 2019.*)

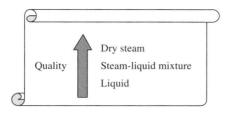

Figure 8-3 The quality of a geothermal resource depends on its phase (and temperature) in the reservoir. The higher the quality, the higher the work potential.

fluid, which is the most common method of disposal. Reinjection may also help to maintain reservoir pressure. Over-ground disposal of geothermal fluid has the potential to pollute rivers and lakes as well as air. Reinjection of geothermal water is a legal requirement in the United States.

A geothermal resource contains geothermal water at a temperature higher than that of the environment. One common classification of geothermal resources is based on the resource temperature.

High-temperature resource: $T > 150°C$

Medium-temperature resource: $90°C < T < 150°C$

Low-temperature resource: $T < 90°C$

The state of geothermal water in the reservoir may be superheated or saturated steam (dry steam), saturated steam-liquid mixture, or liquid (usually compressed liquid). Steam-dominated resources are of higher quality than liquid-dominated resources due to higher enthalpy and exergy (work potential) values (Fig. 8-3). However, most of the world's geothermal resources are liquid dominated. The geysers in California and other locations are rare examples of steam-dominated resources. Liquid-dominated systems can be produced either as brine or as brine-steam mixture, depending on the pressure maintained on the production system. If the pressure is reduced below the saturation pressure at that temperature, some of the brine will flash, and a two-phase mixture will result. If the pressure is maintained above the saturation pressure, the fluid will remain single-phase.

8-2 GEOTHERMAL APPLICATIONS

There are several options for utilizing the thermal energy produced from geothermal energy systems, such as electricity production, space heating and cooling, cogeneration, and geothermal heat pumps. Other applications of geothermal energy include growing plants and crops (greenhouses), drying of lumber, fruits and vegetables, spas, desalination, and fish farming. Ancient people used geothermal energy for heating and bathing. In many parts of the world, hot springs are still used for bathing because of the potential health benefits of minerals in hot geothermal water.

Geothermal energy is most commonly used for base-load electric power generation. The technology for producing power from geothermal resources is well established, and there are numerous geothermal power plants operating worldwide. The temperature of a geothermal resource should be about 150°C or higher for economic power production. However, one geothermal power plant in Northern Nevada used a 120°C liquid resource. Producing electricity from lower temperature geothermal resources is possible, and some novel designs are proposed in the literature, but it becomes difficult to economically justify such investments due to very low conversion efficiencies. The temperature of geothermal

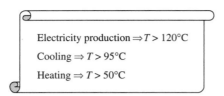

Electricity production $\Rightarrow T > 120°C$

Cooling $\Rightarrow T > 95°C$

Heating $\Rightarrow T > 50°C$

Figure 8-4   Approximate temperature requirements for geothermal applications.

water should be above 95°C for absorption cooling for reasonable coefficient of performance values. For heating, a temperature above 50°C could be used (Fig. 8-4).

A cogeneration system utilizing geothermal energy and producing electricity and heat and/or cooling represents an enhanced use of a resource. This is a cascaded application in which the used geothermal water leaving the power plant is used for heating/cooling before being reinjected back into the ground. Geothermal brine is reinjected at much lower temperatures in cogeneration applications in comparison to single power production. This represents a much higher utilization rate for a given resource, corresponding to higher potential revenues.

In the following sections, we present further information and analysis for common geothermal applications.

8-3 GEOTHERMAL HEATING

A number of residential and commercial districts are effectively heated in winter by low-cost geothermal heat in many parts of the world. Some of the largest district heating installations are in China, Sweden, Iceland, Turkey, and the United States. Almost 90 percent of buildings in Iceland (a relatively small country) are heated in winter by geothermal heat. The annual amount of space heating supplied in the world is estimated to be about 360,000 TJ as of 2015. Noting that 1 terajoule (TJ) = 10^{12} J, this is approximately equivalent to

36×10^{13} kJ

36×10^{13} Btu

36×10^8 therms (1 therm = 100,000 Btu)

7.2×10^9 kg of natural gas at a heating value of 50,000 kJ/kg

10.3×10^9 m³ of natural gas at a heating value of 35,000 kJ/m³

4500×10^6 or $4.5 billion at a natural gas price of $1.25/therm

Geothermal heat is used for space heating mostly in a district heating scheme. Normally, hot geothermal water is not directly circulated to the district due to its undesirable chemical composition and characteristics of geothermal brine. A common operating mode for a geothermal district heating system is shown in Fig. 8-5. Heat exchangers are used to transfer the heat of geothermal water to fresh water, and this heated fresh water is sent to the district. This heat is supplied to the buildings through individual heat exchangers.

District space heating usually involves long distances of piping. This means high values of pumping power are needed to overcome pressure losses. Another concern due to long piping is heat losses. We wish to minimize heat losses from the geothermal site and user district. By using preinsulated pipes, the temperature drop of water can be limited to under 1°C for a few km of pipe length. The selection of optimum pipe diameter is usually a trade-off between reduced pumping power and increased piping cost when larger diameters are selected. A larger diameter pipe also causes greater rates of heat loss due to increased surface area.

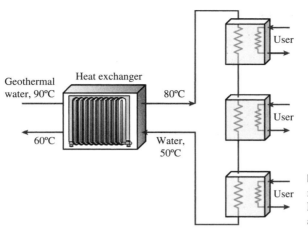

Figure 8-5 A common operating mode for geothermal district space heating systems. Temperature values are representative.

Geothermal heat can also be used for producing hot water for residences, offices, and industrial processes. A possible process heat application of geothermal hot water can be accomplished by preheating water in steam boilers in various industries. The use of geothermal heat for space, water, and process heating and space cooling applications is only viable when the resource is close to the user district. Carrying geothermal water over long distances is not cost-effective. For space and water heating, the resource temperature should be greater than about 50°C.

Geothermal energy is more effective when used directly than when converted to electricity, particularly for moderate and low temperature geothermal resources, since the direct use of geothermal heat for heating and cooling would replace the burning of fossil fuels from which electricity is generated much more efficiently (Kanoğlu and Çengel, 1999a).

When a district is heated by geothermal water, the rate of geothermal heat supplied to the district is determined from

$$\dot{Q}_{\text{heat,useful}} = \dot{m}c_p(T_{\text{supply}} - T_{\text{return}}) \qquad (8\text{-}1)$$

where $\dot{m}$ is the mass flow rate of geothermal water, c_p is the specific heat of geothermal water, and T_{supply} and T_{return} are supply and return temperatures of geothermal water for the district, respectively. Assuming that this heat represents the average rate of heat supplied to the district, the amount of energy supplied for a specified period of time (i.e., operating hours) is determined from

$$\text{Energy consumption} = \frac{\dot{Q}_{\text{heat,useful}} \times \text{Operating hours}}{\eta_{\text{heater}}} \qquad (8\text{-}2)$$

where η_{heater} is the efficiency of the heating equipment. For a geothermal heating application, it can be taken to be unity by neglecting fluid losses and heat losses in fluid lines and the heat exchange system. For a natural gas heater, it typically ranges between 80 and 90 percent. If the efficiency of a natural gas heater is 80 percent, this means that for 100 units of energy provided by natural gas, 80 units are transferred to the district as useful heat, and 20 percent are lost in the heater, mostly by hot exhaust gases. Once the amount of energy consumption is available, the corresponding cost of this energy can be determined from

$$\text{Energy cost} = \text{Energy consumption} \times \text{Unit price of energy} \qquad (8\text{-}3)$$

EXAMPLE 8-1
Potential Revenue from a Geothermal District Heating System

A residential district is currently heated by natural gas heaters in winter with an average efficiency of 85 percent. The price of natural gas is $1.30/therm (1 therm = 100,000 Btu). It is proposed to heat this district with geothermal water. On an average winter day, hot geothermal water is supplied to the district at 200°F at a rate of 55 lbm/s and returns at 130°F after giving its heat to the district. How much revenue can be generated per year if the winter period can be taken to be equivalent to 2800 h of these average conditions, and geothermal heat is sold at a discount of 25 percent with respect to natural gas?

SOLUTION We use the specific heat of water at room temperature: $c_p = 1.0$ Btu/lbm·°F. The rate of geothermal heat supplied to the district is determined from

$$\dot{Q}_{\text{heat,useful}} = \dot{m}c_p(T_{\text{supply}} - T_{\text{return}}) = (55 \text{ lbm/s})(1.0 \text{ Btu/lbm} \cdot °\text{F})(200 - 130)°\text{F} = 3850 \text{ Btu/s}$$

This is the rate of useful heat supplied to the space. Noting that natural gas heaters have an average efficiency of 85 percent (for 100 units of energy provided by natural gas, 85 units are transferred to the district as useful heat), the corresponding consumption of natural gas during a winter period of 2800 h is

$$\text{Natural gas consumption} = \frac{\dot{Q}_{\text{heat,useful}} \times \text{Operating hours}}{\eta_{\text{heater}}}$$

$$= \frac{(3850 \text{ Btu/s})(2800 \text{ h/yr})}{0.85}\left(\frac{3600 \text{ s}}{1 \text{ h}}\right)\left(\frac{1 \text{ therm}}{100,000 \text{ Btu}}\right)$$

$$= 4.566 \times 10^5 \text{ therm/yr}$$

Noting that geothermal heat is sold at a discount of 25 percent, the potential revenue that can be generated by selling geothermal heat is

$$\text{Potential revenue} = (1 - f_{\text{discount}}) \times \text{Natural gas consumption} \times \text{Unit price of natural gas}$$

$$= (1 - 0.25)(4.566 \times 10^5 \text{ therm/yr})(\$1.30/\text{therm})$$

$$= \textbf{\$445,150/yr} \quad \blacktriangle$$

Next, we introduce the degree-day method for annual energy consumption of a building. This method can be used for geothermal heating systems as well as solar, biomass, and other fossil fuel–based systems. The degree-day method is also used to calculate cooling energy consumption in summer. The cost of energy consumption for heating or cooling can be calculated using Eq. (8-3) by multiplying energy consumption by the unit cost of energy.

Degree-Day Method for Annual Energy Consumption

The simplest and most intuitive way of estimating the annual energy consumption of a building is the *degree-day* (or *degree-hour*) *method*, which is a *steady-state* approach. It is based on constant indoor conditions during the heating or cooling season and assumes the efficiency of the heating or cooling equipment is not affected by the variation of outdoor temperature. These conditions will be closely approximated if all the thermostats in a building are set at the same temperature at the beginning of a heating or cooling season and are never changed, and a seasonal average efficiency is used (rather than the full-load or design efficiency) for the heaters or coolers.

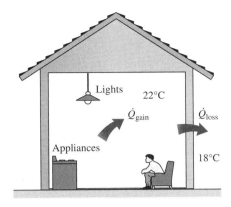

Figure 8-6 The heater of a building will not turn on as long as the internal heat gain makes up for the heat loss from a building (the *balance point* outdoor temperature).

You may think that anytime the outdoor temperature T_o drops below the indoor temperature T_i at which the thermostat is set, the heater will turn on to make up for the heat losses to the outside. However, the internal heat generated by people, lights, and appliances in occupied buildings as well as the heat gain from the sun during the day will be sufficient to compensate for the heat losses from the building until the outdoor temperature drops below a certain value. The *outdoor temperature* above which no heating is required is called the *balance point temperature* $T_{balance}$ (or the *base temperature*) and is determined from Fig. 8-6.

$$K_{overall}(T_i - T_{balance}) = \dot{Q}_{gain} \longrightarrow T_{balance} = T_i - \frac{\dot{Q}_{gain}}{K_{overall}} \tag{8-4}$$

where $K_{overall}$ is the *overall heat transfer coefficient* of the building in W/°C or Btu/h·°F. There is considerable uncertainty associated with the determination of the balance point temperature, but based on the observations of typical buildings, it is usually taken to be 18°C in Europe and 65°F (18.3°C) in the United States for convenience. The rate of energy consumption of the heating system is

$$\dot{Q}_{heating} = \frac{K_{overall}}{\eta_{heater}}(T_{balance} - T_o)^+ \tag{8-5}$$

where η_{heater} is the efficiency of the heating system, which is equal to 1.0 for electric resistance heating systems, the coefficient of performance (COP) for heat pumps, and boiler or heater efficiency (about 0.6 to 0.95) for fuel-burning heaters. If $K_{overall}$, $T_{balance}$, and η_{heater} are taken to be constants, the annual energy consumption for heating can be determined by integration (or by summation over daily or hourly averages) as

$$Q_{heating,year} = \frac{K_{overall}}{\eta_{heater}}\int\left[T_{balance} - T_o(t)\right]^+ dt \cong \frac{K_{overall}}{\eta_{heater}} DD_{heating} \tag{8-6}$$

where $DD_{heating}$ is the *heating degree-days*. The + sign "exponent" in Eqs. (8-5) and (8-6) indicates that only positive values are to be counted, and the temperature difference is to be zero when $T_o > T_{balance}$. The number of degree-days for a heating season is determined from

$$DD_{heating} = (1\ \text{day})\sum_{days}(T_{balance} - T_{o,avg,day})^+ \qquad (\text{°C-day}) \tag{8-7}$$

where $T_{o,avg,day}$ is the *average* outdoor temperature for each day (without considering temperatures above $T_{balance}$), and the summation is performed daily (Fig. 8-7). Similarly, we can also define *heating degree-hours* by using hourly average outdoor temperatures and performing

For a given day:

Highest outdoor temperature: 50°F

Lowest outdoor temperature: 30°F

Average outdoor temperature: 40°F

Degree-days for that day for a
balance-point temperature of 65°F

$$DD = (1 \text{ day})(65 - 40)°F$$
$$= 25°F\text{-day}$$
$$= 600°F\text{-hour}$$

Figure 8-7 The outdoor temperatures for a day during which the heating degree-day is 25°F-day.

the summation hourly. Note that the number of degree-hours is equal to 24 times the number of degree-days. Heating degree-days for each month and the yearly total for a balance point temperature of 65°F are given in Table 8-1 for several cities. *Cooling degree-days* are defined in the same manner to evaluate the annual energy consumption for cooling, using the same balance point temperature.

When heating degree-days are available for a given location, the amount of energy consumption for the entire winter season can be determined from

$$\text{Energy consumption} = \frac{K_{overall}}{\eta_{heater}} DD_{heating} \tag{8-8}$$

The amount of fuel consumption corresponding to this energy consumption can be determined by using the heating value of the fuel as

$$\text{Fuel consumption} = \frac{\text{Energy consumption}}{\text{Heating value of fuel}} \tag{8-9}$$

If the heating is accomplished by a heat pump, η_{heater} needs to be replaced by the heating COP (coefficient of performance) of the heat pump. Also, in this case, energy is consumed in the form of electricity. That is,

$$\text{Electricity consumption} = \frac{K_{overall}}{COP_{heating}} DD_{heating} \tag{8-10}$$

If we are calculating cooling energy consumption, we need to use cooling degree-days and cooling COP of the air conditioner:

$$\text{Electricity consumption} = \frac{K_{overall}}{COP_{cooling}} DD_{cooling} \tag{8-11}$$

In the above equations, the overall heat loss coefficient $K_{overall}$ of the building can be determined from the winter design temperatures and design heating load of the building:

$$\dot{Q}_{design} = UA\Delta T_{design} = UA(T_i - T_o)_{design} = K_{overall}(T_i - T_o)_{design} \tag{8-12}$$

Here, T_o is the design temperature of the location, and it represents a selected minimum winter temperature at which the heating equipment can provide adequate heat to the building, so that indoors can be maintained at the comfort temperature. The outdoor design temperature is usually selected to provide adequate heat to the building for 97.5 percent of the winter period, and thus it is referred to as *97.5% winter design temperature*. This 97.5%

TABLE 8-1   Average Winter Temperatures and Number of Degree-Days for Selected Cities in the United States (ASHRAE, 2008)

State and Station	Average Winter Temp. °F	°C	July	Aug.	Sep.	Oct.	Nov.	Dec.	Jan.	Feb.	March	April	May	June	Yearly Total
Alabama, Birmingham	54.2	12.7	0	0	6	93	363	555	592	462	363	108	9	0	2551
Alaska, Anchorage	23.0	5.0	245	291	516	930	1284	1572	1631	1316	1293	879	592	315	10,864
Arizona, Tucson	58.1	14.8	0	0	0	25	231	406	471	344	242	75	6	0	1800
California, San Francisco	53.4	12.2	82	78	60	143	306	462	508	395	363	279	214	126	3015
Colorado, Denver	37.6	3.44	6	9	117	428	819	1035	1132	938	887	558	288	66	6283
Florida, Tallahassee	60.1	15.9	0	0	0	28	198	360	375	286	202	86	0	0	1485
Georgia, Atlanta	51.7	11.28	0	0	18	124	417	648	636	518	428	147	25	0	2961
Hawaii, Honolulu	74.2	23.8	0	0	0	0	0	0	0	0	0	0	0	0	0
Idaho, Boise	39.7	4.61	0	0	132	415	792	1017	1113	854	722	438	245	81	5809
Illinois, Chicago	35.8	2.44	0	12	117	381	807	1166	1265	1086	939	534	260	72	6639
Indiana, Indianapolis	39.6	4.56	0	0	90	316	723	1051	1113	949	809	432	177	39	5699
Iowa, Sioux City	43.0	1.10	0	9	108	369	867	1240	1435	1198	989	483	214	39	6951
Kansas, Wichita	44.2	7.11	0	0	33	229	618	905	1023	804	645	270	87	6	4620
Kentucky, Louisville	44.0	6.70	0	0	54	248	609	890	930	818	682	315	105	9	4660
Louisiana, Shreveport	56.2	13.8	0	0	0	47	297	477	552	426	304	81	0	0	2184
Maryland, Baltimore	43.7	6.83	0	0	48	264	585	905	936	820	679	327	90	0	4654
Massachusetts, Boston	40.0	4.40	0	9	60	316	603	983	1088	972	846	513	208	36	5634
Michigan, Lansing	34.8	1.89	6	22	138	431	813	1163	1262	1142	1011	579	273	69	6909
Minnesota, Minneapolis	28.3	−1.72	22	31	189	505	1014	1454	1631	1380	1166	621	288	81	8382
Montana, Billings	34.5	1.72	6	15	186	487	897	1135	1296	1100	970	570	285	102	7049
Nebraska, Lincoln	38.8	4.11	0	6	75	301	726	1066	1237	1016	834	402	171	30	5864
Nevada, Las Vegas	53.5	12.28	0	0	0	78	387	617	688	487	335	111	6	0	2709
New York, Syracuse	35.2	2.11	6	28	132	415	744	1153	1271	1140	1004	570	248	45	6756
North Carolina, Charlotte	50.4	10.56	0	0	6	124	438	691	691	582	481	156	22	0	3191
Ohio Cleveland	37.2	3.22	9	25	105	384	738	1088	1159	1047	918	552	260	66	6351
Oklahoma, Stillwater	48.3	9.39	0	0	15	164	498	766	868	664	527	189	34	0	3725
Pennsylvania, Pittsburgh	38.4	3.89	0	9	105	375	726	1063	1119	1002	874	480	195	39	5987
Tennessee, Memphis	50.5	10.6	0	0	18	130	447	698	729	585	456	147	22	0	3232
Texas, Dallas	55.3	13.3	0	0	0	62	321	524	601	440	319	90	6	0	2363
Utah, Salt Lake City	38.4	3.89	0	0	81	419	849	1082	1172	910	763	459	233	84	6052
Virginia, Norfolk	49.2	9.89	0	0	0	136	408	698	738	655	533	216	37	0	3421
Washington, Spokane	36.5	2.83	9	25	168	493	879	1082	1231	980	834	531	288	135	6665

*Based on °F: quantities may be converted to degree-days based on °C by dividing by 1.8. This assumes 18°C corresponds to 65°F.

winter design temperature is commonly listed in tables for various cities. This temperature is used to calculate the required capacity of the heating system for the proper selection of the size of the heating equipment. We normally take design indoor temperature T_i to be 25°C or 77°F. In Eq. (8-12), $\dot{Q}_{design}$ represents the design heating load or capacity of the heating unit corresponding to winter design conditions.

Despite its simplicity, remarkably accurate results can be obtained with the *degree-day method* for most houses and single-zone buildings using a hand calculator. Besides, the degree-days characterize the *severity* of the weather at a location accurately, and the degree-day method serves as a valuable tool for gaining an *intuitive understanding* of annual energy consumption. But when the efficiency of the HVAC equipment changes considerably with the outdoor temperature, or the balance-point temperature varies significantly with time, it may be necessary to consider several bands (or "bins") of outdoor temperatures and to determine the energy consumption for each band using the equipment efficiency for those outdoor temperatures and the number of hours those temperatures are in effect. Then the annual energy consumption is obtained by adding the results of all bands. This modified degree-day approach is known as the *bin method*, and the calculations can still be performed using a hand calculator.

EXAMPLE 8-2
Annual Energy Consumptions for Different Heating Systems

Consider a house with an overall heat loss coefficient of $K_{overall} = 0.5$ kW/°C and heating degree-days of 2500°C-days. Determine the annual heating energy consumption for the following heating systems.

(a) Coal heater, $\eta_{heater} = 0.75$, Heating value of coal = 30,000 kJ/kg
(b) Natural gas heater, $\eta_{heater} = 0.85$
(c) Heat pump, COP = 2.5
(d) Resistance heater, $\eta_{heater} = 1$
(e) Geothermal, $\eta_{heater} = 1$

SOLUTION   The annual heating energy consumption for each heating system is determined as follows:

(a) Coal heater, $\eta_{heater} = 0.75$, Heating value of coal = 30,000 kJ/kg

$$\text{Energy consumption} = \frac{K_{overall}}{\eta_{heater}} DD_{heating}$$

$$= \frac{0.5 \text{ kJ/s·°C}}{0.75}(2500°\text{C-days})\left(\frac{24 \times 3600 \text{ s}}{1 \text{ day}}\right)$$

$$= 1.44 \times 10^8 \text{ kJ}$$

$$\text{Coal consumption} = \frac{\text{Energy consumption}}{\text{Heating value of fuel}} = \frac{1.44 \times 10^8 \text{ kJ}}{30,000 \text{ kJ/kg}} = \textbf{4800 kg}$$

(b) Natural gas heater, $\eta_{heater} = 0.85$

$$\text{Energy consumption} = \frac{K_{overall}}{\eta_{heater}} DD_{heating}$$

$$= \frac{0.5 \text{ kW/°C}}{0.85}(2500°\text{C-days})\left(\frac{24 \text{ h}}{1 \text{ day}}\right)\left(\frac{3412.14 \text{ Btu/h}}{1 \text{ kW}}\right)\left(\frac{1 \text{ therm}}{100,000 \text{ Btu}}\right)$$

$$= \textbf{1204 therms}$$

(*c*) Heat pump, COP = 2.5

$$\text{Electricity consumption} = \frac{K_{\text{overall}}}{\text{COP}} DD_{\text{heating}}$$

$$= \frac{0.5 \text{ kW/}^\circ\text{C}}{2.5} (2500^\circ\text{C-days}) \left(\frac{24 \text{ h}}{1 \text{ day}} \right)$$

$$= \mathbf{12,000 \ kWh}$$

(*d*) Resistance heater, $\eta_{\text{heater}} = 1$

$$\text{Electricity consumption} = \frac{K_{\text{overall}}}{\eta_{\text{heater}}} DD_{\text{heating}}$$

$$= \frac{0.5 \text{ kW/}^\circ\text{C}}{1} (2500^\circ\text{C-days}) \left(\frac{24 \text{ h}}{1 \text{ day}} \right)$$

$$= \mathbf{30,000 \ kWh}$$

(*e*) Geothermal, $\eta_{\text{heater}} = 1$

$$\text{Energy consumption} = \frac{K_{\text{overall}}}{\eta_{\text{heater}}} DD_{\text{heating}}$$

$$= \frac{0.5 \text{ kJ/s}\cdot^\circ\text{C}}{1} (2500^\circ\text{C-days}) \left(\frac{24 \times 3600 \text{ h}}{1 \text{ day}} \right) \left(\frac{1 \text{ MJ}}{1000 \text{ kJ}} \right)$$

$$= \mathbf{108,000 \ MJ} \quad (\text{or } 30,000 \text{ kWh})$$

For a cost comparison, the cost of energy for each system can be obtained by multiplying energy consumption by the unit price of energy. ▲

8-4 GEOTHERMAL COOLING

Geothermal heat may be supplied to an absorption refrigeration system for space cooling applications. A district cooling system utilizing geothermal heat may be feasible depending on the annual cooling load of the district. Geothermal cooling systems are not common due to their high initial cost. A geothermal cooling system installed in the Oregon Institute of Technology was estimated to pay for itself in about 15 years. Next, we provide an overview of absorption cooling systems.

Absorption Cooling System

A form of cooling system that becomes economically attractive when there is a source of inexpensive thermal energy at a temperature of 100 to 200°C is *absorption refrigeration*. Some examples of inexpensive thermal energy sources include geothermal energy, solar energy, and waste heat from cogeneration or process steam plants, and even natural gas when it is available at a relatively low price. A cogeneration plant may involve electricity generation and absorption cooling.

As the name implies, absorption refrigeration systems involve the absorption of a *refrigerant* by a *transport medium*. The most widely used absorption refrigeration system is the ammonia–water system, where ammonia (NH_3) serves as the refrigerant and water (H_2O) as the transport medium. Other absorption refrigeration systems include water–lithium bromide and water–lithium chloride systems, where water serves as the refrigerant. The latter two systems are limited to applications such as air conditioning where the minimum temperature is above the freezing point of water.

To understand the basic principles involved in absorption refrigeration, we examine the NH_3-H_2O system shown in Fig. 8-8. The ammonia–water refrigeration machine was patented by the Frenchman Ferdinand Carre in 1859. Within a few years, machines based on this principle were being built in the United States primarily to make ice and store food.

You will immediately notice from the figure that this system looks very much like the vapor-compression system, except that the compressor has been replaced by a complex absorption mechanism consisting of an absorber, a pump, a generator, a regenerator, a valve, and a rectifier. Once the pressure of NH_3 is raised by the components in the box (this is the only thing they are set up to do), it is cooled and condensed in the condenser by rejecting heat to the surroundings, is throttled to the evaporator pressure, and absorbs heat from the refrigerated space as it flows through the evaporator. So there is nothing new there. Here is what happens in the box:

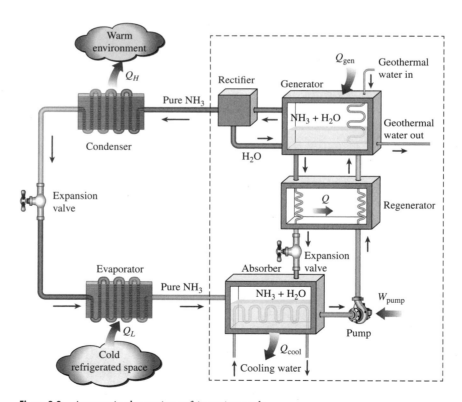

Figure 8-8   Ammonia absorption refrigeration cycle.

Ammonia vapor leaves the evaporator and enters the absorber, where it dissolves and reacts with water to form NH_3-H_2O. This is an exothermic reaction; thus heat is released during this process. The amount of NH_3 that can be dissolved in H_2O is inversely proportional to the temperature. Therefore, it is necessary to cool the absorber to keep its temperature as low as possible, hence to maximize the amount of NH_3 dissolved in water. The liquid NH_3-H_2O solution, which is rich in NH_3, is then pumped to the generator. Heat is transferred to the solution from a source (i.e., geothermal heat) to vaporize some of the solution. The vapor, which is rich in NH_3, passes through a rectifier, which separates the water and returns it to the generator. The high-pressure pure NH_3 vapor then continues its journey through the rest of the cycle. The hot NH_3-H_2O solution, which is weak in NH_3, then passes through a regenerator, where it transfers some heat to the rich solution leaving the pump and is throttled to the absorber pressure.

Compared with vapor-compression systems, absorption refrigeration systems have one major advantage: A liquid is compressed instead of a vapor. The steady-flow work is proportional to the specific volume, and thus the work input for absorption refrigeration systems is very small (on the order of 1 percent of the heat supplied to the generator) and often neglected in the cycle analysis. The operation of these systems is based on heat transfer from an external source. Therefore, absorption refrigeration systems are often classified as *heat-driven systems*.

The absorption refrigeration systems are much more expensive than the vapor-compression refrigeration systems. They are more complex and occupy more space, they are much less efficient, thus requiring much larger cooling towers to reject the waste heat, and they are more difficult to service, since they are less common. Therefore, absorption refrigeration systems should be considered only when the unit cost of thermal energy is low or a free renewable energy source such as geothermal energy is available. An ideal use of an absorption system is when it uses the waste heat of a power plant such as geothermal in a cogeneration scheme. Absorption refrigeration systems are primarily used in large commercial and industrial installations.

The COP of absorption refrigeration systems is defined as

$$\text{COP}_{\text{absorption}} = \frac{Q_{\text{cooling}}}{Q_{\text{gen}} + W_{\text{pump,in}}} \cong \frac{Q_{\text{cooling}}}{Q_{\text{gen}}} \tag{8-13}$$

Air-conditioning systems based on absorption refrigeration, called *absorption chillers,* perform best when the heat source can supply heat at a high temperature with little temperature drop. The absorption chillers are typically rated at an input temperature of 116°C (240°F). The chillers perform at lower temperatures, but their cooling capacity decreases sharply with decreasing source temperature, about 12.5 percent for each 6°C (10°F) drop in the source temperature. For example, the capacity goes down to 50 percent when the supply water temperature drops to 93°C (200°F). In that case, one needs to double the size (and thus the cost) of the chiller to achieve the same cooling.

The COP of the chiller is affected less by the decline of the source temperature. The COP drops by 2.5 percent for each 6°C (10°F) drop in the source temperature. The nominal COP of single-stage absorption chillers at 116°C (240°F) is 0.65 to 0.70. Therefore, for each ton of refrigeration, a heat input of (12,000 Btu/h)/0.65 = 18,460 Btu/h is required. At 88°C (190°F), the COP drops by 12.5 percent, and thus the heat input increases by 12.5 percent for the same cooling effect. Therefore, the economic aspects must be evaluated carefully before any absorption refrigeration system is considered, especially when the source temperature is below 93°C (200°F).

EXAMPLE 8-3
Potential Revenue from
a Geothermal Cooling
System

Geothermal liquid water from a well is available at 95°C at a rate of 150,000 kg/h and is to be used for space cooling using an absorption cooling system. Geothermal water leaves the generator of the absorption system at 80°C. If the COP of the absorption system is 0.65, determine the rate of cooling provided by the system.

This geothermal cooling system is replacing a conventional vapor-compression refrigeration system with an average COP value of 2.5, and the unit price of electricity is $0.125/kWh. Determine the annual potential revenue that can be generated by this geothermal cooling system if geothermal cooling is sold with a discount of 10 percent compared to conventional cooling. Use a seasonal cooling period of 1750 h at the calculated cooling rate.

SOLUTION   The specific heat of water at room temperature is $c_p = 4.18$ kJ/kg·°C. The rate of heat transfer in the generator of the absorption system is

$$\dot{Q}_{gen} = \dot{m}c_p(T_1 - T_2) = (150,000 \text{ kg/h})(4.18 \text{ kJ/kg·°C})(95 - 80)°C = 9.405 \times 10^6 \text{ kJ/h}$$

Using the definition of the COP of an absorption cooling system, the rate of cooling is determined to be

$$COP_{absorption} = \frac{\dot{Q}_{cool}}{\dot{Q}_{gen}} \longrightarrow \dot{Q}_{cool} = COP_{absorption}\dot{Q}_{gen} = (0.65)(9.405 \times 10^6 \text{ kJ/h})\left(\frac{1 \text{ h}}{3600 \text{ s}}\right) = \mathbf{1698 \text{ kW}}$$

If a conventional cooling system with a COP of 2.5 is used for this cooling, the power input would be

$$COP = \frac{\dot{Q}_{cool}}{\dot{W}_{in}} \longrightarrow \dot{W}_{in} = \frac{\dot{Q}_{cool}}{COP} = \frac{1698 \text{ kW}}{2.5} = 679.3 \text{ kW}$$

The electricity consumption for a period of 1750 h is

$$\text{Electricity consumption} = \dot{W}_{in} \times \text{Operating hours} = (679.3 \text{ kW})(1750 \text{ h}) = 1.189 \times 10^6 \text{ kWh}$$

At a discount of 10 percent, the revenue generated by selling the geothermal cooling is

$$\text{Revenue} = (1 - f_{discount}) \times \text{Electricity consumption} \times \text{Unit price of electricity}$$

$$= (1 - 0.1)(1.189 \times 10^6 \text{ kWh})(\$0.125/\text{kWh})$$

$$= \mathbf{\$133,800} \quad \blacktriangle$$

8-5 GEOTHERMAL HEAT PUMP SYSTEMS

Ground-source heat pumps represent the most common use of geothermal energy in terms of the number of units installed. These heat pumps are called geothermal heat pumps as they utilize the heat of the earth. Ground-source heat pumps provide higher values of COP than air-source units. The ground at a few meters depth is at a higher temperature than the ambient air in winter, and it is at a lower temperature than the ambient in summer. These systems use higher ground temperatures in winter for heat absorption (heating mode) and cooler ground temperatures in summer for heat rejection (cooling mode), and this is the reason for higher COPs.

Next, we present an overview of heat pumps, which is followed by the discussion of geothermal or ground-source heat pumps.

Heat Pump Systems

Heat pumps are generally more expensive to purchase and install than other heating systems, but they save money in the long run in some areas because they lower the heating bills. Despite their relatively higher initial costs, the popularity of heat pumps is increasing.

The COP of a heat pump in the heating mode is defined as the heating effect divided by the work input

$$\text{COP} = \frac{Q_{\text{heating}}}{W_{\text{in}}} \tag{8-14}$$

The COP of a heat pump in the cooling mode (called air conditioner) is defined as the cooling effect divided by the work input

$$\text{COP} = \frac{Q_{\text{cooling}}}{W_{\text{in}}} \tag{8-15}$$

The most common energy source for heat pumps is atmospheric air (air-to-air systems), although water and soil are also used. The major problem with air-source systems is *frosting*, which occurs in humid climates when the temperature falls below 2 to 5°C. The frost accumulation on the evaporator coils is highly undesirable since it seriously disrupts heat transfer. The coils can be defrosted, however, by reversing the heat pump cycle (running it as an air conditioner). This results in a reduction in the efficiency of the system. Water-source systems usually use well water from depths of up to 80 m in the temperature range of 5 to 18°C, and they do not have a frosting problem. They typically have higher COPs but are more complex and require easy access to a large body of water such as underground water. Ground-source systems are also rather involved since they require long tubing placed deep in the ground where the soil temperature is relatively constant. The COP of heat pumps usually ranges between 1.5 and 4 depending on the particular system used and the temperature of the source. Recently developed heat pumps that use variable-speed electric motor drives are at least twice as energy efficient as their predecessors.

Both the capacity and the efficiency of a heat pump fall significantly at low temperatures. Therefore, most air-source heat pumps require a supplementary heating system such as electric resistance heaters or an oil or gas furnace. Since water and soil temperatures do not fluctuate much, supplementary heating may not be required for water-source or ground-source systems. However, the heat pump system must be large enough to meet the maximum heating load.

Heat pumps and air conditioners have the same mechanical components. Therefore, it is not economical to have two separate systems to meet the heating and cooling requirements of a building. One system can be used as a heat pump in winter and an air conditioner in summer, and it is just referred to as a heat pump. This is accomplished by adding a reversing valve to the cycle, as shown in Fig. 8-9. As a result of this modification, the condenser of the heat pump (located indoors) functions as the evaporator of the air conditioner in summer. Also, the evaporator of the heat pump (located outdoors) serves as the condenser of the air conditioner. This feature increases the competitiveness of the heat pump. Such dual-purpose units are commonly used in apartment units and motels.

Heat pumps are most competitive in areas that have a large cooling load during the cooling season and a relatively small heating load during the heating season, such as in the southern parts of the United States. In these areas, the heat pump can meet the entire cooling and heating needs of residential or commercial buildings. The heat pump is least competitive in areas where the heating load is very large and the cooling load is small, such as in the northern parts of the United States.

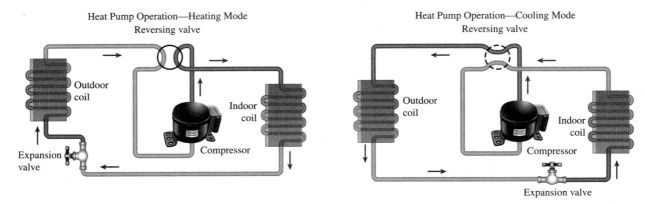

Figure 8-9 A heat pump can be used to heat a house in winter and cool it in summer.

Ground-Source Heat Pump Systems

Ground-source heat pump systems are also known as *geothermal heat pumps* as they use the heat of the earth in their operation. They have higher COPs than ordinary air-source heat pumps because ground is at a higher temperature than ambient air in winter (heating mode) and at a lower temperature than ambient air in summer (cooling mode). Table 8-2 shows how the ground temperature changes with depth in different seasons of the year. The ground temperature increases with depth in winter (see January data) and decreases with depth in summer (see July data). It is clear that the ground temperature is more stable than air temperature throughout the year. It is essentially constant below about 10 m. It starts increasing again at a depth greater than about 60 m.

In the heating mode operation of a ground-source heat pump, heat is absorbed from the ground at T_L, which is higher than the temperature of the ambient air. Heat is supplied to the indoors at T_H by the heat pump. In the cooling mode, heat is absorbed from the indoors at T_L and rejected to the ground at T_H, which is at a lower temperature compared to the temperature of the ambient air (Fig. 8-10).

TABLE 8-2   Variation of Ground Temperature with Depth in a Location for Different Months of the Year (Schöffengrund- Schwalbach GSHP Test Plant, 1985–89)

Depth, m	Ground Temperature, °C			
	January	April	July	October
2	6	5	15	14
4	8	5	10	12
6	9	6	8	11
8	9	7	8	9
10	10	8	8	9
12	10	9	9	9
14	10	9	9	9
16	9	9	9	9

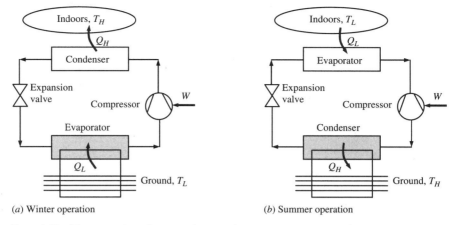

(*a*) Winter operation   (*b*) Summer operation

Figure 8-10 The operation of a ground-source heat pump in winter and in summer.

Now, let us compare performances of an air-source heat pump and a ground-source heat pump with the following data in winter operation:

Air-source, $T_L = 0°C$, $T_H = 25°C$

Ground-source, $T_L = 10°C$, $T_H = 25°C$

We calculate the Carnot COPs of these heat pumps for comparison purposes to be

Air-source: $$\text{COP}_{HP,Carnot} = \frac{T_H}{T_H - T_L} = \frac{298\ \text{K}}{298\ \text{K} - 273\ \text{K}} = 11.9$$

Ground-source: $$\text{COP}_{HP,Carnot} = \frac{T_H}{T_H - T_L} = \frac{298\ \text{K}}{298\ \text{K} - 283\ \text{K}} = 19.9$$

The maximum COP of the ground-source heat pump is 67 percent greater than that of the air-source heat pump. Even though actual COP values are significantly lower than the calculated COPs above, the comparison is still applicable in actual cases. Despite their much higher initial cost as compared to air-source units, ground-source heat pump units are preferred due to their higher COPs. The actual COP of ground-source heat pumps ranges between about 3 and 5, while the actual COP of air-source heat pumps ranges between about 1.5 and 3.

Ground-source heat pumps can be classified according to the configuration of piping and heat source as follows:

Horizontal loop heat pump

It involves horizontal underground piping in 1.2 to 2.0 m depths (Fig. 8-11). It is suitable when there is sufficient area for pipe burial, such as the relatively large backyard of a house.

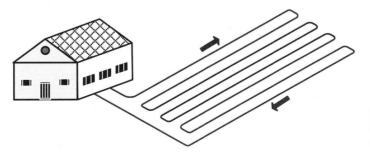

Figure 8-11 Schematic of horizontal loop ground-source heat pump.

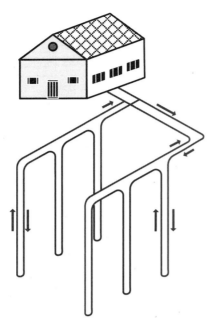

Figure 8-12 Schematic of vertical loop ground-source heat pump.

Vertical loop heat pump

It is also called a borehole loop heat pump. Vertical piping in 10 to 250 m depths is used (Fig. 8-12). It can be installed almost anywhere because a small field allowing vertical drilling is sufficient. This type of heat pump is selected for meeting the heating and cooling requirements of a medium-size university campus with several buildings. Vertical piping is more expensive than horizontal piping for a given heat transfer surface area. These systems require virtually no regular maintenance, and they are safe. However, the capacity is limited per borehole, and the ground temperature is relatively low. A borehole heat pump application in Sweden involves 33 boreholes each 160 m deep, and the capacity is 225 kW. Another example in Istanbul involves 208 boreholes with an average depth of 88 m. The system provides heating in winter and cooling in summer with a capacity of 1 MW (Sanner, 2008).

Ground water wells heat pump

Underground water is circulated through the evaporator of the heat pump unit (Fig. 8-13). Heat is transferred from the water to the refrigerant flowing in the evaporator. The cooler

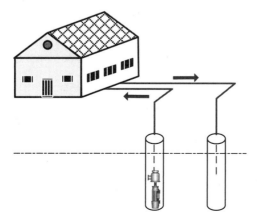

Figure 8-13 Schematic of ground water wells ground-source heat pump.

water leaving the evaporator is dumped back to the ground at a different location. The water well has a depth of 5 to 50 m. These systems can provide high capacities with relatively low cost. The temperature of water is relatively high compared to horizontal loop and borehole heat pumps. An aquifer is needed for sufficient water yield. The wells need to be maintained, and water quality should be monitored.

Ground-source heat pumps can also be classified in terms of the fluid type circulating in the evaporator and the condenser. Usually water or brine is circulated to transfer heat from the ground to the refrigerant in the evaporator by means of a horizontal-loop or borehole heat exchanger. A circulation pump is located in the horizontal or vertical piping to circulate the water between the ground and the evaporator. For example, a heat pump is called a *water-to-air* heat pump if water is circulating through the evaporator and air is circulating through the condenser. It is called a *water-to-water* heat pump if the heat pump is used to produce hot water by circulating water through the condenser and an underground water is circulated in the evaporator. A borehole heat pump may be modified such that a heat pipe is used in place of a borehole heat exchanger. The ground circuit may also be designed such that refrigerant coming out of the expansion valve flows through the piping in the ground, picking up heat and evaporating before entering the compressor. In this configuration, heat is directly transferred from the ground to the refrigerant without the use of another fluid such as water or brine.

Early ground-source heat pumps date back to 1945 when a horizontal loop heat pump was installed in Indianapolis. The pipes were placed in a garden. A ground water wells heat pump was installed in Portland, Oregon in 1948 with 2.4 MW of heating capacity. The temperature of underground water was 18°C. In Europe, the first ground water heat pump was used around 1970 and the first borehole heat pump was installed around 1980. The number of heat pumps and ground-source ones sold in Europe and elsewhere has increased dramatically in recent years. For example, the number of new geothermal heat pumps increased from 4000 in 2000 to 17,000 in 2015 in Germany (Sanner, 2017). The leading countries are the United States, China, Sweden, Germany, and France (Gehlin, 2015).

EXAMPLE 8-4
An Air-Source Heat Pump

A house in Cleveland, Ohio, is equipped with an air-source heat pump system that is used for space heating in winter. The seasonal COP of the heat pump in winter is estimated to be 1.5. The design heating load of the house is calculated to be 19 kW for an indoor temperature of 25°C (Fig. 8-14). The unit cost of electricity is $0.12/kWh. Determine the annual electricity consumption and its cost for this house.

The annual heating degree-days total for Cleveland, Ohio, is 3528°C-days and the 97.5% winter design temperature of Cleveland is −15°C.

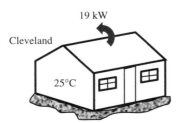

Figure 8-14   Schematic for Example 8-4.

SOLUTION   The annual heating degree-days total for Cleveland, Ohio, is 3528°C-days. The 97.5% winter design temperature of Cleveland is −15°C. The overall heat loss coefficient $K_{overall}$ of the house is determined from

$$\dot{Q}_{design} = UA\Delta T_{design} = UA(T_i - T_o)_{design} = K_{overall}(T_i - T_o)_{design}$$

Substituting,

$$K_{overall} = \frac{\dot{Q}_{design}}{(T_i - T_o)_{design}} = \frac{19 \text{ kW}}{[25 - (-15)]°C} = 0.475 \text{ kW/°C}$$

Then the annual electricity consumption of the house for heating is determined to be

$$\text{Electricity consumption} = \frac{K_{overall}}{\text{COP}} DD_{heating}$$

$$= \frac{0.475 \text{ kW/°C}}{1.5}(3528°C\text{-days})\left(\frac{24 \text{ h}}{1 \text{ day}}\right)$$

$$= \mathbf{26{,}810 \text{ kWh}}$$

The cost of this electricity consumption is

$$\text{Heating cost} = \text{Electricity consumption} \times \text{Unit cost of electricity}$$

$$= (26{,}813 \text{ kWh})(\$0.12/\text{kWh})$$

$$= \mathbf{\$3217} \quad \blacktriangle$$

EXAMPLE 8-5   Reconsider Example 8-4. (*a*) If the house in Cleveland is heated in winter by a ground-source heat
A Ground-Source   pump with a seasonal COP of 3.5, determine the annual electricity consumption and its cost for this
Heat Pump   house. (*b*) If the house in Cleveland is heated in winter by a natural gas furnace that is 88 percent efficient, determine the annual natural gas consumption and its cost in winter. The unit cost of natural gas is $1.10/therm.

SOLUTION   (*a*) The annual electricity consumption of the house for heating is determined to be

$$\text{Electricity consumption} = \frac{K_{overall}}{\text{COP}} DD_{heating}$$

$$= \frac{0.475 \text{ kW/°C}}{3.5}(3528°C\text{-days})\left(\frac{24 \text{ h}}{1 \text{ day}}\right)$$

$$= \mathbf{11{,}490 \text{ kWh}}$$

The cost of this electricity consumption is

$$\text{Heating cost} = \text{Electricity consumption} \times \text{Unit cost of electricity}$$

$$= (11{,}490 \text{ kWh})(\$0.12/\text{kWh})$$

$$= \mathbf{\$1379}$$

(*b*) The annual natural gas consumption of the house for heating is determined to be

$$\text{Gas consumption} = \frac{K_{\text{overall}}}{\eta_{\text{furnace}}} DD_{\text{heating}}$$

$$= \frac{0.475 \text{ kW/°C}}{0.88}(3528°\text{C-days})\left(\frac{24 \text{ h}}{1 \text{ day}}\right)\left(\frac{3412.14 \text{ Btu/h}}{1 \text{ kW}}\right)\left(\frac{1 \text{ therm}}{100,000 \text{ Btu}}\right)$$

$$= 1559 \text{ therms}$$

The cost of this natural gas consumption is

$$\text{Heating cost} = \text{Gas consumption} \times \text{Unit cost of natural gas}$$

$$= (1559 \text{ therms})(\$1.10/\text{therm})$$

$$= \$1715 \quad \blacktriangle$$

8-6 GEOTHERMAL POWER PRODUCTION

Only a fraction of geothermal resources have enough high temperatures to make them suitable for electricity production. Geothermal power plants have been in operation for decades in many parts of the world (Fig. 8-15). The first geothermal power plant was built in Italy in 1904. The first U.S. geothermal plant was built in 1960 in the Geysers in northern California. There are now dozens of geothermal power plants in the United States in California, Nevada, Utah, Idaho, Oregon, and Hawaii.

Figure 8-15 A small-size geothermal power plant in Nevada. The source temperature for this plant is only 120°C. (*Photo by Yunus Çengel.*)

Different thermodynamic cycles can be used for producing power from geothermal resources. See DiPippo (2007) for detailed coverage of geothermal power plants. A case study on an existing binary geothermal power plant is available in Kanoğlu and Çengel (1999b).

The simplest geothermal cycle is the *direct steam* or dry steam cycle. Steam from the geothermal well is passed through a turbine and exhausted to the atmosphere or to a condenser. Flash steam plants are used to generate power from liquid-dominated resources that are hot enough to flash a significant proportion of the water to steam in surface equipment, either at one or two pressure stages. In a *single-flash* plant (Fig. 8-16), the pressure of geothermal water is dropped to a predetermined value in a flash chamber. The resulting two-phase mixture is separated into liquid and vapor in the separator. The vapor is routed to a turbine in which it is expanded to condenser pressure. Steam exiting the turbine is condensed with cooling water obtained in a cooling tower or a spray pond before being reinjected. The liquid geothermal water at state 6 and that in state 5 is reinjected back into the ground.

Thermodynamic analysis of a single-flash geothermal power plant is similar to analysis of the Rankine cycle. Neglecting kinetic and potential energy changes across the turbine, the power output from the turbine is determined from

$$\dot{W}_{\text{out}} = \dot{m}_3(h_3 - h_4) \qquad (8\text{-}16)$$

The thermal efficiency of the plant may be defined as the ratio of the power output to the energy input to the plant:

$$\eta_{\text{th}} = \frac{\dot{W}_{\text{out}}}{\dot{E}_{\text{in}}} \qquad (8\text{-}17)$$

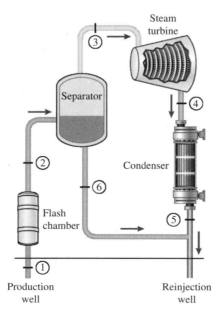

Figure 8-16   Single-flash geothermal power plant.

The energy input to the plant may be taken as the enthalpy difference between the state of the geothermal water at the plant inlet and the enthalpy of liquid water at the environmental state multiplied by the mass flow rate of the geothermal water. This refers to the energy of the geothermal stream at the plant inlet (state 1 in Fig. 8-16) with respect to the environmental state (state 0). Then,

$$\dot{E}_{\text{in}} = \dot{m}_1 (h_1 - h_0) \tag{8-18}$$

The enthalpy at the environmental state can be taken as 1 atm and 25°C, and can be approximated as saturated liquid at 25°C. That is, $h_0 \cong h_{f@25°C}$. Substituting these relations back to Eq. (8-17), we obtain

$$\eta_{\text{th}} = \frac{\dot{W}_{\text{out}}}{\dot{E}_{\text{in}}} = \frac{\dot{m}_3 (h_3 - h_4)}{\dot{m}_1 (h_1 - h_0)} \tag{8-19}$$

The thermal efficiency can also be written using the energy rejected from the plant as

$$\eta_{\text{th}} = \frac{\dot{W}_{\text{out}}}{\dot{E}_{\text{in}}} = 1 - \frac{\dot{E}_{\text{out}}}{\dot{E}_{\text{in}}} \tag{8-20}$$

where the rate of energy rejected from the system is the sum of the energies of stream 6 and stream 4 with respect to the environmental state:

$$\dot{E}_{\text{out}} = \dot{m}_6 (h_6 - h_0) + \dot{m}_4 (h_4 - h_0) \tag{8-21}$$

The flashing process in a flash plant is essentially a constant-enthalpy process as shown in the temperature-enthalpy diagram in Fig. 8-17. The state points refer to Fig. 8-16. Saturated (or compressed) liquid geothermal water (state 1) enters the flash chamber in which its pressure and temperature decrease. The enthalpy of the fluid stream remains constant since the chamber is adiabatic and there is no work interaction. The fluid is a saturated liquid-vapor mixture at the exit of the chamber (state 2). Water vapor (state 3) is separated from the liquid (state 6) in the separator. The water vapor is directed to the turbine while the liquid is sent to a reinjection well.

In a *double-flash* geothermal power plant (Fig. 8-18), the liquid water leaving the separator after the first flashing process is further expanded in a second flash chamber. Additional water vapor resulting from this process is separated and sent to a lower pressure stage of the turbine for additional power production. The rest of the operation is the same as the single-flash plant.

Thermodynamic analysis of a double-flash power plant is very similar to that of a single-flash plant. In this configuration, power is produced from the turbine by the expansion of

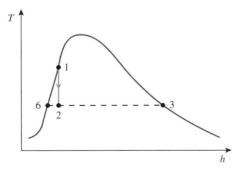

Figure 8-17 Flashing process in temperature-enthalpy diagram.

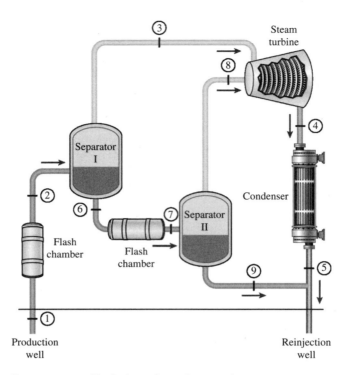

Figure 8-18   Double-flash geothermal power plant.

two fluid streams entering at state 3 and state 8 (Fig. 8-18). Then the thermal efficiency can be expressed as

$$\eta_{th} = \frac{\dot{W}_{out}}{\dot{E}_{in}} = \frac{\dot{m}_3(h_3 - h_4) + \dot{m}_8(h_8 - h_4)}{\dot{m}_1(h_1 - h_0)} \tag{8-22}$$

Binary cycle plants use the geothermal brine from liquid-dominated resources at relatively low temperatures. A binary plant in Alaska uses a geothermal resource at 57°C (Erkan et al., 2008). These plants operate on a Rankine cycle with a binary working fluid (isobutane, pentane, isopentane, R-114, etc.) that has a low boiling temperature. The working fluid is completely vaporized and usually superheated by the geothermal water in a heat exchanger network, as shown in Fig. 8-19. In this configuration, the resulting vapor expands in the turbine and then condenses in an air-cooled condenser (dry cooling tower) before being pumped back to the heat exchangers to complete the cycle.

A temperature-entropy diagram of the binary cycle is given in Fig. 8-20. Common working fluids used in binary geothermal power plants such as isobutane and pentane have a saturation line with positive slope. As a result, the turbine exit (state 4) is always super-heated vapor regardless of the turbine back pressure. The lowest possible pressure at the turbine exit can be used to maximize the power output since we do not have to worry about moisture formation at the later stages of the turbine as in the turbines of steam power plants.

Using the state points in Fig. 8-19, the thermal efficiency of the binary power plant is expressed similar to a flash plant as

$$\eta_{th} = \frac{\dot{W}_{net,out}}{\dot{E}_{in}} = \frac{\dot{W}_{turbine} - \dot{W}_{pump} - \dot{W}_{fan}}{\dot{E}_{in}} \tag{8-23}$$

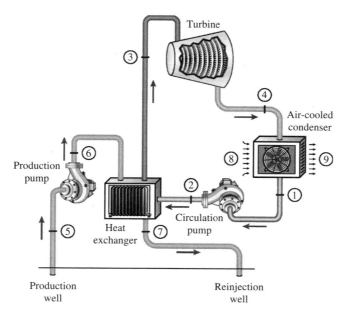

Figure 8-19 Binary cycle geothermal power plant.

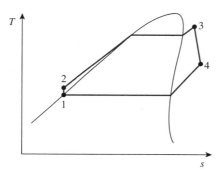

Figure 8-20 Temperature-entropy diagram of the binary cycle.

The energy input is determined from

$$\dot{E}_{in} = \dot{m}_5(h_5 - h_0) \tag{8-24}$$

The power output from the turbine and the power input to the circulation pump are expressed as

$$\dot{W}_{turbine} = \dot{m}_3(h_3 - h_4) \tag{8-25}$$

$$\dot{W}_{pump} = \dot{m}_1(h_2 - h_1) \tag{8-26}$$

The power input to the production and circulation pumps is usually small compared to turbine power. However, the power consumed by the cooling fans in the condenser can be up to 20 percent or more of the turbine power. Ambient temperature has a considerable effect on the power production of air-cooled binary geothermal power plants. As a result of reduced turbine power and increased fan power at higher ambient temperatures, the power output from such a plant decreases by up to 50 percent from winter to summer.

The thermal efficiency of a binary plant can also be expressed based on the energy (heat) input to the binary cycle as

$$\eta_{th} = \frac{\dot{W}_{net,out}}{\dot{Q}_{in}} \tag{8-27}$$

where the heat input is expressible as

$$\dot{Q}_{in} = \dot{m}_6(h_6 - h_7) = \dot{m}_2(h_3 - h_2) \tag{8-28}$$

It should be clear that the thermal efficiency calculated using Eq. (8-27) is greater than that calculated by Eq. (8-23).

The heat exchange between the geothermal water and binary working fluid is shown in Fig. 8-21. The state points refer to Fig. 8-19. Binary fluid should be vaporized completely (states $2a$ to $2b$) and superheated by the geothermal water (states $2b$ to 3) as the water temperature is decreased from T_6 to T_{6a}. Binary fluid is heated from T_2 to T_{2a} as the temperature of geothermal water is decreased from T_{6a} to T_7. To achieve this heat exchange, there must be a temperature difference between the vaporization temperature of the binary fluid (state $2a$) and the temperature of geothermal water at state $6a$. This temperature difference is called pinch-point temperature difference ΔT_{pp}. The state $6a$ is called the pinch-point of geothermal water. An application of the conservation of energy principle on this adiabatic heat exchanger gives the following two equations:

$$\dot{m}_{geo}(h_6 - h_{6a}) = \dot{m}_{binary}(h_3 - h_{2b}) \tag{8-29}$$

$$\dot{m}_{geo}(h_{6a} - h_7) = \dot{m}_{binary}(h_{2a} - h_2) \tag{8-30}$$

Here, $\dot{m}_{geo}$ and $\dot{m}_{binary}$ are the mass flow rates of geothermal brine and binary fluid, respectively, and h is the enthalpy. Note that states $2a$ and $2b$ are the saturated liquid and vapor states at the vaporization temperature of the binary fluid, respectively. That is, $h_{2a} = h_{f@T_{vap}}$ and $h_{2b} = h_{g@T_{vap}}$. Solving these two equations simultaneously gives the mass flow rate of the binary fluid and the exit temperature of geothermal water when the inlet temperatures of geothermal water and binary fluid, the exit temperature of the binary fluid, and the pinch-point temperature difference ΔT_{pp} are specified. The value of ΔT_{pp} is usually between 5 and 10°C.

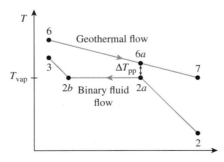

Figure 8-21   Heat exchange process between the geothermal brine and the binary working fluid in the heat exchanger of a binary cycle power plant.

A *combined flash/binary* plant (Fig. 8-22), also called a *hybrid* plant, incorporates both a binary unit and a flashing unit to exploit the advantages associated with both systems. The liquid portion of the geothermal mixture supplies the input heat for the binary cycle, while the steam portion drives a steam turbine to produce power. Power is obtained from both the steam turbine and the binary turbine. The geothermal liquid water is reinjected into the ground at a lower temperature (state 7) than in a single-flash plant. Referring to Fig. 8-22, the thermal efficiency of a combined flash/binary geothermal power plant can be determined from

$$\eta_{th} = \frac{\dot{W}_{net,out}}{\dot{E}_{in}} = \frac{\dot{W}_{turbine} - \dot{W}_{pump} - \dot{W}_{fan}}{\dot{E}_{in}}$$

$$= \frac{\dot{m}_3(h_3 - h_4) + \dot{m}_8(h_8 - h_9) - \dot{W}_{pump} - \dot{W}_{fan}}{\dot{m}_1(h_1 - h_0)}$$

(8-31)

where $\dot{W}_{fan}$ is the power consumed by the fans in the air-cooled condenser. Note that power is produced from both the steam and binary turbines in the plant.

The most common geothermal plant type is single flash, which accounts for more than 5000 MW of installations worldwide. This is followed by dry steam, double flash, and binary plants as shown in Fig. 8-23. Triple flash and combined flash/binary plants also exist but in much smaller capacities.

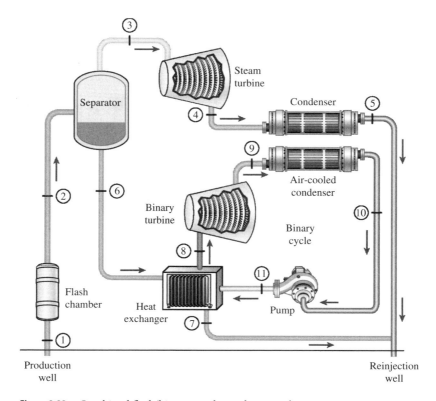

Figure 8-22   Combined flash/binary geothermal power plant.

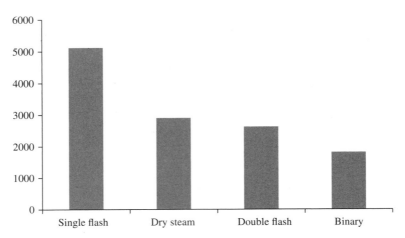

Figure 8-23 Installed capacity of different types of geothermal plants in the world as of 2015. The values are in MW. (*Bertani, 2015.*)

EXAMPLE 8-6
Maximum Power Potential
of a Geothermal Resource

A geothermal site contains geothermal water available at wellhead at a rate of 100 kg/s. The geothermal water is at 180°C with a vapor fraction of 20 percent. The ambient conditions are 25°C and 1 atm.

(*a*)   What is the maximum thermal efficiency of a geothermal power plant using this resource?
(*b*)   What is the maximum amount of power that can be produced from this site?
(*c*)   If a typical second-law efficiency for a geothermal power plant is 30 percent, predict the actual power output from a geothermal power plant at this site and its thermal efficiency.

SOLUTION   (*a*) We use properties of water for geothermal fluid. The maximum thermal efficiency can be determined from the Carnot efficiency relation using the geothermal source and ambient (dead state) temperatures:

$$\eta_{th,max} = 1 - \frac{T_L}{T_H} = 1 - \frac{(25 + 273) \text{ K}}{(180 + 273) \text{ K}} = 0.3422 \text{ or } \textbf{34.2 percent}$$

(*b*) The enthalpy and entropy values of geothermal water at the wellhead state and dead state are obtained from steam tables (Table A-3):

$$\left. \begin{array}{l} T_1 = 180°C \\ x_1 = 0.20 \end{array} \right\} \begin{array}{l} h_1 = h_{f@180°C} + x_1 h_{fg@180°C} = 763.05 + (0.20)(2014.2 \text{ kJ/kg}) = 1165.9 \text{ kJ/kg} \\ s_1 = s_{f@180°C} + x_1 s_{fg@180°C} = 2.1392 + (0.20)(4.4448 \text{ kJ/kg·K}) = 3.0282 \text{ kJ/kg·K} \end{array}$$

$$\left. \begin{array}{l} T_0 = 25°C \\ P_0 = 1 \text{ atm} \end{array} \right\} \begin{array}{l} h_0 \cong h_{f@25°C} = 104.83 \text{ kJ/kg} \\ s_0 \cong s_{f@25°C} = 0.3672 \text{ kJ/kg·K} \end{array}$$

The maximum power potential is equal to the exergy of the geothermal water available at the wellhead:

$$\dot{W}_{max} = \dot{X}_1 = \dot{m}_1 [h_1 - h_0 - T_0(s_1 - s_0)]$$

$$= (100 \text{ kg/s}) \left[(1165.9 - 104.83) \text{kJ/kg} - (298 \text{ K})(3.0282 - 0.3672) \text{kJ/kg·K} \right]$$

$$= \textbf{26,800 kW}$$

(*c*) The second-law efficiency of a power plant is defined as the actual power output divided by the maximum power output. It is also defined as the actual thermal efficiency divided by the maximum (i.e., Carnot) thermal efficiency. Then, for a second-law efficiency of 30 percent, the actual power output and actual thermal efficiency are predicted to be

$$\eta_{II} = \frac{\dot{W}_{actual}}{\dot{W}_{max}} \longrightarrow \dot{W}_{actual} = \eta_{II}\dot{W}_{max} = (0.30)(26{,}800 \text{ kW}) = \mathbf{8040 \text{ kW}}$$

$$\eta_{II} = \frac{\eta_{th,actual}}{\eta_{th,max}} \longrightarrow \eta_{th,actual} = \eta_{II}\eta_{th,max} = (0.30)(0.3422) = 0.1027 \text{ or } \mathbf{10.3\%} \quad \blacktriangle$$

EXAMPLE 8-7
Thermodynamic and
Economic Analysis of a
Single-Flash Geothermal
Power Plant

Geothermal liquid water at 200°C is extracted from a geothermal well at a rate of 100 kg/s. This water is flashed to a pressure of 500 kPa in the flash chamber of a single-flash geothermal power plant (Fig. 8-24). The condenser is maintained at a pressure of 10 kPa. The isentropic efficiency of the turbine is 83 percent.

(a) What is the mass flow rate of water vapor at the turbine inlet?

(b) Determine the power output from the turbine and the thermal efficiency of the plant.

(c) Assume this single-flash plant is retrofitted to operate as a double-flash plant (see Fig. 8-18). The first and second flash pressures are selected as 500 kPa and 150 kPa, respectively. What are the power output and thermal efficiency of this double-flash plant? Other parameters remain the same.

(d) Consider the single-flash plant again. Instead of retrofitting the plant to a double-flash design, the liquid water leaving the separator is to be used for district space heating of buildings. Assume the return temperature of geothermal water from the district is 60°C. Geothermal space heating will be replacing heating by natural gas heaters. Geothermal heat will be sold to the district at the same price as the natural gas heating. What is the annual potential revenue from the selling of geothermal heat to the district? The unit price of natural gas is $1.25/therm (1 therm = 100,000 Btu = 105,500 kJ) and the efficiency of a natural gas heater is 83 percent. Take the annual heating hours to be 3000 h.

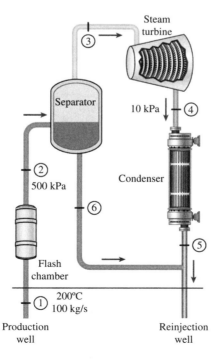

Figure 8-24   Single-flash geothermal power plant considered in Example 8-7.

Compare this revenue to the selling of electricity from the plant. The selling price of electricity is $0.075/kWh. Take the annual operating hours of the plant to be 8000 h.

SOLUTION   (a) We neglect kinetic and potential energy changes and use water properties for geothermal fluid. Steam properties are obtained from Tables A-3 through A-5. First, the properties of water at the flash chamber and separator are

$$\left. \begin{array}{l} T_1 = 200°C \\ x_1 = 0 \end{array} \right| \begin{array}{l} P_1 = 1554.9 \text{ kPa} \\ h_1 = 852.26 \text{ kJ/kg} \end{array}$$

$$\left. \begin{array}{l} P_2 = 500 \text{ kPa} \\ h_2 = h_1 = 852.26 \text{ kJ/kg} \end{array} \right| \begin{array}{l} T_2 = 151.83°C \\ h_6 = h_f = 640.09 \text{ kJ/kg} \\ h_3 = h_g = 2748.1 \text{ kJ/kg} \end{array}$$

$$\left. \begin{array}{l} P_6 = 500 \text{ kPa} \\ x_6 = 0 \end{array} \right\} h_6 = 640.09 \text{ kJ/kg}$$

$$\left. \begin{array}{l} P_3 = 500 \text{ kPa} \\ x_3 = 1 \end{array} \right| \begin{array}{l} h_3 = 2748.1 \text{ kJ/kg} \\ s_3 = 6.8207 \text{ kJ/kg·K} \end{array}$$

The enthalpy at the turbine exit for the isentropic process is

$$\left. \begin{array}{l} P_4 = 10 \text{ kPa} \\ s_4 = s_3 = 6.8207 \text{ kJ/kg·K} \end{array} \right\} h_{4s} = 2160.2 \text{ kJ/kg}$$

Using the definition of the turbine isentropic efficiency, the actual enthalpy at the turbine exit is determined as

$$\eta_{turb} = \frac{h_3 - h_4}{h_3 - h_{4s}} \longrightarrow 0.83 = \frac{2748.1 - h_4}{2748.1 - 2160.2} \longrightarrow h_4 = 2260.1 \text{ kJ/kg}$$

The quality of water at state 2 is

$$x_2 = \frac{h_2 - h_f}{h_{fg}} = \frac{852.26 - 640.09}{2108.0} = 0.1006$$

The mass flow rate of liquid water at state 6 is

$$\dot{m}_6 = (1 - x_2)\dot{m}_1 = (1 - 0.1006)(100 \text{ kg/s}) = 89.94 \text{ kg/s}$$

Note that about 90 percent of the water extracted from the reservoir is sent to reinjection at a temperature of 151.8°C. The mass flow rate of steam at the turbine inlet is then

$$\dot{m}_3 = x_2 \dot{m}_1 = (0.1006)(100 \text{ kg/s}) = \textbf{10.06 kg/s}$$

(b) The power output from the turbine is

$$\dot{W}_{out} = \dot{m}_3(h_3 - h_4) = (10.06 \text{ kg/s})(2748.1 - 2260.1) \text{ kJ/kg} = \textbf{4909 kW}$$

The enthalpy of saturated liquid water at 25°C is $h_0 = 104.83$ kJ/kg. The energy input to the power plant is

$$\dot{E}_{in} = \dot{m}_1(h_1 - h_0) = (100 \text{ kg/s})(852.26 - 104.83) \text{ kJ/kg} = 74{,}743 \text{ kW}$$

The thermal efficiency of the plant is then

$$\eta_{th} = \frac{\dot{W}_{out}}{\dot{E}_{in}} = \frac{4909 \text{ kW}}{74{,}743 \text{ kW}} = 0.0657 \text{ or } \mathbf{6.57\%}$$

(c) If this power plant is retrofitted to operate as a double-flash design, the properties at other states (see Fig. 8-18) would be

$$\left. \begin{array}{l} T_1 = 200°C \\ x_1 = 0 \end{array} \right\} \begin{array}{l} P_1 = 1554.9 \text{ kPa} \\ h_1 = 852.26 \text{ kJ/kg} \end{array}$$

$$\left. \begin{array}{l} P_7 = 150 \text{ kPa} \\ h_7 = h_6 = 640.09 \text{ kJ/kg} \end{array} \right\} \begin{array}{l} T_8 = 111.35°C \\ h_9 = h_{f\,@\,150\,kPa} = 467.13 \text{ kJ/kg} \\ h_8 = h_{g\,@\,150\,kPa} = 2693.1 \text{ kJ/kg} \end{array}$$

The quality of water at state 7 is

$$x_7 = \frac{h_7 - h_f}{h_{fg}} = \frac{640.09 - 467.13}{2226.0} = 0.07770$$

The mass flow rate of saturated steam at state 8 is

$$\dot{m}_8 = x_7 \dot{m}_7 = (0.07770)(89.94 \text{ kg/s}) = 6.988 \text{ kg/s}$$

The additional power output from the turbine due to expansion of this additional steam in the turbine is

$$\dot{W}_{out,additional} = \dot{m}_8 (h_8 - h_4) = (6.988 \text{ kg/s})(2693.1 - 2260.1) \text{kJ/kg} = 3026 \text{ kW}$$

The thermal efficiency of the plant is then

$$\eta_{th} = \frac{\dot{W}_{out}}{\dot{E}_{in}} = \frac{(4909 + 3026) \text{ kW}}{74{,}743 \text{ kW}} = 0.106 \text{ or } \mathbf{10.6\%}$$

(d) We note that the geothermal water is supplied to the district for space heating at 151.8°C (state 6) and returns at 60°C. The enthalpy of water at 60°C is

$$h_{return} = 251.18 \text{ kJ/kg}$$

This corresponds to a heating rate of

$$\dot{Q}_{heat} = \dot{m}_6 (h_6 - h_{return}) = (89.94 \text{ kg/s})(640.09 - 251.18) \text{kJ/kg} = 34{,}979 \text{ kW}$$

The amount of heat for a heating period of 3000 h, in therm, is

$$Q_{heat} = \dot{Q}_{heat} \Delta t = (34{,}979 \text{ kW})(3000 \text{ h}) \left(\frac{3600 \text{ kJ}}{1 \text{ kWh}} \right) \left(\frac{1 \text{ therm}}{105{,}500 \text{ kJ}} \right) = 3.581 \times 10^6 \text{ therm}$$

The natural gas heater is 83 percent efficient, and this means that when 100 units of heat are supplied to the heater by burning natural gas, 83 units will be supplied by the heater as the useful space heat. Then the amount of natural gas consumed is determined to be

$$\text{Gas consumption} = \frac{Q_{heat}}{\eta_{heater}} = \frac{3.581 \times 10^6 \text{ therm}}{0.83} = 4.314 \times 10^6 \text{ therm}$$

The annual potential revenue in selling the geothermal heat is then

$$\text{Revenue from heat} = \text{Gas consumption} \times \text{Unit price of natural gas}$$

$$= (4.314 \times 10^6 \text{ therm})(\$1.25/\text{therm})$$

$$= \mathbf{\$5.39 \times 10^6}$$

The plant produces 4909 kW of electricity. Noting that the annual operating period is 8000 h and the price of electricity is $0.075/kWh, the annual revenue potential from selling electricity is

$$\text{Revenue from electricity} = \dot{W}_{out} \times \Delta t \times \text{Unit price of electricity}$$

$$= (4909 \text{ kW})(8000 \text{ h})(\$0.075/\text{kWh})$$

$$= \mathbf{\$2.95 \times 10^6}$$

Therefore, the plant owner can make an additional $5.4 million from selling geothermal heat by using otherwise reinjected geothermal brine. This example shows that utilization of geothermal energy for heating has greater revenue potential than producing and selling electricity. ▲

8-7 GEOTHERMAL COGENERATION

Cogeneration is *the production of more than one useful form of energy (such as process heat and electric power) from the same energy source.* A combination of power production and cooling can also be used in a cogeneration scheme. It is called *trigeneration* if three useful forms of energy (such as electric power, process heat, and cooling) are produced from the same energy source. A steam-turbine (Rankine) cycle, a gas-turbine (Brayton) cycle, a combined cycle (combination of Rankine and Brayton cycles), an internal combustion engine, or any other power producing plant (such as a geothermal power plant) can be used as the power cycle in a cogeneration plant. Cogeneration systems utilizing internal combustion engines and gas turbines in open cycle are the most utilized technologies worldwide.

The "cascading" of energy use from high- to low-temperature uses often distinguishes cogeneration systems from conventional separate electrical and thermal energy systems (e.g., a power plant and an industrial boiler), and from simple heat recovery strategies. The principal technical advantage of cogeneration systems is their ability to improve the efficiency of fuel use in the production of electrical and thermal energy. Less fuel is required to produce a given amount of electrical and thermal energy in a single cogeneration unit than is needed to generate the same quantities of both types of energy with separate, conventional technologies. This is because heat from the turbine-generator set, which uses a substantial quantity of fuel to fire the turbine, becomes useful thermal energy (e.g., process steam) in a cogeneration system rather waste heat (Benelmir and Feidt, 1998; Wilkinson and Barnes, 1980).

The technical advantages of cogeneration lead to significant energy savings and corresponding environmental advantages. That is, the increase in efficiency and corresponding decrease in fuel use by a cogeneration system, compared to other conventional processes for thermal and electrical energy production, normally yield large reductions in energy use and greenhouse gas emissions. These reductions can be as large as 50 percent in some situations, while the same thermal and electrical services are provided. When cogeneration is used in a renewable energy scheme, the environmental benefits are greater with respect to fossil fuel–based cogeneration units. Renewable cogeneration systems make the most use of a resource, and this translates into a greater replacement of fossil fuels.

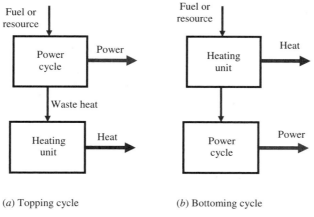

(a) Topping cycle   *(b)* Bottoming cycle

Figure 8-25   Topping and bottoming cycles.

The use of cogeneration dates to the beginning of this century when power plants were integrated into a community to provide district heating, that is, space, hot water, and process heating for residential and commercial buildings. The district heating systems lost their popularity in the 1940s owing to low fuel prices. However, the rapid rise in fuel prices in the 1970s brought about renewed interest in district heating. Cogeneration plants have proved to be economically very attractive. Consequently, more and more such plants have been installed in recent years, and more are being installed.

Cogeneration systems can generally be arranged as topping cycles and bottoming cycles, as shown in Fig. 8-25. In a *topping cycle*, the fuel is burned or an energy source such as geothermal is utilized in a power cycle and power is generated first. The waste heat output from the power cycle is then used to produce a useful heat output in a boiler or heat exchanger. This scheme of cogeneration is most commonly used. It is particularly suitable when the electric demand of the facility is greater than the thermal demand and the process heat temperature requirements are not high. In a *bottoming cycle*, heat is produced first in a boiler or heat exchanger, and electricity is generated by using the waste heat of the fluid stream leaving the process heat equipment and/or the boiler. This scheme of cogeneration is less common than the topping cycle. It is particularly suitable when the plant needs process heat at a high temperature. Some examples include steel reheat furnaces, clay and glass kilns, and aluminum remelt furnaces (Thumann and Mehta, 2008).

A geothermal resource can be used for single uses of power production, heating, and cooling. It can also be used for cogeneration and trigeneration schemes where two or three of these applications can be applied in a cascaded design. For example, the waste geothermal water of a power plant can be used for a space cooling application if the temperature of geothermal water is sufficiently high (i.e., greater than about 95°C). If the temperature of geothermal water is greater than about 50°C, a space heating application can be cascaded into a geothermal power unit or geothermal cooling application.

The efficiency of a cogeneration plant is usually expressed by a utilization factor. The utilization factor of a cogeneration plant involving an electricity producing unit and an absorption cooling system may be expressed as

$$\varepsilon_u = \frac{\text{Net power output} + \text{Rate of cooling delivered}}{\text{Total rate of heat input}} = \frac{\dot{W}_{\text{net}} + \dot{Q}_{\text{cooling}}}{\dot{Q}_{\text{in}}} \tag{8-32}$$

Similarly, the utilization factor of a cogeneration plant involving an electricity producing unit and a heating system may be expressed as

$$\varepsilon_u = \frac{\text{Net power output} + \text{Rate of heating delivered}}{\text{Total rate of heat input}} = \frac{\dot{W}_{net} + \dot{Q}_{heating}}{\dot{Q}_{in}} \qquad (8\text{-}33)$$

If a cogeneration plant provides electricity, heat, and cooling at the same time (i.e., trigeneration), the utilization factor may be expressed as

$$\varepsilon_u = \frac{\dot{W}_{net} + \dot{Q}_{heating} + \dot{Q}_{cooling}}{\dot{Q}_{in}} \qquad (8\text{-}34)$$

This scheme with power production, heating, and cooling may represent the best use of a geothermal resource, and a high utilization factor is a measure of this.

EXAMPLE 8-8

Comparison of Potential Revenues from Geothermal Energy for Various Uses

A geothermal resource can be used for single uses of power production, heating, and cooling. It can also be used for cogeneration and trigeneration schemes where two or three of these applications can be applied in a cascaded design. For example, the waste geothermal water of a power plant can be used for space cooling if the temperature of geothermal water is sufficiently high (i.e., greater than about 95°C). If the temperature of geothermal water is greater than about 50°C, a space heating application can be cascaded into a geothermal power unit or geothermal cooling application.

Compare potential revenues for three possible uses of a geothermal resource at 165°C that supplies water at a rate of 10 kg/s. Consider the following options under the specified conditions. The second and third options are cogeneration applications.

1. Power generation by a binary geothermal power plant with a conversion efficiency of 12 percent. Geothermal water enters at 165°C and leaves the plant at 90°C. Annual operating hours are 8000 h. The unit price of electricity generated is $0.10/kWh.

2. Power generation by a binary power plant with a conversion efficiency of 12 percent and space heating with a conversion efficiency of 100 percent. Geothermal water enters the power plant at 165°C and leaves at 90°C. It enters the heating system at 90°C and is discharged at 50°C. Annual operating hours are 8000 h for power generation and 3000 h for space heating. The unit price of heat supplied is $1.20/therm (1 therm = 105,500 kJ = 29.31 kWh).

3. Power generation by a binary power plant with a conversion efficiency of 12 percent and space cooling by an absorption system that has a COP of 0.7. Geothermal water enters the power plant at 165°C and leaves at 115°C. It enters the absorption system at 115°C and is discharged at 95°C. Annual operating hours are 8000 h for power generation and 3000 h for space cooling. The unit price of cooling provided is $0.05/kWh.

SOLUTION   We provide the analysis for each option in Table 8-3.

The results show that using the same geothermal resource for a cogeneration application (power + heating) can provide potential revenues of $506,000 per year compared to $301,000 when it is used for power generation only. A cogeneration scheme involving power generation and absorption cooling can only provide potential revenues of $310,000, which is a little higher than those for a power generation system. However, the temperature of geothermal water at the exit of the absorption cooling system is still high (90°C), which makes it suitable for heating applications. Note that space heating cannot be added to the space cooling in this trigeneration scheme since no space heating is needed when space cooling is used in summer. A potential solution is to operate the cogeneration system with power generation and space heating in winter and with power generation and space cooling

TABLE 8-3   **Potential Revenues from Geothermal Energy for Various Uses**

Parameter	Geothermal Power Generation	Power Generation + Heating		Power Generation + Cooling	
		Power Generation	Heating	Power Generation	Cooling
Rate of geothermal water, $\dot{m}$	10 kg/s	10 kg/s	10 kg/s	10 kg/s	10 kg/s
Source temperature of geothermal brine, T_s	165°C	165°C	90°C	165°C	115°C
Discharge temperature of geothermal brine, T_d	90°C	90°C	50°C	115°C	90°C
Rate of geothermal heat utilized, $\dot{Q}_u = \dot{m}c_p(T_s - T_d)$ where $c_p = 4.18$ kJ/kg.°C	3135 kW	3135 kW	1672 kW	2090 kW	1045 kW
Annual operating period, Δt	8000 h	8000 h	3000 h	8000 h	3000 h
Amount of geothermal heat utilized, $Q_u = \dot{Q}_u \Delta t$	2.51×10^7 kWh	2.51×10^7 kWh	5.016×10^6 kWh or 171,162 therm	1.67×10^7 kWh	3.14×10^6 kWh
Conversion efficiency, η_c	0.12	0.12	1.0	0.12	0.7
Amount of commodity (electricity, heating, or cooling) delivered, $E_d = \eta_c Q_u$	3.01×10^6 kWh	3.01×10^6 kWh	171,162 therm	2.01×10^6 kWh	2.19×10^6 kWh
Unit price of electricity, heat, and cooling, UP	$0.10/kWh	$0.10/kWh	$1.20/therm	$0.10/kWh	$0.05/kWh
Potential revenue, $PR = E_d \times UP$	$300,960	$300,960	$205,395	$200,640	$109,725
Total potential revenue	**$301,000**	**$506,000**		**$310,000**	

in summer. When this is the case, the annual potential revenue becomes $PR = \$301,000$ (power) + \$205,000 (heating) + \$110,000 (cooling) = \$616,000. The final values for total potential revenues are rounded to three significant digits in Table 8-3, and a comparison of all four cases is shown in Fig. 8-26.

Geothermal direct use applications such as district space heating and cooling are characterized by high initial costs. Therefore, a life cycle cost analysis should be performed to investigate the feasibility of each cogeneration and trigeneration option. A more detailed discussion on this example can be found in Kanoğlu and Çengel (1999a).

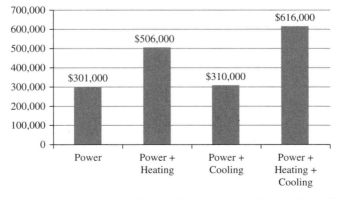

Figure 8-26   Comparison of potential revenues (in $) from geothermal energy for various uses.   ▲

REFERENCES

ASHRAE. 2008. American Society of Heating, Refrigeration, and Air-Conditioning Engineers. In *Handbook: HVAC Systems and Equipment*. Atlanta, GA.

Benelmir Rand Feidt M. 1998. "Energy Cogeneration Systems and Energy Management Strategy." *Energy Conversion and Management*, 39(16–18): 1791–1802.

Bertani R. 2015. "*Geothermal Power Generation in the World 2010-2014 Update Report.*" Proceedings World Geothermal Congress. Melbourne, Australia. 19–25 April, 2015.

Çengel, YA, Boles MA, and Kanoğlu M. 2019. *Thermodynamics: An Engineering Approach*, 9th ed. New York: McGraw Hill.

Çengel YA and Ghajar AJ. 2020. *Heat and Mass Transfer: Fundamentals and Applications*, 6th ed. New York: McGraw Hill.

NEED Project P.O. Box 10101, Manassas, VA 20108.www.need.org.

DiPippo R. 2007. *Geothermal Power Plants: Principles, Applications, and Case Studies*, 2nd ed. Oxford, UK: Elsevier.

Erkan K, Holdmann G, Benoit W, and Blackwell D. 2008. "Understanding the Chena Hot Flopë Springs, Alaska, Geothermal System using Temperature and Pressure Data." *Geothermics*, 37 (6): 565–585.

Gehlin S. 2015. Feel Green—Not Blue. IEA Heat Pump Centre. Vol. 33, No. 3. p. 3. Boras, Sweden.

Geothermal Worldwide. 2019. http://geothermalworldwide.com/egs.html. Access date: Mar. 2, 2019.

Kanoğlu M and Çengel YA. 1999a. "Economic Evaluation of Geothermal Power Generation, Heating, and Cooling." *Energy*, 24 (6): 501–509.

Kanoğlu M and Çengel YA. 1999b. "Improving the Performance of an Existing Binary Geothermal Power Plant: A Case Study." *Transactions of the ASME, Journal of Energy Resources Technology*, 121 (3): 196–202.

Sanner B. 2017. *Ground Source Heat Pumps—History, Development, Current Status and Future Prospects*. 12th IEA Heat Pump Conference, Rotterdam, Holland.

Schöffengrund-Schwalbach GSHP test plant, 1985–1989.

Thumann A and Mehta DP. 2008. *Handbook of Energy Engineering*, 6th ed. Lilburn, GA: The Fairmont Press.

Wilkinson BW and Barnes RW. 1980. *Cogeneration of Electricity and Useful Heat*. Boca Raton, FL: CRC Press.

PROBLEMS

INTRODUCTION

8-1   What is geothermal energy? Why is it a renewable energy source?

8-2   Describe different layers of earth's interior, and explain how a geothermal resource is formed.

8-3   Classify geothermal resources. Briefly describe them. Which ones are currently exploited?

8-4   Classify geothermal resources based on temperature.

8-5   Describe the states of geothermal water produced from a reservoir and compare their quality.

8-6   The thermal energy within the earth's interior is called

(*a*) Inner energy   (*b*) Magma energy   (*c*) Geothermal energy

(*d*) Biomass energy   (*e*) Ground energy

8-7   Which is not a classification of geothermal resources?

(*a*) Hydrothermal   (*b*) Magma   (*c*) Enhanced   (*d*) Mantle   (*e*) Geopressurized

8-8   Which geothermal fields are currently exploited?

(*a*) Geopressurized   (*b*) Hydrothermal   (*c*) Molten rock   (*d*) Enhanced   (*e*) Mantle

GEOTHERMAL APPLICATIONS

8-9   List the main geothermal applications. Which application is most common?

8-10   What are typical geothermal resource temperature requirements for electricity production, cooling, and heating?

8-11   Consider a liquid geothermal resource at a temperature of 110°C. An investor is considering a power plant on this site, and the investor is asking for your opinion whether this is an attractive investment. What would you say? Explain.

8-12   Which is not an application of geothermal energy?

(*a*) Heating   (*b*) Cooling   (*c*) Electricity generation

(*d*) Heat pump   (*e*) None of these

8-13   Which geothermal sources are of the highest quality?
(*a*) Compressed liquid   (*b*) Saturated vapor   (*c*) Superheated vapor
(*d*) Saturated mixture   (*e*) Saturated liquid

8-14   Which application of geothermal energy requires the highest temperature?
(*a*) Electricity generation   (*b*) Space heating   (*c*) Space cooling
(d) Process heating   (*e*) Greenhouse

GEOTHERMAL HEATING

8-15   How is the optimum pipe diameter selected for geothermal heating applications? Explain.

8-16   How are the amount and cost of energy (fuel and electricity) used for heating and cooling of an existing building determined? If a house uses natural gas for space and water heating, how can you determine the natural gas consumption for space heating?

8-17   What is the degree-day method used for? What assumptions are involved?

8-18   What is the balance point temperature? Is it normally below or above the temperature of indoors? Why?

8-19   Someone claims that the °C-days for a location can be converted to °F-days by simply multiplying °C-days by 1.8. But another person insists that 32 must be added to the result because of the formula $T(°F) = 1.8T(°C) + 32$. Which person do you think is right?

8-20   Consider city A and city B. The degree-day in city A is 2000°C-day and that in city B is 3000°C-day. For which city will the percent fuel savings be greater for a 5°C temperature setback in both cities? Why?

8-21   A residential district is to be heated by geothermal water in winter. On an average winter day, geothermal water enters the main heat exchanger network of the district at 90°C at a rate of 70 kg/s and leaves the heat exchangers at 50°C. How much revenue will be generated if the total winter period can be taken to be equivalent to 2500 h of these average conditions and geothermal heat is sold at a price of $1.20/10^5 kJ?

8-22   A commercial district is currently heated by coal-burning heaters in winter with an average efficiency of 75 percent. The heating value of the coal is 25,000 kJ/kg and its unit price is $0.25/kg. It is proposed to heat this district by geothermal water. On an average winter day, hot geothermal water is supplied to the district at 80°C at a rate of 20 kg/s and returns at 60°C. How much revenue can be generated per year if the winter period can be taken to be equivalent to 3500 h of these average conditions and geothermal heat is sold at the same price as coal?

8-23   What is the number of heating degree-days for a winter day during which the average outdoor temperature was 10°C, and it never went above 18°C?

8-24   What is the number of heating degree-days for a winter month during which the average outdoor temperature was 54°F, and it never went above 65°F?

8-25   Suppose you have moved to Syracuse, New York, in August, and your roommate, who is short of cash, offered to pay the heating bills during the upcoming year (starting January 1) if you pay the heating bills for the current calendar year until December 31. Is this a good offer for you?

8-26   Consider a family in Atlanta, Georgia, whose average heating bill has been $600 a year. The family now moves to an identical house in Denver, Colorado, where the fuel and electricity prices are also the same. How much do you expect the annual heating bill of this family to be in their new home?

8-27   Using indoor and outdoor winter design temperatures of 70°F and −10°F, respectively, the design heat load of a 2500-ft² house in Billings, Montana, is determined to be 83,000 Btu/h. The house is to be heated by natural gas that is burned in a 95 percent efficient, high-efficiency furnace. If the unit cost of natural gas is $1.10/therm, estimate the annual gas consumption of this house and its cost.

If this house is heated by geothermal energy, determine the annual geothermal energy consumption and its cost. Take the efficiency of geothermal heating to be 100 percent neglecting heat losses in the geothermal heating system and assume that geothermal heat is sold at a price of $1.0/10^5 Btu.

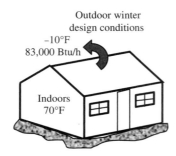

Figure P8-27

8-28 The design heating load of a new house in Cleveland, Ohio, is calculated to be 10 kW for an indoor temperature of 23°C. If the house is to be heated by geothermal heat that is 97 percent efficient due to heat losses in the geothermal system, predict the annual geothermal energy consumption of this house, in MJ (megajoules). Assume the entire house is maintained at indoor design conditions at all times.

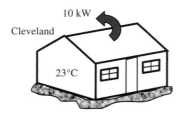

Figure P8-28

8-29 For an indoor temperature of 22°C, the design heating load of a residential building in Charlotte, North Carolina, is determined to be 28 kW. The house is to be heated by geothermal energy, which costs $13/10^6$ kJ. The efficiency of geothermal heating can be taken as 100 percent. Determine (a) the heat loss coefficient of the building, in kW/°C, and (b) the annual energy usage for heating and its cost.

8-30 In a typical residential building, most energy is consumed for
(a) Space heating and water heating (b) Space heating and lighting
(c) Water heating and refrigerators (d) Refrigerators and lighting
(e) Lighting and air conditioning

8-31 In a typical commercial building, most energy is consumed for
(a) Space heating and water heating (b) Space heating and lighting
(c) Water heating and refrigerators (d) Refrigerators and lighting
(e) Lighting and air conditioning

8-32 For buildings that are at the design or construction stage, the evaluation of annual energy consumption does not involve
(a) Heat transfer through the building envelope (b) Efficiency of furnace
(c) COP of cooling or heat pump system (d) Cost of fuel and electricity
(e) Energy consumed by the distribution system

8-33   The balance point temperature is usually taken to be
(a) 12°C   (b) 15°C   (c) 18°C   (d) 20°C   (e) 23°C

8-34   The average outdoor temperature in a location in a given day is 3°C. The heating degree-day in that day is
(a) 12°C-day   (b) 15°C-day   (c) 18°C-day   (d) 20°C-day   (e) 23°C-day

8-35   In a geothermal heating system, the amount of energy supplied can be determined from: (Energy Consumption) = (Rate of useful heat)(Operating hours)/(Heater efficiency). If the unit of (Energy Consumption) is kJ, what is the unit of (Rate of useful heat)?
(a) kJ   (b) kW   (c) kWh   (d) kJ/h   (e) kW/h

GEOTHERMAL COOLING

8-36   Which system is used for cooling a space by geothermal energy?

8-37   What drives an absorption system? Give some examples.

8-38   What is the most common system for absorption refrigeration? What are the other systems? Which system is recommended if the minimum temperature is below 0°C (32°F)?

8-39   What is the main advantage of absorption cooling systems compared to conventional vapor-compression systems?

8-40   What are the disadvantages of absorption refrigeration systems? Are they suitable for small-scale applications?

8-41   How is the coefficient of performance of an absorption refrigeration system defined?

8-42   Can the COP of an absorption refrigeration system be greater than 1? Can it be smaller than 1?

8-43   An absorption refrigeration system provides 15 kW of cooling by receiving heat in the generator at a rate of 21 kW. What is the COP of this system?

8-44   Geothermal liquid water from a well is available at 210°F at a rate of 86,000 lbm/h and is to be used for space cooling using an absorption refrigeration system. Geothermal water leaves the generator of the absorption system at 180°F. If the COP of the absorption system is 0.70, determine the rate of cooling provided by the system. This geothermal cooling system is replacing a conventional cooling system (vapor-compression refrigeration system) with a COP of 2.3, and the price of electricity is $0.14/kWh. Determine the potential revenue that can be generated by this geothermal cooling system if geothermal cooling is sold with a discount of 20 percent compared to conventional cooling. Use an operating cooling period of 2000 h at the calculated cooling rate.

8-45   Geothermal liquid water from a well is available at 120°C with a vapor fraction of 10 percent at a rate of 18 kg/s and is to be used for space cooling using an absorption cooling system. Geothermal water leaves the generator of the absorption system at 110°C as a saturated liquid. (a) If the COP of the absorption system is 0.75, determine the rate of cooling provided by the system. (b) If the geothermal cooling is to be sold at a rate of $5.0/10^6 kJ, determine the annual potential revenue that can be generated by this geothermal cooling system. Use a seasonal cooling period of 2400 h at the calculated cooling rate.

8-46   Which system is used for cooling applications of geothermal energy?
(a) Thermoelectric refrigeration system   (b) Absorption refrigeration system
(c) Vapor-compression refrigeration system   (d) Evaporative cooling system
(e) Air-source heat pump

8-47   Which one is not a component of an absorption refrigeration system?
(a) Compressor   (b) Evaporator   (c) Condenser
(d) Generator   (e) Expansion valve

8-48   Which statement is not correct for absorption refrigeration systems?
(a) The COP can be greater than 1.   (b) The COP can be smaller than 1.
(c) The pump work is negligible.   (d) It is more suitable for small-scale applications.
(e) It is a heat-driven system.

8-49   Which of the following is not a characteristic of absorption refrigeration systems?
(a) High initial cost   (b) High coefficient of performance values
(c) Difficult to service   (d) Suitable for large industrial applications
(e) Uses waste heat

8-50   Which equation is not correct?
(a) COP of absorption cooling system in summer = (Cooling load)/(Work input)
(b) COP of a heat pump in summer = (Cooling load)/(Work input)
(c) COP of a heat pump in winter = (Heating load)/(Work input)
(d) Thermal efficiency of geothermal power plant = (Net work)/(Energy input to the plant)
(e) Thermal efficiency of geothermal power plant = 1 − (Heat output)/(Energy input to the plant)

8-51   An absorption refrigeration system provides 24 kW of cooling while receiving 18 kW of solar heat input. The COP of this system is
(a) 0.5   (b) 0.75   (c) 1.0   (d) 1.33   (e) 4.0

GEOTHERMAL HEAT PUMP

8-52   What is the difference between a ground-source heat pump and a geothermal heat pump? Explain. Why do geothermal heat pumps have higher COPs than air-source heat pumps?

8-53   Why are ground-source heat pumps more efficient than air-source heat pumps? What is the disadvantage of ground-source heat pumps?

8-54   Consider two identical heat pumps, both maintaining a room at the same temperature in winter. Heat pump A operates in an ambient at 5°C while heat pump B operates in an ambient at 10°C. Which heat pump consumes more electricity for the same rate of heat supply to the room? Why?

8-55   Most air-source heat pumps require a supplementary heating system such as electric resistance heaters. Why? How about water-source or ground-source heat pumps?

8-56   Consider a split air conditioner that operates as a heat pump in winter and as an air conditioner in summer. Do you recommend using this air conditioner in (a) a mild climate with a large cooling load and a small heating load or (b) a cold climate with a large heating load and a small cooling load? Why?

8-57   Describe advantages and disadvantages of vertical loop heat pump systems.

8-58   Consider an air-source heat pump that operates between the ambient air at 0°C and the indoors at 20°C. The working fluid refrigerant-134a evaporates at −10°C in the evaporator. The refrigerant is saturated vapor at the compressor inlet and saturated liquid at the condenser exit. The refrigerant condenses at 39.4°C. Heat is lost from the indoors at a rate of 48,000 kJ/h. Determine (a) the power input and (b) the COP. (c) Also, determine the COP if a ground-source heat pump is used with a ground temperature of 15°C and the evaporating temperature in this case is 5°C. Take everything else the same. What is the percent increase in COP when a ground-source heat pump is used instead of an air-source heat pump?

For air-source heat pump, the properties of R-134a at the compressor inlet and exit and at the condenser exit states are obtained from R-134a tables, respectively, to be $h_1 = 244.51$ kJ/kg, $h_2 = 287.42$ kJ/kg, $h_3 = 107.37$ kJ/kg.

For a ground-source heat pump, the properties of R-134a at the compressor inlet and exit and at the condenser exit states are obtained from R-134a tables, respectively, to be $h_1 = 253.34$ kJ/kg, $h_2 = 281.31$ kJ/kg, $h_3 = 107.37$ kJ/kg.

8-59   Consider a building whose annual air-conditioning load is estimated to be 120,000 kWh in an area where the unit cost of electricity is $0.10/kWh. Two air conditioners are considered for the building. Air conditioner A has a seasonal average COP of 3.2 and costs $5500 to purchase and install. Air conditioner B has a seasonal average COP of 5.0 and costs $7000 to purchase and install. If all else is equal, determine how long it will take for air conditioner B to pay for its price differential from the electricity it saves.

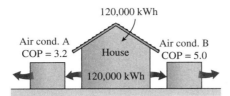

120,000 kWh

Air cond. A
COP = 3.2

House

Air cond. B
COP = 5.0

120,000 kWh

Figure P8-59

8-60 Repeat Prob. 8-59 if the seasonal COP of air-conditioner B is 3.8.

8-61 A house in Cleveland, Ohio, is equipped with a heat pump system that is used for space heating in winter. The seasonal COP of the heat pump in winter is estimated to be 1.5. The design heating load of the house is calculated to be 65,000 Btu/h for an indoor temperature of 75°F and outdoor temperature of 5°F. The unit cost of electricity is $0.12/kWh. Determine the annual electricity consumption and its cost for this house.

8-62 Reconsider Prob. 8-61. (*a*) If the house in Cleveland is heated in winter by a ground-source heat pump with a seasonal COP of 3.5, determine the annual electricity consumption and its cost for this house. (*b*) If the house in Cleveland is heated in winter by a natural gas furnace that is 88 percent efficient, determine the annual natural gas consumption and its cost in winter. The unit cost of natural gas is $1.10/therm.

8-63 A building in a location is equipped with a heat pump system that is used for space heating in winter and cooling in summer. The seasonal COPs of the heat pump in winter and summer are estimated to be 2.2 and 2.8, respectively. The design heating load of the house is calculated to be 750,000 kJ/h for an indoor temperature of 25°C. The unit cost of electricity is $0.09/kWh. The annual heating degree-days total in this location is 2400°C-days, and the 97.5% winter design temperature is −9°C. The cooling degree-days are 1100°C-days. Determine the annual electricity consumption and its cost for heating and cooling of this house. Assume the same overall heat loss coefficient for winter and summer.

8-64 Reconsider Prob. 8-63. (*a*) The existing air-source heat pump system is to be replaced by a ground-source heat pump system. The seasonal COPs of this ground-source heat pump in winter and summer are estimated to be 4.0 and 3.5, respectively. Determine the annual electricity consumption and its cost for heating and cooling of this house. (*b*) It is estimated that the cost and installation of the ground-source heat pump is $40,000. Determine the simple payback period of this investment in comparison to the existing air-source heat pump system.

8-65 The management of a new building is considering two options for space heating and cooling of the building:

Option A: Space heating in winter by a natural gas boiler whose average efficiency is 85 percent. Cooling in summer by an air-source air conditioner whose seasonal COP is 2.2.

Option B: A ground-source heat pump that provides heating in winter and cooling in summer. The seasonal COP for both winter and summer operation is 3.5.

   The heating period is 3000 h and the cooling period is 2000 h. The unit costs of natural gas and electricity are $1.20/therm and $0.11/kWh, respectively. The average heating load in winter is 60 kW, and the average cooling load in summer is 30 kW.

(*a*)   Determine the amount of energy consumed and the monetary cost for heating and cooling of this building for each option.

(*b*)   An initial cost analysis indicates that the initial cost of option B is greater than option A by $15,000. Determine how long it will take for option B to pay for its price differential from the energy it saves.

8-66 The COP of a heat pump is 2.5 in cooling mode. For the same amounts of heat absorption Q_L and heat rejection Q_L, what is the COP of this heat pump in heating mode?
(*a*) 1.25   (*b*) 1.5   (*c*) 2.0   (*d*) 2.5   (*e*) 3.5

8-67   Which of the following is not a characteristic of a geothermal heat pump?
(*a*) Its COP is higher than that of an air-source heat pump.
(*b*) It is also called a ground-source heat pump.
(*c*) It is the most common use of geothermal energy in terms of the number of units installed.
(*d*) Lower initial cost compared to an air-source heat pump.
(*e*) It operates on a vapor-compression refrigeration cycle.

8-68   Heat pumps are most competitive in places that have
(*a*) Small cooling load, small heating load   (*b*) Large cooling load, large heating load
(*c*) Large cooling load, small heating load   (*d*) Small cooling load, large heating load

8-69   A heat pump is called a water-to-air heat pump if water is circulating through the evaporator and air is circulating through the
(*a*) Borehole loop   (*b*) Compressor   (*c*) Vertical piping
(*d*) Condenser   (*e*) Expansion valve

8-70   The operation of a ground-source heat pump in winter involves heat absorption from the ground in
(*a*) Evaporator   (*b*) Condenser   (*c*) Compressor   (*d*) Expansion valve
(*e*) None of these

8-71   The operation of a ground-source heat pump in summer involves heat rejection to the ground in
(*a*) Evaporator   (*b*) Condenser   (*c*) Compressor   (*d*) Expansion valve
(*e*) None of these

8-72   Which is not a type of ground-source heat pump system?
(*a*) Horizontal loop   (*b*) Vertical loop   (*c*) Borehole loop
(*d*) Cascade   (*e*) Ground water wells

8-73   A horizontal loop heat pump involves horizontal underground piping in _____ depths.
(*a*) 0.2–1 m   (*b*) 1.2–2 m   (*c*) 2–3 m   (*d*) 3–6 m   (*e*) 10–20 m

GEOTHERMAL POWER PRODUCTION

8-74   What is the purpose of the flashing process in geothermal power plants? Which property remains constant during a flashing process? What happens to pressure and temperature during a flashing process?

8-75   A geothermal resource contains compressed liquid water at 160°C. Which cycle(s) is best suited for this resource?

8-76   A geothermal resource contains compressed liquid water at 240°C. Which cycle(s) is best suited for this resource?

8-77   A geothermal power plant produces 11.5 MW of power utilizing a liquid resource at 170°C. If the mass flow rate of geothermal water is 210 kg/s and the environment temperature is 20°C, determine the thermal efficiency of this power plant.

8-78   A geothermal site contains geothermal water available at wellhead at a rate of 100 lbm/s. The geothermal water is at 300°F with a vapor fraction of 35 percent. Determine the maximum thermal efficiency and the maximum amount of power that can be produced from this site. Take the dead state temperature to be 80°F.

8-79   Geothermal water enters a flash chamber at a temperature of 210°C as a saturated liquid. The pressure of water is decreased to 600 kPa at the exit of the flash chamber. Determine the temperature and the fractions of liquid and vapor phases after the flashing process.

8-80   Reconsider Prob. 8-79. The flow rate of geothermal water at the flash chamber inlet is 50 kg/s. The vapor resulting from the flashing process is routed to a steam turbine whose isentropic efficiency is 88 percent. The steam leaves the turbine and enters a condenser maintained at 20 kPa. Determine the power output from the turbine.

8-81   A single-flash geothermal power plant uses geothermal liquid water at 150°C at a rate of 420 kg/s as the heat source and produces 15.8 MW of net power in an environment at 1 atm and 25°C. Determine the thermal efficiency, the second-law efficiency, and the total rate of exergy destroyed in this power plant.

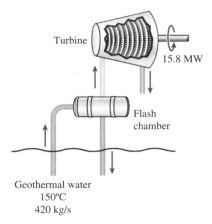

Figure P8-81

8-82 The schematic of a single-flash geothermal power plant with state numbers is given in Fig. P8-82. The geothermal resource exists as saturated liquid at 230°C. The geothermal liquid is withdrawn from the production well at a rate of 230 kg/s, and is flashed to a pressure of 500 kPa by an essentially isenthalpic flashing process where the resulting vapor is separated from the liquid in a separator and directed to the turbine. The steam leaves the turbine at 10 kPa with a moisture content of 10 percent and enters the condenser, where it is condensed and routed to a reinjection well along with the liquid coming off the separator. Determine (*a*) the mass flow rate of steam through the turbine, (*b*) the isentropic efficiency of the turbine, (*c*) the power output of the turbine, and (*d*) the thermal efficiency of the plant (the ratio of the turbine work output to the energy of the geothermal fluid relative to standard ambient conditions).

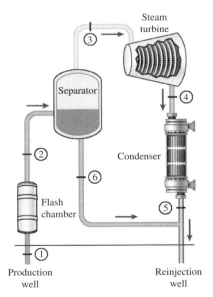

Figure P8-82

8-83 Reconsider Prob. 8-82. Now, it is proposed that the liquid water coming out of the separator be routed through another flash chamber maintained at 150 kPa, and the steam produced be directed to a lower stage of the same turbine (Fig. P8-83). Both streams of steam leave the turbine at the same state of 10 kPa and 90 percent quality. Determine (*a*) the temperature of steam at the outlet of the second flash chamber, (*b*) the power produced by the lower stage of the turbine, and (*c*) the thermal efficiency of the plant.

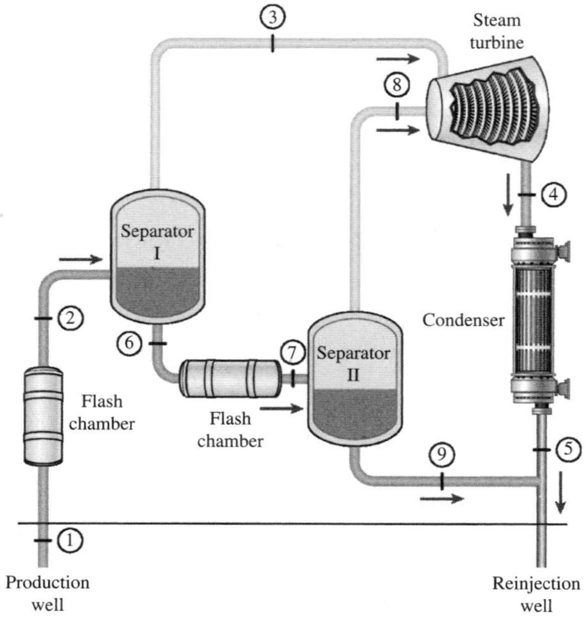

Figure P8-83

8-84 Reconsider Prob. 8-82. Now, it is proposed that the liquid water coming out of the separator be used as the heat source in a binary cycle with isobutane as the working fluid (Fig. P8-84). Geothermal liquid water leaves the heat exchanger at 90°C while isobutane enters the turbine at 3.25 MPa and 145°C and leaves at 80°C and 400 kPa. Isobutane is condensed in an air-cooled condenser and then

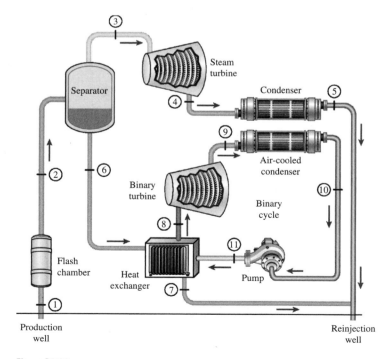

Figure P8-84

pumped to the heat exchanger pressure. Assuming an isentropic efficiency of 90 percent for the pump, determine (*a*) the mass flow rate of isobutane in the binary cycle, (*b*) the net power outputs of both the flashing and the binary sections of the plant, and (*c*) the thermal efficiencies of the binary cycle and the combined plant. The properties of isobutane at various states in the cycle are $h_8 = 755.05$ kJ/kg, $h_9 = 691.01$ kJ/kg, $h_{10} = 270.83$ kJ/kg, and $h_{11} = 276.65$ kJ/kg.

8-85   Consider a binary geothermal power plant with isobutane as the working fluid using geothermal liquid water at 150°C at a rate of 150 kg/s as the heat source (Fig. P8-85). Geothermal liquid water leaves the heat exchanger at 90°C while isobutane enters the turbine at 3.25 MPa and 145°C and leaves at 80°C and 400 kPa. Isobutane is condensed in an air-cooled condenser and then pumped to the heat exchanger pressure. Assuming an isentropic efficiency of 90 percent for the pump, determine (*a*) the mass flow rate of isobutane in the binary cycle, (*b*) the net power output from the plant if 15 percent of the turbine power output is used for the fans of the condenser, and (*c*) the thermal efficiency of the binary cycle. The properties of isobutane at various states in the cycle are $h_1 = 270.83$ kJ/kg, $h_2 = 276.65$ kJ/kg, $h_3 = 755.05$ kJ/kg, and $h_4 = 691.01$ kJ/kg.

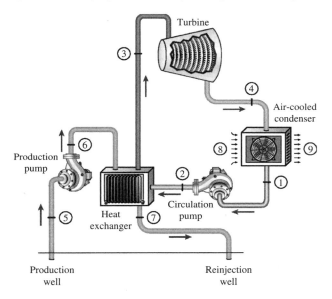

Figure P8-85

8-86   Which system is not used in geothermal power plants?
(*a*) Brayton   (*b*) Direct steam   (*c*) Single-flash   (*d*) Double-flash   (*e*) Binary

8-87   In a _____ geothermal power plant, the working fluid is not steam.
(*a*) Single-flash cycle   (*b*) Binary cycle   (*c*) Double-flash cycle
(*d*) Direct steam cycle without a condenser   (*e*) Direct steam cycle with a condenser

8-88   The maximum thermal efficiency of a power plant using a geothermal source at 180°C in an environment at 25°C is
(*a*) 43.1%   (*b*) 34.2%   (*c*) 19.5%   (*d*) 86.1%   (*e*) 12.7%

8-89   Geothermal liquid water is available at a site at 150°C at a rate of 10 kg/s in an environment at 20°C. The maximum amount of power that can be produced from this site is
(*a*) 606 kW   (*b*) 692 kW   (*c*) 740 kW   (*d*) 883 kW   (*e*) 955 kW

8-90   Geothermal water enters a flash chamber at 230°C as a saturated liquid. The pressure of water is decreased to 750 kPa at the exit of the flash chamber. The vapor fraction of geothermal water after the flashing process is
(*a*) 0.092   (*b*) 0.117   (*c*) 0.137   (*d*) 0.155   (*e*) 0.184

8-91   A binary geothermal power plant produces 5 MW of power using a geothermal resource at 175°C at a rate of 110 kg/s. If the temperature of geothermal water is decreased to 90°C in the heat exchanger after giving its heat to binary fluid, the thermal efficiency of the binary cycle is
(*a*) 7.5%   (*b*) 16.3%   (*c*) 14.4%   (*d*) 12.5%   (*e*) 9.7%

GEOTHERMAL COGENERATION

8-92   What is cogeneration? What is trigeneration?

8-93   What cycles can be used as the power cycle in a cogeneration plant? Which ones are most common?

8-94   What is the advantage of cogeneration compared to separate electrical and process heat systems?

8-95   What are a topping cycle and a bottoming cycle? Which one is more commonly used? For which applications are they more suitable?

8-96   Can a wind turbine be used as the power-producing system of a cogeneration plant? Explain.

8-97   A trigeneration plant produces 25 MW of electricity, 12 MW of process heat, and 9 MW of cooling. If the rate of heat input to the plant is 66 MW, determine the utilization factor of this plant.

8-98   A trigeneration plant produces 180 kW of electricity, 780,000 Btu/h of process heat, and 80 tons of cooling. If the rate of heat input to the plant is 2,700,000 Btu/h, determine the utilization factor of this plant.

8-99   A geothermal-based trigeneration plant produces 2.5 MW of electricity, 1.8 MW of process heat, and 1.4 MW of cooling. Electricity is produced for a period of 7500 h, heating for 2500 h, and cooling for 2000 h per year. The geothermal energy consumed in the plant is 4×10^7 kWh during an entire year. Determine the annual average utilization factor of this plant.

8-100   Geothermal water enters a single-flash geothermal power plant at 230°C as a saturated liquid at a rate of 50 kg/s (state 1), as shown in Fig. P8-82. It is flashed (throttled) to a pressure of 500 kPa, during which enthalpy remains constant. The resulting vapor (state 3) is directed to a steam turbine and is expanded to a condenser pressure of 50 kPa (state 4). The isentropic efficiency of the turbine is 82 percent. The liquid separated from vapor in the flash chamber (state 6) is normally reinjected back to the ground. It is proposed to use this water for space heating of commercial buildings in winter. This is done by routing geothermal water to a heat exchanger in which the heat of geothermal water is transferred to fresh water. The heated fresh water circulates in the buildings for space heating. Geothermal water leaves the heat exchanger at 90°C.

(*a*)   Determine the power produced and the rate of space heating.

(*b*)   Determine the thermal efficiency of the geothermal power plant and the utilization factor of the cogeneration plant.

8-101   Reconsider Prob. 8-100. The price of electricity and space heating are $0.07/kWh and $0.9/therm (1 therm = 105,500 kJ), respectively.

(*a*)   Determine the annual potential revenues due to selling electricity and space heating. Assume an annual period of 8500 h for electricity production and 3000 h for space heating.

(*b*)   Assume that all geothermal water is used for space heating during winter, and no electricity is generated. In this mode, geothermal water enters the heat exchanger at 230°C as a saturated liquid at a rate of 50 kg/s and leaves at 90°C. In non-winter times, geothermal water is only used to generate electricity. Assume an annual period of 3000 h for space heating and 5500 h for electricity production. Determine the total potential revenue and compare to that calculated in part (*a*).

8-102   Consider a binary geothermal cogeneration plant like that in Fig. P8-102 using geothermal water at 165°C with isobutane as the working fluid. The mass flow rate of geothermal water is 175 kg/s and the net power output is 6.9 MW. It is determined that geothermal water leaves the plant at 85°C (state 6) and leaves the heater at 65°C (state 7). The fresh liquid water, heated to 75°C (state 9), is used for space heating and returns to the cogeneration plant at 50°C (state 8).

(*a*)   Determine the rate of space heating provided by the system and the mass flow rate of water used for space heating.

(*b*)   Determine the thermal efficiency of the power plant and the utilization factor for the entire cogeneration plant.

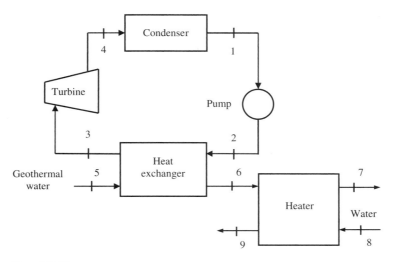

Figure P8-102

8-103 Reconsider Prob. 8-102. The space heating provided by this cogeneration plant will replace natural gas heating. How much money does this plant make by selling geothermal heat at a discount rate of 25 percent with respect to natural gas? The efficiency of a natural gas furnace is 85 percent, and the price of natural gas is \$1.3/therm (1 therm = 105,500 kJ). Assume a winter operation period of 4200 h under these average conditions.

8-104 An absorption refrigeration system is powered by a geothermal power plant whose net power output is 5000 kW, and its thermal efficiency is 8 percent. Geothermal water enters the power plant at 180°C as a liquid and leaves the plant at 120°C. It then enters the generator of the absorption cooling system and leaves at 100°C. The COP of the absorption cooling system is 0.75.

(*a*) Determine the mass flow rate of geothermal water, the rate of cooling provided by the absorption cooling system, and the utilization factor of this cogeneration plant.

(*b*) The absorption cooling system is replacing a conventional vapor-compression system whose COP is 2.5. The price of absorption cooling is 20 percent less than the price of conventional cooling. The unit price of electricity is \$0.085/kWh. Determine the revenues generated by selling electricity and absorption cooling. The power plant operates 8200 h a year, but the cooling is supplied for 4000 h a year.

8-105 Which one cannot be used as the power producing system in a cogeneration plant?
(*a*) Gas turbine   (*b*) Steam turbine   (*c*) Wind turbine
(*d*) Geothermal   (*e*) Internal combustion engine

8-106 Which one cannot be used as the power producing system in a renewable-based cogeneration plant?
(*a*) Biomass plant   (*b*) Geothermal plant   (*c*) Solar thermal plant
(*d*) Hydroelectric power plant

8-107 A geothermal cogeneration plant produces 10 MW of electricity and supplies 2 MW of district heating. If the energy input to the plant is 80 MW, the utilization factor of this plant is
(*a*) 0.125   (*b*) 0.15   (*c*) 0.20   (*d*) 0.025   (*e*) 1.0

8-108 A cogeneration plant produces 2 MW of electricity and supplies 500 kW of cooling with a heat input of 4000 kW. The utilization factor of this plant is
(*a*) 0.125   (*b*) 0.375   (*c*) 0.50   (*d*) 0.625   (*e*) 1.6

CHAPTER 9

Biomass Energy

9-1 BIOMASS RESOURCES

Biomass is an organic renewable energy source. It is mostly produced from agriculture and forest products and residues, energy crops, and algae. The organic component of municipal and industrial wastes and the fuel produced from food processing waste such as used cooking oil are also considered biomass. Despite the lengthy periods needed to grow crops and trees, they can be regrown by planting, and therefore biomass is considered to be a renewable energy source. It is estimated that about half of all renewable energy consumed in the United States is biomass.

Before coal, oil, and natural gas replaced it as primary fuels, wood was the primary fuel for space heating in winter. Wood is still used in many parts of the developing world for space heating. Unfortunately, the burning rate of wood is generally higher than the growth rate of trees in these parts of the world. The net result is deforestation with well known negative consequences.

Liquid and gaseous fuels are generally more convenient forms of fuel compared to solid fuels. Therefore, crops and forest products are usually converted to liquid and gaseous fuels through some engineering processes. The growing of crops and trees as well as the conversion to liquid and gaseous fuels involves the consumption of energy in the form of electricity and heat.

Biomass can be obtained from a variety of resources called feedstocks. Biomass resources can be listed as follows (DOE/EERE, 2018):

Dedicated Energy Crops. These *herbaceous energy crops* are perennials that are harvested after reaching maturity. These include such grasses as switchgrass, miscanthus, bamboo, sweet sorghum, tall fescue, kochia, wheatgrass, and others.

Agricultural Crops. These include corn, soybeans, wheat, and other vegetables. They generally yield sugars, oils, and extractives.

Agriculture Crop Residues. Biomass materials consisting primarily of stalks and leaves not used for commercial use such as corn stover (stalks, leaves, husks, and cobs), wheat straw, and rice straw are included in this resource. Approximately 80 million acres of corn is planted annually.

Forestry Residues. These are biomass not harvested or used in commercial forest processes, including materials from dead and dying trees.

Aquatic Crops. Aquatic biomass resources include algae, giant kelp, other seaweed, and marine microflora.

Biomass Processing Residues. Byproducts and waste streams produced by biomass processing are called residues, and they represent an additional biomass resource.

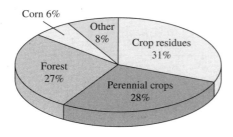

Figure 9-1 Contribution of various feedstocks to biomass production in the United States (*DOE/EERE, 2018*).

Municipal Waste. Plant-based organic material generated from industrial, residential, and commercial waste represents an important biomass source. Some examples include waste paper, wood waste, yard waste, and cooking oil.

Animal Waste. Animal wastes consist of organic materials and are generated from farms and animal-processing operations. Animal waste is used as a heating fuel in some parts of the world.

Most of the biomass feedstock in the United States comes from crop residues (31 percent), perennial crops (28 percent), and forest resources (27 percent), as indicated in Fig. 9-1. Corn has a contribution of only 6 percent, while the remaining 8 percent come from other sources (Perlack et al., 2005).

9-2 CONVERSION OF BIOMASS TO BIOFUEL

Biomass can be converted into liquid or gaseous fuels through biochemical- and thermochemical-based conversion processes. In *biochemical conversion processes*, enzymes and microorganisms are used as biocatalysts to convert biomass or biomass-derived compounds into desirable products. In *thermochemical conversion processes*, heat energy and chemical catalysts are used to break down biomass into intermediate compounds or products. In gasification, biomass is heated in an oxygen-starved environment to produce a gaseous fuel. Solvents, acids, and bases can be used to fractionate biomass into an array of products including sugars, cellulosic fibers, and lignin.

Research is underway on *photobiological conversion processes*. Photobiological conversion processes use the natural photosynthetic activity of organisms to produce biofuels directly from sunlight. For example, the photosynthetic activities of bacteria and green algae have been used to produce hydrogen from water and sunlight.

Various pathways for the conversion of biomass into gaseous and liquid fuels are shown in Fig. 9-2. The conversion to gaseous fuels can be achieved by two major methods: biological process and thermal process. Anaerobic digestion is used as the biological process and the resulting gaseous fuel is biogas. The thermal gasification process yields producer gas. Both biogas and producer gas can be upgraded to other gaseous fuels such as hydrogen, methane, ammonia, and dimethyl ether.

Liquid fuels produced from biomass is mostly used as transportation fuel. The sources of liquid fuels include substrates derived from biomass through physical, chemical, and thermal processes. They can be classified into four categories as carbohydrates, triglycerides, syngas, and bio-oil/biocrude. The main liquid fuel from carbohydrates is ethanol, but butanol, isoprenes, furans, and alkanes can also be obtained. Biodiesel (methyl or ethyl ester) is the main product from triglycerides, but alkanes and propane may also be generated. Fischer-Tropsch liquids, methanol, ethanol, and MTB (methanol-to-gasoline) may be obtained from syngas. Finally, bio-oil and biocrude may be upgraded to produce gasoline, diesel fuel, and aviation fuel (Kreith and Kreider, 2011).

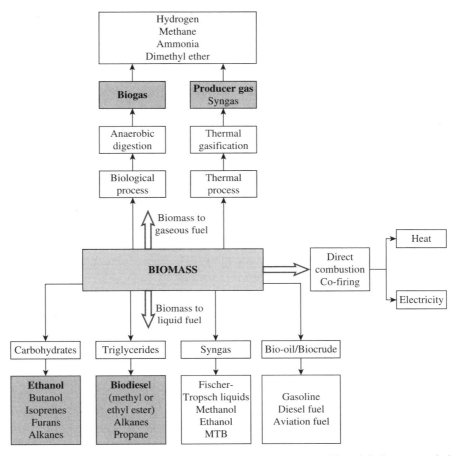

Figure 9-2 Pathways for the conversion of biomass to gaseous and liquid fuels. Heat and electricity production from biomass by direct combustion and co-firing is also shown.

A third main method of using biomass is heat and electricity generation through direct combustion of biomass or co-firing. *Co-firing* refers to replacing part of fossil fuel in a conventional energy system by biomass. Heat production is normally done in boilers by generating a hot fluid or steam. Electricity production is done by various power plant systems including steam power plants, gas turbines, internal combustion engines, and fuel cells. Another pathway to electricity generation is first producing gaseous and liquid fuels from biomass and burning these fuels in power plants.

A major product of biomass is biofuels, which are a replacement for petroleum-based fuels. Biofuels can be liquid or gas. They are mostly used for transportation as engine fuel but are also used for heating and electricity generation. The two most common biofuels are ethanol and biodiesel. Other products include methanol, pyrolysis oil, biogas, producer gas, and synthesis gas.

Biomass is primarily used to produce biofuels such as ethanol and biodiesel, but other products made from fossil fuels can also be made by biomass. Some of these products include antifreeze, plastics, glues, artificial sweeteners, and gel for toothpaste.

The extensive use of farmland for energy crops to produce biofuels means that less land will be available for food production. Using a certain farmland for fuel production may be feasible from an economic and environmental point of view but it may be morally wrong if it contributes to shortage of food.

9-3 ANAEROBIC DIGESTION

Anaerobic digestion is a biological process, and it refers to the decomposition of waste biomass into gaseous fuels by bacteria action without the presence of oxygen. Organic matter is broken into well-known simpler compounds. The main output of anaerobic digestion is *biogas* that consists of methane (CH_4), carbon dioxide (CO_2), and small amounts of other gases.

The biological anaerobic digestion process takes place in multiple stages, and it consists of three steps, as shown in Fig. 9-3 (Klass, 1998):

Hydrolysis. Hydrolytic and fermentative bacteria break down proteins, carbohydrates, and fats into simple organics, long-chain acids, amino acids, alcohols, hydrogen, carbon dioxide, and sugar units.

Acidification. This is a transitional digestion step, and it is achieved by acid-forming bacteria to produce acetate (acetic acid, propionic acid, C_4+ acid), hydrogen, and carbon dioxide.

Methanogenesis. Methane-forming bacteria action converts the acetate formed in the previous step into methane.

The final desired product of the three steps of anaerobic digestion is biogas, but a stream of sludge also forms. Biogas contains sulfur compounds like hydrogen sulfide, and these

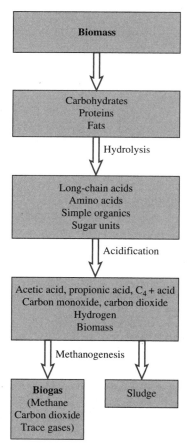

Figure 9-3 Anaerobic digestion steps. (*Adapted from Klass, 1998.*)

compounds need to be removed before biogas is substituted for natural gas. Otherwise, combustion products contain sulfur dioxide, a very undesirable pollutant.

Thermodynamic performance of an anaerobic digestor may be expressed by the fraction of energy content of dry waste that is converted into the energy content of biogas. This efficiency is about 50-70 percent in typical digestors. Carbon dioxide has no energy value, and thus the energy content of biogas is mainly due to methane. Volumetric methane yield per mass of the volatile solid waste is also used as a performance parameter. This parameter ranges greatly depending on the type of the feedstock and the anaerobic reactor design. Most digestors produce between 0.2 and 0.3 m³ methane per kg of volatile solids. Up to 90 percent thermodynamic efficiency is achieved in laboratory applications. Note that the volatile matter in the biomass evaporates to a gas at moderate temperatures (about 400°C) in a nonoxidizing environment.

Some of the operating parameters affecting digestor performance are pretreatment, preheating, mixing, nutrient addition, specialized bacteria edition, and pH adjustment for the waste (Kreith and Kreider, 2011). The feedstock to the digestor is heated to a certain temperature for bacteria action. A liquid is heated in a boiler and the energy of this hot liquid is transferred to the biomass feedstock using a heat exchanger. Minimizing heat requirement of the process is an effective method of improving the digestor efficiency.

Common feedstocks for anaerobic digestion are various organic wastes high in moisture content such as sewage sludge, municipal solid waste (MSW), food scraps, animal manures, fats, oils, greases, industrial organic residuals, and agricultural waste. Anaerobic digesters perform better when they are designed for a certain type of waste.

Different types of anaerobic digesters exist. One classification is based on the method of biomass feed: batch, intermittent, or steady-flow biomass feed (Fig. 9-4). The basic function in each type is maintaining an oxygen-free environment for methane-forming microorganisms to grow. The two outflow streams are biogas and sludge. Some digester designs include a simple batch digester, complete mix digester, plug flow digester, anaerobic contact digester, fixed film digester, upflow anaerobic sludge blanket digester, induced bed reactor, and anaerobic sequencing batch reactor. A main objective in digester design is to increase biomass contact with active bacteria.

Biogas

Biogas, also called swamp gas, landfill gas, or digester gas, usually consists of 50 to 80 percent methane (CH_4) and 20 to 50 percent carbon dioxide (CO_2) by volume. It also contains small amounts of hydrogen, carbon monoxide, and nitrogen. Noting that the higher heating value

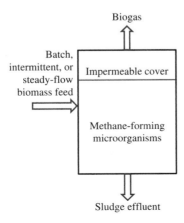

Figure 9-4 Anaerobic digesters can be designed for batch, intermittent, or steady-flow biomass feed.

(HHV) of methane is 55,530 kJ/kg, the HHV of biogas with 50 percent methane by volume is 14,800 kJ/kg and that of biogas with 80 percent methane is 32,900 kJ/kg.

Biogas can be produced from biological waste such as animal manure, food waste, and agricultural waste. The process is called *anaerobic digestion*, which is the decomposition of organic waste into a gaseous fuel by bacterial action without the presence of oxygen. It is possible to produce 200 to 400 m³ of biogas from 1000 kg of organic waste with 50 to 75 percent methane by volume (Boyle, 2004).

Biogas is basically a gaseous fuel similar to natural gas but with a lower energy content due to its significant carbon dioxide fraction. Biogas can be fed to a natural gas pipeline after carbon dioxide is removed. It can be easily burned in a boiler for space, process, and water heating applications. It can also be used to generate steam in a steam power plant to produce electricity. Many municipalities in different countries have solid waste treatment facilities in which they produce biogas from the waste and use the biogas for electricity generation. Some facilities produce both electricity and heat from biogas (i.e., cogeneration). A gas-turbine or an internal combustion engine can be used as the power-producing unit.

An existing biogas-burning combined power plant and its schematic are shown in Fig. 9-5. This plant produces power using a gas turbine cycle. An additional useful output from the plant is liquid fertilizers. The feedstock for biogas production is chicken manure. The steps from chicken manure inlet to biogas outlet are shown on the right side of the figure. Chicken manure is fermented as raw material. Protein, fat, and carbohydrates are mixed with water and decomposed into acetic acid, propionic acid, and butyric acid by acid-producing bacteria. Methanogenic bacteria convert these acids into biogas. The process takes place at an average temperature of about 35°C. The resulting biogas from chicken manure is 60 to 70 percent methane, 30 to 40 percent carbon dioxide, and trace amounts of hydrogen, hydrogen sulfide, ammonia, water, and nitrogen. About 70 percent of the methane released in this process comes from the methyl groups of acids, and the rest comes from the reaction of carbon dioxide and hydrogen.

The operation of the gas-turbine plant is shown on the left side of Fig. 9-5b. Atmospheric air is compressed in a compressor. This high-pressure air is heated by the hot exhaust gases leaving the gas turbine before entering the combustion chamber. Biogas is burned with air in the combustion chamber and the resulting combustion gases are expanded in the turbine producing power. Mechanical power output from the turbine is converted to electrical power in a generator connected to the turbine. The power output from this gas turbine unit is 4000 kW. It is estimated that 31 percent of the energy content of biogas is converted to power in the power plant (Arslan and Yilmaz, 2022).

EXAMPLE 9-1
Heating Value
of Biogas

A certain biogas consists of 75 percent methane (CH_4) and 25 percent carbon dioxide (CO_2) by volume. If the higher heating value (HHV) of methane is 55,530 kJ/kg and the lower heating value (LHV) is 50,050 kJ/kg, what are the HHV and LHV of this biogas?

SOLUTION The molar masses of C, H_2, and O_2 are 12 kg/kmol, 2 kg/kmol, and 32 kg/kmol, respectively (Table A-1). The mole fractions of CH_4 and CO_2 are equal to their volume fractions:

$$y_{CH_4} = 0.75$$

$$y_{CO_2} = 0.25$$

(a)

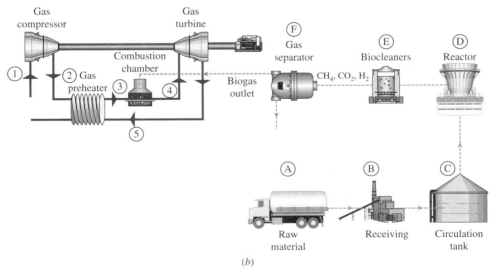

(b)

Figure 9-5   An existing biogas production facility and the gas turbine cycle. (a) Picture of the plant, (b) Schematic of operation. (*Arslan and Yilmaz, 2022.*)

We consider 100 kmol of biogas mixture. In this mixture, there are 75 kmol of CH_4 and 25 kmol of CO_2. The masses of CH_4 and CO_2 are

$$m_{CH_4} = N_{CH_4} M_{CH_4} = (75 \text{ kmol})(16 \text{ kg/kmol}) = 1200 \text{ kg}$$

$$m_{CO_2} = N_{CO_2} M_{CO_2} = (25 \text{ kmol})(44 \text{ kg/kmol}) = 1100 \text{ kg}$$

The total mass of the mixture is

$$m_{total} = m_{CH_4} + m_{CO_2} = 1200 + 1100 = 2300 \text{ kg}$$

The mass fraction of CH_4 is

$$\text{mf}_{CH_4} = \frac{m_{CH_4}}{m_{total}} = \frac{1200 \text{ kg}}{2300 \text{ kg}} = 0.5217$$

The heating value of CO_2 is zero. Therefore, the higher and lower heating values of biogas are

$$\text{HHV}_{biogas} = \text{mf}_{CH_4} \times \text{HHV}_{CH_4} = (0.5217)(55,530 \text{ kJ/kg}) = \mathbf{28,970 \text{ kJ/kg}}$$

$$\text{LHV}_{biogas} = \text{mf}_{CH_4} \times \text{LHV}_{CH_4} = (0.5217)(50,050 \text{ kJ/kg}) = \mathbf{26,110 \text{ kJ/kg}} \quad \blacktriangle$$

9-4 THERMAL GASIFICATION

Thermal gasification is a thermal process in which biomass is exposed to high temperatures (750 to 800°C) to turn solid biomass into gaseous fuel, known as *producer gas*. Producer gas consists of hydrogen, carbon monoxide, methane, nitrogen, carbon dioxide, and small amounts of other hydrocarbons. Thermal gasification can also yield synthetic gas (or syngas) whose composition is mainly carbon monoxide and hydrogen. Heating biomass to high temperatures requires significant amounts of heat input. This heat may be supplied externally or most commonly by burning part of the biomass. The amount of air used is typically less than 25 percent of stoichiometric amount for complete combustion of biomass. Biogas production through anaerobic digestion is more common than producer gas production through thermal gasification. This is partly due to high temperature requirement and the corresponding high cost of gasification. In addition to the heat and electricity production, the producer gas can also be used to produce liquid fuels and chemicals.

Historically, thermal gasification has been used for converting coal to gaseous fuels. It is still used in some parts of the world. However, we are interested in using biomass as the feedstock. Common biomass feedstocks include cellulosic biomass such as wood chips, wood pellets, wood powder, and agricultural product or wastes such as straw and husks. The moisture content of feedstock is usually less than 20 percent.

There are multiple steps in a thermal gasification process, as shown in Fig. 9-6 (Kreith and Kreider, 2011):

Heating and drying. Biomass is heated to remove moisture content by evaporation of the water. An energy input is required for heating and drying. This heat is normally supplied by burning part of the biomass input. This can be controlled by admitting a limited amount of air to allow partial combustion of the biomass. This should provide enough heat for the heating and drying as well as the required temperature levels for the endothermic reactions in the further steps of the gasification.

Pyrolysis. This is the most important step of the gasification. Pyrolysis is the exposure of biomass to high temperatures in the absence of air or oxygen. Pyrolysis starts at 300 to 400°C. The product of pyrolysis are the gases including carbon monoxide (CO), carbon dioxide (CO_2), hydrogen (H_2), and water vapor (H_2O), and some other hydrocarbons. Relative amounts of these gases depend on the composition of the feedstock and the

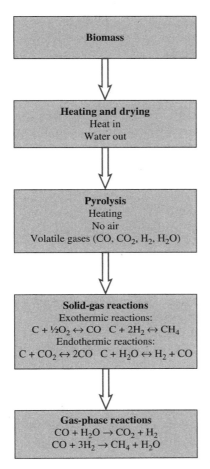

Figure 9-6 Thermal gasification steps. (*Adapted from Kreith and Kreider, 2011.*)

temperature during the process. The volatile matter in the biomass is roughly equal to the gaseous yield after the pyrolysis. The remaining yield consists of condensable vapors such as water, methanol, acetic acid, and acetone as well as solid carbon particles and ash.

Solid-gas reactions. Solid carbon in char is converted into gaseous products including CO, H_2, and CH_4. Following reactions take place (Higman and Burgt, 2008):

Exothermic reactions:

Carbon-oxygen reaction:	$C + \frac{1}{2}O_2 \Leftrightarrow CO$
Boudouard reaction:	$C + CO_2 \Leftrightarrow 2CO$

Endothermic reactions:

Carbon-water reaction:	$C + H_2O \Leftrightarrow H_2 + CO$
Hydrogenation reaction:	$C + 2H_2 \Leftrightarrow CH_4$

Comparing the two exothermic reactions, more carbon reacts with oxygen than with hydrogen. As a result, the first exothermic reaction supplies more heat compared to the second reaction for drying, pyrolysis, and the two endothermic solid-gas reactions.

Gas-phase reactions. This is the last step of the thermal gasification, and it results in the final composition of the producer gas. The reactions are as follows:

Water-gas shift reaction:	$CO + H_2O \Leftrightarrow CO_2 + H_2$
Methanation:	$CO + 3H_2 \Leftrightarrow CH_4 + H_2O$

The final product is called producer gas consisting of hydrogen, carbon monoxide, methane, nitrogen, carbon dioxide, and some other light hydrocarbons. More hydrogen and carbon monoxide are formed at higher reaction temperatures while more methane is formed at low reaction temperatures. The composition of producer gas depends on the reaction time and temperature as well as the amount of air used for partial combustion.

Another product of gasification is a viscous liquid known as *tar*, consisting of heavy hydrocarbons. It is an undesirable product as it can block valves and filters and interfere with the reactions. Also, tar in the producer gas causes various operating problems when producer gas is used as the fuel in different power production systems such as internal combustion engines and fuel cells. Steam injection and catalysts are sometimes used to minimize tar yield. Particulate matter consisting of carbon particles and ash resulting from the gasification process can deposit on the components of the power system and degrade the performance (Kreith and Kreider, 2011).

Certain practical methods are available to clean the gas and remove tar and particulate matter from the producer gas. One method is scrubbing gas stream with a fine mist of water or oil to remove tar and solid particles. Another method is called *cracking*, which is a process of converting higher-molecular-weight component into lower-molecular-weight components. In this process, the gas is heated to very high temperatures for converting tar into carbon monoxide and hydrogen. The required temperature is about 1000°C without a catalyst and about 600 to 800°C with the use of a catalyst.

For a better gasifier performance, the amount of air used for partial combustion of biomass should be as low as possible. Nitrogen in the air eventually becomes part of the producer gas at the end of the gasification process, and this lowers the heating value of the producer gas. One possible solution is using pure oxygen instead of air, but this solution is too costly to be implemented in most practical gasifiers.

The efficiency of a thermal gasification process may be defined as the total energy content of the producer gas to the chemical energy of the input biomass. The total energy content of the producer gas consists of chemical energy (heating value of the constituents of the producer gas) and the sensible energy due to high temperature of the gas at the exit of the gasifier. This efficiency may be called as *hot gas efficiency*, and it typically ranges between 70 and 90 percent. The sensible heat of the gas can be used for various uses such as space heating, process heating, and drying. In most power producing applications, the producer gas is cooled, and thus the sensible heat is lost. In such cases, it is more proper to define the efficiency considering only the heating value of the producer gas (*cold gas efficiency*). This efficiency is typically between 50 and 70 percent.

Common biomass gasifiers include fixed bed, fluidized bed, and entrained flow reactors (Fig. 9-7). Fixed bed reactors could be updraft (countercurrent) or downdraft (concurrent). Updraft gasifiers admit biomass from above and some air enters from the below. Partial combustion of biomass occurs at the bottom section and provides necessary temperatures for drying, pyrolysis, and reduction reactions. The producer gas leaving from the top contains considerable levels of tar. In downdraft gasifiers, the biomass enters from the top and the producer gas leaves at the bottom. Air enters the reactor in a middle section where the combustion forms a bed in the throat. The condensable gases resulting from pyrolysis flow through this hot bed, and this allows the cracking of tar to lower-molecular-weight components. As a result, less tar is produced compared to updraft gasifiers (Kreith and Kreider, 2011).

In fluidized bed gasifiers, biomass fuel must be injected as small particles such as sawdust and fine wood shavings. Combustion takes place with suspended biomass fuel using air entering from the bottom. These systems involve high initial and operating costs, and therefore they are preferred for large gasification systems. Entrained flow gasifiers have been

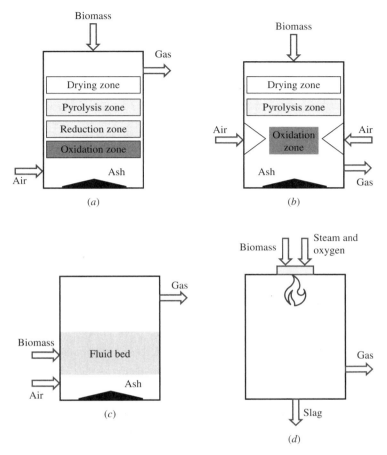

Figure 9-7   Various biomass thermal gasification systems. (*a*) Updraft fixed bed, (*b*) Downdraft fixed bed, (*c*) Fluidized bed, (*d*) Entrained flow.

developed to gasify coal, and it is a candidate for the gasification of pretreated biomass. The fuel enters from the top along with steam and oxidant. They involve very high temperatures (1200 to 1500°C), and the output is a high-quality synthesis gas (syngas) without tar. The high-temperature operation also melts the coal ash into slag.

Producer Gas

Producer gas is produced by thermal gasification, which is the partial oxidation of a solid biomass at high temperatures into a gaseous fuel. Steam and oxygen interact with solid biomass such as wood during a gasification process. Most practical gasification systems can convert 70 to 80 percent of the heat of the biomass into energy of producer gas. The resulting producer gas consists of CO, H_2, CH_4, N_2, and CO_2. Note that except for N_2 and CO_2, other ingredients in producer gas are fuels. For example, CO is a fuel with a heating value of 10,100 kJ/kg. (Is this LHV or HHV?) The composition of producer gas varies greatly. The heating value of producer gas depends on percentages of ingredient gases, and it varies between 15 percent and 50 percent of the heating value for natural gas. Note that natural gas is primarily methane with small amounts of other HCs. Producer gas can be used as a feedstock for liquid fuels or it can be burned directly in a furnace.

Synthesis Gas

Synthesis gas is also called *biosynthesis gas* or *syngas*, and it is produced by thermal gasification using oxygen. It consists of CO and H_2. If a synthesis gas has 50 percent CO and 50 percent H_2 fraction by volume, its LHV and HHV will be 17,430 kJ/kg and 18,880 kJ/kg, respectively. Synthesis gas is commonly produced from natural gas, coal, and heavy diesel. However, we are more interested in its production from biomass feedstock. Wood and other solid biomass can be used to produce syngas. In addition to being used as the fuel for conversion to heat and electricity, synthesis gas can be used to make plastics and acids, which can then be used to make photographic films, textiles, and synthetic fabrics.

EXAMPLE 9-2
Efficiency of a Biomass Gasifier

A biomass gasifier converts 100 kg of dry wood powder into 500 m^3 of producer gas at 1 atm and 400°C. Determine the hot gas efficiency and the cold gas efficiency for this thermal gasification process. Use the following data:

Heating value of wood powder, HV_{wood} = 19,000 kJ/kg

Heating value of producer gas, HV_{gas} = 5800 kJ/m^3

Density of producer gas at standard atmospheric conditions (1 atm, 25°C), ρ = 1.29 kg/m^3

The average specific heat of producer gas for sensible heat calculation, c_p = 1.1 kJ/kg·°C

SOLUTION   Using the given density of the producer gas at standard atmospheric conditions [1 atm (101.3 kPa), 25°C], the gas constant is determined from the ideal gas equation to be

$$R = \frac{P}{\rho T} = \frac{101.3 \text{ kPa}}{(1.29 \text{ kg/m}^3)[(25+273) \text{ K}]} = 0.2635 \text{ kJ/kg·K}$$

The mass of the producer gas at the gasifier exit can be determined from the ideal gas relation using the given pressure and temperature of the gas:

$$m_{gas} = \frac{PV}{RT} = \frac{(101.3 \text{ kPa})(500 \text{ m}^3)}{(0.2635 \text{ kJ/kg·K})[(400+273) \text{ K}]} = 285.6 \text{ kg}$$

The mass of the producer gas is much more than the mass of the wood due to addition of air into the gasifier. A large portion of the producer gas is nitrogen coming with the air.

The heating value of the producer gas is given per unit volume and the volume is based on the standard atmospheric conditions (1 atm, 25°C). The heating value of the gas per unit mass is determined using the density at the standard atmospheric conditions:

$$HV_{gas} = \frac{HV}{\rho} = \frac{5800 \text{ kJ/m}^3}{1.29 \text{ kg/m}^3} = 4496 \text{ kJ/kg}$$

The sensible heat of the producer gas due to its temperature above the standard atmospheric temperature of 25°C (298 K) is estimated from

$$Q_{sensible} = mc_p\Delta T = (285.6 \text{ kg})(1.1 \text{ kJ/kg·K})[(673-298) \text{ K}] = 117,810 \text{ kJ}$$

This sensible heat value is approximate as we used a constant value of specific heat. A more accurate value for the sensible heat can be determined after obtaining enthalpy values of the producer gas at 400 and 25°C, and using the relation $Q_{sensible} = m\Delta h$.

The hot gas efficiency and the cold gas efficiency are

$$\eta_{\text{hot gas}} = \frac{m_{\text{gas}}HV_{\text{gas}} + Q_{\text{sensible}}}{m_{\text{wood}}HV_{\text{wood}}} = \frac{(285.6 \text{ kg})(4496 \text{ kJ/kg}) + 117{,}810 \text{ kJ}}{(100 \text{ kg})(19{,}000 \text{ kJ/kg})} = 0.738 = \textbf{73.8\%}$$

$$\eta_{\text{cold gas}} = \frac{m_{\text{gas}}HV_{\text{gas}}}{m_{\text{wood}}HV_{\text{wood}}} = \frac{(285.6 \text{ kg})(4496 \text{ kJ/kg})}{(100 \text{ kg})(19{,}000 \text{ kJ/kg})} = 0.676 = \textbf{67.6\%}$$

The sensible heat is less than 10 percent of the total energy content of the producer gas.   ▲

9-5 ETHANOL

The most common liquid fuels obtained from biomass are ethanol and biodiesel (methyl or ethyl ester). Ethanol is a good substitute for gasoline in spark-ignition engines while biodiesel is more appropriate for use as a fuel in compression-ignition engines.

Ethanol is easily produced from carbohydrates derived from biomass. Carbohydrates are basically sugar molecules found in sugars, starch, and cellulose. Common biomass feedstock for ethanol production are starch and sugar crops such as sugarcane, sugar beets, and sweet sorghum. The sugars in sugar crops can be fermented directly into ethanol by yeasts or bacteria. Sugarcane is the main source of ethanol production in Brazil.

Common starch crops for ethanol feedstock are grain and potatoes. Corn is the main source of ethanol production in the United States. Starch consists of long chains of glucose with hydrogen bonds. Breaking these bonds results in simple sugars, which then can be fermented into ethanol. *Hydrolysis* of starch involves breaking down of the carbohydrate fractions of starch to smaller sugars by heating starch granules in a dilute acid solution or heating the starch and reacting with thermophilic (high-temperature tolerant) enzymes. Inulin is another source of carbohydrates. It consists of chains of sugar. A hydrolysis process converts them to simple sugars using acids or enzymes.

Ethanol production by corn is usually facilitated in dry grind or wet-milling plants. Dry grind plants grind whole corn kernels. In wet-milling plants, water and acid are added to grain, and corn germ, fiber, gluten, and starch are separated before mechanical grinding (Brown, 2003). A dry grind plant involves four major process steps (Fig. 9-8):

- Pretreatment
- Cooking
- Fermentation
- Distillation

Corn is mixed with water, ammonia, and enzymes during pretreatment. The resulting mixture is heated (cooking), and bacteria levels are reduced. Then, it is cooled and fermented for about 40 h or more. Distillation removes water and stillage from ethanol, and molecular sieves are used to obtain pure ethanol.

The performance of a dry grind plant is usually expressed as the amount of ethanol produced per bushel of corn. One bushel of corn is equal to approximately 35.2 L (or 25.4 kg). A dry grind plant produces about 2.7 to 2.8 gal of ethanol per bushel of corn or about 0.3 kg of ethanol per 1 kg of corn.

Lignocellulose is a composite of cellulose fibers, and it is essential part of woody cell walls of plants. Producing ethanol from lignocellulosic biomass involves pretreatment, enzymatic hydrolysis, fermentation, and distillation steps. Before the enzymatic hydrolysis process, cellulose, hemicellulose, and lignin fractions must be separated from

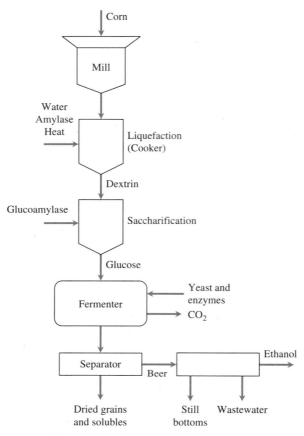

Figure 9-8 Dry milling plant operation for ethanol production from corn. (*Adapted from Kreith and Kreider, 2011.*)

lignocellulosic materials. Cellulase and hemicellulase enzymes break down the carbohydrate fractions of cellulose and hemicellulose to five- and six-carbon sugars during the process of hydrolysis. Yeast and bacteria then ferment the sugars into ethanol. Lignin is separated from the product and used separately as fuel.

Common lignocellulosic feedstock includes hybrid poplar, corn stover, corn cobs, and wheat straw. Conversion efficiencies from carbohydrates to ethanol for various feedstock are expressed in liter of ethanol per kilogram of feedstock, and some representative values are listed in Table 9-1.

Ethanol is made primarily from the starch in corn. Corn, sugar beets, sugar cane, and even cellulose (wood and paper) are some of the sources of ethanol. The feedstock used for ethanol should be high in sugar content. First, the feedstock is converted to sugar, and the sugar (glucose) is fermented into ethanol through the following reaction (Hodge, 2010)

$$C_6H_{12}O_6 \rightarrow 2C_2H_5OH + 2CO_2 \qquad (9\text{-}1)$$

This reaction is exothermic, and thus heat is also released. When ethanol is burned with air, heat is released, and the combustion products include CO_2, water vapor, and nitrogen (N_2)

$$C_2H_5OH + 3(O_2 + 3.76N_2) \rightarrow 2CO_2 + 3H_2O + 11.28N_2 \qquad (9\text{-}2)$$

TABLE 9-1   Conversion Efficiencies of Various Feedstocks to Ethanol (Kreith and Kreider, 2011)

Feedstock	Conversion efficiency from carbohydrate to ethanol, L/ton
Sugar crops, sweet sorghum	80
Sugar beets	90–100
Sugarcane	75
Corn	350–400
Wheat	400
Hybrid poplar	400
Corn stover	450
Corn cobs	510
Wheat straw	490

The source of glucose in biomass is solar energy by means of photosynthesis process in which carbon dioxide and water react in the presence of solar energy to produce glucose

$$6CO_2 + 6H_2O + \text{solar light} \rightarrow C_6H_{12}O_6 + 6O_2 \tag{9-3}$$

The cost of producing ethanol is relatively high due to the costs of growing corn and the manufacturing and processing involved. Some studies suggest that the energy consumed during the production of ethanol (plowing, planting, harvesting, fermenting, and delivery) can be quite high per unit mass of the ethanol produced, and it is sometimes comparable to the energy content of ethanol itself (O'Donnell, 1994; Pimentel, 2003). The ratio of the energy content of ethanol to the total energy input is estimated to be 1.4 in the United States and 9.2 in Brazil. This very favorable value for Brazil is due to multiple advantages of Brazil such as favorable climate, use of sugarcane as feedstock, efficiency of overall process, and the minimum use of energy consuming machinery (Dunlap, 2019).

Ethanol emits less hydrocarbon (HC) than gasoline and is commonly added to gasoline to reduce engine emissions. Its use also represents a renewable replacement for gasoline. Adding ethanol to gasoline increases the octane number of gasoline, allowing higher compression ratios and corresponding higher efficiencies for the engine. A significant problem with ethanol-gasoline blends is that water is easily absorbed by ethanol. When water leaks into an ethanol tank or pipeline, a layer rich in water forms at the bottom. This causes various operational problems in an internal combustion engine.

Two common uses of ethanol for automobiles in the United States include E10 (also called gasohol) and E85. Gasohol is a gasoline-ethanol mixture with 10 percent ethanol, while E85 contains 85 percent ethanol. The vehicles that can operate on E85 are labelled as flexible fuel vehicles (FFV) that can operate on gasoline and its blend with up to 85 percent ethanol. The 15 percent gasoline is added to eliminate operating problems with the use of pure ethanol. About half of the gasoline used in the United States includes 5 to 10 percent ethanol while E85 is available in about 4000 gas stations (Fig. 9-9). Brazil is the leading user of ethanol with more than 70 percent of the cars operating a mix of ethanol and gasoline.

The main advantage of ethanol as a transportation fuel is that it exists as a liquid at room temperature. Ethanol or ethyl alcohol (C_2H_5OH) has a higher heating value (HHV)

Figure 9-9   A gas station in San Diego, CA, providing a mixture of gasoline and ethanol to drivers. (*Photo by Sinan Demir.*)

of 29,670 kJ/kg and a lower heating value (LHV) of 26,810 kJ/kg. This is significantly lower than those of gasoline (HHV = 47,300 kJ/kg, LHV = 44,000 kJ/kg). This also means that a full tank of ethanol gets less mileage than one of gasoline (see Table 9-2).

The most common liquid fuel produced from carbohydrates is ethanol, but butanol, isoprenes, furans, esters, fatty alcohols, and alkanes can also be obtained by using alternative fermentations. Butanol has an advantage over ethanol in that its heating value on volumetric basis is about 25 percent more, as shown in Table 9-2. A higher volumetric heating value allows better compatibleness with gasoline. Methanol, ethanol, and butanol all have higher octane numbers and lower theoretical or stoichiometric air-fuel ratios compared to gasoline. An engine using a high-octane fuel can be designed for a higher compression ratio, and this results in higher fuel efficiency. When the air-fuel ratio is low (but still equal to higher than stoichiometric), more power can be generated from a given displacement volume of the engine because more fuel can be burned. A significant disadvantage of butanol over ethanol is that butanol yield during its fermentation (about 20 g/L) is much less than the yield from ethanol fermentations (about 150 g/L).

TABLE 9-2   Properties of Common and Alternative Fuels for Internal Combustion Engines

Fuel	Formula	Density, kg/m³	Octane number (anti-knock index)	Theoretical air-fuel ratio	Lower heating value, kJ/kg	Lower heating value, kJ/L
Methanol	CH_3OH	0.792	99	6.5	20,050	15,900
Ethanol	C_2H_5OH	0.789	100	9.0	26,810	21,200
Butanol	C_4H_9OH	0.810	97	11.2	33,100	26,800
Gasoline	C_8H_{15}	0.680	85 – 98	14.6	44,000	29,920
Light diesel	$C_{12.3}H_{22.2}$	0.820	—	14.5	42,500	34,900
Biodiesel	$C_{12}H_{23}$	0.880	—	12.5	39,000	34,300

EXAMPLE 9-3
Ethanol Production from Sugar Beet Roots

In the production of ethanol, the feedstock high in sugar content is first converted to sugar, and the sugar (glucose) is fermented into ethanol. Consider 100 kg of sugar beet roots whose sugar content represents 30 percent of total mass. How much ethanol can be produced from these sugar beet roots?

SOLUTION   The molar masses of C, H_2, and O_2 are 12 kg/kmol, 2 kg/kmol, and 32 kg/kmol, respectively (Table A-1). The reaction for conversion of glucose to ethanol is

$$C_6H_{12}O_6 \rightarrow 2\ C_2H_5OH + 2\ CO_2$$

The molar masses of ethanol C_2H_5OH and sugar $C_6H_{12}O_6$ are

$$M_{C_2H_5OH} = 2 \times 12 + 3 \times 2 + 0.5 \times 32 = 46\ \text{kg/kmol}$$

$$M_{C_6H_{12}O_6} = 6 \times 12 + 6 \times 2 + 3 \times 32 = 180\ \text{kg/kmol}$$

The mass of sugar in 100 kg of sugar beet roots is

$$m_{sugar} = \text{mf}_{sugar} \times m_{beet} = (0.30)(100\ \text{kg}) = 30\ \text{kg}$$

The ratio of the mass of ethanol to the mass of sugar is obtained from the chemical reaction as

$$\frac{m_{ethanol}}{m_{sugar}} = \frac{2M_{C_2H_5OH}}{M_{C_6H_{12}O_6}} = \frac{(2 \times 46)\ \text{kg}}{180\ \text{kg}} = 0.5111$$

Then the mass of the ethanol produced becomes

$$m_{ethanol} = m_{sugar}\frac{m_{ethanol}}{m_{sugar}} = (30\ \text{kg})(0.5111) = \textbf{15.3 kg}\quad \blacktriangle$$

EXAMPLE 9-4
Maximum Power from an Ethanol Engine

A 1.6-L gasoline-fueled internal combustion engine produces a maximum power of 100 kW when the fuel is burned stoichiometrically with air. The thermal efficiency of the engine at this maximum power is 38 percent. Estimate the maximum power from this engine if ethanol (C_2H_5OH or C_2H_6O) is used under stoichiometric combustion with the same thermal efficiency. Determine the rates of gasoline and ethanol consumption for maximum power operation. The stoichiometric air-fuel ratio for gasoline is 14.6. Assume a combustion efficiency of 100 percent.

SOLUTION The molar masses of C, H_2, and O_2 are 12 kg/kmol, 2 kg/kmol, and 32 kg/kmol, respectively (Table A-1). The lower heating values of gasoline and ethanol are 44,000 kJ/kg and 26,810 kJ/kg, respectively (Table A-7). The balanced reaction equation for stoichiometric air is

$$C_2H_6O + a_{th}[O_2 + 3.76N_2] \longrightarrow 2\,CO_2 + 3\,H_2O + a_{th} \times 3.76\,N_2$$

The stoichiometric coefficient a_{th} is determined from an O_2 balance:

$$0.5 + a_{th} = 2 + 1.5 \longrightarrow a_{th} = 3$$

Substituting,

$$C_2H_6O + 3[O_2 + 3.76N_2] \longrightarrow 2\,CO_2 + 3\,H_2O + 11.28\,N_2$$

Therefore, the air-fuel (AF) ratio for this stoichiometric reaction is

$$AF = \frac{m_{air}}{m_{fuel}} = \frac{(3 \times 4.76 \times 29)\ \text{kg}}{(2 \times 12 + 6 \times 1 + 1 \times 16)\ \text{kg}} = \frac{414.1\ \text{kg}}{46\ \text{kg}} = 9.0$$

For a given engine size (i.e., volume), the power produced is proportional to the heating value of the fuel and inversely proportional to the air-fuel ratio. The ratio of power produced by the gasoline engine to that of the ethanol engine is expressed as

$$\frac{\dot{W}_{gasoline}}{\dot{W}_{ethanol}} = \frac{\text{LHV}_{gasoline}}{\text{LHV}_{ethanol}} \times \frac{\text{AF}_{ethanol}}{\text{AF}_{gasoline}} = \left(\frac{44{,}000\ \text{kJ/kg}}{26{,}810\ \text{kJ/kg}} \right)\left(\frac{9.0}{14.6} \right) = 1.012$$

Then the maximum power from the engine with ethanol as the fuel becomes

$$\dot{W}_{ethanol} = \frac{\dot{W}_{gasoline}}{1.012} = \frac{100\ \text{kW}}{1.012} = \mathbf{98.81\ kW}$$

The thermal efficiency of this engine is defined as the net power output divided by the rate of heat input, which is equal to the heat released by the combustion of fuel:

$$\eta_{th} = \frac{\dot{W}_{out}}{\dot{Q}_{in}} = \frac{\dot{W}_{out}}{\dot{m}_{fuel} \times \text{LHV}}$$

Note that the lower heating value is used in the analysis of internal combustion engines since the water in the exhaust is normally in the vapor phase. Also, the combustion efficiency is given to be 100 percent. Solving the above equation for the rates of gasoline and ethanol consumption, we obtain

$$\dot{m}_{gasoline} = \frac{\dot{W}_{gasoline}}{\eta_{th} \times \text{LHV}_{gasoline}} = \frac{100\ \text{kJ/s}}{(0.38)(44{,}000\ \text{kJ/kg})}\left(\frac{60\ \text{s}}{1\ \text{min}} \right) = \mathbf{0.359\ kg/min}$$

$$\dot{m}_{ethanol} = \frac{\dot{W}_{out}}{\eta_{th} \times \text{LHV}_{ethanol}} = \frac{98.81\ \text{kJ/s}}{(0.38)(26{,}810\ \text{kJ/kg})}\left(\frac{60\ \text{s}}{1\ \text{min}} \right) = \mathbf{0.582\ kg/min}$$

The rate of ethanol consumption is about 60 percent greater than the rate of gasoline consumption when the engine produces the maximum power. ▲

9-6 BIODIESEL

Biodiesel is ethyl or methyl ester that is produced through a process that combines organically derived oils with ethanol or methanol in the presence of a catalyst. Triglycerides are a type of lipid (fat), and they are used to produce biodiesel fuel from biomass. Triglycerides are long-chain carboxylic acids. Fats have high fractions of saturated acids and oils have

TABLE 9-3   Approximate Oil Production Values per Year (in Liter per Hectare Farmland) from Various Plants

Plant	Oil production, L/ha
Cotton seed	150
Soybean	400
Peanut oil	850
Rapeseed	1100
Jatropha	1600
Oil palm	4000
Algae	20,000

high fractions of unsaturated fatty acids. Fats are usually solid at room temperature while oils are liquid and normally derived from plants.

The feedstock for biodiesel production includes animal fats, grease, vegetable oils, cooking oils, recycled restaurant greases, and algae. Soybean oil, corn oil, and recycled cooking oils are common sources of biodiesel production in the United States. The composition and fuel characteristics of biodiesel show slight variations depending on the type of feedstock. Glycerin is a major byproduct in the process of converting triglycerides to biodiesel.

Certain plant seeds are common sources of triglycerides. Oil production is about 150 L/ha (1 ha = 1 hectare = 10,000 m^2) for cotton seed, 400 L/ha for soybeans, 850 L/ha for peanut oil, and 4000 L/ha for oil palm (Table 9-3). Alternative oil seed crops that could be grown in waste land or saline soils are proposed to prevent forest destruction and replacement of food crops such as Jatropha curcas, Salicornia curcas, and Chinese tallow tree (Openshaw, 2000). Algae is another potential source of triglycerides, and it can be grown in brackish water, seawater, or waste land with easy access to water. More than half of the algae mass can be converted to lipid.

In order to extract the oil from biomass, oil seeds are mechanically crushed. If the oil fraction is less than 20 percent, solvent extraction is used. The remains of the seed are typically used as animal feed. Various processes involving biodiesel production from biomass is shown in Fig. 9-10.

The heating value of oils (around 37,000 kJ/kg for vegetable oil) and fat are relatively high, but they have high viscosity and low volatility. This makes it unsuitable as the fuel in a compression ignition engine. However, the properties of biodiesel are compatible with petroleum diesel including heating value, viscosity, density, cetane number, and flash point. Biodiesel can be used in compression ignition engines as a single fuel or added to conventional diesel fuel. The HHV of biodiesel is about 40,700 kJ/kg (17,500 Btu/lbm), which is about 12 percent less than that of petroleum diesel (HHV = 46,100 kJ/kg). Biodiesel can also be used as a heating fuel in boilers for space and process heating.

The most common biodiesel mixture used in the United States is B20, which is 20 percent biodiesel and 80 percent conventional diesel. Biodiesel is also used as a single fuel in compression ignition engines, called B100. Whereas B20 can be used in a diesel engine without any modification, B100 usually requires the use of special material for parts such as hoses and gaskets. Due to its lower energy content, B100 provides less power from the engine. Some maintenance issues can also arise, and some manufacturers may not cover biodiesel use in their warranties. One percent petroleum diesel is added to biodiesel in

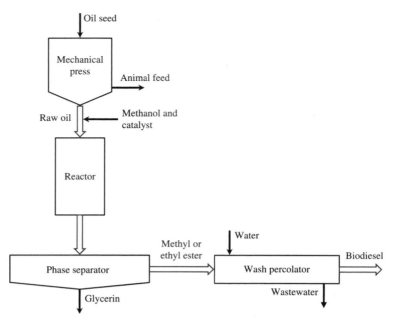

Figure 9-10 Biodiesel production from oil seed. (*Adapted from Kreith and Kreider, 2011.*)

some parts of the world (B99) to prevent mold growth. B100 and B99 could increase nitrogen oxide emissions while significantly reducing hydrocarbon, sulfur, particulate, and carbon monoxide emissions.

9-7 LIQUID FUELS FROM SYNGAS, PYROLYSIS OIL, AND BIOCRUDE

Synthetic gas (or syngas) is produced from biomass by thermal gasification process. Syngas is a mixture of carbon monoxide and hydrogen. Various biomass feedstock including wood, sugarcane, straw, switchgrass, and corn are used for syngas production. Syngas can be used directly for heat and electricity production, but it can also be converted to liquid fuels. Ethanol can be produced from syngas using iron-based catalysts through the reaction

$$CO + 3H_2 \rightarrow CH_3CH_2OH \qquad (9\text{-}4)$$

Actual applications yield various liquid fuels with a high rate of alcohols but the challenge of producing pure ethanol by this method remains. Another method of producing ethanol from syngas is *syngas fermentation* during which microorganisms consume CO and H_2 to form products such as alcohols, esters, and carboxylic acids.

Another liquid product from syngas is methanol, and it can be produced from syngas by the exothermic reaction

$$CO + 2H_2 \rightarrow CH_3OH \qquad (9\text{-}5)$$

For methanol production from syngas, a fixed catalytic bed or a liquid-phase slurry reactor may be used. Pressure levels in the reactor are high (60–100 bar) and temperatures are relatively low (around 250°C). In these reactors, large amounts of heat released during the exothermic reaction must be removed effectively (Cybulski, 1994).

Methanol or methyl alcohol (CH_3OH) has an HHV of 22,660 kJ/kg and an LHV of 19,920 kJ/kg. This is less than half of the values for gasoline. Pure methanol and its blend with gasoline have been extensively tested as alternatives to gasoline. Two common mixtures are M85 (85 percent methanol, 15 percent gasoline) and M10 (10 percent methanol, 90 percent gasoline). There is no noticeable emission reduction due to the use of M10 in engines, but M85 reduces HC and carbon monoxide (CO) emissions significantly while also replacing more of gasoline consumption. Methanol can be produced from fossil sources or biomass. Coal, oil, and natural gas represent fossil sources. Natural gas is the main feedstock in the United States for methanol production. Synthesis gas produced from biomass can replace natural gas for this process.

Fischer-Tropsch process refers to producing liquid hydrocarbons fuels (called Fischer-Tropsch liquids) from syngas. Metal catalysts such as iron and cobalt are used in the reactions with temperature levels of 150 to 300°C and pressure levels of 10 to 40 bar. This process has been used historically to produce liquid fuels from coal and natural gas. The generalized reaction with the syngas is

$$CO + 2H_2 \rightarrow CH_2 + H_2O$$

In addition to Fischer-Tropsch liquids, hydrocarbon gases and alcohols are also produced. The composition of liquids depends on the composition of syngas, the type and characteristics of catalyst, and the reaction temperature and pressure. Higher reaction temperatures tend to favor the production of gasoline in liquids and methane gas. Distillation can be used to obtain various byproducts from Fischer-Tropsch liquids including gasoline and diesel fuel. Gasoline yield is valuable but its fraction in the products is typically less than 50 percent.

Fast pyrolysis involves the rapid heating (less than 2 s) of biomass to high temperatures (around 500°C) in an oxygen free environment to decompose the biomass and then applying the rapid cooling at the end of the process. The resulting fuel product is called *pyrolysis liquid*, *pyrolysis oil*, or *bio-oil*. Bio-oil is a mixture of organic compounds such as alcohols, aldehydes, carbohydrates, carboxylic acids, phenols, and lignin with low viscosity and up to 20 percent water fraction. Syngas, ash, and char are other components in the products.

Pyrolysis oil is produced when biomass is exposed to high temperatures without the presence of air, causing it to decompose. A possible reaction involves heating of cellulosic feedstock in grain form for a short period (less than half a second) to a temperature of 400 to 600°C and quenching it. The product is highly oxygenated and has considerable amounts of water. This makes these liquids corrosive and unstable with a low heating value. Pyrolysis oil is not suitable as a replacement for conventional fuels such as gasoline or diesel. Further processes are needed to make this fuel compatible with conventional HC fuels. A chemical called phenol can be extracted from pyrolysis oil, and it is used to make wood adhesives, molded plastic, and foam insulation.

Wood and agricultural waste are commonly used as the feedstock for fast pyrolysis process. About 60 to 70 percent of biomass mass is converted to bio-oil in this process. Bio-oil is considered as a low-quality fuel. Upgrading to higher quality fuels such as gasoline or diesel fuel are desirable. This can be done by distillation and cracking processes, as done in oil refineries. Distillation involves heating and separating components at different levels of temperatures while cracking is breaking of higher molecular weight components into lower molecular weight components.

Heating of wet biomass at relatively high pressure to produce liquid hydrocarbons, carbohydrate, and gaseous products is called *hydrothermal processing*. The resulting liquid hydrocarbons are called *biocrude*. The temperature range during hydrothermal processing is 300 to 350°C and the pressure range is 120 to 180 bar. The pressure should be high enough to prevent boiling of water in the biomass. A feasible use of biocrude is its upgrade to gasoline, diesel fuel, or aviation fuel through distillation and cracking processes.

9-8 ELECTRICITY AND HEAT PRODUCTION BY BIOMASS

The production of electricity and heat from biomass is called *biopower*. There are three technologies used to convert biomass energy to heat and electricity: direct combustion, co-firing, and anaerobic digestion (DOE/EERE, 2018).

Biomass consisting of waste wood products (i.e., wood pellet) can be burned in *direct combustion* in conventional boilers to generate steam or hot water. This steam is run through a turbine coupled with a generator to produce electricity. *Co-firing* refers to replacing only a portion of fossil fuel in coal-fired boilers with biomass. This technology has been successfully demonstrated in most boiler technologies, including pulverized coal, cyclone, fluidized bed, and spreader stoker units. Sulfur dioxide emissions of coal-fired power plants can be reduced considerably by co-firing biomass.

Anaerobic digestion, or *methane recovery*, is a common technology used to convert organic waste to methane and heat. In this process, organic matter is decomposed by bacteria in the absence of oxygen to produce natural gas consisting primarily of methane and other byproducts such as carbon dioxide. This gas can be used for space and water heating or for electricity production.

Wood constitutes a considerable fraction of biomass use. The combustion of wood releases carbon dioxide to the environment, thus contributing to the global greenhouse gas emission. In order for this process to be carbon neutral, equal number of trees must be planted as trees and other living organic matter sequester carbon dioxide (CO_2). However, this is not happening with resulting global scale deforestation. Wood combustion is associated with the production of some pollutants including nitrogen oxides (NO_x) and particulate matter. When dead trees are left in their environment, they emit methane (CH_4) as they decay. The alternative is collecting the dead trees and burning them for energy production. This results in CO_2 emission. However, both CH_4 and CO_2 are greenhouse gases but the global warming potential of CH_4 is about 25 times more than CO_2 (Dunlap, 2019).

The heating value of wood ranges between 14,000 and 21,000 kJ/kg. This is considerably less than high quality coal whose heating value can be as high as 31,000 kJ/kg. Wood is mostly used as the fuel for space heating in winter in many parts of the world. There are also some power plants generating electricity by burning wood. For example, a wood-burning power plant in Sweden generates 1500 MW electricity and another one in Austria generates 500 MW.

EXAMPLE 9-5
Replacing a Propane Furnace with a Pellet Furnace

A homeowner is currently using a propane furnace for heating the house in winter, but he is considering replacing this system with a furnace burning wood pellets (Fig. 9-11). The homeowner currently pays an average of $2500 per year for space heating. Using the values below, determine how much the homeowner saves from heating per year if he replaces the existing furnace with a pellet-burning furnace.

Propane furnace: HV = 21,640 Btu/lbm, $\eta_{furnace} = 0.85$, Unit price = $0.80/lbm

Wood pellet: HV = 8500 Btu/lbm, $\eta_{furnace} = 0.80$, Unit price = $0.12/lbm

Figure 9-11 A wood pellet–fired boiler for residential use.

SOLUTION The unit price of useful heat supplied by the furnace in the case of propane, in $/Btu, is

$$\text{Unit price of heat (propane)} = \frac{\text{Unit price of propane}}{\eta_{\text{furnace}} \times \text{HV}} = \frac{\$0.80/\text{lbm}}{(0.85)(\$21{,}640 \text{ Btu/lbm})} = \$4.349 \times 10^{-5}/\text{Btu}$$

The unit price of useful heat supplied by the furnace in the case of pellet, in $/Btu, is

$$\text{Unit price of heat (pellet)} = \frac{\text{Unit price of pellet}}{\eta_{\text{furnace}} \times \text{HV}} = \frac{\$0.12/\text{lbm}}{(0.80)(8500 \text{ Btu/lbm})} = \$1.765 \times 10^{-5}/\text{Btu}$$

Using these unit prices, the annual fuel cost in the case of pellet furnace would be

$$\text{Annual pellet cost} = \text{Annual propane cost} \times \frac{\text{Unit price of heat (pellet)}}{\text{Unit price of heat (gas)}} = (\$2500)\frac{\$1.765 \times 10^{-5}/\text{Btu}}{\$4.349 \times 10^{-5}/\text{Btu}} = \$1015$$

Therefore, the savings due to this replacement would be

$$\text{Savings} = \text{Annual propane cost} - \text{Annual pellet cost} = \$2500 - \$1015 = \textbf{\$1485}$$

9-9 MUNICIPAL SOLID WASTE

An important class of biomass is produced by households as trash or garbage. This is referred to as *municipal solid waste* (MSW). MSW includes mostly organic materials such as paper, food scraps, wood, and yard trimmings, but some fossil content such as plastic also exists. Most MSW comes from residences (55 to 65 percent) and 35 to 45 percent comes from businesses, schools, and hospitals. MSW does not include industrial, hazardous, or construction waste.

There are four general methods of dealing with MSW:

- Recycling
- Composting
- Landfilling
- Combustion with energy recovery

Recycling refers to recovery of useful materials such as paper, glass, plastic, and metals from the trash to use to make new products. Recycling reduces raw material use and associated energy consumption and greenhouse gas emissions, which cause global warming. Air and water pollution associated with making new products are also avoided.

Composting, on the other hand, refers to storing organic waste such as food scraps and yard trimmings under certain conditions to help it break down naturally. The resulting product can be used as a natural fertilizer.

In addition to recycling and composting, waste can also be reduced by *waste prevention*, which is the design of products to minimize the production of waste and making the resulting waste less toxic.

Landfilling refers to disposal of waste to a landfill. This is the most common method of dealing with MSW.

Combustion with energy recovery involves burning of combustible portion of MSW in a boiler to generate heat. The resulting heat may be used directly for various heating purposes or used to generate steam to drive a turbine-generator unit for the generation of electricity.

A common use of MSW is biogas production by the decomposition of organic material at landfill sites. The resulting biogas which is a mixture of methane and carbon dioxide is usually burned in a combustion chamber and the resulting hot combustion gases drive a gas turbine to generate power.

Another method of utilizing MSW is to convert the combustible portion of MSW into refuse-derived fuel (RDF) by separating noncombustibles from the combustible portion of MSW. RDF should be obtained in a form such that it can be burned directly in a boiler such as in the form of pellet. RDF can also be heated in a thermal gasification process to yield a gas (producer gas) consisting of mostly methane and hydrogen. This gas can be used for heat and electricity generation.

The U.S. Environmental Protection Agency (EPA) collects and reports data on the generation and disposal of waste in the United States. This data is used to measure the success of waste reduction and recycling programs across the country (EPA, 2021). Materials in MSW and their percentages are given in Fig. 9-12. Organic materials represent the largest component of MSW. Paper and paperboard account for 23.1 percent while yard trimmings and food scraps together account for another 33.7 percent. Plastics, metals, rubber, leather, and textiles together make up 29.9 percent. Wood and glass follow at 6.2 and 4.2 percent, respectively.

About 292.4 million tons of MSW was generated in the United States in 2018. Note that EPA's ton in their data is in the English unit system [1 ton = 2000 pound-mass (lbm) and 1 lbm = 0.4536 kg]. An average American produced 4.90 lbm (2.22 kg) of solid waste per day and only recycled 23.6 percent (1.16 lbm or 0.53 kg) of this waste. Total MSW

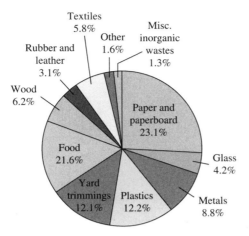

Figure 9-12 Materials in municipal solid waste and their percentages in the United States. (*EPA, 2021.*)

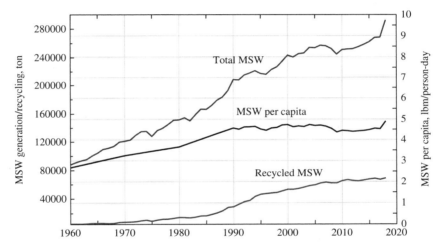

Figure 9-13 Municipal solid waste generation from 1960 to 2018 in the United States. Total MSW generation is in tons, recycled MSW is in tons, and MSW generation per capita basis is in lbm/person-day. 1 ton = 2000 lbm and 1 lbm = 0.4536 kg. (*EPA, 2021.*)

production increased steadily since 1960 but started to decrease between 2005 and 2010. Per capita MSW production increased from 1960 until 2000. Its value ranged between 4.37 and 4.90 lbm per person per day between 2000 and 2018 (Fig. 9-13).

The recycling culture in the United States improved significantly after 1985, when only 10.1 percent of MSW was recycled. After dramatic increases between 1985 and 1995 in recycling rates, there was a steady increase in recycling after 1995 (Fig. 9-13). In 2018, 69.1 million tons of MSW was recycled and 24.9 million tons of MSW was composted. The highest recycling rates are achieved in paper and paperboard, yard trimmings, and metals. Approximately 67 percent of the paper and paperboard were recycled and 52 percent of yard trimmings were composted. It is estimated that the 69.1 million tons of MSW recycled in 2018 saved 165 million tons of carbon dioxide emissions. This is equivalent to the removal of 30 million cars from the roads.

It is remarkable that disposal of waste to a landfill has decreased from 89 percent of the amount generated in 1980 to 56.1 percent of MSW in 2018. The remaining 11.8 percent (34.6 million tons) is combusted for energy recovery (Fig. 9-14). The number of U.S. landfills has steadily declined over the years, but the average landfill size has increased.

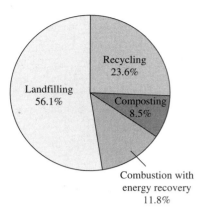

Figure 9-14   Use of municipal solid waste in the United States in 2018 (*EPA, 2021*).

About 15 percent of renewable electricity generation excluding hydroelectric power comes from municipal waste facilities in the United States. Energy can be recovered in the form of electricity by both burying and burning options.

What is the best use of municipal solid waste? Is it better to burn or bury waste when trying to recover energy and minimize emissions?

An EPA research study compared two options for producing electricity from MSW. The first option is known as *waste to energy* (WTE), where waste is burned directly for generating steam. This steam is run through a turbine to generate electricity. The second option is known as *landfill-gas-to-energy* (LFGTE), and it involves harvesting biogas (mostly methane) from the buried waste as it decomposes. Biogas is then used as the fuel in an internal combustion engine or gas turbine to generate electricity. The research indicates that burning waste through the WTE method can produce up to 10 times more electricity than burying the same amount of waste through the LFGTE method. It is also determined

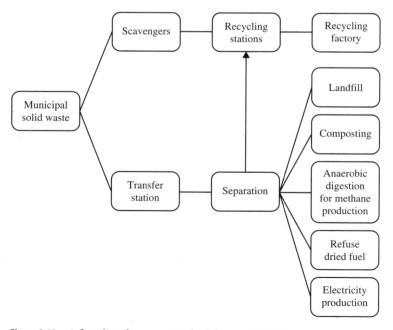

Figure 9-15   A flowchart for a municipal solid waste (MSW) treatment system.

that greenhouse gas emissions per unit of electricity produced are two to six times higher in landfills than in waste burning plants (Kaplan et al., 2009).

Note that WTE plants involve combustion of both *biogenic* (nonfossil materials in the waste) and fossil materials of waste, while methane from landfills only results from anaerobic breakdown of biogenic materials.

Figure 9-15 gives a flowchart for an MSW treatment system. Various options are available for the treatment of MSW including recycling, composting, landfill burying, anaerobic digestion to produce methane gas, electricity production, and refuse dried fuel. Note that only organic waste materials are accepted in composting facilities.

EXAMPLE 9-6
Steam and Power
Generation from MSW

Fifty tons of MSW is burned directly in a boiler to generate superheated steam at 800 kPa and 400°C (Fig. 9-16). Water enters the boiler as a saturated liquid at 50 kPa. The heating value of MSW is 18,000 kJ/kg and the boiler efficiency is 80 percent. The generated steam is run through a turbine-generator unit to generate electricity. The steam pressure at the turbine exit is 50 kPa. The turbine isentropic efficiency is 90 percent and the generator efficiency is 97 percent. Determine (*a*) the amount of steam generated and (*b*) the amount of electricity produced.

SOLUTION (*a*) The properties of water at the boiler inlet (state 1), boiler exit or turbine inlet (state 2), and turbine exit (state 3) are obtained from steam tables (Tables A-3 through A-5)

$$\left.\begin{array}{l} P_1 = 50 \text{ kPa} \\ x_1 = 0 \text{ (sat. liq.)} \end{array}\right\} \quad h_1 = 340.54 \text{ kJ/kg}$$

$$\left.\begin{array}{l} P_2 = 800 \text{ kPa} \\ T_2 = 400°C \end{array}\right| \begin{array}{l} h_2 = 3267.7 \text{ kJ/kg} \\ s_2 = 7.5735 \text{ kJ/kg·K} \end{array}$$

$$\left.\begin{array}{l} P_3 = 50 \text{ kPa} \\ s_3 = s_2 = 7.5735 \text{ kJ/kg·K} \end{array}\right\} \quad h_{3s} = 2638.3 \text{ kJ/kg}$$

The actual enthalpy at the turbine exit is determined from the definition of isentropic efficiency as

$$\eta_T = \frac{h_2 - h_3}{h_2 - h_{3s}} \quad \rightarrow \quad 0.90 = \frac{3267.7 - h_3}{3267.7 - 2638.3} \quad \rightarrow \quad h_3 = 2701.2 \text{ kJ/kg}$$

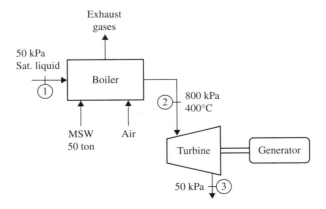

Figure 9-16 Schematic for Example 9-6.

The boiler efficiency is defined as the useful heat delivered to the steam divided by the energy released from the burning of the fuel. Then, the amount of steam produced is determined to be

$$\eta_{boiler} = \frac{Q_{useful}}{Q_{in}} = \frac{m_{steam}(h_2 - h_1)}{m_{MSW} \times HV_{MSW}}$$

$$0.80 = \frac{m_{steam}(2701.2 - 340.54)kJ/kg}{(50,000\ kg)(18,000\ kJ/kg)}$$

$$m_{steam} = \textbf{305,000 kg}$$

(*b*) The amount of work output from the turbine is

$$W_{turbine} = m_{steam}(h_2 - h_3) = (305,000\ kg)(3267.7 - 2701.22)\ kJ/kg\left(\frac{1\ kWh}{3600\ kJ}\right) = 47,995\ kWh$$

Then the amount of electricity from the generator is

$$W_{electric} = \eta_{generator}W_{electric} = (0.97)(47,995\ kWh) = \textbf{46,560 kWh} \quad \blacktriangle$$

REFERENCES

Arslan M and Yilmaz C. 2022. "Thermodynamic Optimization and Thermoeconomic Evaluation of Afyon Biogas Plant Assisted by Organic Rankine Cycle for Waste Heat Recovery." *Energy*, 248(2022): 123487.

Boyle G. 2004. *Renewable Energy*. Oxford, UK: Oxford University Press.

Brown RC. 2003. *Biorenewable Resources: Engineering New Products from Agriculture*. Iowa State Press. Ames, IA.

Cybulski A. 1994. "Liquid-Phase Methanol Synthesis: Catalysts, Mechanism, Kinetics, Chemical Equilibria, Vapor-Liquid Equilibria, and Modeling–A Review." *Catalysis Review*, 36(4): 557–615.

DOE/EERE. 2018. Department of Energy, Energy Efficiency and Renewable Energy. www.eere.energy.gov

Dunlap RA, 2019. *Sustainable Energy*. 2nd ed. Cengage Learning. Boston, MA.

EPA (Environmental Protection Agency). 2021. National Overview: Facts and Figures on Materials, Wastes and Recycling. Washington, DC. https://www.epa.gov/facts-and-figures-about-materials-waste-and-recycling/national-overview-facts-and-figures-materials. Accessed on Jan. 2, 2022.

Hodge BK. 2010. *Alternative Energy Systems and Applications*. New York: Wiley.

Higman C and Burgt VD. 2008. *Gasification*. Gulf Professional Publishing, Burlington, MA.

Kaplan PO, Decarolis J, and Thornelow S. 2009. "Is It Better to Burn or Bury Waste for Clean Electricity Generation?" *Environmental Science & Technology*, 43(6): 1711–1717.

Klass D. 1998. *Biomass for Renewable Energy, Fuels, and Chemicals*. Academia Press, San Diego, CA.

Kreith F and Kreider JF. 2011. *Principles of Sustainable Energy*. New York: Taylor & Francis.

O'Donnell J. 1994. "Gasoline Allies." *Autoweek*, pp. 16–18, February 1994.

Openshaw K. 2000. "A Review of *Jatropha curcas*: An Oil Plant of Unfulfilled Promise." *Biomass and Bioenergy*, 19(1):1–15.

Perlack RD, Wright LL, Turhollow AF, Graham RL, Stokes BJ, and Erbach DC. 2005. "Biomass as Feedstock for a Bioenergy and Bioproducts Industry: The Technical Feasibility of a Billion-Ton Annual Supply," DOE/GO-102005-2136, ORNL/TM-2005/66.

Pimentel D. 2003. "Ethanol Fuels: Energy Balance, Economics, and Environmental Impacts are Negative." *Natural Resources Research*, 12(2): 127–134.

PROBLEMS

BIOMASS RESOURCES

9-1   What are the sources of biomass energy? Why is it a renewable energy source? Explain.

9-2   List biomass feedstocks.

9-3   What are the most common sources of biomass feedstock in the United States?

9-4   Which is not a source of biomass energy?
(*a*) Forest   (*b*) Oil   (*c*) Agriculture   (*d*) Energy crops   (*e*) Municipal waste

9-5   Which is not one of the four major feedstocks in the United States?
(*a*) Municipal waste   (*b*) Crop residues   (*c*) Corn   (*d*) Forest resources
(*e*) Perennial crops

CONVERSION OF BIOMASS TO BIOFUEL

9-6   What is the purpose of biochemical and thermochemical conversion processes? What is the difference between them?

9-7   What is a photobiological conversion process?

9-8   What are the two major methods of converting biomass to gaseous fuels? Name the resulting gaseous fuels?

9-9   What are the two most common liquid fuels from biomass? Which substrates derived from biomass are used for their production?

9-10   What is the most common application of liquid fuels obtained from biomass? What are the four types of substrates derived from biomass, which are used to produce liquid fuels?

9-11   Heat energy and chemical catalysts are used to break down biomass into intermediate compounds or products. This process is called
(*a*) Biochemical conversion   (*b*) Anaerobic digestion   (*c*) Hydrolysis
(*d*) Photobiological conversion   (*e*) Thermochemical conversion

9-12   Enzymes and microorganisms are used as biocatalysts to convert biomass or biomass-derived compounds into desirable products. This process is called
(*a*) Biochemical conversion   (*b*) Solid-gas reactions   (*c*) Pyrolysis
(*d*) Photobiological conversion   (*e*) Thermochemical conversion

9-13   Which of the following are the two major gaseous fuels resulting from the anaerobic digestion and thermal gasification processes?
  I. Water gas
 II. Biogas
III. Ethanol
IV. Producer gas
(*a*) I and II   (*b*) I and III   (*c*) II and III   (*d*) II and IV   (*e*) I and IV

9-14   Which of the following are the two most common liquid fuels from biomass?
  I. Ethanol
 II. Biogas
III. Biodiesel
IV. Methanol
(*a*) I and II   (*b*) I and III   (*c*) II and III   (*d*) II and IV   (*e*) I and IV

9-15   The sources of liquid fuels include substrates derived from biomass. Which of the following is not one of these substrates?
(*a*) Triglycerides   (*b*) Bio-oil/biocrude   (*c*) Syngas   (*d*) Carbohydrates   (*e*) Alkanes

9-16   Which of the following is the substrate derived from biomass used for biodiesel production?
(*a*) Triglycerides   (*b*) Bio-oil/biocrude   (*c*) Syngas   (*d*) Carbohydrates   (*e*) Alkanes

ANAEROBIC DIGESTION

9-17   Define anaerobic digestion process. What is the main product gas of anaerobic digestion and what does it consist of?

9-18   What are the three steps of the anaerobic digestion process?

9-19   Describe two methods of expressing thermodynamic performance of an anaerobic digestor?

9-20   Which operating parameters affect performance of an anaerobic digestor?

9-21   What are the common feedstocks for anaerobic digestion?

9-22   What are the main constituents of biogas? What are the sources of biogas?

9-23   Compare the heating value of biogas to that of natural gas. Which is higher? Why?

9-24   A certain biogas consists of 65 percent methane (CH_4) and 35 percent carbon dioxide (CO_2) by volume. If the higher heating value (HHV) of methane is 55,530 kJ/kg, what is the HHV of this biogas?

9-25   Decomposition of waste biomass into gaseous fuels by bacteria action without the presence of oxygen is called
(*a*) Solid-gas reactions   (*b*) Anaerobic digestion   (*c*) Acidification
(*d*) Hydrolysis   (*e*) Methanogenesis

9-26   Which of the following is not a step of the anaerobic digestion process?
(*a*) Pyrolysis   (*b*) Hydrolysis   (*c*) Acidification   (*d*) Methanogenesis

9-27   Which of the following is not one of the common feedstocks for anaerobic digestion?
(*a*) Food scraps   (*b*) Sewage sludge   (*c*) Wood powder
(*d*) Animal manures   (*e*) Municipal solid waste

9-28   Producing biogas from the municipal waste and using the biogas for electricity production is done in _____ plants.
(*a*) Pyrolysis conversion   (*b*) Biochemical conversion   (*c*) Solid waste treatment
(*d*) Thermochemical conversion   (*e*) Anaerobic digestion

9-29   A certain biogas consists of 75 percent methane (CH_4) and 25 percent carbon dioxide (CO_2) by volume. If the higher heating value (HHV) of methane is 55,530 kJ/kg, the HHV of this biogas is
(*a*) 25,100 kJ/kg   (*b*) 29,000 kJ/kg   (*c*) 33,900 kJ/kg
(*d*) 41,400 kJ/kg   (*e*) 55,200 kJ/kg

THERMAL GASIFICATION

9-30   Define thermal gasification process. What is the main product gas of anaerobic digestion and what does it consist of?

9-31   What are the four steps of the thermal gasification process? Which step is most important?

9-32   What is a pyrolysis process?

9-33   What are the required temperatures for a pyrolysis process?

9-34   How do you compare anaerobic digestion and thermal gasification in terms of moisture content in their feedstocks?

9-35   Which factors does the composition of producer gas depend on?

9-36   What is cracking? Why is it used?

9-37   Provide two definitions for expressing thermodynamic efficiency of the thermal gasification process?

9-38   List common types of thermal biomass gasifiers.

9-39   How is producer gas obtained? What does it consist of?

9-40   How is synthesis gas obtained? What does it consist of?

9-41   A synthesis gas consists of 50 percent CO and 50 percent H_2 by volume. This gas is burned with the stoichiometric amount of air. Write the balanced chemical reaction and determine the air-fuel ratio.

9-42   A synthesis gas consists of 40 percent CO and 60 percent H_2 by volume. Determine its higher and lower heating values. The higher and lower heating values of hydrogen are 141,800 kJ/kg and 120,000 kJ/kg, respectively, and the heating value of carbon monoxide is 10,100 kJ/kg.

9-43   A producer gas consists of 30 percent CO, 25 percent H_2, 20 percent CH_4, 15 percent N_2, and 10 percent CO_2 by volume. Determine the higher heating value (HHV) of this gas. The HHVs of CO, H_2, and CH_4 are 10,100 kJ/kg, 141,800 kJ/kg, and 55,530 kJ/kg, respectively.

9-44   A biomass gasifier converts 200 lbm of biomass into 18,000 ft³ of producer gas at 1 atm and 500°F. Determine the hot gas efficiency and the cold gas efficiency for this thermal gasification process. Use the following data:

Heating value of biomass, $HV_{biomass}$ = 9500 Btu/lbm
Heating value of producer gas, HV_{gas} = 165 Btu/ft³
Density of producer gas at standard atmospheric conditions (1 atm, 77°F), ρ = 0.0805 lbm/ft³
The average specific heat of producer gas for sensible heat calculation, c_p = 0.27 Btu/lbm·°F

9-45   A biomass gasifier converts 500 kg of biomass into producer gas. The gas is cooled to the atmospheric temperature and contained in a tank whose volume is 2200 m³. Determine the efficiency for this thermal gasification process. Use the following data:

Heating value of biomass, $HV_{biomass}$ = 17,500 kJ/kg
Heating value of producer gas, HV_{gas} = 8500 kJ/m³
Density of producer gas at standard atmospheric conditions (1 atm, 25°C), ρ = 1.29 kg/m³

9-46   A steady-flow biomass gasifier converts wheat straw entering at a rate of 50 kg/min and producer gas leaves the gasifier at a rate of 280 m³/min at 1 atm and 500°C. Determine the hot gas efficiency and the cold gas efficiency for this thermal gasification process. Use the following data:

Heating value of straw, HV_{straw} = 15,600 kJ/kg
Heating value of producer gas, HV_{gas} = 5500 kJ/m³
Density of producer gas at standard atmospheric conditions (1 atm, 25°C), ρ = 1.29 kg/m³
The average specific heat of producer gas for sensible heat calculation, c_p = 1.1 kJ/kg·°C

9-47   Which of the following is not a step of the thermal gasification process?
(*a*) Pyrolysis   (*b*) Gas-phase reactions   (*c*) Solid-gas reactions
(*d*) Heating and drying   (*e*) Hydrolysis

9-48   Biomass is exposed to high temperatures without the presence of air, causing it to decompose. This process is called
(*a*) Biochemical conversion   (*b*) Hydrolysis   (*c*) Pyrolysis
(*d*) Photobiological conversion   (*e*) Anaerobic digestion

9-49   Which component is not among main outputs after the pyrolysis step of a thermal gasification process?
(*a*) CO   (*b*) CO_2 (*c*) H_2 (*d*) CH_4 (*e*) H_2O

9-50   Which of the following components in producer gas do not contribute to the heating value of the gas?
  I. CO
 II. CO_2
III. CH_4
IV. N_2
(*a*) I and II   (*b*) I and III   (*c*) II and III   (*d*) II and IV   (*e*) I and IV

9-51   Which biomass product is produced by thermal gasification using oxygen and consists of CO and H_2?
(*a*) Synthesis gas   (*b*) Ethanol   (*c*) Biodiesel   (*d*) Producer gas   (*e*) Biogas

9-52   Which biomass product usually consists of 50 to 80 percent methane and 20 to 50 percent carbon dioxide by volume?
(*a*) Synthesis gas   (*b*) Ethanol   (*c*) Biodiesel   (*d*) Producer gas   (*e*) Biogas

9-53   Producer gas consists of carbon monoxide, hydrogen, methane, nitrogen and carbon dioxide. Which of the following choices include fuels only?
(*a*) Methane, nitrogen   (*b*) Carbon monoxide, carbon dioxide
(*c*) Carbon dioxide, nitrogen   (*d*) Hydrogen, carbon monoxide
(*e*) Hydrogen, nitrogen

ETHANOL

9-54   What are the two most common biofuels used in internal combustion engines?

9-55   List common biomass feedstocks for ethanol production? What are the main sources of ethanol production in Brazil and the United States?

9-56   What are the four major process steps in dry grind plants for ethanol production?

9-57   What are the main advantages and disadvantages of butanol compared to ethanol?

9-58   What are gasohol and E85 fuels?

9-59   How much ethanol can be produced from 100 kg of glucose ($C_6H_{12}O_6$)?

9-60   How much ethanol can be produced from 100 kg of biomass with a glucose fraction of 60 percent by mass?

9-61   In the production of ethanol, the feedstock high in sugar content is first converted to sugar, and the sugar (glucose) is fermented into ethanol through the reaction $C_6H_{12}O_6 \rightarrow 2\ C_2H_5OH + 2\ CO_2$. Consider 100 kg of sugar beet roots whose sugar content represents 20 percent of total mass. How much ethanol can be produced from these sugar beet roots?

9-62   Determine the ratio of power produced from an ethanol (C_2H_5OH or C_2H_6O) burning internal combustion engine to that produced from a methanol (CH_3OH or CH_4O) burning engine. The lower heating values of ethanol and methanol are 26,810 kJ/kg and 19,920 kJ/kg, respectively.

9-63   A 2.4-L gasoline-fueled internal combustion engine produces maximum power of 200 hp when the fuel is burned stoichiometrically with air. The thermal efficiency of the engine at this maximum power is 40 percent. Estimate the maximum power from this engine if methanol (CH_3OH or CH_4O) is used under stoichiometric combustion with the same thermal efficiency. Determine the rates of gasoline and methanol consumption for maximum power operation. The stoichiometric air-fuel ratio for gasoline is 14.6. Assume a combustion efficiency of 100 percent. The lower heating values of gasoline and methanol are 18,900 Btu/lbm and 8570 Btu/lbm, respectively.

9-64   Order the following major process steps in dry grind plants for ethanol production?
  I. Fermentation
 II. Cooking
III. Distillation
IV. Pretreatment
(*a*) I, II, III, IV   (*b*) II, IV, I, III   (*c*) IV, I, III, II   (*d*) IV, II, I, III   (*e*) II, I, IV, III

9-65   Order the following fuels from a higher heating value to a lower one on a volumetric basis?
  I. Butanol
 II. Methanol
III. Ethanol
IV. Gasoline
(*a*) I, II, III, IV   (*b*) II, IV, I, III   (*c*) IV, I, III, II   (*d*) IV, II, I, III   (*e*) II, I, IV, III

9-66   Adding ethanol to gasoline increases the octane number of gasoline and this allows higher _____ corresponding to higher efficiencies for the engine.
(*a*) Engine volume   (*b*) Power output   (*c*) Compression ratio
(*d*) Torque   (*e*) Exhaust temperature

9-67   Consider 100 kg of sugar beet roots whose sugar content represents 15 percent of total mass. The sugar is converted into ethanol through the reaction $C_6H_{12}O_6 \rightarrow 2\,C_2H_5OH + 2\,CO_2$. The amount of ethanol produced is

(a) 15 kg   (b) 13 kg   (c) 9.8 kg   (d) 7.7 kg   (e) 5.5 kg

9-68   Determine the ratio of power produced from an ethanol (C_2H_5OH or C_2H_6O) burning internal combustion engine to that produced from a gasoline burning engine. The lower heating values of ethanol and gasoline are 26,810 kJ/kg and 44,000 kJ/kg, respectively. Air-fuel ratios for ethanol and gasoline are 9.0 and 15, respectively.

(a) 0.50   (b) 0.61   (c) 0.98   (d) 1.02   (e) 1.67

BIODIESEL

9-69   List common biomass feedstocks for biodiesel production? What are the main sources of biodiesel production in the United States?

9-70   Which properties of oils derived from plants make them unsuitable for use in diesel engines?

9-71   Consider two cars with one using ethanol and the other using biodiesel as the fuel. For a given full tank of liquid fuel, which car gets more mileage? Why?

9-72   Consider two cars with one using petroleum diesel and the other using biodiesel as the fuel. For a given full tank of liquid fuel, which car gets more mileage? Why? Also, compare the two fuels in terms of emissions.

9-73   Which properties of oils derived from plants make them unsuitable for use in diesel engines?
(a) High viscosity, low volatility   (b) High viscosity, high volatility
(c) Low viscosity, high volatility   (d) Low viscosity, low volatility

9-74   Common sources of biodiesel include which of the following?
 I. Used vegetable oils
 II. Sugar beets
III. Corn
IV. Animal fats
(a) I and II   (b) II and III   (c) I and III   (d) I and IV   (e) III and IV

9-75   Which fuel produces most power in an internal combustion engine?
(a) Petroleum diesel   (b) Ethanol   (c) Biogas   (d) Biodiesel   (e) Methanol

LIQUID FUELS FROM SYNGAS, PYROLYSIS OIL, AND BIOCRUDE

9-76   Which liquid fuels can be produced from syngas?

9-77   What is a Fischer-Tropsch process? What are Fischer-Tropsch liquids?

9-78   Describe a fast pyrolysis process. What is the product of fast pyrolysis?

9-79   Describe hydrothermal processing. What is the product of hydrothermal processing?

9-80   Which is not a parameter affecting the composition of the liquids after Fischer-Tropsch process?
(a) Type of catalyst   (b) Reaction temperature   (c) Composition of syngas
(d) Reaction temperature   (e) None of these

9-81   Which processes are used to obtain gasoline, diesel fuel, and aviation fuel from bio-oil and biocrude?
(a) Pyrolysis   (b) Distillation and cracking   (c) Fast pyrolysis
(d) Hydrothermal processing   (e) Syngas fermentation

ELECTRICITY AND HEAT PRODUCTION BY BIOMASS

9-82   What are the three technologies used to convert biomass energy into heat and electricity? Briefly describe each one of them.

9-83   Under what condition is the burning of wood a carbon neutral process?

9-84   Which undesired emissions are produced when wood is burned?

9-85   Compare leaving the dead trees in their environment versus collecting and burning them for energy production.

9-86   An industrial facility is considering switching from a natural gas boiler to wood pellet boiler for space heating in winter. In a typical winter the facility consumes 5000 therms of natural gas (1 therm = 100,000 Btu) with a unit price of $1.5/therm. The price of wood pellet is $0.1/kg and its heating value is 12,000 kJ/kg. Determine the amount of pellet consumption and the cost savings to the facility in a typical winter season due to switching to pellet boiler. Assume the boiler efficiency is the same for both boiler systems.

9-87   A homeowner is currently using a natural gas furnace for heating the house in winter, but he is considering replacing this system with a furnace burning wood pellets. The homeowner currently pays an average of $3000 per year for space heating. Using the values below, determine how much the homeowner will save per year if he replaces the existing furnace with the pellet-burning furnace.

Natural gas furnace: $\eta_{boiler} = 0.90$, Unit price = $1.35/therm (1 therm = 105,500 kJ)
Wood pellet: HV = 20,000 kJ/kg, $\eta_{boiler} = 0.80$, Unit price = $0.15/kg

9-88   A building owner is currently using a coal boiler for heating the building in winter, but he is considering replacing this system with a biogas boiler. The building owner currently pays an average of $40,000 per year for space heating. Using the values below, determine how much the homeowner will pay for heating if he replaces the existing coal boiler with the biogas boiler.

Coal boiler: HV = 26,000 kJ/kg, $\eta_{boiler} = 0.70$, Unit price = $0.30/kg
Biogas boiler: HV = 22,000 kJ/m³, $\eta_{boiler} = 0.90$, Unit price = $0.40/m³

9-89   A building is heated in winter by a boiler burning wood pellets. The boiler is rated at 40 kW. The heating value of pellet is 18,000 kJ/kg, and the average boiler efficiency is 80 percent. Taking the total winter period to be 4000 h and assuming that the boiler operates at an average load of 60 percent, determine the amount of wood pellets consumed in the winter season.

9-90   A house is heated in winter by a biogas boiler. The boiler is rated at 70,000 Btu/h. The heating value of biogas is 13,000 Btu/h, and the average boiler efficiency is 87 percent. Taking the total winter period to be 3000 h and assuming that the boiler operates at an average load of 75 percent, determine the amount of biogas consumed in therm (1 therm = 100,000 Btu) in the winter season. If the biogas is sold at a unit price of $1.20/therm, determine the cost of this biogas to the homeowner.

9-91   Organic matter is decomposed by bacteria in the absence of oxygen to produce natural gas consisting primarily of methane and other byproducts such as carbon dioxide. This process is called
(*a*) Anaerobic digestion   (*b*) Direct combustion   (*c*) Co-firing
(*d*) Photobiological conversion   (*e*) Biochemical conversion

9-92   Replacing only a portion of fossil fuel in coal-fired boilers with biomass is called
(*a*) Anaerobic digestion   (*b*) Direct combustion   (*c*) Co-firing
(*d*) Photobiological conversion   (*e*) Biochemical conversion

9-93   Which is not a technology used to convert biomass energy into heat and electricity?
(*a*) Anaerobic digestion   (*b*) Direct combustion   (*c*) Co-firing
(*d*) Biochemical conversion   (*e*) Methane recovery

9-94   A house is heated in winter by a pellet boiler. The boiler is rated at 20 kW. The heating value of pellet is 20,000 kJ/kg, and the average boiler efficiency is 85 percent. Taking the total winter period to be 3000 h and assuming that the boiler operates at an average load of 50 percent, determine the amount of wood pellets consumed in the winter season.
(*a*) 6350 kg   (*b*) 8500 kg   (*c*) 4940 kg   (*d*) 3280 kg   (*e*) 5810 kg

MUNICIPAL SOLID WASTE

9-95   What is municipal solid waste (MSW)? List some of the components of MSW. Are industrial, hazardous, and construction waste considered MSW?

9-96   What is the difference between recycling and composting?

9-97   Define waste prevention.

9-98   Describe waste to energy (WTE) and landfill-gas-energy (LFGTE) options for producing electricity from municipal solid waste.

9-99   Ten tons of municipal solid waste (MSW) is burned directly in a boiler to generate saturated steam at 200°C. Water enters the boiler at 20°C. The heating value of MSW is 18,000 kJ/kg, and the boiler efficiency is 75 percent. Determine the amount of steam generated. The generated steam is run through a turbine-generator unit to generate electricity. The steam pressure at the turbine exit is 100 kPa. If the turbine isentropic efficiency is 85 percent and the generator efficiency is 95 percent, determine the amount of electricity produced. How much revenue can be generated if the electricity is sold at a price of $0.11/kWh?

9-100   A 75-percent-efficient steam boiler is used to generate saturated water vapor at 800 kPa at a rate of 10 kg/s. Water enters the boiler at the same pressure as a saturated liquid. Municipal solid waste (MSW) is burned in the boiler, and the heating value of MSW is 21,000 kJ/kg. The generated steam is run through an ideal turbine-generator unit to generate electricity. The pressure at the turbine outlet is 75 kPa. Determine the rate of MSW consumption and the power generated in the generator.

9-101   A 78-percent-efficient steam boiler is used to generate saturated water vapor at 350°F. Water enters the boiler at the steam pressure as a saturated liquid. Municipal solid waste (MSW) is burned at a rate of 60 lbm/min in the boiler, and the heating value of MSW is 9000 Btu/lbm. The generated steam is run through an ideal turbine-generator unit to generate electricity. The pressure at the turbine outlet is 10 psia. Determine the rate of steam generation and the power output from the generator.

9-102   Which is not included in municipal solid waste?
(*a*) Food scraps   (*b*) Construction waste   (*c*) Paper
(*d*) Yard trimmings   (*e*) Wood

9-103   Recovery of useful materials such as paper, glass, plastic, and metals from the trash to use to make new products is called
(*a*) Waste to energy   (*b*) Landfill gas to energy   (*c*) Waste prevention
(*d*) Composting   (*e*) Recycling

9-104   Storing organic waste such as food scraps and yard trimmings under certain conditions to help it break down naturally is called
(*a*) Waste to energy   (*b*) Landfill gas to energy   (*c*) Waste prevention
(*d*) Composting   (*e*) Recycling

9-105   The design of products to minimize the production of waste and making the resulting waste less toxic is called
(*a*) Waste to energy   (*b*) Landfill gas to energy   (*c*) Waste prevention
(*d*) Composting   (*e*) Recycling

9-106   Which is not an advantage of recycling?
(*a*) Reduces air pollution   (*b*) Reduces greenhouse gas emission
(*c*) Reduces energy consumption   (*d*) Reduces raw material use
(*e*) None of these

CHAPTER 10

Ocean Energy

10-1 INTRODUCTION

Tremendous amounts of energy are available in oceans and seas due to waves and tides, which are mechanical forms of energy related to kinetic and potential energy of the ocean water. Oceans also have thermal energy due to solar heating of ocean water. In this chapter, we study ocean thermal energy conversion, wave, and tidal energies.

10-2 OCEAN THERMAL ENERGY CONVERSION

As a result of solar energy absorption, the water at the sea or ocean surface is warmer, and the water at a deeper location is cooler. In tropical climates, surface temperatures can reach 28°C while the temperatures are as low as 4°C about 1 km below the surface.

The ocean can be considered to be a large heat engine with a source temperature of 28°C and a sink temperature of 4°C. This heat engine can be operated to use the surface warm water as the heat source and deep cold water as the heat sink for the conversion of heat to power. This principle of power production is called *ocean thermal energy conversion* (OTEC). The main disadvantage of an OTEC plant is the low thermal efficiency. The maximum thermal efficiency can be determined from the Carnot relation using the above source and sink temperatures to be

$$\eta_{\text{th,max}} = 1 - \frac{T_L}{T_H} = 1 - \frac{(4+273)\text{ K}}{(28+273)\text{ K}} = 0.080 \text{ or } 8.0\%$$

Actual thermal efficiencies of OTEC systems are only around 3 percent. Experiments have been performed using the OTEC principle, but the results have not been promising due to the large installation cost and low thermal efficiency. The equipment in OTEC systems must be very large for significant amounts of power production. For example, for a power output of 100 kW, the heat transfer rating of the heat exchanger of an OTEC system should be 3300 kW at a thermal efficiency of 3 percent. This corresponds to a very large surface area for the heat exchanger and corresponding high capital and pumping costs. For example, consider a closed system OTEC application whose evaporator operates at a log mean temperature difference of 5°C and an overall heat transfer coefficient of 1.5 kW/m²·°C. The required surface area of this evaporator for a heat transfer rating of 3300 kW is determined from

$$\dot{Q} = UA_s \Delta T_{\text{lm}} \longrightarrow A_s = \frac{\dot{Q}}{U\Delta T_{\text{lm}}} = \frac{3300 \text{ kW}}{(1.5 \text{ kW/m}^2 \cdot °\text{C})(5°\text{C})} = 440 \text{ m}^2$$

For 1 MW power output, the heat exchanger will have a surface area of 4400 m².

Tropical and equatorial waters are more suitable for OTEC installations due to relatively high surface temperatures. Surface water temperatures vary with season and latitude. In tropical waters, surface temperatures vary from 29°C in August to 24°C in February. Hawaii, island nations in the South Pacific Ocean, and Gulf nations are some of the known locations for OTEC applications.

A pilot OTEC plant operated between 1993 and 1998 in Hawaii with average surface and deep-water temperatures of 26 and 6°C, respectively, and a net power output of 103 kW. About half of the power produced by the plant was used internally for pumps and parasitic work (Vega, 2002).

Two basic designs can be used for OTEC systems: *open* and *closed* systems. Open systems operate on the *Claude cycle* and closed systems operate on the *Anderson cycle*.

The schematic of an open-system OTEC plant is given in Fig. 10-1. The system is similar to the flash cycle of geothermal power plants. A chamber (i.e., evaporator) is maintained at a subatmospheric (i.e., vacuum) pressure by a vacuum pump. Warm surface water (state 1) flows into this chamber, where its pressure is reduced. The enthalpy of water remains constant during this pressure reducing process since the chamber is adiabatic and there is no work interaction, as shown in the temperature-enthalpy diagram in Fig. 10-2. As the pressure of warm water decreases, its temperature also decreases, resulting in a liquid-vapor mixture (state 2).

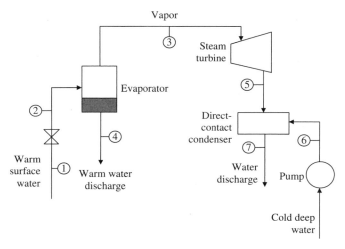

Figure 10-1   Operation of an open-system OTEC plant.

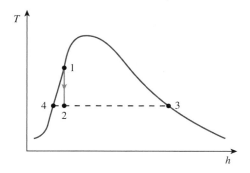

Figure 10-2   Constant-enthalpy process in the evaporator of open-system OTEC plant.

Low-pressure vapor (state 3) is directed to a steam turbine while the liquid is discharged from the chamber (state 4). The vapor (state 5) exits to a direct-contact condenser, which is maintained at a much lower pressure. Cold deep water (state 6) is supplied to the condenser by a pump, and mixing of this cold water with the vapor from the turbine outlet (state 5) turns the vapor to the liquid (state 7), which is discharged.

EXAMPLE 10-1
Thermodynamic Analysis
of an Open-System
OTEC Plant

An open-system OTEC plant operates with a surface water temperature of 30°C and a deep-water temperature of 10°C. The evaporator pressure is 3 kPa and condenser pressure is 1.5 kPa. The mass flow rate of warm surface water entering the evaporator is 100 kg/s, and the turbine has an isentropic efficiency of 85 percent. Determine (*a*) the mass and volume flow rates of steam at the turbine inlet, (*b*) the turbine power output and the thermal efficiency of the plant, and (*c*) the mass flow rate of the cold deep water. Neglect pumping power and other internal or auxiliary power consumption in the plant.

SOLUTION   (*a*) We neglect kinetic and potential energy changes and use ordinary water properties for ocean water. Water properties are obtained from Tables A-3 through A-5. We refer to Fig. 10-1 for state numbers. First, the properties of water at the evaporator are

$$\left. \begin{array}{l} T_1 = 30°C \\ x_1 = 0 \end{array} \right\} \ h_1 \cong h_{f @ 30°C} = 125.74 \ \text{kJ/kg}$$

$$\left. \begin{array}{l} P_2 = 3 \ \text{kPa} \\ h_2 = h_1 = 125.74 \ \text{kJ/kg} \end{array} \right\} \ \begin{array}{l} T_2 = 24.08°C \\ h_4 = h_{f @ 3 \ \text{kPa}} = 100.98 \ \text{kJ/kg} \\ h_3 = h_{g @ 3 \ \text{kPa}} = 2544.8 \ \text{kJ/kg} \\ s_3 = s_{g @ 3 \ \text{kPa}} = 8.5765 \ \text{kJ/kg·K} \\ v_3 = v_{g @ 3 \ \text{kPa}} = 45.654 \ \text{m}^3/\text{kg} \end{array}$$

The quality of water at state 2 is

$$x_2 = \frac{h_2 - h_f}{h_{fg}} = \frac{125.74 - 100.98}{2544.8 - 100.98} = 0.010132$$

The mass flow rate of steam at the turbine inlet is

$$\dot{m}_3 = x_2 \dot{m}_1 = (0.01013)(100 \ \text{kg/s}) = \textbf{1.0132 kg/s}$$

The volume flow rate of steam at the turbine inlet is

$$\dot{V}_3 = v_3 \dot{m}_3 = (45.654 \ \text{m}^3/\text{kg})(1.0132 \ \text{kg/s}) = \textbf{46.26 m}^3\textbf{/s}$$

(*b*) The enthalpy at the turbine exit for the isentropic process is

$$\left. \begin{array}{l} P_5 = 1.5 \ \text{kPa} \\ s_{5s} = s_3 = 8.5765 \ \text{kJ/kg·K} \end{array} \right\} \ \begin{array}{l} x_{5s} = 0.97098 \\ h_{5s} = 2453.1 \ \text{kJ/kg} \\ T_{5s} = 13.02°C \end{array}$$

Using the definition of the turbine isentropic efficiency, the actual enthalpy at the turbine exit is determined as

$$\eta_{\text{turb}} = \frac{h_3 - h_5}{h_3 - h_{5s}} \longrightarrow 0.85 = \frac{2544.8 - h_5}{2544.8 - 2453.1} \longrightarrow h_5 = 2466.9 \ \text{kJ/kg}$$

The specific volume at the turbine exit is

$$\left. \begin{array}{l} P_5 = 1.5 \ \text{kPa} \\ h_5 = 2466.9 \ \text{kJ/kg} \end{array} \right\} \ v_5 = 85.90 \ \text{m}^3/\text{kg}$$

The power output from the turbine is

$$\dot{W}_{out} = \dot{m}_3(h_3 - h_5) = (1.0132 \text{ kg/s})(2544.8 - 2466.9)\text{kJ/kg} = \textbf{78.93 kW}$$

The energy or heat input to the power plant can be expressed as the energy difference between states 1 and 4. Then,

$$\dot{Q}_{in} = \dot{m}_1 h_1 - \dot{m}_4 h_4$$

$$= (100 \text{ kg/s})(125.74 \text{ kJ/kg}) - (100 - 1.0132 \text{ kg/s})(100.98 \text{ kJ/kg})$$

$$= 2578 \text{ kW}$$

The thermal efficiency of the plant is then

$$\eta_{th} = \frac{\dot{W}_{out}}{\dot{Q}_{in}} = \frac{78.93 \text{ kW}}{2578 \text{ kW}} = 0.0306 \text{ or } \textbf{3.06\%}$$

(c) A direct-contact condenser is basically a mixing chamber. The steam leaving the turbine is mixed with cold deep water so that the water at state 7 is saturated liquid at 1.5 kPa pressure. The enthalpies of liquid water at states 6 and 7 are

$$h_6 \cong h_{f@10°C} = 42.022 \text{ kJ/kg}$$

$$h_7 \cong h_{f@1.5 \text{ kPa}} = 54.688 \text{ kJ/kg}$$

Then the mass flow rate of cold deep water is determined from an energy balance to be

$$\dot{m}_5 h_5 + \dot{m}_6 h_6 = \dot{m}_7 h_7$$

$$(1.0132 \text{ kg/s})(2466.9 \text{ kJ/kg}) + \dot{m}_6(42.022 \text{ kJ/kg}) = (\dot{m}_6 + 1.0132 \text{ kg/s})(54.688 \text{ kJ/kg})$$

Solving this equation gives

$$\dot{m}_6 = \textbf{193 kg/s}$$

In Example 10-1, the specific volumes at the turbine inlet and exit are 45.65 and 85.90 m³/kg, respectively. These values are very high compared to those found in turbines of fossil fuel–burning conventional steam power plants. For example, the specific volume of steam is only 0.033 m³/kg (at 10 MPa and 500°C) at the inlet of a conventional steam turbine. Therefore, the steam turbines of OTEC systems should be very large to handle high-volume flow rates, and their technology needs to be somewhat different than conventional steam turbines.

The schematic of a closed-cycle OTEC system, called the *Anderson cycle*, is given in Fig. 10-3. This is similar to the binary cycle of geothermal power plants. Closed-cycle plants operate on a Rankine cycle with a binary working fluid such as propane that has a low boiling temperature. The working fluid is completely vaporized in the evaporator by the warm surface water. The resulting vapor (state 3) expands in the turbine and then condenses by transferring its heat to the cool deep water flowing through the condenser (states 7 to 8). The condensed working fluid (state 1) is pumped to the evaporator to complete the closed cycle. The surface water is discharged (state 6) after transferring its heat to the working fluid.

Referring to the closed-cycle OTEC plant shown in Fig. 10-3, the power output from the turbine can be determined from

$$\dot{W}_{out} = \dot{m}_{wf}(h_3 - h_4) \tag{10-1}$$

where $\dot{m}_{wf}$ is the mass flow rate of working fluid, and h_3 and h_4 are enthalpies of the working fluid at the turbine inlet and exit, respectively. The energy or heat input to the power plant

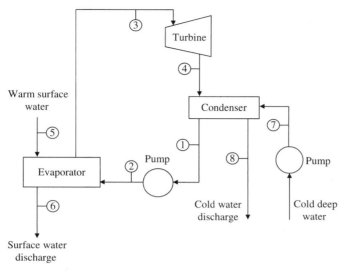

Figure 10-3 Operation of a closed-cycle OTEC plant.

can be expressed as the energy transferred to the working fluid in the evaporator, which is given by

$$\dot{Q}_{in} = \dot{m}_{wf}(h_3 - h_2) \tag{10-2}$$

where h_2 and h_3 are enthalpies of the working fluid at the evaporator inlet and exit, respectively. The heat picked up by the working fluid is equal to heat given by the warm surface water. Therefore, the rate of heat input can also be expressed as

$$\dot{Q}_{in} = \dot{m}_{ww}(h_5 - h_6) \tag{10-3}$$

where $\dot{m}_{ww}$ is the mass flow rate of warm water, and h_5 and h_6 are enthalpies of the warm water at the evaporator inlet and exit, respectively. The thermal efficiency of the plant is then expressed as

$$\eta_{th} = \frac{\dot{W}_{out}}{\dot{Q}_{in}} = \frac{\dot{m}_{wf}(h_3 - h_4)}{\dot{m}_{wf}(h_3 - h_2)} = \frac{h_3 - h_4}{h_3 - h_2} \tag{10-4}$$

In this analysis, we neglected the work required for pumping working fluid in the closed cycle and the pump work of deep cold water. Also, other auxiliary works in the plant are not considered. Therefore, Eq. (10-1) gives the gross power of the plant, and Eq. (10-4) is the expression of the gross thermal efficiency.

10-3 WAVE ENERGY

Like most other renewable energy sources, *wave energy* is ultimately caused by solar energy. Ocean and sea waves are caused by wind, and wind is caused by uneven solar heating of earth and water bodies and the resulting temperature fluctuations. Rotation of the earth is also a contributing factor to wind and thus wave energy. Wave energy is also referred to as *wave power* as we try to extract power from the energy of waves as a sustainable method of power generation.

The interest in wave energy has been around since the energy crisis of the 1970s due to the renewable nature of this energy and its huge potential, as well as the fact that this energy is free. One can easily see the enormous potential of producing power from ocean and sea waves considering that an average of some 3 million waves break onto the coast every year. The technologies for wave energy conversion are still in the developmental stage because of technical, practical, and economic complications. Some of the problems of wave power conversion are (El-Wakil, 1984):

- There are many sites of large wave activity around the world, but the access to these sites is often limited.
- The sites of wave energy are often far from populated and industrial districts and sometimes far from power grids. This makes it difficult and expensive to transmit generated electricity.
- The equipment for converting wave energy to electricity is large and complicated. High structural and mechanical strength are needed to withstand strong motions in the sea.
- For reasonable amounts of power production, very large equipment is necessary. Hence, wave power systems involve high capital investment as well as high maintenance expenses. As a result, wave power production is not cost-competitive to conventional fossil fuel or renewable power systems.
- The effect of wave power installations on marine life could be a problem.

One traveling around the world can observe that some ocean and sea sites involve more waves and higher wave heights than others. These locations typically attract wind surfers. Some such places include

- Molokai and Alenuihaha channels in the Hawaiian islands
- Pacific coast of North America
- Arabian Sea off Pakistan and India
- North Atlantic coast of Scotland
- Coast of New England in the United States

These sites involve wave heights of 1 to 5 m. A study of these sites reveals about half of the time more than 2-m-high waves are observed with a period of about 6 s (Hogben and Lamb, 1967).

Power Production from Waves

How much energy is available in ocean and sea waves? How much power can be produced from wave energy? In this section we try to answer these questions following the treatment of El-Wakil (1984). For a wave, the relationship between wavelength and period is

$$\lambda = 1.56\tau^2 \tag{10-5}$$

where the wavelength λ is in meters and period τ is in seconds. In English units, $\lambda = 5.12\tau^2$ where λ is in feet and τ is in seconds. A traveling wave can be expressed as

$$y = a\sin\left(\frac{2\pi}{\lambda}x - \frac{2\pi}{\tau}t\right) \tag{10-6}$$

where y is height above mean sea level in meters, a is amplitude in meters, and t is time in seconds. These parameters are indicated by a typical traveling wave in Fig. 10-4. In Eq. 10-6,

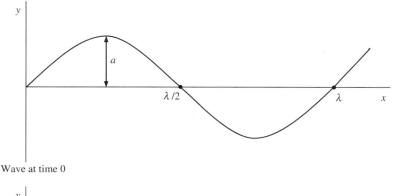

Wave at time 0

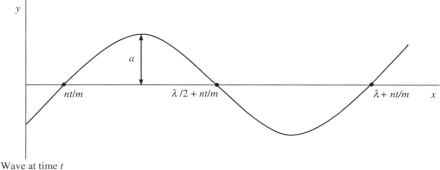

Wave at time t

Figure 10-4 A typical traveling wave and its parameters at time 0 and time t.

$n = 2\pi/\tau$ represents phase rate in 1/s, $m = 2\pi/\lambda$ and $mx - nt = 2\pi(x/\lambda - t/\tau)$ represents phase angle, which is dimensionless. After an initial time of $t = 0$, the wave is repeated after a period of $t = \tau$.

The wave motion is in the horizontal direction (x direction), and the wave velocity is given by

$$V = \frac{\lambda}{\tau} \tag{10-7}$$

Water does not flow exactly with a wave. Instead, a water droplet rotates in place in an elliptical path in the plane of wave propagation with horizontal and vertical semiaxes. The paths of water particles are shown in Fig. 10-5. The horizontal and vertical semiaxes of the ellipses are given, respectively, by (Lamb, 1932; Milne-Thompson, 1955)

$$\alpha = a\frac{\cosh m\eta}{\sinh mh} \tag{10-8}$$

$$\beta = a\frac{\sinh m\eta}{\sinh mh} \tag{10-9}$$

where h is depth of water in meters, and η is distance from the bottom. The amplitude is half of water height: $2a = h$. Note that horizontal semiaxis α is normally greater than vertical semiaxis β. Also,

when $\eta = 0$, $\beta = 0$ (bottom of water)

when $\eta = h$, $\beta = a$ (surface of water)

for large depths, $\alpha \approx \beta \approx a$

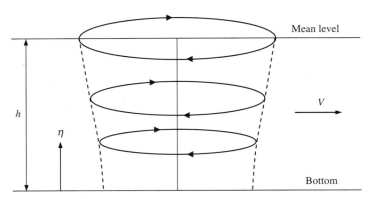

Figure 10-5   Paths of water particles at different heights.

A wave has speed and height, and therefore a wave has both kinetic and potential energies, and the total energy of a wave is the sum of its potential and kinetic energies. The potential energy is due to elevation of the water with respect to $y = 0$ in Fig. 10-4. For a differential volume of $yL\,dx$, its mean height is $y/2$. Then, the potential energy is expressed as

$$d\text{PE} = m\frac{yg}{2} = (\rho yL\,dx)\frac{yg}{2} = \frac{\rho L}{2}y^2 g\,dx \tag{10-10}$$

where $m = y\,dx$ is mass of liquid in kilograms, g is acceleration due to gravity in m/s², ρ is density in kg/m³, and L is the arbitrary width of the two-dimensional wave, perpendicular to the direction of wave propagation x in meters. Combining Eq. (10-10) with Eq. (10-6) and integrating gives

$$\text{PE} = \frac{1}{4}\rho a^2 \lambda Lg \qquad \text{(J)} \tag{10-11}$$

Potential energy per unit area (pe) can be obtained by dividing Eq. (10-11) by the area $A = \lambda L$, resulting

$$\text{pe} = \frac{1}{4}\rho a^2 g \qquad \text{(J/m}^2\text{)} \tag{10-12}$$

The kinetic energy of the wave is due to the velocity of the water contained in one complete wavelength. From hydrodynamic theory, the kinetic energy is given by (Lamb, 1932)

$$\text{KE} = \frac{1}{4}i\rho Lg\int \omega\,d\varpi \qquad \text{(J)} \tag{10-13}$$

where ω is a complex potential given by

$$\omega = \frac{ac}{\sinh mh}\cos(mz - nt) \tag{10-14}$$

Here, z is distance measured from a reference point. The integral in Eq. (10-13) over the entire water body gives

$$\text{KE} = \frac{1}{4}\rho a^2 \lambda Lg \qquad \text{(J)} \tag{10-15}$$

Kinetic energy per unit area (ke) can be obtained by dividing Eq. (10-15) by the area $A = \lambda L$, resulting in

$$\text{ke} = \frac{1}{4}\rho a^2 g \qquad (\text{J/m}^2) \qquad\qquad (10\text{-}16)$$

Therefore, kinetic and potential energies of a traveling wave are identical, and the total wave energy per unit area (i.e., energy density) is

$$e_{\text{wave}} = \text{pe} + \text{ke} = \frac{1}{2}\rho a^2 g \qquad (\text{J/m}^2) \qquad\qquad (10\text{-}17)$$

This is in fact the energy potential of a wave or available work from a wave:

$$w_{\text{available}} = \frac{1}{2}\rho a^2 g \qquad (\text{J/m}^2) \qquad\qquad (10\text{-}18)$$

Finally, the wave energy per unit time (i.e., power per unit area or power density) can be obtained by multiplying Eq. (10-17) by the frequency f. The frequency is defined as the reciprocal of period $f = 1/\tau$. Then,

$$\dot{e}_{\text{wave}} = \frac{1}{2}f\rho a^2 g = \frac{1}{2\tau}\rho a^2 g \qquad (\text{W/m}^2) \qquad\qquad (10\text{-}19)$$

This is the power potential of a wave or available power from a wave:

$$\dot{w}_{\text{available}} = \frac{1}{2\tau}\rho a^2 g \qquad (\text{W/m}^2) \qquad\qquad (10\text{-}20)$$

EXAMPLE 10-2
Work and Power Potentials
of an Ocean Wave

An ocean wave is 3 m high and lasts for a period of 5 s. The depth of water is 75 m. Determine (a) the wavelength and the wave velocity, (b) the horizontal and vertical semiaxes for water motion at the surface, and (c) the work and power potentials per unit area. Take the density of seawater to be 1025 kg/m³.

SOLUTION (a) First the wavelength and wave velocity are determined from

$$\lambda = 1.56\tau^2 = 1.56(5 \text{ s})^2 = \mathbf{39 \text{ m}}$$

$$V = \frac{\lambda}{\tau} = \frac{39 \text{ m}}{5 \text{ s}} = \mathbf{7.8 \text{ m/s}}$$

(b) The water height is $2a = 3$ m, and thus the amplitude is

$$a = 1.5 \text{ m}$$

and

$$m = \frac{2\pi}{\lambda} = \frac{2\pi}{39 \text{ m}} = 0.1611 \text{ m}^{-1}$$

Also, at the surface,

$$\eta = h = 75 \text{ m}$$

The horizontal and vertical semiaxes are determined from

$$\alpha = a\frac{\cosh m\eta}{\sinh mh} = (1.5 \text{ m})\frac{\cosh(0.1611 \times 75)}{\sinh(0.1611 \times 75)} = 1.5 \text{ m}$$

$$\beta = a\frac{\sinh m\eta}{\sinh mh} = (1.5 \text{ m})\frac{\sinh(0.1611 \times 75)}{\sinh(0.1611 \times 75)} = 1.5 \text{ m}$$

The semiaxes are equal due to the large depth, and this indicates the circular motion.
(c) The work potential of the wave per unit area is

$$w_{\text{available}} = \frac{1}{2}\rho a^2 g = \frac{1}{2}(1025 \text{ kg/m}^3)(1.5 \text{ m})^2(9.81 \text{ m/s}^2)\left(\frac{1 \text{ N}}{1 \text{ kg·m/s}^2}\right)\left(\frac{1 \text{ J}}{1 \text{ N·m}}\right) = \mathbf{11,310 \ J/m^2}$$

The wave power potential per unit area is

$$\dot{w}_{\text{available}} = \frac{1}{2\tau}\rho a^2 g$$

$$= \frac{1}{2\,(5 \text{ s})}(1025 \text{ kg/m}^3)(1.5 \text{ m})^2(9.81 \text{ m/s}^2)\left(\frac{1 \text{ N}}{1 \text{ kg·m/s}^2}\right)\left(\frac{1 \text{ J}}{1 \text{ N·m}}\right)\left(\frac{1 \text{ W}}{1 \text{ J/s}}\right)$$

$$= \mathbf{2262 \ W/m^2} \quad \blacktriangle$$

Wave Power Technologies

Various systems and technologies have been proposed for wave energy conversion. One such application involves transferring the energy of wave water to air and using the compressed air to drive a turbine. A company uses oscillating water column technology. The system operates as follows (Fig. 10-6): Water moves into a hollow container by the wave motion. This compresses air that was in the container. The compressed air drives the turbine, which is connected to a generator where electricity is produced. When the wave moves in the opposite direction, air fills the container from the top, which is open to the atmosphere. Air flows through the turbine from both sides, depending on whether the device is "breathing" in or out. The process is repeated with the next wave. The system uses a special rotor geometry so that there is no need to change the blade angles or the direction of rotation. These systems must be large, and many of them should be installed to produce significant amounts of electricity. Oscillating water column technology has been installed off the coast of western Scotland.

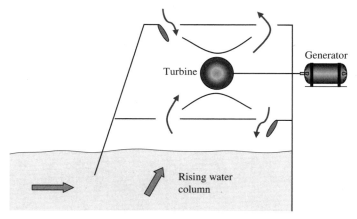

Figure 10-6   Oscillating water column technology for wave power conversion. (*Adapted from Hodge, 2010 and www.wavegen.co.uk.*)

Oscillating water column technology can be applied in three types of coastal wave power plants.

1. *Shoreline power plants:* The plant is installed on the coast with a collector structure open to the sea. Air in the chamber is compressed and calmed by the moving and returning waves. The power outputs of shoreline power plants can range from a few hundred kilowatts to a few megawatts, depending on the size of the waves and the size of installations.

2. *Near-shore power plants:* In this case, the plant is installed a few hundred meters from the coastline, which is about 10 m deep. The collector is connected to the land by a dam. The power ratings range from 10 to 100 MW depending on the size of the waves and power plant. With government incentives, these plants can operate economically.

3. *Breakwater power plants:* In order to reduce the cost of wave power plants, the plant is integrated into a coastal structure such as a harbor breakwater or a coastal protection project. Power ratings more than 10 MW can be installed this way.

The first breakwater power plant with a rating of 300 kW was installed in 2011 on the Spanish Atlantic coast at Mutriku. The plant was built into the breakwater around the harbor. The power plant has 16 identical turbines. Since the cost of the breakwater is handled by the local municipality, the cost of the plant is relatively low.

Another commercial technology for wave energy conversion involves using wave water to drive a power machine, known as the Pelamis Wave Energy Converter. The machine consists of five tube sections connected by joints which allow flexing in two directions (Fig. 10-7). The machine floats submerged on the surface of the water. It faces the direction of moving waves. The power systems are housed inside each tube joint. As waves move in the bent tubes, the moving water drives hydraulic power take-off systems. The generated power is transferred to shore using subsea cables and equipment. Hydraulic cylinders drive the power take-off system. Hydraulic cylinders resist the water motion and pump fluid into high-pressure accumulators. This way power generation is continuous.

In this wave power technology, the machine is located about 2 to 10 km from the coast where the water is 50 m deep or more. A standard machine is rated at 750 kW, but the average production is about 25 to 40 percent of this rated power. This machine is 150 m long, and the tube diameter is 3.5 m. One machine is estimated to produce 2.7 GWh (gigawatt hours) of electricity per year at a capacity rate of 40 percent. A Pelamis machine was first connected to the U.K. grid in 2004. These machines were also tested on the coast of Portugal in 2009 and Scotland in 2014.

Currently, this technology needs significant incentives to make it cost-competitive to conventional power generation. Some of the areas that need improvements are optimization

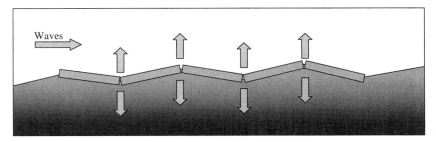

Figure 10-7   Five tube sections of a wave power machine.

Figure 10-8   The test unit of Wavestar technology located on the western coast of Denmark. The test unit has two floats, but the full commercial unit will have 20 floats. (*Courtesy of Wavestar.*)

of the system configuration to maximize the efficiency of energy conversion, more economical structural materials, standardized production of machine components, and improved control systems.

A large-scale test system for wave power was installed near Hanstholm on the western coast of Denmark in 2009. The unit produces power from ocean waves that are 5 to 10 s apart. This is achieved with a row of submerged floats which rise and fall as the wave passes the length of the machine. The up and down motion of the floats is transferred by a hydraulic system into a generator to generate electricity.

The test unit has two floats placed on one side, whereas the full commercial converter will have 20 floats with 10 floats on each side (Fig. 10-8). The maximum wave height in the test site is 6 m, with water depth of 5 to 8 m. The dimensions of the main structure are 32 m × 17 m × 6.5 m. The unit has a maximum power potential of 110 kW. In tests, a power output of 32.4 kW has been measured with a wave height of 2.6 m. The test results indicate an estimated power production of 46 MWh. The system is designed to shut down the operation when the wave height exceeds certain limits. In this case, the floats are pulled out of the water by retracting hydraulic cylinders (Wavestar, 2013).

10-4 TIDAL ENERGY

Power can be produced using the energy in tides. A reservoir can be charged by the high tide and discharged by the low tide. As the water flows in and out of the reservoir, it runs through a hydraulic turbine to produce power. This is similar to a hydroelectric power plant since both systems use the potential energy of water.

The tidal motion of ocean and seawater is due to the gravitational force of the moon and the sun. These forces balance the centrifugal force on the water due to rotation of the earth. The tides are not continuous, and they do not have consistent patterns. Their timings and heights vary from day to day. Also, they vary with location on the earth. Some coastal regions have more tidal movements with greater tide heights and some have fewer tidal movements.

As with other ocean energy systems, the energy is free but the systems should be very large to yield significant amounts of power. This corresponds to a very large capital investment.

The tidal schedule is based on moon motion around the earth, which lasts 24 h and 50 min. The tides rise and fall twice in a lunar day, and thus a full tidal cycle lasts 12 h and 25 min. The tidal range (the difference in water elevation between high tide and low tide) varies during a lunar month, which is 29.5 days. Figure 10-9 shows how the tidal range varies in a typical lunar month. During new moon and full moon, the range is maximum, which is called *spring tide*. Around first and third quarters the range is minimum, which is called *neap tide*. An average range is about one-third of spring tide.

Tidal ranges vary with location of the earth, profile of the shoreline, and water depth. High-range locations are more suitable for tidal power generation to better justify the installation cost. Some of the known locations with high tide ranges are

- Bay of Fundy, Canada
- River Severn Estuary, UK
- Puerto Rio Gallegos, Argentina
- Bay of Mezen, Russia
- Sea of Okhotsk, Russia
- La Rance, France
- Port of Granville, France

These coastlines have tidal ranges typically greater than 10 m. The recoverable tidal power potential on earth is estimated to be about 1.5 million MW. Several tidal systems are operational with a total power rating of over 260 MW (Gorlov, 2001). The largest one is located in La Rance, France, with a power rating of 240 MW and an average tide height of 8.6 m. It has 24 identical turbines. The plant was built in 1967 and is still operating, generating about 600 million kWh of electricity per year. This corresponds to a capacity factor of about 30 percent. The second largest one is installed in Annapolis, Canada, with a power rating of 18 MW and an average tide height of 6.4 m.

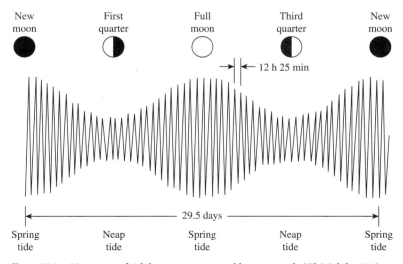

Figure 10-9 Variation of tidal range in a typical lunar month (*El-Wakil, 1984*).

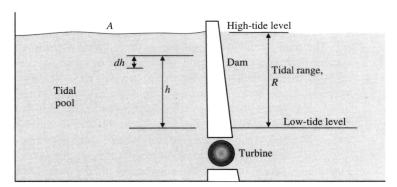

Figure 10-10   A single-pool tidal system for tidal power generation.

In order to estimate the power potential of tidal energy, consider a *simple single-pool tidal system*, as shown in Fig. 10-10 based on the procedure given in El-Wakil (1984). The potential energy or work potential of a differential element dm is

$$dW = ghdm \tag{10-21}$$

where h is the height or head of the differential element and

$$dm = -\rho A dh \tag{10-22}$$

where ρ is the density of water and A is the surface area of the pool. Substituting,

$$dW = -g\rho Ahdh \tag{10-23}$$

Integrating this expression during a full emptying or filling period gives the total work potential

$$W = \int_{R}^{0} -g\rho Ah\, dh = -g\rho A \int_{R}^{0} h\, dh$$

which gives

$$W_{\text{available}} = \frac{1}{2} g\rho AR^2 \qquad \text{(J)} \tag{10-24}$$

where R is tidal range. This is the theoretical amount of work that can be produced by tidal energy with a range R. The power potential or available power is available work over the time period:

$$\dot{W}_{\text{available}} = \frac{W}{\Delta t} = \frac{1}{2\Delta t} g\rho AR^2 \qquad \text{(W)} \tag{10-25}$$

Each full tidal cycle lasts 12 h and 25 min, and each emptying or filling process takes 6 h and 12.5 min or 6.2083 h, or 22,350 s. Also, the density of seawater can be taken to be 1025 kg/m³. Therefore, the average power potential is expressed as

$$\dot{W}_{\text{available}} = \frac{1}{2\Delta t} g\rho AR^2 = \frac{1}{2(22{,}350\text{ s})}(9.81 \text{ m/s}^2)(1025 \text{ kg/m}^3)AR^2 \qquad \text{(W)}$$

or

$$\dot{W}_{available} = 0.225AR^2 \qquad \text{(W)} \qquad (10\text{-}26)$$

Based on the units used to obtain Eq. (10-26), when area A is in m² and range R is in meters, this equation gives power in watts. For a unit surface area and a tidal range of 10 m, the power would be

$$\dot{W}_{available} = 0.225AR^2 = 0.225(1 \text{ m}^2)(10 \text{ m})^2 = 22.5 \text{ W}$$

Therefore, the power potential of a 10-m-high tide is only 22.5 W/m². Note that the power generation in a tidal system is not continuous. The system generates power only during the emptying period of the tidal pool.

The Bay of Fundy in Canada has an area of 13,000 km² with an average range of 8 m. The tidal power potential is then

$$\dot{W}_{available} = 0.225AR^2 = 0.225(13{,}000 \times 10^6 \text{ m}^2)(8 \text{ m})^2 = 1.872 \times 10^{11} \text{ W}$$

or 187,200 MW. This is a huge amount of power potential. Because of the frictional losses and turbine and generator inefficiencies, actual power will be less. Assuming a conversion efficiency of 30 percent, the actual power potential becomes 56,000 MW, which is still very high.

A good strategy for emptying the pool is to do this slowly over time. This is called a *modulated single-pool tidal system*. The average head in this case is less than the tidal range, and the instantaneous power output is reduced in comparison to fast emptying of the pool. This strategy reduces power fluctuations, and as a result, smaller turbine-generator units can be used with longer power generation periods. According to this operation module, the work potential for a tidal cycle is evaluated from

$$W_{available} = g\rho AR^2 \left\{ 0.988a \left[\cos\left(\frac{\pi t_1}{6.2083}\right) - \cos\left(\frac{\pi t_2}{6.2083}\right) \right] - \frac{a^2}{2}(t_2^2 - t_1^2) \right\} \qquad \text{(J)} \qquad (10\text{-}27)$$

where a is a constant in 1/s and it controls the slope of the pool filling, and t_1 and t_2 are initial and final time of work production in hours, respectively. In Eq. (10-27), the value 6.2083 h represents the period of a filling or emptying process of the pool.

It can be shown that the work production given for a modulated single-pool system [Eq. (10-27)] is much lower than that given by a simple single-pool system [Eq. (10-24)]. However, Eq. (10-24) is applicable for fast emptying of the pool with a high power rating, but the period of power production is very short. Large turbines and generators should be installed to match the high power rating. This also has a negative effect on the power grid. In the slow emptying of a tidal pool, as represented by Eq. (10-27), less power is produced over longer periods of time.

Equations (10-24) and (10-27) give work potential or available work for simple and modulated single-pool systems, respectively. The actual work output can be obtained by defining tidal system efficiency as

$$\eta_{tidal} = \frac{W_{actual}}{W_{available}} \qquad (10\text{-}28)$$

Efficiency of tidal power systems can be taken to be around 30 percent.

EXAMPLE 10-3
Power Production from Tidal Energy

A modulated single-pool tidal system has a tidal range of 10 m, an area of 1 km², the parameter a is 0.08 h⁻¹, and work is produced between $t_1 = 1$ h and $t_2 = 4$ h. Using an overall efficiency of 30 percent, determine the actual work and power outputs. Also, determine the actual work output for the case of a simple single-pool system. Take the density of water to be 1025 kg/m³.

SOLUTION The work output given for a modulated single-pool system is determined to be

$$W_{available} = g\rho AR^2 \left\{ 0.988a \left[\cos\left(\frac{\pi t_1}{6.2083}\right) - \cos\left(\frac{\pi t_2}{6.2083}\right) \right] - \frac{a^2}{2}(t_2^2 - t_1^2) \right\}$$

$$= (9.81 \text{ m/s}^2)(1025 \text{ kg/m}^3)(1 \times 10^6 \text{ m}^2)(10 \text{ m})^2$$

$$\left\{ 0.988 \times 0.08 \text{ h}^{-1} \left[\cos\left(\frac{\pi (1 \text{ h})}{6.2083}\right) - \cos\left(\frac{\pi (4 \text{ h})}{6.2083}\right) \right] - \frac{(0.08 \text{ h}^{-1})^2}{2} \left((4 \text{ h})^2 - (1 \text{ h})^2 \right) \right\}$$

$$= 5.606 \times 10^{10} \text{ J}$$

The actual work output is

$$W_{actual} = \eta_{tidal} W_{available} = (0.30)(5.606 \times 10^{10} \text{ J}) = \mathbf{1.682 \times 10^{10} \text{ J}}$$

This is equivalent to 4672 kWh since 1 kWh = 3600 kJ. The available average power output during this 3-h power generation period is

$$\dot{W}_{available} = \frac{W}{t_2 - t_1} = \frac{5.606 \times 10^{10} \text{ J}}{(4 \text{ h}) - (1 \text{ h})} \left(\frac{1 \text{ h}}{3600 \text{ s}}\right) = 5.191 \times 10^6 \text{ W}$$

The actual power output is

$$\dot{W}_{actual} = \eta_{tidal} \dot{W}_{available} = (0.30)(5.191 \times 10^6 \text{ W}) = \mathbf{1.557 \times 10^6 \text{ W}}$$

This is equivalent to 1557 MW since 1 MW = 10⁶ W.

The available work output for the case of a simple single-pool system is determined from

$$W_{available} = \frac{1}{2} g\rho AR^2$$

$$= \frac{1}{2}(9.81 \text{ m/s}^2)(1025 \text{ kg/m}^3)(1 \times 10^6 \text{ m}^2)(10 \text{ m})^2 \left(\frac{1 \text{ N}}{1 \text{ kg·m/s}^2}\right)\left(\frac{1 \text{ J}}{1 \text{ N·m}}\right)$$

$$= 5.028 \times 10^{11} \text{ J}$$

Using the same efficiency value, the actual work output in this case becomes

$$W_{actual} = \eta_{tidal} W_{available} = (0.30)(5.028 \times 10^{11} \text{ J}) = \mathbf{1.508 \times 10^{11} \text{ J}}$$

The work output in the simple system is about 9 times the work output by a modulated system. However, the time period for power generation in this mode is much shorter, and the turbines must be very large to accommodate the high power rating. ▲

REFERENCES

El-Wakil MM. 1984. *Power Plant Technology*. New York: McGraw Hill.

Gorlov AM. 2001. *Tidal Energy*. Encyclopedia of Ocean Sciences, London: Academic Press, 2955–2960.

Hodge BK. 2010. *Alternative Energy Systems and Applications*. New York: Wiley.

Hogben N and FE Lamb. 1967. "Ocean Wave Statistics." National Physics Laboratory, London, Her Majesty's Stationary Office.

Lamb H. 1932. *Hydrodynamics*, 6th ed. New York: Dover Publications.

Milne-Thompson LM. 1955. *Theoretical Hydrodynamics*. 3rd ed. New York: Macmillan Publishing.

Vega LA. 2002. "Ocean Thermal Energy Conversion Primer." *Marine Technology Society Journal*, 6(4): 25–35.

Wavestar. 2013. "Wavestar Prototype at Roshage, Performance Data for Forsk VE." Project no 2009-1-10305 phase 1 & 2. www.wavestarenergy.com.

PROBLEMS

OCEAN THERMAL ENERGY CONVERSION

10-1   What are mechanical and thermal energies available in oceans?

10-2   What are the typical surface and deep-water temperatures in tropical regions?

10-3   Describe the principle of OTEC (ocean thermal energy conversion).

10-4   What is the difference between a solar pond and an OTEC plant?

10-5   What factors affect surface water temperatures of oceans? Which regions are more suitable for OTEC applications?

10-6   An OTEC plant is to be built in an ocean that operates between a surface temperature of 26°C and a deep-water temperature of 6°C. What is the maximum thermal efficiency this power plant can have?

10-7   An inventor claims to have developed an OTEC plant that has an efficiency of 11 percent when the plant operates between a surface temperature of 32°C and a deep-water temperature of 5°C. Is this claim possible or reasonable?

10-8   Consider a closed-cycle OTEC application whose evaporator operates at a log mean temperature difference of 6.5°C and an overall heat transfer coefficient of 1.25 kW/m²·°C. Determine the required surface area of this evaporator for a heat transfer rating of 2000 kW.

10-9   A closed-cycle OTEC plant produces 1000 kW of power with a thermal efficiency of 2.5 percent. What is the heat transfer rating of the evaporator?

10-10   An OTEC power plant built in Hawaii in 1987 was designed to operate between the temperature limits of 30°C at the ocean surface and 5°C at a depth of 640 m. About 50 m³/min of cold seawater was to be pumped from deep ocean through a 1-m-diameter pipe to serve as the cooling medium or heat sink. If the cooling water experiences a temperature rise of 3.3°C and the thermal efficiency is 2.5 percent, determine the power generated. Take the density of seawater to be 1025 kg/m³.

10-11   An open-system OTEC power plant operates with a surface water temperature of 27°C and a deep-water temperature of 13°C. The evaporator is maintained at a saturation pressure of 3.17 kPa and a saturation temperature of 25°C, and condenser pressure and temperature at saturation condition are 1.706 kPa and 15°C, respectively. The mass flow rate of warm surface water entering the evaporator is 1000 kg/s, and the turbine has an isentropic efficiency of 80 percent. Determine (*a*) the mass flow rate of steam at the turbine inlet, (*b*) the volume flow rates of steam at the turbine inlet and outlet, (*c*) the turbine power output and the thermal efficiency of the cycle, and (*d*) the mass flow rate of cold deep water. Neglect pumping power and other internal or auxiliary power consumption in the plant.

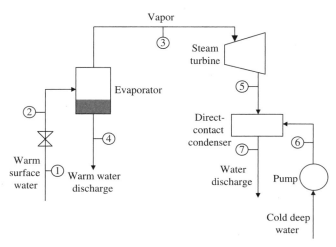

Figure P10-11

10-12 An OTEC power plant operates on an open cycle with a surface water temperature of 30°C and a deep-water temperature of 6°C. The evaporator is maintained at a saturation pressure of 3 kPa, and condenser pressure is 1.5 kPa. The turbine is isentropic and produces 200 kW of power. Determine (a) the mass flow rate of warm surface water, (b) the thermal efficiency of the cycle, and (c) the mass flow rate of cold deep water. Neglect pumping power and other internal or auxiliary power consumption in the plant.

10-13 An OTEC power plant operates on a closed cycle with a surface water temperature of 30°C and a deep-water temperature of 5°C. The working fluid is propane. Propane leaves the evaporator as a saturated vapor at 20°C and condenses at 10°C. The surface water leaves the evaporator at 25°C, and the cold water leaves the condenser at 8°C. The turbine has an isentropic efficiency of 85 percent, and the power output from the turbine is 100 kW. Determine (a) the mass flow rates of surface water, deep water, and propane and (b) the thermal efficiency of the plant. Neglect pumping power and other internal or auxiliary power consumption in the plant. The enthalpies of propane are: Pump and evaporator inlet $h_2 \cong h_1 = 223.32$ kJ/kg, evaporator outlet $h_3 = 581.85$ kJ/kg, turbine outlet $h_4 = 570.73$ kJ/kg.

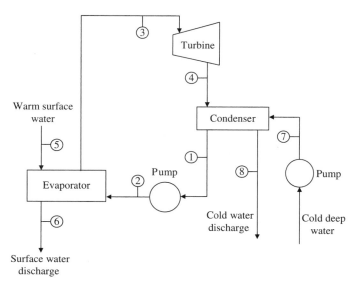

Figure P10-13

10-14   Repeat Prob. 10-13 if the working fluid is refrigerant-134a. The enthalpies of R-134a are: Pump and evaporator inlet $h_2 \cong h_1 = 65.43$ kJ/kg, evaporator outlet $h_3 = 261.59$ kJ/kg, turbine outlet $h_4 = 256.00$ kJ/kg.

10-15   Reconsider Prob. 10-14. The evaporator and condenser of the plant operate at a log mean temperature difference of 4.5°C and an overall heat transfer coefficient of 1.5 kW/m²·°C. Determine the required surface areas of the evaporator and condenser.

10-16   What are the typical surface and deep-water temperatures in tropical regions, respectively?
(*a*) 80°C, 20°C   (*b*) 55°C, 15°C   (*c*) 28°C, 4°C   (*d*) 45°C, 5°C   (*e*) 33°C, 18°C

10-17   An OTEC plant is to be built in an ocean that operates between a surface temperature of 25°C and a deep-water temperature of 9°C. What is the maximum thermal efficiency this power plant can have?
(*a*) 64%   (*b*) 25%   (*c*) 9.0%   (*d*) 7.1%   (*e*) 5.4%

10-18   A closed-cycle OTEC plant has a gross thermal efficiency of 3.2 percent, and the rate of heat transfer in the evaporator is 3500 kW. What is the power output from the plant?
(*a*) 32 kW   (*b*) 112 kW   (*c*) 84 kW   (*d*) 175 kW   (*e*) 350 kW

WAVE ENERGY

10-19   What causes wave energy? Explain.

10-20   List at least two problems of wave power conversion.

10-21   Why is wind power not cost-competitive with conventional power systems?

10-22   What are the two main technologies of wave energy conversion?

10-23   Describe the oscillating water column technology of wave power conversion.

10-24   List three types of coastal wave power plants, and describe breakwater power plants.

10-25   An ocean wave is 4 m high and lasts for a period of 3.5 s. Determine (*a*) the wavelength and the wave velocity and (*b*) the work and power potentials per unit area. Take the density of seawater to be 1025 kg/m³.

10-26   An ocean wave is 10 ft high and lasts for a period of 6 s. Determine (*a*) the wavelength and the wave velocity and (*b*) the work and power potentials per unit area, in Btu/ft² and kW/ft², respectively. Take the density of seawater to be 64 lbm/ft³.

10-27   The average wave height of an ocean area is 2.5 m high, and each wave lasts for an average period of 7 s. Determine (*a*) the energy and power of the waves per unit area and (*b*) the work and average power output of a wave power plant on this site with a plant efficiency of 35 percent and a total ocean wave area of 1 km². Take the density of seawater to be 1025 kg/m³.

10-28   Which is not considered a problem of wave energy conversion?
(*a*) The access to sites of large wave activity is often limited.
(*b*) Wave power systems involve high capital investment as well as high maintenance expenses.
(*c*) For reasonable amounts of power production, very large equipment is necessary.
(*d*) The efficiency of wave power systems is limited by Carnot efficiency.
(*e*) The devices for converting wave energy to electricity are large and complicated.

10-29   Which is not a suitable wave power site in the world?
(*a*) Molokai and Alenuihaha channels in the Hawaiian Islands
(*b*) Mediterranean coast of Greece and Italy
(*c*) Pacific coast of North America
(*d*) North Atlantic coast of Scotland
(*e*) Arabian Sea off Pakistan and India

10-30   What are the typical wave heights of suitable wave energy sites on the earth?
(*a*) 1 to 5 m   (*b*) 5 to 10 m   (*c*) 10 to 15 m   (*d*) 15 to 20 m   (*e*) 10 to 20 m

10-31   The energy of a 3-m-high ocean wave per unit area is (the density of seawater is 1025 kg/m³)
(*a*) 24,500 J/m²   (*b*) 19,800 J/m²   (*c*) 16,200 J/m²   (*d*) 13,600 J/m²   (*e*) 11,300 J/m²

10-32   The wave power potential of a 2-m-high ocean wave that lasts for a period of 6 s per unit area is (the density of seawater is 1025 kg/m³)
(*a*) 838 W/m²   (*b*) 940 W/m²   (*c*) 1275 W/m²   (*d*) 1850 W/m²   (*e*) 2090 W/m²

TIDAL ENERGY

10-33   How can power be produced by the energy of tides? Explain.

10-34   What causes tidal motion? Explain.

10-35   What factors cause variations in tidal motion?

10-36   What is a tidal range? How does the tidal range vary during a lunar month according to the shape of the moon?

10-37   What are spring tide and neap tide?

10-38   Determine the work potential of a 13-m-high tide in kWh for an area of 1000 m². Take the density of water to be 1025 kg/m³.

10-39   Determine the average power potential of a 30-ft-high tide in W per unit ft² area. Take the density of water to be 64 lbm/ft³.

10-40   The average theoretical power for tidal energy is expressed as

$$\dot{W}_{available} = 0.225AR^2$$

where the power is in watts, *A* is in m², and *R* is in meters. Obtain the counterpart of this equation in English units with power in watts, *A* in ft², and *R* in feet. Take the density of water to be 64 lbm/ft³.

10-41   A modulated single-pool tidal system has a tidal range of 8 m, a total area of 3 km², the parameter *a* is 0.065 h⁻¹, and work is produced between 2 and 4 h. Determine the work potential in kWh and power potential in kW. Take the density of water to be 1025 kg/m³.

10-42   A modulated single-pool tidal system has a tidal range of 6 m, a total area of 1 km², the parameter *a* is 0.0625 h⁻¹, and work is produced between 1 and 4 h. Determine the work potential in kWh and the average power potential during the total period of 6.2083 h. Take the density of water to be 1025 kg/m³.

10-43   The Bay of Rance in France has a total basin area of 22 ×10⁶ m². Tidal power is to be generated using a modulated single-pool tidal system with an average tidal range of 11 m. The parameter *a* is 0.07 h⁻¹, and work is produced between 1 and 3 h. Using an overall efficiency of 30 percent, determine the actual power potential from this site. Take the density of water to be 1025 kg/m³. Also, determine the potential revenue from this site during this 2-h period if electricity is sold at a rate of $0.12/kWh.

10-44   How long does a tidal cycle last?
(*a*) 12 h   (*b*) 12 h 25 min   (*c*) 24 h   (*d*) 24 h and 50 min   (*e*) 6 h

10-45   Which is not a factor affecting tidal range?
(*a*) Location of the earth   (*b*) Profile of the shoreline   (*c*) Shape of the moon
(*d*) Water depth   (*e*) None of these

10-46   Which renewable energy application has the lowest efficiency in power production?
(*a*) Geothermal   (*b*) Wind   (*c*) OTEC   (*d*) Wave   (*e*) Tidal

10-47   Which renewable energy application has the highest efficiency in power production?
(*a*) Hydropower   (*b*) Wind   (*c*) Solar   (*d*) Tidal   (*e*) Geothermal

10-48   Determine the work potential of a 7-m-high tide for a unit area. Take the density of water to be 1025 kg/m³.
(*a*) 1100 kJ   (*b*) 670 kJ   (*c*) 415 kJ   (*d*) 246 kJ   (*e*) 105 kJ

10-49   Determine the power potential of a 5-m-high tide for a unit area. Take the density of water to be 1025 kg/m³.
(*a*) 22.5 W   (*b*) 11.3 W   (*c*) 5.6 W   (*d*) 41.0 W   (*e*) 94.1 W

CHAPTER 11

Hydrogen and Fuel Cells

11-1 HYDROGEN: AN ENERGY CARRIER

One major problem with renewable energy sources is the inability to store the produced energy in a viable manner. This is not a problem with biomass since the fuels produced from biomass such as ethanol and biodiesel can be stored and used anytime. However, the electricity produced from solar systems, hydroelectric dams, geothermal power plants, and wind turbines cannot be stored for later use. Batteries are not a viable option in today's technology due to their limited capacity. One possible solution to this problem is production of hydrogen from renewable electricity by the electrolysis of water. Once produced, hydrogen can be stored and used anytime.

Hydrogen is a colorless, odorless, nonmetallic, tasteless, highly flammable diatomic gas with the molecular formula H_2. It is also the lightest element, with a molecular mass of 2.016 kg/kmol. Hydrogen is a *fuel* with a higher heating value of 141,800 kJ/kg and a lower heating value of 120,000 kJ/kg. Note, however, that hydrogen is not an energy source like coal, oil, and natural gas since there are no hydrogen reserves in the earth. Although hydrogen is the most plentiful element in the universe, making up about three-quarters of all matter, free hydrogen is scarce. Hydrogen must be produced from other fuels such as natural gas or from water through electrolysis by consuming electricity. Therefore, hydrogen should be called an *energy carrier* rather than an *energy source*.

Currently, most hydrogen is produced from natural gas through a *steam reforming* process. This reaction can be written as

$$CH_4 + H_2O \rightarrow CO + 3H_2 \tag{11-1}$$

The steam reforming reaction can be followed by a *water-gas shift reaction* to obtain more hydrogen:

$$CO + H_2O \rightarrow CO_2 + H_2 \tag{11-2}$$

The combination of the two reactions gives

$$CH_4 + 2H_2O \rightarrow CO_2 + 4H_2 \tag{11-3}$$

Therefore, for 1 kmol (about 16 kg) of methane used in this reaction, 4 kmol (about 8 kg) of hydrogen is produced. The steam reforming process is endothermic, requiring 206,000 kJ/kmol H_2 of energy input, while the water-gas shift reaction is exothermic, with an energy output of 41,000 kJ/kmol H_2. The net result is 165,000 kJ/kmol H_2 energy input. This is equivalent to 81,850 kJ (or 22.73 kWh) energy input per kg of hydrogen produced. Hydrogen is also produced from oil, coal, and biomass.

EXAMPLE 11-1
Hydrogen Production by Steam Reforming

Hydrogen is to be produced by a steam reforming and water-gas shift reaction through $CH_4 + 2H_2O \rightarrow CO_2 + 4H_2$. (a) How much hydrogen can be produced from 100 kg of natural gas (approximated as methane)? (b) How much energy is consumed during this process, in kWh?

SOLUTION   The molar masses of methane (CH_4) and hydrogen (H_2) are 16.043 and 2.016 kg/kmol, respectively (Table A-1). The reaction for this process is

$$CH_4 + 2H_2O \rightarrow CO_2 + 4H_2$$

According to this reaction, 4 kmol hydrogen is produced for 1 kmol of methane:

$$\frac{m_{H_2}}{m_{CH_4}} = \frac{(4 \text{ kmol})(2.016 \text{ kg/kmol})}{(1 \text{ kmol})(16.043 \text{ kg/kmol})} = 0.5026 \text{ kg } H_2/\text{kg } CH_4$$

Therefore, for 100 kg of natural gas, we obtain

$$m_{H_2,\text{total}} = m_{CH_4,\text{total}} \frac{m_{H_2}}{m_{CH_4}} = (100 \text{ kg } CH_4)(0.5026 \text{ kg } H_2/\text{kg } CH_4) = \textbf{50.26 kg } H_2$$

81,850 kJ energy input is required per kg of hydrogen produced in the steam reforming process (from the text). For 50.26 kg of hydrogen production, the energy required is

$$E = e \times m_{H_2,\text{total}} = (81,850 \text{ kJ/kg } H_2)(50.26 \text{ kg } H_2)\left(\frac{1 \text{ kWh}}{3600 \text{ kJ}}\right) = \textbf{1142 kWh}$$

That is, 1142 kWh of energy input is required to obtain 50.3 kg of hydrogen from 100 kg of natural gas.   ▲

The steam reforming is a cheaper method of producing hydrogen compared with water electrolysis. However, using a fossil fuel source (natural gas) for producing hydrogen is not a sustainable path. We are primarily interested in water electrolysis since it can use electricity generated from renewable energies. Nuclear electricity can also be used. The water electrolysis reaction is

$$H_2O \rightarrow H_2 + \tfrac{1}{2}O_2 \tag{11-4}$$

This reaction produces 1 kmol of hydrogen (about 2 kg) and 0.5 kmol of oxygen (about 16 kg) when 1 kmol (about 18 kg) of water is used. The minimum work required for this endothermic reaction is given by the Gibbs function of water at 25°C, which is equal to 237,180 kJ per kmol of liquid water entering the reaction (Table A-6). This is equivalent to 117,650 kJ (or 32.68 kWh) per kg of hydrogen produced by the reaction. Note that this is very close to the lower heating value of hydrogen (120,000 kJ/kg). An actual electrolyzer involves irreversibilities, and the actual electricity consumption will be greater than this minimum value. The efficiency of a typical electrolyzer is about 80 percent. Therefore, there is no thermodynamic advantage to producing hydrogen through water electrolysis by consuming electricity. The reason is that we cannot store renewable electricity effectively by other methods.

EXAMPLE 11-2
Hydrogen Production by
Water Electrolysis

Hydrogen is to be produced by a water electrolysis process through $H_2O \rightarrow H_2 + \frac{1}{2}O_2$. (*a*) How much hydrogen can be produced from 100 kg of water? (*b*) How much energy is consumed during this process, in kWh, if the process is ideal?

SOLUTION   The molar masses of water (H_2O) and hydrogen (H_2) are 18.015 and 2.016 kg/kmol, respectively. The reaction for this process is

$$H_2O \rightarrow H_2 + \frac{1}{2}O_2$$

According to this reaction, 1 kmol of hydrogen is produced for 1 kmol of water:

$$\frac{m_{H_2}}{m_{H_2O}} = \frac{(1 \text{ kmol})(2.016 \text{ kg/kmol})}{(1 \text{ kmol})(18.015 \text{ kg/kmol})} = 0.1119 \text{ kg } H_2/\text{kg } H_2O$$

Therefore, for 100 kg of water, we obtain

$$m_{H_2, \text{total}} = m_{H_2O, \text{total}} \frac{m_{H_2}}{m_{H_2O}} = (100 \text{ kg } H_2O)(0.1119 \text{ kg } H_2/\text{kg } H_2O) = \textbf{11.19 kg } \textbf{H}_2$$

117,650 kJ of energy input is required per kg of hydrogen produced in an ideal water electrolysis process (from the text). For 11.19 kg hydrogen production, the energy required is

$$E = e \times m_{H_2, \text{total}} = (117,650 \text{ kJ/kg } H_2)(11.19 \text{ kg } H_2)\left(\frac{1 \text{ kWh}}{3600 \text{ kJ}}\right) = \textbf{365.7 kWh}$$

That is, a minimum of 365.7 kWh of energy input is required to obtain 11.2 kg hydrogen from 100 kg of water. ▲

Once produced, hydrogen can be used as a fuel for an internal combustion engine. It can be burned just like gasoline, or a fuel cell can be used to convert the energy of hydrogen into electricity. One of the greatest advantages of hydrogen is that the exhaust of a hydrogen engine does not contain carbon monoxide, sulfur, hydrocarbon, or carbon dioxide emissions. The exhaust of a fuel cell using hydrogen is water. Hydrogen is the primary fuel to power fuel cells. Fuel cell–powered cars, bikes, stationary and portable generators, and electronic devices such as computers and cell phones can all use hydrogen as the fuel. A mobile phone powered by a fuel cell using hydrogen as the fuel can have several months of battery life compared to just several days with current batteries. Hydrogen can be used in a variety of applications, including electricity generation plants and various industrial, commercial, and residential uses (Fig. 11-1).

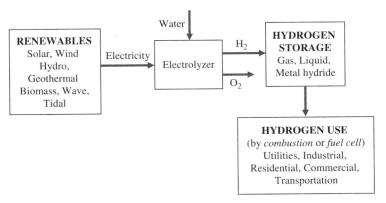

Figure 11-1   Production, storage, and utilization of hydrogen.

Hydrogen can be combusted with air with concentrations ranging from 4 to 75 percent. Due to this wide flammability range, hydrogen must be handled with care. In the fuel cell, the hydrogen flow to the stack can be shut off in case of a fire. The high diffusivity of hydrogen and fast venting of storage tanks can help reduce the potential for spreading flames (IAHE, 2014).

Production, storage, transportation, refueling, and use of hydrogen present a number of technical, economical, and practical problems, and much research is underway to tackle these problems. For example, the storage of hydrogen in liquid form requires very low temperatures (about 20 K or −253°C), which are expensive to achieve and maintain. Its storage in gas form at high pressure, on the other hand, involves low density, and a full tank will have a limited range. Extensive tests indicate that pressures as high as 70 MPa can be safely used for compressed hydrogen storage in commercial vehicles. Hydrogen can also be stored in a metal hydride solid. A hydride can absorb about 3.5 percent of its mass as hydrogen. Cooling a hydride allows absorption of hydrogen, while heating it slightly releases hydrogen as a gas. At present, hydrogen cars used in most engineering student competitions use this technique to store hydrogen.

An alternative to hydrogen-fueled fuel cell cars is electric cars powered by lithium-ion batteries. In recent years, electric automobiles from various companies were introduced to the market. Some major car companies are planning to introduce their fuel cell vehicles fueled by hydrogen to the market. A comparison of electric cars and fuel cell cars follows (IAHE, 2014):

Hydrogen has a lower heating value of 120,000 kJ/kg, while lithium metal has a specific energy of 41,000 kJ/kg (NIST, 2014). Lithium metal is unstable, and safe storage is needed to prevent runaway thermal events leading to combustion. For both hydrogen and lithium batteries, the storage is challenging and the containers are very heavy compared to the weight of hydrogen and lithium themselves. Considering the total weight of the battery, 1 kg of a current lithium-ion battery contains 840 kJ of electrical energy (Panasonic, 2011). It is estimated that a carbon fiber tank weighing 127.6 kg can store only 5.8 kg of compressed hydrogen at 70 MPa (Ahluwalia et al., 2013). This corresponds to a specific energy of 5455 kJ/kg. Considering an approximate 100 kg weight for a fuel cell stack in an automobile and an approximate operating voltage of 0.7 V (Mench, 2008), the specific energy for a fuel cell system becomes 1700 kJ/kg. This is only 1.42 percent of the heating value of hydrogen but about twice that of a lithium-ion battery. Inclusion of other necessary units in a fuel cell vehicle further reduces the available specific energy, making it comparable to that of a lithium-ion battery. The energy density of a commercial lithium-ion battery is currently about 2500 kJ/L. Note for comparison that the energy density of gasoline is about 32,000 kJ/L.

The above discussion only considers the storage of hydrogen as a compressed gas since this method is normally used for fuel cell automobiles. Storing hydrogen in liquid form at cryogenic temperatures and in metal hydrides is not as commercially ready as compressed hydrogen gas. Refueling periods for compressed hydrogen by means of refueling stations is much shorter than those for batteries. On the fuel cell side, there are significant challenges to clean methods of hydrogen production. Manufacturing of carbon fiber hydrogen tanks requires energy input. Also, possible operational problems of commercial fuel cell vehicles remain to be seen. On the other hand, mining and manufacturing processes for batteries are energy intensive, and some environmental concerns exist for their disposal (IAHE, 2014).

To power hydrogen-fueled buses, automobiles, bikes, and equipment, a number of hydrogen refueling stations are now available across China, Europe, Japan, and North America. As of 2022, the total number in the world reached 685. China has 250 public hydrogen refueling stations, Japan has 161, Germany has 92, and the United States has 54 (www.H2stations.org).

A possible long-term solution to hydrogen distribution is the formation of national hydrogen pipelines. This can only be justified economically with widespread use of fuel cell automobiles. For locations without any hydrogen fuel infrastructure or fueling stations, local electrolysis and reformation-based technologies can be used.

Battery, hybrid, and fuel cell technologies are all expected to play important roles in the future automobile market as evidenced by the fact that major automobile companies continue to invest in all these technologies. A widespread hydrogen distribution structure and large-scale renewable-based hydrogen production are essential if fuel cell automobiles will have a good share among other competing technologies.

EXAMPLE 11-3
Amount and Cost of Hydrogen by Electrolysis

Hydrogen is to be produced by the water electrolysis process using excess renewable electricity produced by solar cells. The unit cost of this renewable electricity is $0.09/kWh. If the efficiency of electrolysis is 80 percent, determine the amount of hydrogen produced for 1000 kWh of electricity consumed. What is the unit cost of producing hydrogen in $/kg H_2?

SOLUTION It takes a minimum of 117,650 kJ work to produce 1 kg of hydrogen (from the text). The efficiency of an electrolyzer can be defined as the minimum work consumption to produce 1 kg of hydrogen divided by the actual work consumption. Using this definition, the actual work is determined to be

$$\eta_{electrolyzer} = \frac{w_{min}}{w_{actual}} \longrightarrow w_{actual} = \frac{w_{min}}{\eta_{electrolyzer}} = \frac{117,650 \text{ kJ/kg}}{0.80} = 147,060 \text{ kJ/kg}$$

Then, the amount of hydrogen produced is determined as

$$m_{hydrogen} = \frac{W_{actual}}{w_{actual}} = \frac{1000 \text{ kWh}}{147,060 \text{ kJ/kg}}\left(\frac{3600 \text{ kJ}}{1 \text{ kWh}}\right) = \textbf{24.48 kg } \mathbf{H_2}$$

The cost of consumed electricity divided by the amount of hydrogen produced gives the unit cost of hydrogen:

$$\text{Cost}_{hydrogen} = \frac{\text{Cost of electricity}}{m_{hydrogen}} = \frac{(1000 \text{ kWh})(\$0.09/\text{kWh})}{24.48 \text{ kg } H_2} = \textbf{\$3.68/kg } \mathbf{H_2}$$

The cost of hydrogen is quite high by electrolysis. However, this is a sustainable operation since renewable electricity is used. ▲

EXAMPLE 11-4
Hydrogen Production by Electricity from Wind Turbine

Consider a wind turbine with a blade diameter of 250 ft. It is installed in a location where average wind velocity is 20 ft/s. The overall efficiency of the wind turbine is 35 percent. The produced electricity is to be used for producing hydrogen by water electrolysis with an efficiency of 85 percent. Determine the rate of hydrogen production in lbm/s. Take the density of air to be 0.078 lbm/ft³.

SOLUTION We assume that wind flows steadily at the specified speed. The blade span area is

$$A = \pi D^2/4 = \pi(250 \text{ ft})^2/4 = 49,087 \text{ ft}^2$$

The wind power potential is

$$\dot{W}_{\text{available}} = \frac{1}{2}\rho A V^3 = \frac{1}{2}(0.078 \text{ lbm/ft}^3)(49{,}087 \text{ ft}^2)(20 \text{ ft/s})^3 \left(\frac{1 \text{ kW}}{23{,}730 \text{ lbm·ft}^2/\text{s}^3}\right) = 645.4 \text{ kW}$$

The electric power generated is

$$\dot{W}_{\text{electric}} = \eta_{\text{wt,overall}} \dot{W}_{\text{available}} = (0.35)(645.4 \text{ kW}) = 225.9 \text{ kW}$$

This power input is used for the water electrolysis process. The minimum work consumption for producing hydrogen by electrolysis is 117,650 kJ/kg H_2 (from the text). The actual work consumption for electrolysis is

$$\eta_{\text{electrolyzer}} = \frac{w_{\text{min}}}{w_{\text{actual}}} \longrightarrow w_{\text{actual}} = \frac{w_{\text{min}}}{\eta_{\text{electrolyzer}}} = \frac{117{,}650 \text{ kJ/kg}}{0.85} = 138{,}412 \text{ kJ/kg}$$

The rate of hydrogen production is

$$\dot{m}_{\text{H2}} = \frac{\dot{W}_{\text{actual}}}{w_{\text{actual}}} = \frac{225.9 \text{ kJ/s}}{138{,}412 \text{ kJ/kg}}\left(\frac{2.2046 \text{ lbm}}{1 \text{ kg}}\right) = \mathbf{0.00359 \text{ lbm/s}}$$

That is, hydrogen is produced at a rate of 0.00359 lbm/s when using electricity from this wind turbine. ▲

11-2 FUEL CELLS

Fuels like natural gas (primarily methane), coal, and oil are commonly burned to provide thermal energy at high temperatures for use in heat engines. For example, when methane (CH_4) at 25°C is burned with 50 percent excess air at 25°C adiabatically, the products will be at 1789 K. A second-law analysis of this process reveals that the exergy of the reactants (818 MJ/kmol CH_4) decreases by 288 MJ/kmol as a result of the irreversible adiabatic combustion process alone. That is, the exergy of the hot combustion gases at the end of the adiabatic combustion process is 818 − 288 = 530 MJ/kmol CH_4. In other words, the work potential of the hot combustion gases is about 65 percent of the work potential of the reactants. It seems that when methane is burned, 35 percent of the work potential is lost before we even start using the thermal energy (Fig. 11-2).

Thus, the second law of thermodynamics suggests that there should be a better way of converting the chemical energy to work. The better way is, of course, the less irreversible way, the best being the reversible case. In chemical reactions, the irreversibility is due to uncontrolled electron exchange between the reacting components. The electron exchange can be controlled by replacing the combustion chamber with electrolytic cells, like car batteries. (This is analogous to replacing unrestrained expansion of a gas in mechanical systems with restrained expansion.) In electrolytic cells, the electrons are exchanged through conductor wires connected to a load, and the chemical energy is directly converted to electric energy. The energy conversion devices that work on this principle are called fuel cells. Fuel cells are not heat engines, and thus their efficiencies are not limited by the Carnot efficiency. They convert chemical energy to electric energy essentially in an isothermal manner.

A fuel cell functions like a battery, except that it produces its own electricity by combining a fuel with oxygen in a cell electrochemically without combustion, and discards the waste heat. A fuel cell consists of two electrodes separated by an electrolyte such as a solid oxide, phosphoric acid, or molten carbonate. The electric power generated by a single fuel cell is usually too small to be of any practical use. Therefore, fuel cells are usually stacked in practical applications. This modularity gives the fuel cells considerable flexibility in applications: The same design can be used to generate a small amount of power for a

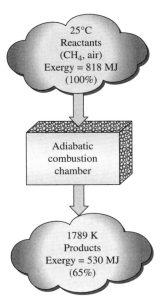

Figure 11-2 The exergy (work potential) of methane decreases by 35 percent as a result of an irreversible combustion process (*Çengel et al., 2019*).

remote switching station or a large amount of power to supply electricity to an entire town. Therefore, fuel cells are termed the "microchip of the energy industry."

The operation of a hydrogen-oxygen fuel cell is illustrated in Fig. 11-3. Hydrogen is ionized at the surface of the anode, and hydrogen ions flow through the electrolyte to the cathode. There is a potential difference between the anode and the cathode, and free electrons flow from the anode to the cathode through an external circuit (such as a motor or a generator). Hydrogen ions combine with oxygen and the free electrons at the surface of the cathode, forming water. Therefore, the fuel cell operates like an electrolysis system working in reverse. In steady operation, hydrogen and oxygen continuously enter the fuel cell as reactants, and water leaves as the product. Therefore, the exhaust of the fuel cell is drinkable quality water.

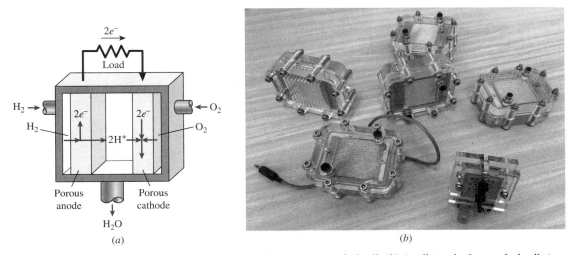

Figure 11-3 (*a*) The operation of a hydrogen-oxygen fuel cell. (*b*) Small-size hydrogen fuel cells in a university lab.

The fuel cell was invented by William Groves in 1839, but it did not receive serious attention until the 1960s, when they were used to produce electricity and water for the Gemini and Apollo spacecraft during their missions to the moon. Today they are used for the same purpose in the space shuttle missions. Despite the irreversible effects, such as internal resistance to electron flow, fuel cells have great potential for much higher conversion efficiencies. Currently fuel cells are available commercially, but they are competitive only in some niche markets because of their higher cost. Fuel cells produce high-quality electric power efficiently and quietly while generating low emissions using a variety of fuels such as hydrogen, natural gas, propane, and biogas.

A single fuel cell produces between 0.5 and 0.9 V of electricity. For usable voltage and reasonable power outputs, fuel cells are combined into stacks. Many fuel cells have been installed to generate electricity. For example, a remote police station in Central Park in New York City is powered by a 200-kW phosphoric acid fuel cell that has an efficiency of 40 percent with negligible emissions (it emits 1 ppm NO_x and 5 ppm CO).

Major car manufacturers have intense research and development programs for *fuel cell electric vehicles* (FCEVs), thus nearly doubling the efficiency from about 30 percent for gasoline engines to up to 60 percent for fuel cells. Fuel cell vehicles produce no harmful emissions because they only emit water vapor and warm air. These vehicles require a hydrogen infrastructure to fuel them. Hydrogen fueling stations can fuel in less than 4 minutes providing a driving range close to 500 km. Automobile manufacturers have already been offering a growing number of FCEVs in special markets, subject to the availability of hydrogen fueling stations.

FCEVs use the electrical power output from a fuel cell stack to power an electric motor (Fig. 11-4). The energy input to the fuel cell is supplied by hydrogen fuel. A battery that captures braking energy is used to provide additional power during acceleration. The battery also provides power for low power needs during which the fuel cell could be turned off. The power rating of the vehicle is the sum of the electric power from the fuel cell and the battery (DOE, 2022).

Hybrid power systems (HPS) that combine high-temperature fuel cells and gas turbines have the potential for very high efficiency in converting natural gas (or even coal) to electricity. The primary objective of a fuel cell is to generate electricity, but heat is also

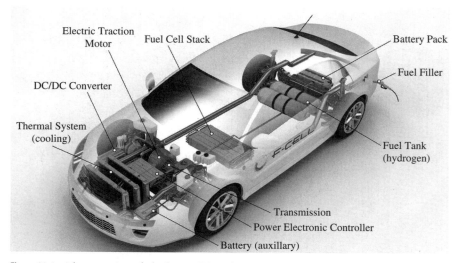

Figure 11-4   The operation of a hydrogen FCEV. (*DOE, 2022.*)

produced. This heat can be used for space and water heating for buildings. The efficiency of a combined heat and power plant that operates on a fuel cell can reach a utilization efficiency of 90 percent.

A technology of fuel cells involves dual operation of the system. The fuel cell is normally used to produce electricity. However, the operation can be reversed so that the fuel cell produces hydrogen (and oxygen) by consuming electricity. This electricity should come from excess electricity produced from renewable energy sources such as solar or wind systems. The hydrogen produced will be used when electricity demand is high. The fuel cells that operate on this principle are called *regenerative fuel cells* or *reversible fuel cells*.

Fuel cells are usually classified according to the electrolyte used. The main types are briefly described based on Hodge (2010) and the Department of Energy, Energy Efficiency and Renewable Energy website (DOE/EERE, 2018).

Proton exchange membrane fuel cell (PEMFC): It is also called a polymer electrolyte membrane fuel cell. This fuel cell contains a thin plastic polymer membrane, which is coated by platinum particles acting as the electrolyte. Their operating temperatures are low (close to 100°C). An important advantage of PEMFC is its small size for a given power output. The power output can be varied easily depending on the demand, which makes these fuel cells the best candidates for automobiles. Stationary power generation is another good application for PEMFC.

Direct-methanol fuel cell (DMFC): This fuel cell uses a polymer membrane as the electrolyte. It uses methanol directly in the cell. These fuel cells can be used for powering laptop computers and battery rechargers such as in cell phones.

Alkaline fuel cell (AFC): This type of fuel cell uses potassium hydroxide or an alkaline membrane as the electrolyte. They operate between 20 and 250°C with efficiencies of up to 60 percent at low temperatures. Alkaline fuel cells require pure hydrogen as the fuel, which is expensive. As a result, their application is mostly limited to spacecraft to produce electricity and water. Portable power production is a new application for alkaline fuel cells.

Phosphoric acid fuel cell (PAFC): Phosphoric acid is used as the electrolyte while platinum is used as the catalyst, which is expensive. Porous carbon electrodes are used in both cathode and anode. The operating temperatures are about 200°C. Their current thermal efficiencies are between 35 and 45 percent. PAFCs are mostly used in stationary power generation, with power outputs of greater than 400 kW. Ideal applications include hotels, stores, hospitals, and office buildings for which rejected heat from the fuel cell can also be used for heating purposes.

Molten carbonate fuel cell (MCFC): The advantage of MCFC is that it can use various fuels such as hydrogen, natural gas, propane, diesel, and gasified coal. When natural gas or biogas is used, the conversion from natural gas or biogas to hydrogen is accomplished within the cell. Molten carbonate salt mixtures (lithium and potassium salts) immobilized in a porous matrix are used as the electrolyte. It can reach up to 60 percent efficiency with operating temperatures of about 600°C. MCFCs are currently used in medium- and large-scale stationary applications at high efficiency.

Solid oxide fuel cell (SOFC): Phosphoric acid and molten carbonate are highly corrosive, which causes material problems and reduces cell life. A thin layer of solid ceramic material is used as the electrolyte in an SOFC, allowing reliable design and high temperatures. They can use carbon monoxide, hydrogen, and gasified coal as the fuel. Natural gas can also be reformed into hydrogen within the cell. An SOFC operates at temperatures up to 1000°C with efficiencies up to 60 percent. The fuel cell can be combined with a gas turbine to provide efficiencies up to 75 percent. Solid oxide fuel cells are good candidates for stationary power production and auxiliary power devices such as heavy-duty trucks.

Thermodynamic Analysis of Fuel Cells

A hydrogen-oxygen fuel cell reaction can be written as

$$H_2 + \tfrac{1}{2}O_2 \rightarrow 2e^- + 2H^+ + O \rightarrow H_2O \tag{11-5}$$

with the half-cell reactions of

$$\text{Anode:} \qquad H_2 \rightarrow 2e^- + 2H^+$$

$$\text{Cathode:} \qquad 2H^+ + 2e^- + \tfrac{1}{2}O_2 \rightarrow H_2O$$

The energy or heat released during this reaction is simply the difference between the enthalpy of the reactants entering and the enthalpy of the products leaving the reaction,

$$\Delta \bar{H} = \sum H_r - \sum H_p = \sum N_r (\bar{h}_f^o + \bar{h} - \bar{h}^o)_r - \sum N_p (\bar{h}_f^o + \bar{h} - \bar{h}^o)_p \tag{11-6}$$

where $\bar{h}_f^o$ is the enthalpy of formation and $\bar{h} - \bar{h}^o$ is the enthalpy difference between the given state of the reactants or products and the standard environmental state (1 atm, 25°C). In a fuel cell reaction, ideally both the reactants and the products are at environment temperature. Then, the $\bar{h} - \bar{h}^o$ term becomes zero, and the reaction reduces to

$$\Delta \bar{H} = \sum N_r (\bar{h}_f^o)_r - \sum N_p (\bar{h}_f^o)_p \tag{11-7}$$

Using the enthalpy of formation values from Table A-6, we obtain

$$\Delta \bar{H} = N_{H_2} (\bar{h}_f^o)_{H_2} - N_{O_2} (\bar{h}_f^o)_{O_2} - N_{H_2O} (\bar{h}_f^o)_{H_2O}$$

$$= (1)(0) - (0.5)(0) - (1 \text{ kmol})(-241{,}820 \text{ kJ/kmol})$$

$$= 241{,}820 \text{ kJ/kmol H}_2$$

Note that we used the enthalpy of formation $\bar{h}_f^o$ value from Table A-6 for water vapor. If we use the $\bar{h}_f^o$ value for liquid water ($-285{,}830$ kJ/kmol), we obtain 285,830 kJ/kmol H_2. Dividing the energy output by the molar mass of hydrogen (2.016 kg/kmol), we obtain the lower heating value (LHV$_{H_2}$ = 120,000 kJ/kg) and the higher heating value (HHV$_{H_2}$ = 141,800 kJ/kg) for hydrogen. In other words, the energy output per unit mass of the fuel during a fuel cell reaction is equal to the heating value of the fuel.

The maximum work that can be done during a chemical reaction is given by

$$W_{max} = \sum N_r (\bar{h}_f^o + \bar{h} - \bar{h}^o - T_0 \bar{s})_r - \sum N_p (\bar{h}_f^o + \bar{h} - \bar{h}^o - T_0 \bar{s})_p \tag{11-8}$$

When both the reactants and the products are at the temperature of the surroundings T_0, the relation becomes

$$W_{max} = \Delta \bar{G} = \sum N_r \bar{g}_{0,r} - \sum N_p \bar{g}_{0,p} \tag{11-9}$$

since the Gibbs function at T_0 is defined as $\bar{g}_0 = (\bar{h} - T_0\bar{s})_{T_0}$. This relation can also be written in terms of Gibbs function of formation $\bar{g}_f^o$ when the reactants, products, and surroundings are at 25°C. That is,

$$W_{max} = \Delta\bar{G} = \sum N_r \bar{g}_{f_r}^o - \sum N_p \bar{g}_{f_p}^o \qquad (11\text{-}10)$$

This equation gives the maximum work output during a chemical reaction. For the hydrogen-oxygen fuel cell reaction described in Eq. (11-5), the application of this equation gives

$$W_{max} = \Delta\bar{G} = N_{H_2}(\bar{g}_f^o)_{H_2} - N_{O_2}(\bar{g}_f^o)_{O_2} - N_{H_2O}(\bar{g}_f^o)_{H_2O}$$

$$= (1)(0) - (0.5)(0) - (1\ \text{kmol})(-228{,}590\ \text{kJ/kmol})$$

$$= 228{,}590\ \text{kJ/kmol}\ H_2$$

We used the Gibbs function of formation value $\bar{g}_f^o$ for water vapor from Table A-6. Using the $\bar{g}_f^o$ value for liquid water, we obtain 237,180 kJ/kmol H_2. Therefore, the maximum power that can be generated by a fuel cell at the environmental temperature and pressure is equal to the Gibbs function of formation for water. Dividing this result per kmol by the molar mass of hydrogen (2.016 kg/kmol), the maximum work becomes

Water is vapor in the products: $W_{max} = \Delta\bar{G} = 228{,}590\ \text{kJ/kmol}\ H_2$
$$\text{or } 113{,}390\ \text{kJ/kg}\ H_2$$

Water is liquid in the products: $W_{max} = \Delta\bar{G} = 237{,}180\ \text{kJ/kmol}\ H_2$
$$\text{or } 117{,}650\ \text{kJ/kg}\ H_2$$

The enthalpy difference, heating value, and Gibbs function difference values for the hydrogen-oxygen fuel cell reaction at 25°C and 1 atm are given in Table 11-1 for water as vapor and liquid in the products. Since the terms $\Delta\bar{H}$ and $\Delta\bar{G}$ used in this section are per kmol of fuel, these terms can be replaced by $\Delta\bar{h}$ and $\Delta\bar{g}$, respectively.

A fuel cell converts the chemical energy of hydrogen into electricity. Since a fuel cell is not limited by the Carnot efficiency, the values in Table 11-1 represent the upper limits for electricity generated by a hydrogen fuel cell. An actual fuel cell produces less electricity than the maximum, and thus the *second-law efficiency* of a fuel cell can be expressed as

$$\eta_{\text{fuel cell, II}} = \frac{w_{\text{actual}}}{w_{\text{max}}} = \frac{w_{\text{actual}}}{\Delta\bar{g}} \qquad (11\text{-}11)$$

TABLE 11-1   **Enthalpy Difference, Heating Value, and Gibbs Function Difference for Hydrogen-Oxygen Fuel Cell Reaction at 25°C and 1 atm**

The values are given for water vapor and liquid in the products per unit kmol of hydrogen and per unit mass of hydrogen. The reaction is $H_2 + \frac{1}{2}O_2 \rightarrow H_2O$.

	Water is vapor	Water is liquid
$\Delta\bar{h}$	241,820 kJ/kmol	285,830 kJ/kmol
Δh or HV	120,000 kJ/kg (LHV)	141,800 kJ/kg (HHV)
$\Delta\bar{g}$ or w_{max}	228,590 kJ/kmol	237,180 kJ/kmol
Δg or w_{max}	113,390 kJ/kg	117,650 kJ/kg

Similarly, a *first-law* or *thermal efficiency* for a fuel cell can be expressed as the ratio of actual work output to the enthalpy difference between the reactants and products. That is,

$$\eta_{\text{fuel cell, I}} = \frac{w_{\text{actual}}}{\Delta\overline{h}} \tag{11-12}$$

Here, the denominator represents the heating value of the fuel if the fuel cell reaction occurs isothermally at 25°C. The maximum first-law efficiency of a hydrogen-oxygen fuel cell can be expressed as

$$\eta_{\text{fuel cell, I, max}} = \frac{w_{\text{max}}}{\Delta\overline{h}} = \frac{\Delta\overline{g}}{\Delta\overline{h}} = \frac{\Delta g}{\Delta h} \tag{11-13}$$

Note that the parameter with an overbar is the quantity per unit mole and the parameter without the overbar is the quantity per unit mass. For example, the unit for $\Delta\overline{g}$ is kJ/kmol and that for Δg is kJ/kg. The maximum first-law efficiency of a hydrogen-oxygen fuel cell at 25°C is determined to be

$$\eta_{\text{fuel cell, I, max}} = \frac{w_{\text{max}}}{\Delta\overline{h}} = \frac{\Delta\overline{g}}{\Delta\overline{h}} = \frac{\Delta g}{\text{LHV}} = \frac{113{,}390 \text{ kJ/kg}}{120{,}000 \text{ kJ/kg}} = 0.945 \text{ or } 94.5\%$$

Using the higher heating value and the Gibbs function of formation for liquid water, the maximum first-law efficiency becomes

$$\eta_{\text{fuel cell, I, max}} = \frac{w_{\text{max}}}{\Delta\overline{h}} = \frac{\Delta\overline{g}}{\Delta\overline{h}} = \frac{\Delta g}{\text{HHV}} = \frac{117{,}650 \text{ kJ/kg}}{141{,}800 \text{ kJ/kg}} = 0.830 \text{ or } 83.0\%$$

It should be noted that the fuel cell has a higher maximum first-law efficiency (94.5%) based on the lower heating value than that based on the higher heating value (83.0%). The high-efficiency potential is the main reason why fuel cells are very popular.

The ideal first-law efficiency of a fuel cell can also be expressed as

$$\eta_{\text{fuel cell, I, max}} = \frac{w_{\text{max}}}{\Delta\overline{h}} = \frac{\Delta\overline{g}}{\Delta\overline{h}} = 1 - \frac{T\Delta\overline{s}}{\Delta\overline{h}} = 1 - \frac{T\Delta s}{\Delta h} = 1 - \frac{q_{\text{rev}}}{\Delta h} \tag{11-14}$$

Here, q_{rev} represents the heat rejected during the isothermal (constant temperature) operation of an ideal fuel cell. This rejected heat can potentially be used for other purposes such as heating.

EXAMPLE 11-5
First-Law and Second-Law
Efficiencies of a Fuel Cell

A hydrogen-oxygen fuel cell stack consumes hydrogen at a rate of 0.005 kg/s while generating electricity at a rate of 475 kW. (*a*) Determine the rate of water produced. (*b*) Determine the first-law efficiency and second-law efficiency of this fuel cell if the water in the products is vapor.

SOLUTION   (*a*) The molar masses of water (H_2O) and hydrogen (H_2) are 18.015 and 2.016 kg/kmol, respectively. The hydrogen-oxygen fuel cell reaction is

$$H_2 + \tfrac{1}{2}O_2 \rightarrow H_2O$$

According to this reaction, 1 kmol of water is produced for 1 kmol of hydrogen:

$$\frac{m_{H_2O}}{m_{H_2}} = \frac{(1 \text{ kmol})(18.015 \text{ kg/kmol})}{(1 \text{ kmol})(2.016 \text{ kg/kmol})} = 8.936 \text{ kg } H_2O/\text{kg } H_2$$

Therefore, the rate of water produced is

$$\dot{m}_{H_2O} = \dot{m}_{H_2} \frac{m_{H_2O}}{m_{H_2}} = (0.005 \text{ kg/s } H_2)(8.936 \text{ kg } H_2O/\text{kg } H_2) = \mathbf{0.04468 \text{ kg/s } H_2O}$$

(b) Using the heating value and Gibbs function difference in Table 11-1, the first-law efficiency and the second-law efficiency are determined as

$$\eta_{\text{fuel cell, I}} = \frac{\dot{W}_{\text{actual}}}{\dot{m}_{H_2} \times \text{LHV}} = \frac{475 \text{ kJ/s}}{(0.005 \text{ kg/s})(120,000 \text{ kJ/kg})} = 0.792 = \mathbf{79.2\%}$$

$$\eta_{\text{fuel cell, II}} = \frac{\dot{W}_{\text{actual}}}{\dot{m}_{H_2} \times \Delta g} = \frac{475 \text{ kJ/s}}{(0.005 \text{ kg/s})(113,390 \text{ kJ/kg})} = 0.838 = \mathbf{83.8\%}$$

That is, this fuel cell stack converts 79.2 percent of its heating value and 83.8 percent of its power potential to electricity. ▲

In thermodynamics, the second $T\,d\overline{s}$ relation per kmol basis is expressed as (overbar represents the property per unit mole)

$$T\,d\overline{s} = d\overline{h} - \overline{v}\,dP \tag{11-15}$$

Using the definition of the Gibbs function $\overline{g} = \overline{h} - T\overline{s}$, and differentiating both sides of the equation, and noting that the fuel cell operates at constant temperature,

$$d\overline{g} = d\overline{h} - T\,d\overline{s} - \overline{s}\,dT = d\overline{h} - T\,d\overline{s} \tag{11-16}$$

Substituting into Eq. (11-15), we obtain

$$d\overline{g} = \overline{v}\,dP \tag{11-17}$$

Using the ideal gas equation state for each component in the fuel cell reaction $\overline{v} = R_u T/P$, where R_u is the universal gas constant, and substituting into Eq. (11-17),

$$d\overline{g} = R_u T \frac{dP}{P} \tag{11-18}$$

Integrating this equation between a reference state (subscript 0) at P_0 and a general state P gives

$$\overline{g} = \overline{g}_0 + R_u T \ln P \tag{11-19}$$

For a general fuel cell reaction,

$$a\,A + b\,B \rightarrow c\,C + d\,D \tag{11-20}$$

Using partial pressures of gases (assumed to be ideal gases) in the reaction,

$$\Delta\overline{g} = \Delta\overline{g}_0 + R_u T \ln\left(\frac{P_A^a P_B^b}{P_C^c P_D^d}\right) \tag{11-21}$$

Here, $\Delta\overline{g}$ can be expressed as

$$\Delta\overline{g} = N_e F_c V \tag{11-22}$$

where N_e is the number of moles of electrons per kmol of reacting fuel, F_c is Faraday's constant (9.6487×10^7 C/kmol electron or 96,487 kJ/V·kmol electron), and V is the *cell voltage*, also called *open-circuit voltage*. Note that 1 kmol of electrons contains 6.022×10^{26} electrons, each with a charge of 1.602×10^{-19} C. Faraday's constant is the product of these two values. Solving Eq. (11-22) for the voltage,

$$V = \frac{\Delta \overline{g}}{N_e F_c} \qquad (11\text{-}23)$$

Substituting Eq. (11-22) into Eq. (11-21) and solving for the open-circuit voltage, we obtain

$$V = V_0 - \frac{R_u T}{N F_c} \ln\left(\frac{P_C^c P_D^d}{P_A^a P_B^b}\right) \qquad (11\text{-}24)$$

This equation is known as the *Nernst equation*. For a hydrogen-oxygen fuel cell reaction at 25°C and 1 atm, the ideal voltage is determined to be

$$V_0 = \frac{\Delta \overline{g}}{N_e F_c} = \frac{237{,}180 \text{ kJ/kmol}}{(2 \text{ kmol electron/kmol})(96{,}487 \text{ kJ/V·kmol electron})} = 1.229 \text{ V}$$

We used the Gibbs function value with water in the liquid phase in the products. The cell voltage with water in the vapor phase can be calculated to be

$$V_0 = \frac{\Delta \overline{g}}{N_e F_c} = \frac{228{,}590 \text{ kJ/kmol}}{(2 \text{ kmol electron/kmol})(96{,}487 \text{ kJ/V·kmol electron})} = 1.185 \text{ V}$$

Note that $N_e = 2$ kmol of electrons are produced per kmol of fuel, as indicated in the fuel cell reaction [Eq. (11-5)].

If air at 1 atm is used to provide oxygen to a fuel cell, the partial pressure of oxygen is 0.21 atm since the mole fraction of oxygen in air is 21 percent. Then the cell voltage is calculated from Eq. (11-22) to be

$$V = V_0 - \frac{R_u T}{N F_c} \ln\left(\frac{P_C^c P_D^d}{P_A^a P_B^b}\right)$$

$$= 1.185 \text{ V} - \frac{(8.314 \text{ kJ/kmol·K})(298 \text{ K})}{(2 \text{ kmol electron/kmol})(96{,}487 \text{ kJ/V·kmol electron})}$$

$$\ln\left[\frac{(1 \text{ atm})^1}{(1 \text{ atm})^1 (0.21 \text{ atm})^{0.5}}\right]$$

$$= 1.175 \text{ V}$$

The maximum work output for a hydrogen-air fuel cell can be determined from Eq. (11-21) to be

$$\Delta \overline{g} = \Delta \overline{g}_0 + R_u T \ln\left(\frac{P_A^a P_B^b}{P_C^c P_D^d}\right)$$

$$= 228{,}590 \text{ kJ/kmol} + (8.314 \text{ kJ/kmol·K})(298 \text{ K}) \ln\left[\frac{(1 \text{ atm})^1 (0.21 \text{ atm})^{0.5}}{(1 \text{ atm})^1}\right]$$

$$= 226{,}660 \text{ kJ/kmol}$$

TABLE 11-2   Maximum Work and Maximum Thermal Efficiency Values for Hydrogen-Oxygen and Hydrogen-Air Fuel Cell Reactions at 25°C and 1 atm

	Hydrogen-oxygen fuel cell		Hydrogen-air fuel cell	
	Water is vapor	Water is liquid	Water is vapor	Water is liquid
$\Delta\bar{g}$ or w_{max}	228,590 kJ/kmol	237,180 kJ/kmol	226,660 kJ/kmol	235,250 kJ/kmol
Δg or w_{max}	113,390 kJ/kg	117,650 kJ/kg	112,430 kJ/kg	116,690 kJ/kg
$\eta_{fuel\ cell,\ I,\ max}$	94.5%	83.0%	93.7%	82.3%

Water is assumed to be vapor in the products. If water is liquid, the maximum work output is

$$\Delta\bar{g} = \Delta\bar{g}_0 + R_u T \ln\left(\frac{P_A^a P_B^b}{P_C^c P_D^d}\right)$$

$$= 237,180 \text{ kJ/kmol} + (8.314 \text{ kJ/kmol·K})(298 \text{ K}) \ln\left[\frac{(1 \text{ atm})^1 (0.21 \text{ atm})^{0.5}}{(1 \text{ atm})^1}\right]$$

$$= 235,250 \text{ kJ/kmol}$$

For a hydrogen-air fuel cell, the maximum first-law efficiency based on the lower and higher heating values becomes, respectively,

$$\eta_{fuel\ cell,\ I,\ max} = \frac{w_{max}}{\Delta\bar{h}} = \frac{\Delta\bar{g}}{\Delta\bar{h}} = \frac{\Delta\bar{g}/M}{LHV} = \frac{(226,660 \text{ kJ/kmol})/(2.016 \text{ kg/kmol})}{120,000 \text{ kJ/kg}} = 0.937 \text{ or } 93.7\%$$

$$\eta_{fuel\ cell,\ I,\ max} = \frac{w_{max}}{\Delta\bar{h}} = \frac{\Delta\bar{g}}{\Delta\bar{h}} = \frac{\Delta\bar{g}/M}{HHV} = \frac{(235,250 \text{ kJ/kmol})/(2.016 \text{ kg/kmol})}{141,800 \text{ kJ/kg}} = 0.823 \text{ or } 82.3\%$$

The analysis above shows that the maximum work and maximum first-law efficiency values are within 1 percent of each other for hydrogen-oxygen and hydrogen-air fuel cell reactions (Table 11-2).

EXAMPLE 11-6
A Hydrogen-Oxygen Fuel Cell

Consider a hydrogen-oxygen fuel cell whose overall reaction is

$$H_2 + \tfrac{1}{2}O_2 \rightarrow 2e^- + 2H^+ + O \rightarrow H_2O\ (g)$$

The fuel cell operates at 400 K. Determine the maximum work per unit kmol of fuel, the maximum first-law efficiency, the heat transfer, and the ideal open-circuit voltage. Use the enthalpy and entropy values from the following table.

Substance	$\bar{h}_{400\ K} = \bar{h}_f^o + \bar{h} - \bar{h}^o$ kJ/kmol	$\bar{s}_{400\ K}$ kJ/kmol·K
H_2	2958	139.10
O_2	3027	213.76
$H_2O\ (g)$	−238,358	198.67

SOLUTION The maximum work output is equal to the Gibbs function difference and can be determined from Eq. (11-8) to be

$$
\begin{aligned}
w_{max} = \Delta\bar{g} &= \sum N_r(\bar{h}_f^o + \bar{h} - \bar{h}^o - T_0\bar{s})_r - \sum N_p(\bar{h}_f^o + \bar{h} - \bar{h}^o - T_0\bar{s})_p \\
&= \sum N_r(\bar{h} - T\bar{s})_r - \sum N_p(\bar{h} - T\bar{s})_p \\
&= N_{H_2}(\bar{h} - T\bar{s})_{H_2} + N_{O_2}(\bar{h} - T\bar{s})_{O_2} - N_{H_2O}(\bar{h} - T\bar{s})_{H_2O} \\
&= (1)[2958 - (400)(139.10)] + (0.5)[3027 - (400)(213.76)] - (1)[-238,358 - (400)(198.67)] \\
&= \mathbf{223{,}910\ kJ/kmol\ H_2}
\end{aligned}
$$

The energy input to the system is expressed as

$$
\begin{aligned}
\Delta\bar{h} &= \sum N_r(\bar{h}_f^o + \bar{h} - \bar{h}^o)_r - \sum N_p(\bar{h}_f^o + \bar{h} - \bar{h}^o)_p \\
&= \sum (N\bar{h})_r - \sum (N\bar{h})_p \\
&= (N\bar{h})_{H_2} + (N\bar{h})_{O_2} - (N\bar{h})_{H_2O} \\
&= (1)(2958) + (0.5)(3027) - (1)(-238,358) \\
&= 242{,}830\ kJ/kmol\ H_2
\end{aligned}
$$

The maximum first-law efficiency is determined from Eq. (11-13) to be

$$
\eta_{fuel\ cell,\ I,\ max} = \frac{w_{max}}{\Delta\bar{h}} = \frac{\Delta\bar{g}}{\Delta\bar{h}} = \frac{223{,}910\ kJ/kmol}{242{,}830\ kJ/kmol} = 0.922
$$

Therefore, the upper limit for the first-law efficiency of this fuel cell is 92.2 percent. The heat transfer for this ideal operation is

$$
q_{rev} = \Delta\bar{h} - \Delta\bar{g} = 242{,}830 - 223{,}910 = \mathbf{19{,}000\ kJ/kmol\ H_2}
$$

In this reaction, $N_e = 2$ since 2 kmol of electrons are released per kmol of fuel. Then, the ideal open-circuit voltage is determined from Eq. (11-23) to be

$$
V = \frac{\Delta\bar{g}}{N_e F_c} = \frac{223{,}910\ kJ/kmol}{(2\ kmol\ electron/kmol)(96{,}487\ kJ/V\cdot kmol\ electron)} = \mathbf{1.160\ V} \quad \blacktriangle
$$

The fuel cell is essentially an isothermal device, and it operates ideally at the environmental temperature. If we repeat the calculations in Example 11-6 at different temperatures, we obtain Fig. 11-5. It is clear that the efficiency of a fuel cell decreases as the cell operating temperature increases. This change is significant for a hydrogen-oxygen fuel cell, as shown in Fig. 11-5, but small for methane-oxygen and carbon-oxygen fuel cells (Wark, 1995). The ideal cell voltage also decreases with increasing temperature.

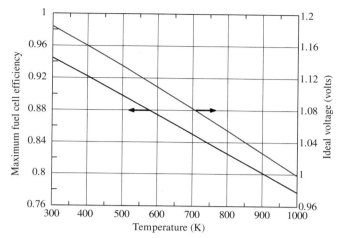

Figure 11-5 Effect of temperature on the maximum fuel cell first-law efficiency and ideal open-circuit voltage for a hydrogen-oxygen fuel cell.

EXAMPLE 11-7
Second-Law Efficiency of a Hydrogen-Oxygen Fuel Cell

Reconsider Example 11-6. Determine the second-law efficiency if this fuel cell produces 2.7 kW of electricity by consuming 2.2 g of hydrogen per minute.

SOLUTION   In Example 11-6, the maximum work output or the Gibbs function difference was determined to be 223,910 kJ/kmol. The maximum power output for a hydrogen consumption of 2.2 g/min is

$$\dot{W}_{max} = \dot{m}_{H_2} \frac{\Delta \overline{g}}{M_{H_2}} = (0.0022/60 \text{ kg/s})\frac{223{,}910 \text{ kJ/kmol}}{2.016 \text{ kg/kmol}} = 4.07 \text{ kW}$$

The second-law efficiency is determined from its definition in Eq. (11-11) as

$$\eta_{\text{fuel cell, II}} = \frac{\dot{W}_{\text{actual}}}{\dot{W}_{max}} = \frac{2.7 \text{ kW}}{4.07 \text{ kW}} = \textbf{0.663}$$

That is, this fuel cell produces 66.3 percent of its power-generating potential.   ▲

REFERENCES

Ahluwalia RK, Hua TQ, Peng J-K, and Roh HS. 2013. *System Level Analysis of Hydrogen Storage Options. Project ID: ST001.* DOE Hydrogen and Fuel Cells Program Review Arlington, VA.

Çengel, YA, Boles MA, and Kanoğlu M. 2019. *Thermodynamics: An Engineering Approach,* 9th ed. New York: McGraw Hill.

DOE/EERE. 2018. Department of Energy, Energy Efficiency and Renewable Energy. www.eere.energy.gov.

DOE (U.S. Department of Energy). 2022. How Do Fuel Cell Electric Vehicles Work Using Hydrogen? https://afdc .energy.gov/vehicles/how-do-fuel-cell-electric-cars-work

Hodge BK. 2010. *Alternative Energy Systems and Applications.* New York: Wiley.

IAHE. 2014. "Is Elon Musk Correct?" International Association for Hydrogen Energy. Electronic Newsletter, Vol. 6, Issue 1.

Mench MM. 2008. *Fuel Cell Engines.* Hoboken, NJ: John Wiley & Sons, Inc.

NIST Chemistry WebBook. 2014. http://webbook.nist.gov/chemistry (accessed Feb 26, 2014).

Panasonic. 2011. "Energy Catalog: A Comprehensive Guide by Product Grouping and Part Number for Panasonic OEM Batteries." p. 1.

Wark K. 1995. *Advanced Thermodynamics for Engineers.* New York: McGraw Hill.

PROBLEMS

HYDROGEN: AN ENERGY CARRIER

11-1 Why are we interested in hydrogen energy?

11-2 Is hydrogen an energy source? Explain.

11-3 Describe steam reforming and water-gas shift reactions.

11-4 Why are we interested in hydrogen production by water electrolysis instead of steam reforming from natural gas? Explain.

11-5 What are the applications of hydrogen?

11-6 What methods are used to store hydrogen? Briefly describe them.

11-7 Hydrogen is to be produced by a steam reforming process through the reaction $CH_4 + H_2O \rightarrow CO + 3H_2$. (a) How much hydrogen can be produced from 100 kg of natural gas (approximated as methane)? (b) How much energy is consumed during this process, in MJ?

11-8 Hydrogen is to be produced by a steam reforming and water-gas shift reaction through $CH_4 + 2H_2O \rightarrow CO_2 + 4H_2$. A total of 10,000 MJ of energy is consumed in the process. Determine the amounts of natural gas consumption and hydrogen production.

11-9 Hydrogen is to be produced by a water electrolysis process through $H_2O \rightarrow H_2 + \frac{1}{2}O_2$. (a) How much hydrogen can be produced from 100 lbm of water? (b) How much energy is consumed during this process, in kWh, if the process is ideal?

11-10 Hydrogen is to be produced by a water electrolysis process using electricity produced by solar cells. If the efficiency of electrolysis process is 70 percent, determine the amount of electricity consumed (in kWh) for 100 kg of hydrogen production.

11-11 Hydrogen is to be produced by the water electrolysis process using excess renewable electricity produced by solar cells. The unit cost of this renewable electricity is $0.12/kWh. If the electrolysis process is ideal, determine the amount of hydrogen produced for 1000 kWh of electricity consumed. What is the unit cost of producing hydrogen, in $/kg H_2?

11-12 Hydrogen is to be produced by a water electrolysis process using geothermal electricity. Consider a resource at 150°C supplying geothermal liquid water at a rate of 100 kg/s. Determine the rate of hydrogen production if the geothermal power plant and the electrolyzer operate ideally (no irreversibilities). Take the dead state to be 25°C and 1 atm.

11-13 Consider a wind turbine with a blade diameter of 25 m. It is installed in a location where average wind velocity is 6 m/s. The overall efficiency of the wind turbine is 34 percent. The produced electricity is to be used for producing hydrogen by water electrolysis with an efficiency of 75 percent. Determine the rate of hydrogen production in g/s. Take the density of air to be 1.3 kg/m³.

11-14 Choose the wrong statement.
(a) Hydrogen is a fuel.   (b) Hydrogen is an energy source.
(c) Hydrogen is an energy carrier.   (d) Hydrogen can be obtained from natural gas.
(e) Hydrogen can be produced from water.

11-15 Which is a hydrogen storage method?
 I. Liquid
 II. Gas
III. Metal hydride
(a) Only I   (b) Only II   (c) I and II   (d) II and III   (e) I, II, and III

11-16 At what temperature is hydrogen normally stored as a liquid?
(a) 0 K   (b) 20 K   (c) 50 K   (d) 100 K   (e) 298 K

11-17 A metal hydride can absorb about _____ of its mass as hydrogen.
(a) 20%   (b) 14%   (c) 10%   (d) 7.5%   (e) 3.5%

11-18   Hydrogen is to be produced by a steam reforming process through the reaction $CH_4 + H_2O \rightarrow CO + 3H_2$. How much hydrogen can be produced from 10 kg of natural gas (approximated as methane)?
(a) 1 kg   (b) 8 kg   (c) 3.75 kg   (d) 10 kg   (e) 6 kg

11-19   Hydrogen is to be produced by a water electrolysis process through $H_2O \rightarrow H_2 + \frac{1}{2}O_2$. How much hydrogen can be produced from 10 kg of water?
(a) 0.05 kg   (b) 1.1 kg   (c) 2 kg   (d) 0.18 kg   (e) 0.5 kg

11-20   Hydrogen is to be produced by a water electrolysis process using electricity produced by solar cells. If the electrolysis process is ideal, determine the amount of electricity consumed for 10 kg of hydrogen production. The minimum work consumption for producing hydrogen is 117,650 kJ/kg H_2.
(a) 224 kWh   (b) 485 kWh   (c) 78 kWh   (d) 327 kWh   (e) 119 kWh

FUEL CELLS

11-21   Why is a fuel cell more efficient than a conventional heat engine?

11-22   Is the fuel cell a heat engine? Are their efficiencies limited by Carnot efficiency? Explain.

11-23   Describe the operation of a fuel cell.

11-24   Describe the operation of a regenerative fuel cell.

11-25   Describe the characteristics and applications of alkaline fuel cells.

11-26   A hydrogen-oxygen fuel cell stack consumes hydrogen at a rate of 0.2 lbm/h while producing electricity at a rate of 2.5 kW. (a) Determine the rate of water produced. (b) Determine the first-law efficiency and second-law efficiency of this fuel cell if the water in the products is liquid.

11-27   A hydrogen-air fuel cell stack consumes hydrogen at a rate of 0.005 kg/s while generating electricity at a rate of 475 kW. (a) Determine the rate of water produced. (b) Determine the first-law efficiency and second-law efficiency of this fuel cell if the water in the products is liquid.

11-28   A hydrogen-air fuel cell operates at 400 K and 1 atm. Determine the ideal cell voltage and the maximum work output, in kJ/kmol, if the water in the products is vapor. Use the enthalpy and entropy values from the following table.

Substance	$\bar{h}_{400\,K} = \bar{h}_f^o + \bar{h} - \bar{h}^o$ kJ/kmol	$\bar{s}_{400\,K}$ kJ/kmol·K
H_2	2958	139.10
O_2	3027	213.76
$H_2O\ (g)$	−238,358	198.67

11-29   Consider a hydrogen-oxygen fuel cell whose overall reaction is

$$H_2 + \tfrac{1}{2}O_2 \rightarrow 2e^- + 2H^+ + O \rightarrow H_2O\ (g)$$

The fuel cell operates at 360 K. Determine the maximum work per unit kmol of fuel, the maximum first-law efficiency, and the ideal open-circuit voltage.

Substance	$\bar{h}_{360\,K} = \bar{h}_f^o + \bar{h} - \bar{h}^o$ kJ/kmol	$\bar{s}_{360\,K}$ kJ/kmol·K
H_2	1794	136.04
O_2	1829	210.60
$H_2O\ (g)$	−239,732	195.08

11-30   Reconsider Prob. 11-29. Determine the heat rejected from this ideal fuel cell. Determine the work potential of this heat transfer. Compare this work potential to the difference between maximum work outputs of the fuel cell at 25°C (298 K) and 360 K.

11-31   Reconsider Prob. 11-29. Determine the second-law efficiency and the first-law efficiency if this fuel cell produces 13.5 kW of electricity by consuming 9.5 g of hydrogen per minute.

11-32   Consider two hydrogen-oxygen fuel cell reactions, one operating at 300 K and the other operating at 600 K. Determine the changes in maximum first-law efficiency and ideal voltage between the two fuel cells.

11-33   Which is not a type of fuel cell?
(*a*) Direct methanol   (*b*) Alkaline   (*c*) Direct combustion
(*d*) Proton exchange membrane   (*e*) Molten carbonate

11-34   Which type of fuel cell operates at temperatures up to 1000°C with efficiencies up to 60 percent?
(*a*) Direct methanol   (*b*) Alkaline   (*c*) Molten carbonate
(*d*) Proton exchange membrane   (*e*) Solid oxide

11-35   Which type of fuel cell uses potassium hydroxide or an alkaline membrane as the electrolyte; requires pure hydrogen as the fuel; and is mostly limited to space applications to produce electricity and water?
(*a*) Direct methanol   (*b*) Alkaline   (*c*) Molten carbonate
(*d*) Proton exchange membrane   (*e*) Solid oxide

11-36   A hydrogen-oxygen fuel cell stack consumes hydrogen at a rate of 0.004 kg/min while generating electricity at a rate of 6 kW. What is the first-law efficiency of this fuel cell if the water in the product is vapor?
(*a*) 0.71   (*b*) 0.73   (*c*) 0.75   (*d*) 0.80   (*e*) 0.84

11-37   A hydrogen-oxygen fuel cell stack consumes hydrogen at a rate of 0.005 kg/min while generating electricity at a rate of 8 kW. What is the second-law efficiency of this fuel cell if the water in the product is vapor?
(*a*) 0.82   (*b*) 0.79   (*c*) 0.76   (*d*) 0.74   (*e*) 0.71

11-38   Hydrogen is produced by an ideal water electrolysis process by consuming 10 kW of electricity. This hydrogen is now used in an ideal fuel cell to generate electricity. How much electricity is to be produced?
(*a*) 5 kW   (*b*) 7.5 kW   (*c*) 8 kW   (*d*) 9 kW   (*e*) 10 kW

CHAPTER 12

Economics of Renewable Energy

12-1 ENGINEERING ECONOMICS

A technical analysis for a renewable energy application is normally accompanied by an economic analysis. Renewable energy applications replace conventional fossil fuel–based systems while providing the same desired output. A renewable energy project certainly provides savings from fossil fuel use, but it may or may not provide cost savings compared to fossil fuel systems. A proper account of cost savings can be handled by means of *engineering economics.*

An economic analysis can also provide a calculation of total cost of a project over its lifetime, and it allows for a cost comparison of competing projects and technologies. Another useful application of economic analysis is the determination of payback period for a project and specifically for a renewable energy application. A decision whether to implement a project is made based on the results of an economic analysis that complements a technical analysis.

A project may involve a *capital* or *initial cost* and *operating and maintenance (O&M) costs.* The capital cost is an account of the initial investment, and it may include labor and other expenses occurring at the start of the project. Operating and maintenance costs are those needed to keep the system running in proper condition over its lifetime. O&M costs can be fixed (such as regular maintenance costs) or variable depending on the use of a commodity (such as the electricity cost of operating an electric motor). The sale value of the system at the end of its lifetime is called *salvage value* or scrap value. The total cost of a system is the sum of the capital and O&M costs. This total should be obtained considering the changes in the value of the money over time.

The cash flow during the life of a project can be indicated by a cash flow diagram as shown in Fig. 12-1. The horizontal line gives the time series in years, and the vertical arrows show the cash flows. There is *n* number of years. The negative arrows indicate expenditures and positive arrows indicate incomes. At time zero, the capital cost occurs. Annual expenditures such as operating and maintenance costs occur at each time interval. The salvage value represents an income that occurs at the end of the lifetime of the system.

The main concepts and principles of engineering economics are covered in Blank and Tarquin (2011) and Rubin (2001), and this chapter applies these economic principles to renewable energy projects.

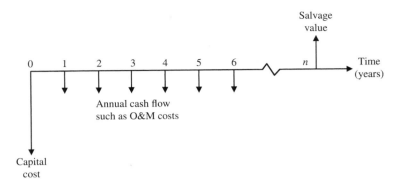

Figure 12-1   A cash flow diagram indicating various costs during the lifetime of a project.

12-2 THE TIME VALUE OF MONEY

The value of money changes with time due to interest and inflation, among other factors. This is also due to the fact that money is always in limited supply with respect to what people want to buy with money. For example, having $1000 today is preferable to having $1000 in 1 year. This is because $1000 today becomes $1100 in 1 year if deposited in a bank with an interest rate of 10 percent. In a given energy project, the cash flow can occur at different times in the lifetime of the project. If one simply adds the incomes and subtracts the expenditures, this neglects the time value of the money. An appropriate economic analysis considers how the value of the money changes with time. The tools for such an analysis are provided next.

Let P represent the *present value of money* and F the single *future value of money*. What is the value of a present amount of money in a future time? Let i represent the *interest rate*. The interest rate is sometimes called the *discount rate*. The value of this money in 1 year is

$$F = P\,(1 + i)$$

At the end of 2 years,

$$F = P\,(1 + i)(1 + i) = P\,(1 + i)^2$$

Let n represent the number of time periods. Then, at the end of n years,

$$F = P(1 + i)^n \tag{12-1}$$

This equation is known as the *compound interest* formula. The money grows exponentially over the time. This is known as *compounding*. The compounding can be done yearly, monthly, weekly, daily, or even hourly. However, in engineering calculations, yearly compounding is commonly used.

As an example of the use of Eq. (12-1), assume you deposit $1000 in a bank account that pays an annual interest rate of 6 percent. What is the value of this amount in 4 years? Application of Eq. (12-1) gives

$$F = P(1 + i)^n = (\$1000)(1 + 0.06)^4 = \$1262$$

Therefore, $1000 today is worth $1262 in 4 years at an interest rate of 6 percent. What happens if the money is compounded monthly instead of annually? In this case, the monthly interest rate is $i = 0.06/12 = 0.005$, and the number of time periods is $n = 4 \times 12 = 48$. Then, from Eq. (12-1),

$$F = P(1 + i)^n = (\$1000)(1 + 0.005)^{48} = \$1270$$

The difference in the future value of money for annual and monthly compounding cases is negligible for most applications. Equation (12-1) can be solved for the present value P when the future value F is known, yielding

$$P = \frac{F}{(1 + i)^n} \qquad (12\text{-}2)$$

EXAMPLE 12-1
Present Value of a
Future Amount

You need to replace the existing natural gas boiler in your home in 6 years. It is estimated that the new pellet boiler will cost $5000 at that time. How much money do you need to deposit in a bank that pays 7 percent annual interest?

SOLUTION   This question can easily be answered by the application of Eq. (12-2):

$$P = \frac{F}{(1 + i)^n} = \frac{\$5000}{(1 + 0.07)^6} = \textbf{\$3332}$$

That is, if you deposit $3332 into a bank account, it will be compounded into $5000 in 6 years, allowing you to buy the new pellet boiler.   ▲

In many instances, the incomes and expenditures occur in each time period. Salary deposits, mortgage payments, and car lease payments are some examples. They can be on a yearly or monthly basis. The periodic income or expenditure in this case is called a *uniform series amount*, denoted by U. It may be expressed as a function of the present value as

$$U = P\left[\frac{i}{1 - (1 + i)^{-n}}\right] \qquad (12\text{-}3)$$

The U in this equation represents the fixed amount over the number of time periods n (Fig. 12-2). If the uniform series amount is given, the corresponding present value may be expressed by rewriting Eq. (12-3) as

$$P = U\left[\frac{1 - (1 + i)^{-n}}{i}\right] \qquad (12\text{-}4)$$

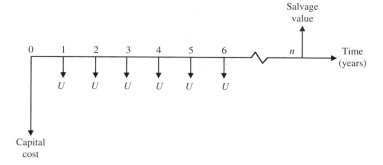

Figure 12-2   A cash flow diagram indicating uniform series amount U.

EXAMPLE 12-2
Uniform Series Amount of
a Present Value

A flat-plate solar collector is to be installed in a facility for hot water production. The collector costs $10,000. You borrow $10,000 from a bank at an annual interest rate of 12 percent compounded on a monthly basis. You will have to pay back the bank in 36 months. What is the amount of the monthly payments?

SOLUTION The monthly interest rate is $i = 0.12/12 = 0.01$. Your monthly payment to the bank can be calculated from Eq. (12-3) as

$$U = P\left[\frac{i}{1-(1+i)^{-n}}\right] = \$10,000\left[\frac{0.01}{1-(1+0.01)^{-36}}\right] = \mathbf{\$332}$$

The total amount you pay during the 36-month period is $11,957. You pay an additional $1957 to the bank for the interest. ▲

The uniform series amount U can also be expressed by a future amount F by combining Eqs. (12-1) and (12-3), yielding

$$U = F\left[\frac{i}{(1+i)^n - 1}\right] \tag{12-5}$$

Solving Eq. (12-5) for F, we obtain

$$F = U\left[\frac{(1+i)^n - 1}{i}\right] \tag{12-6}$$

The six equations forming the basis for economic calculations are summarized in Table 12-1. The representation of each term is given in a cash flow diagram in Fig. 12-3.

TABLE 12-1   Summary of Basic Equations Used in Economic Analysis

Notation	Equation
P = present value	$F = P(1+i)^n$
F = future value	$P = \dfrac{F}{(1+i)^n}$
U = uniform series amount	$U = P\left[\dfrac{i}{1-(1+i)^{-n}}\right]$
i = interest rate	$P = U\left[\dfrac{1-(1+i)^{-n}}{i}\right]$
n = number of time periods	$U = F\left[\dfrac{i}{(1+i)^n - 1}\right]$
	$F = U\left[\dfrac{(1+i)^n - 1}{i}\right]$

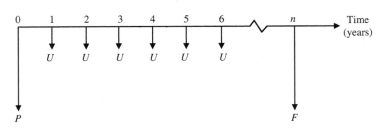

Figure 12-3   A cash flow diagram indicating present value P, future value F, and uniform series amount U.

Effect of Inflation and Taxation on Interest Rate

In the preceding discussion, the time value of money was determined by the interest rate. However, the buying power of money also changes because the price of goods and services increases over time. This is called *inflation* and is denoted by e. If the interest and inflation occur at the same time at constant rates, an *inflation-adjusted interest rate* can be defined as

$$i_{\text{adjusted}} = \frac{i - e}{1 + e} \qquad \text{(adjusted for inflation)} \qquad (12\text{-}7)$$

For example, if the interest rate is 8 percent and the inflation rate is 3 percent, the inflation-adjusted interest rate is

$$i_{\text{adjusted}} = \frac{i - e}{1 + e} = \frac{0.08 - 0.03}{1 + 0.03} = 0.0485$$

The calculated value is different from a simple difference between the interest rate and inflation rate, $i = 0.08 - 0.03 = 0.05$. Therefore, Eq. (12-7) should be used.

In addition to the effect of inflation, the *taxation* (denoted by t) on an investment with interest decreases the amount of money. Assuming that interest, inflation, and a taxation on the interest occur over the same period at constant rates, the *inflation-taxation-adjusted interest rate* can be expressed as

$$i_{\text{adjusted}} = \frac{(1 - t)i - e}{1 + e} \qquad \text{(adjusted for inflation and taxation)} \qquad (12\text{-}8)$$

For example, if the interest rate is 8 percent, the inflation rate is 3 percent, and the tax rate on the interest is 2 percent, the inflation-taxation-adjusted interest rate is

$$i_{\text{adjusted}} = \frac{(1 - t)i - e}{1 + e} = \frac{(1 - 0.02)0.08 - 0.03}{1 + 0.03} = 0.0470$$

The taxation decreases the true value of the interest rate from 4.85 to 4.70 percent.

EXAMPLE 12-3
Comparison of Two Investment Options

Company A installs a wind turbine that costs $15,000. Calculations indicate that this wind turbine will save $3000 per year from the electricity it saves for a period of 10 years. The salvage value of the turbine at the end of 10 years is $1500. Company B takes a different route and deposits $15,000 in a savings account. Taking the inflation-taxation-adjusted interest rate for both applications to be 5 percent, determine which company will have more money after 10 years.

SOLUTION The total future value of cost savings that occur for a period of 10 years is [from Eq. (12-6)]:

$$F = U \left[\frac{(1 + i)^n - 1}{i} \right] = \$3000 \left[\frac{(1 + 0.05)^{10} - 1}{0.05} \right] = \$37{,}733$$

The salvage value is already given in the future amount, and it should be added to this amount:

$$F = \$37{,}733 + \$1500 = \mathbf{\$39{,}200}$$

We used three significant digits to give the final answer. If the money saved each year was not deposited into the bank, the total savings in 10 years would be $30,000.

For company B, $15,000 deposited in the bank will provide a savings of [from Eq. (12-1)]

$$F = P(1 + i)^n = (\$15{,}000)(1 + 0.05)^{10} = \mathbf{\$24{,}400}$$

It is clear that company A will have $14,800 more money available to them at the end of 10 years. This analysis neglects the operating and maintenance expenses associated with the wind turbine. ▲

12-3 LIFE CYCLE COST ANALYSIS

Equations (12-1) through (12-6) are very powerful, and they form the basis for energy economics calculations. One can use these equations to convert expenses occurring at different times to a desired time so that the total cost of the project can be expressed by a single value. It also allows the comparison of competing projects and options. This comparison can be done by calculating the total cost of a project, known as *life cycle cost analysis*.

The life cycle cost can be evaluated in different ways. The *net present value* method is based on expressing all expenses and benefits that will occur over the lifetime of the project on the present time basis *P*. The formulation given in Sec.12-2 is used to express the expenses and benefits occurring at different times. Summing all the benefits and subtracting from the expenses on the present time gives the net present value of the project. The greater the positive value or the smaller the negative value, the more desirable the project from an economic point of view.

The life cycle cost of a project can also be calculated using the *levelized annual cost* (or *levelized annual value*) method. The net cost (or benefit) of the project is expressed by equal annual amounts over the lifetime of the project. Each benefit/expense of the project occurring at different times is expressed by a uniform series amount *U*. The net value of *U* is calculated by adding benefits and subtracting expenses on an annual basis.

Cost-Benefit Analysis

In general, a project option may involve costs and benefits in economic terms. Some projects only involve costs with no monetary benefits. For example, the installation of an environmental control technology requires cost consumption, but it may not yield any monetary benefit. The benefits in this case are due to the reduction in certain pollutants. Of course, the indirect economic benefits of reducing pollutants are not accounted for. A life cost analysis results in the total cost of the project expressed in the present term or annual cost term. Renewable energy projects are beneficial to the environment, but they are also intended to be financially attractive to the investors. Therefore, the economic benefits are expected to outweigh the cost of the project.

A comparison of benefits and costs associated with a project can be made using a *cost-benefit analysis*. The net present value calculated by adding all the benefits and subtracting from all the expenses on a present time basis is used to determine whether the project is acceptable.

Net present value (NPV) = Total benefits in present time − Total costs in present time   (12-9)

If the net present value is greater than zero, the project yields a net benefit, and therefore, it is acceptable. The cost-benefit method can also be expressed using the *benefit-cost ratio*,

$$\text{Benefit-cost ratio} = \frac{\text{Total benefits in present time}}{\text{Total costs in present time}} \qquad (12\text{-}10)$$

If this ratio is greater than unity, the project is acceptable. If it is less than unity, it is unacceptable. In general, the projects whose benefit-cost ratio is sufficiently greater than unity receive a green light.

EXAMPLE 12-4
Net Present Value of a
Geothermal Heating
System

A geothermal space heating system is to be installed for a commercial building to support natural gas heaters. The initial cost of the system is $50,000. It is estimated that the system will save the building $10,000 per year from the natural gas it saves. It will take $2000 per year to operate and maintain the geothermal system. The system will require a major cleaning after 6 years at a cost of $7000. The lifetime of the system is 12 years, after which the geothermal system will have a salvage value of $3000 (Fig. 12-4). Determine the net present value of this project. Take the inflation-adjusted interest rate as 4 percent. Also, calculate the benefit-cost ratio.

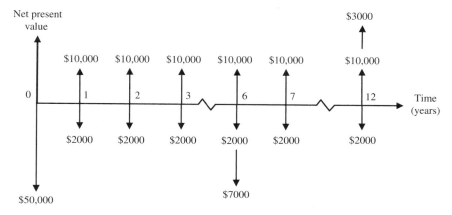

Figure 12-4   Cash flow diagram for Example 12-4.

SOLUTION   The cash flow diagram in Fig. 12-4 gives all the cash flows during the lifetime of the project. The capital cost is already expressed in present time:

$$P_1 = \$50,000$$

The present value of annual savings is [Eq. (12-4)]

$$P_2 = U\left[\frac{1-(1+i)^{-n}}{i}\right] = \$10,000\left[\frac{1-(1+0.04)^{-12}}{0.04}\right] = \$93,851$$

The present value of the operating and maintenance expenses is

$$P_3 = U\left[\frac{1-(1+i)^{-n}}{i}\right] = \$2000\left[\frac{1-(1+0.04)^{-12}}{0.04}\right] = \$18,770$$

The major cleaning cost that will occur after 6 years can be expressed in present time as [Eq. (12-2)]

$$P_4 = \frac{F}{(1+i)^n} = \frac{\$7000}{(1+0.04)^6} = \$5532$$

The salvage value that will occur after 12 years is expressed in present time as

$$P_5 = \frac{F}{(1+i)^n} = \frac{\$3000}{(1+0.04)^{12}} = \$1874$$

The net present value is determined by adding benefits and subtracting expenses from it:

$$\text{Net present value} = \text{Benefits} - \text{Expenses}$$
$$= (P_2 + P_5) - (P_1 + P_3 + P_4)$$
$$= (\$93,851 + \$1874) - (\$50,000 + \$18,770 + \$5532)$$
$$= \$95,725 - \$74,302$$
$$= \mathbf{\$21,400}$$

The installation of this geothermal heating system will provide a monetary benefit of $21,400 over the entire lifetime of the project expressed in present time value. If we did not consider the time value of money and just added benefits and subtracted expenses, the net value of the project would be

$$\text{Net value} = (\$10,000 \times 12 + \$3000) - (\$50,000 + \$2000 \times 12 + \$7000) = \$42,000$$

This is twice the amount calculated above, indicating the importance of considering the time value of money.

The benefit-cost ratio is calculated from Eq. (12-10) to be

$$\text{Benefit-cost ratio} = \frac{\$95,725}{\$74,302} = \mathbf{1.29}$$

This ratio is significantly greater than unity, and thus this project should receive a green light.   ▲

EXAMPLE 12-5
**Levelized Annual Value
of a Geothermal Heating
System**

Reconsider Example 12-4. This time, determine the levelized annual value (cost or benefit) of this project.

SOLUTION   First, the capital cost given in present time value will be converted to a uniform series payment [Eq. (12-3)]

$$U_1 = P\left[\frac{i}{1-(1+i)^{-n}}\right] = \$50,000\left[\frac{0.04}{1-(1+0.04)^{-12}}\right] = \$5328$$

Annual savings and the O&M expenses are already given in terms of a uniform series payment:

$$U_2 = \$10,000$$

$$U_3 = \$2000$$

The cleaning cost occurs once after 6 years. In order to find its uniform series payment, we need to first obtain its present value using a time period of 6 years ($n = 6$) and then calculate the uniform series payment applicable for 12 years ($n = 12$). That is,

$$P_4 = \frac{F}{(1+i)^n} = \frac{\$7000}{(1+0.04)^6} = \$5532$$

$$U_4 = P_4\left[\frac{i}{1-(1+i)^{-n}}\right] = \$5532\left[\frac{0.04}{1-(1+0.04)^{-12}}\right] = \$589$$

The salvage value is expressed on an annual benefit basis as [Eq. (12-5)]

$$U_5 = F\left[\frac{i}{(1+i)^n - 1}\right] = \$3000\left[\frac{0.04}{(1+0.04)^{12} - 1}\right] = \$200$$

The levelized annual value is determined by adding benefits and subtracting expenses from it:

$$\text{Levelized annual value} = \text{Benefits} - \text{Expenses}$$
$$= (U_2 + U_5) - (U_1 + U_3 + U_4)$$
$$= (\$10,000 + \$200) - (\$5328 + \$2000 + \$589)$$
$$= \$10,200 - \$7917$$
$$= \mathbf{\$2283}$$

The installation of the geothermal heating system will provide an annual monetary benefit of $2283 for the entire lifetime of the project. The same result could also be obtained using the net present value calculated in Example 12-4 as follows:

$$U = P\left[\frac{i}{1-(1+i)^{-n}}\right] = \$21,423\left[\frac{0.04}{1-(1+0.04)^{-12}}\right] = \$2283$$ ▲

Unit Product Cost

In many manufacturing and other facilities, the cost of producing a unit of product is of prime interest. This may be expressed in terms of the levelized annual cost as

$$\text{Unit product cost (UPC)} = \frac{\text{Levelized annual cost}}{\text{Annual production}} \qquad (12\text{-}11)$$

Unit product cost is different from specific cost (or specific energy cost). Specific cost refers to the energy cost of a unit product, while unit product cost refers to the total cost (including initial cost, energy cost, operating and maintenance cost, and salvage value) expressed on a levelized annual cost basis per unit product.

Annual production can be in different units depending on the type of production. A power plant produces electricity in kWh, and the UPC is expressed in $/kWh. For a facility producing corn flakes, the unit of UPC is $/kg. For a facility producing a notebook computer, the unit of UPC could be $/notebook. In the calculation of UPC, the levelized annual cost should include all costs, including energy cost. A renewable energy application may decrease the amount of energy consumption, corresponding to a reduction in the unit product cost.

Comparison of Projects Based on Life Cycle Cost Analysis

The life cycle cost analysis based on the *net present value method* and the *levelized annual cost method* can be used to compare different technology options. If the present value method is used for the comparison, the total cost of the project over the entire lifetime of each option is calculated in present time. The result sometimes represents the amount of investment needed today to carry out the project for its entire lifetime. This method requires that each option has the same lifetime.

If the levelized annual cost method is used for comparison, the total cost of the project over its entire lifetime is expressed as equal annual payments. This method does not require that the lifetime of each option have the same lifetime. If an option has a shorter lifetime, it is assumed that the project will be replaced by an equivalent project at the end of its life.

In both methods, the project with the smallest total cost is selected. If the project yields a net positive benefit instead of expenditure, the project with the highest benefit should be selected. Note that both methods are equivalent, and they indicate the same project as the economically viable choice.

EXAMPLE 12-6
Comparison of Three Projects Based on Their Net Present Values

An electric motor is to be purchased for use in a renewable energy application, and there are three options. Compare the total costs of standard motor, high-efficiency motor, and premium-efficiency motor. Use the net present value method for comparison, and consider the following data:

All three motors provide the same mechanical output of 45 kW.

The efficiencies of the standard, high-efficiency, and premium-efficiency motors are 88, 91, and 94 percent, respectively. The efficiency is defined as the actual mechanical power output over the electricity input.

The initial costs of the standard, high-efficiency, and premium-efficiency motors are $11,000, $13,000, and $14,500, respectively.

The lifetime is 10 years for each motor.

The motor operates 4000 hours a year at an average load factor of 65%.

The price of electricity is \$0.10/kWh.

The operating and maintenance expenses are \$500 per year.

The interest rate is 8 percent.

The motors have no salvage value at the end of 10 years.

SOLUTION   We determine the total costs of each motor over its lifetime by calculating the present values. First, we consider the standard motor. The initial cost at the present time is

$$P_1 = \$11,000$$

The present value of the operating and maintenance expenses is

$$P_2 = U\left[\frac{1-(1+i)^{-n}}{i}\right] = \$500\left[\frac{1-(1+0.08)^{-10}}{0.08}\right] = \$3355$$

Next, we need to determine annual electricity consumption and its cost.

$$\text{Electricity consumption} = \text{Mechanical power} \times \text{Load factor} \times \text{Operating hours} \times \frac{1}{\eta_{\text{motor}}}$$

$$= (45 \text{ kW})(0.65)(4000 \text{ h/yr})\left(\frac{1}{0.88}\right)$$

$$= 132,955 \text{ kWh/yr}$$

$$\text{Electricity cost} = \text{Electricity consumption} \times \text{Unit cost of electricity}$$

$$= (132,955 \text{ kWh/yr})(\$0.10/\text{kWh})$$

$$= \$13,296/\text{yr}$$

The present value of the annual electricity cost is

$$P_3 = U\left[\frac{1-(1+i)^{-n}}{i}\right] = \$13,296\left[\frac{1-(1+0.08)^{-10}}{0.08}\right] = \$89,216$$

Then the net present value of total costs becomes (with four significant digits)

$$\text{Net present value} = P_1 + P_2 + P_3 = \$11,000 + \$3355 + \$89,216 = \$103,571 \cong \mathbf{\$103,600}$$

Now, we repeat the calculations for the high-efficiency motor with an efficiency of 91 percent:

$$P_1 = \$13,000$$
$$P_2 = \$3355$$

$$\text{Electricity consumption} = \text{Mechanical power} \times \text{Load factor} \times \text{Operating hours} \times \frac{1}{\eta_{\text{motor}}}$$

$$= (45 \text{ kW})(0.65)(4000 \text{ h/yr})\left(\frac{1}{0.91}\right)$$

$$= 128,572 \text{ kWh/yr}$$

$$\text{Electricity cost} = \text{Electricity consumption} \times \text{Unit cost of electricity}$$

$$= (128,572 \text{ kWh/yr})(\$0.10/\text{kWh})$$

$$= \$12,857/\text{yr}$$

$$P_3 = U\left[\frac{1-(1+i)^{-n}}{i}\right] = \$12,857\left[\frac{1-(1+0.08)^{-10}}{0.08}\right] = \$86,270$$

$$\text{Net present value} = P_1 + P_2 + P_3 = \$13,000 + \$3355 + \$86,270 = \$102,625 \cong \mathbf{\$102,600}$$

TABLE 12-2 Summary of Cost Results for Each Motor Considered in Example 12-6

Motor	Present Value of Costs	Total Cost
Standard motor efficiency = 88%	Initial cost = $11,000 O&M costs = $3355 Electricity cost = $89,216	$103,600
High-efficiency motor efficiency = 91%	Initial cost = $13,000 O&M costs = $3355 Electricity cost = $86,270	$102,600
Premium-efficiency motor efficiency = 94%	Initial cost = $14,500 O&M costs = $3355 Electricity cost = $83,519	$101,400

Finally, we repeat the calculations for the premium-efficiency motor with an efficiency of 94 percent:

$$P_1 = \$14,500$$

$$P_2 = \$3355$$

$$\text{Electricity consumption} = \text{Mechanical power} \times \text{Load factor} \times \text{Operating hours} \times \frac{1}{\eta_{\text{motor}}}$$

$$= (45 \text{ kW})(0.65)(4000 \text{ h/yr})\left(\frac{1}{0.94}\right)$$

$$= 124,469 \text{ kWh/yr}$$

$$\text{Electricity cost} = \text{Electricity consumption} \times \text{Unit cost of electricity}$$

$$= (124,469 \text{ kWh/yr})(\$0.10/\text{kWh})$$

$$= \$12,447/\text{yr}$$

$$P_3 = U\left[\frac{1 - (1+i)^{-n}}{i}\right] = \$12,447\left[\frac{1 - (1+0.08)^{-10}}{0.08}\right] = \$83,519$$

$$\text{Net present value} = P_1 + P_2 + P_3 = \$14,500 + \$3355 + \$83,519 = \$101,374 \cong \textbf{\$101,400}$$

The results for each motor are summarized in Table 12-2. The total cost of the premium-efficiency motor is lower than that of the high-efficiency motor by about $1250, and that of the standard motor by about $2200. ▲

12-4 PAYBACK PERIOD ANALYSIS

In most renewable energy projects, it is very important to know how long it will take for the investment to pay for itself. This is called the *payback period*. The payback period should be less than the life of the project, but this does not guarantee a green light for the project. The payback period should be sufficiently shorter than the life of the project. In today's economic environment, it is difficult to get support for projects whose payback periods are longer than 3 or 4 years. However, renewable energy projects also provide environmental benefits, and this may justify longer payback periods.

Consider a project that requires a total investment in the amount of P in present time and it provides an annual savings in the amount of U. How long will it take for the investment to pay for itself from the savings it provides? You may have figured out that the answer to this question is given by Eq. (12-4):

$$P = U\left[\frac{1-(1+i)^{-n}}{i}\right]$$

Of course, the unknown here is the number of years n, which is equal to the payback period n_{dpb}. Solving for the payback period, we obtain

$$n_{dpb} = \frac{\log\left[1-\left(\frac{P}{U}\right)i\right]^{-1}}{\log(1+i)} \qquad (12\text{-}12)$$

Here, P is the present value of the total cost of the project, and its calculation, in addition to capital cost, may involve operating and maintenance expenses and salvage value of the investment. U represents annual savings due to the project, and i is the interest rate. This payback period is called *discounted payback period* n_{dpb} because the time value of money is considered.

In most engineering applications, the time value of the money is neglected, and a *simple payback period* is calculated from

$$n_{spb} = \frac{\text{Investment}}{\text{Annual savings}} = \frac{\text{Investment}}{U} \qquad \left(\frac{\$}{\$/\text{yr}} = \text{yr}\right) \qquad (12\text{-}13)$$

Here, "Investment" represents the total cost of the project, and it is the simple sum of all the costs associated with the project over its lifetime. Usually only initial cost is considered. Again, U represents annual savings. It can be shown that the simple payback period n_{spb} is shorter than the discounted payback period n_{dpb}, and thus it underestimates the actual value. As the interest rate gets smaller, the difference between the two payback periods also gets smaller. The simple payback period n_{spb} is primarily used in engineering calculations, and the discounted payback period n_{dpb} is rarely used.

The payback period analysis is sometimes used for comparing competing options. In this analysis, the project with a shorter payback period is selected. However, this may be misleading because an option with a longer payback period may have better net economic benefits for the entire lifetime of the project. For example, consider two projects with the same life periods. Option A has a net present value (benefit) of $10,000 and a payback period of 2 years. Option B has a net present value of $15,000 with a payback period of 3 years. Clearly, option B provides higher monetary benefits, and thus it is economically more attractive even though it has a longer payback period.

EXAMPLE 12-7
Payback Period of a Renewable Energy Project

A renewable energy project for an existing building involves a solar photovoltaic system installation and construction of a Trombe wall on the south side of the building. The building is heated by a natural gas boiler in winter and cooled by air conditioners in summer. It is estimated that the Trombe wall will reduce the natural gas consumption by 7500 therm in winter, and the photovoltaic system will reduce the electricity consumption by 50,000 kWh in summer. The costs of the photovoltaic system and the Trombe wall are $35,000 and $25,000, respectively, including the materials and labor. The unit costs of natural gas and electricity are $1.2/therm and $0.11/kWh, respectively. Taking the interest rate to be 6 percent, determine the discounted and simple payback periods for each application. Neglect operating and maintenance cost and the salvage value in your analysis.

SOLUTION   This project involves neither operating and maintenance cost nor the salvage value. The total cost is equal to the cost of purchase and installation of the insulation. The cost savings due to the reduction in natural gas and electricity consumption are

Electricity cost savings = Electricity savings × Unit cost of electricity

$$= (50,000 \text{ kWh/yr})(\$0.11/\text{kWh})$$

$$= \$5500/\text{yr}$$

$$\text{Natural gas cost savings} = \text{Natural gas savings} \times \text{Unit cost of natural gas}$$
$$= (7500 \text{ therm/yr})(\$1.2/\text{therm})$$
$$= \$9000/\text{yr}$$

$$\text{Total annual savings} = \$5500/\text{yr} + \$9000/\text{yr} = \$14{,}500/\text{yr}$$

For the photovoltaic system, the discounted payback period is determined from

$$n_{dpb} = \frac{\log\left[1 - \left(\dfrac{P}{U}\right)i\right]^{-1}}{\log(1+i)} = \frac{\log\left[1 - \left(\dfrac{\$35{,}000}{\$5500/\text{yr}}\right)0.06\right]^{-1}}{\log(1+0.06)} = \textbf{8.25 yr}$$

The simple payback period is

$$n_{spb} = \frac{\text{Investment}}{U} = \frac{\$35{,}000}{\$5500/\text{yr}} = \textbf{6.36 yr}$$

For the Trombe wall, the discounted payback period and the simple payback period are

$$n_{dpb} = \frac{\log\left[1 - \left(\dfrac{P}{U}\right)i\right]^{-1}}{\log(1+i)} = \frac{\log\left[1 - \left(\dfrac{\$25{,}000}{\$9000/\text{yr}}\right)0.06\right]^{-1}}{\log(1+0.06)} = \textbf{3.13 yr}$$

$$n_{spb} = \frac{\text{Investment}}{U} = \frac{\$25{,}000}{\$9000/\text{yr}} = \textbf{2.78 yr}$$

The photovoltaic system has a longer payback period than the Trombe wall. The simple payback period underestimates the actual payback period by about 23 months for the photovoltaic system and about 4 months for the Trombe wall.   ▲

In this chapter, basic concepts of engineering economics are provided, and applications are given for renewable energy projects. It should be noted that it is difficult to quantify all of the costs and benefits associated with a project, and therefore economic analysis involves a certain degree of uncertainty. Sometimes, some benefits of a project are difficult to express in monetary terms, and their existence can be a deciding factor for a green light. This is particularly true for renewable energy projects because they reduce or eliminate pollution and greenhouse emissions with positive consequences for humanity. The method of obtaining money for an investment (loan from a bank, equity, etc.) as well as tax regulations affect the outcome of an economic evaluation. Every project has its own benefits and costs and requires a particular analysis. In previous chapters, the technical treatment of various renewable energy applications was presented with simple calculations of cost savings and simple payback periods. However, once the technical analysis of a measure is complete, one can easily perform the economic analysis using the methods described in this chapter.

REFERENCES

Blank L and Tarquin A. 2011. *Engineering Economy*. 7th ed. New York: McGraw Hill.
Rubin ES. 2001. *Introduction to Engineering and the Environment*. New York: McGraw Hill.

PROBLEMS

ENGINEERING ECONOMICS

12-1   What are the purposes of an engineering economic analysis for renewable energy projects?

12-2   List the costs involved in a project and define them.

12-3   What is salvage value? Does it represent an expenditure or income for the project?

12-4   Draw a cash flow diagram indicating various costs during the lifetime of a project.

12-5   Which one is not a purpose of an engineering economic analysis for renewable energy projects?
(*a*) Calculating cost savings   (*b*) Calculating energy savings
(*c*) Calculating total cost of project   (*d*) Cost comparison of competing project options
(*e*) Calculating payback period

12-6   Which one is not part of the total cost of a project?
(*a*) Initial cost   (*b*) Capital cost   (*c*) Salvage cost
(*d*) Maintenance cost   (*e*) Operating cost

12-7   Which cash flow occurs at the end of the lifetime of the system?
(*a*) Annual savings   (*b*) Capital cost   (*c*) Initial cost
(*d*) Operating and maintenance cost   (*e*) Salvage value

12-8   Which cash flow occurs at the start of a project?
(*a*) Capital cost   (*b*) Annual savings   (*c*) Scrap value
(*d*) Operating and maintenance cost   (*e*) Salvage value

THE TIME VALUE OF MONEY

12-9   Why does the value of money change with time?

12-10   Why should the time value of money be considered in a project?

12-11   What is compounding? Which compounding period is commonly used in engineering calculations?

12-12   Draw a cash flow diagram and indicate the present value of money, future value of money, and uniform series amount.

12-13   What is inflation?

12-14   If you deposit $10,000 today in a bank that pays 5 percent annual interest, how much money will you have after 7 years?

12-15   You deposit $10,000 today in a bank that pays 5 percent annual interest. How much money will you have after 7 years if the money is compounded monthly?

12-16   You will need to replace the existing windows in your house in 10 years. It is estimated that the new energy-efficient windows will cost $15,000 at that time. In order to raise money for this, how much money do you need to deposit in a bank that pays 4 percent annual interest?

12-17   You have $3000 today, and you want to deposit it in a bank. What is the minimum rate of annual interest so that you will have $5000 after 6 years, if the money is compounded (*a*) yearly and (*b*) weekly?

12-18   You need to pay $8000 per year to bank A for a period of 10 years for your mortgage payments. Bank B offers a mortgage with a payment of $5000 for a period of 17 years instead. Which one is a better option? The annual interest rate is 3 percent for both cases.

12-19   The compressor in your facility stopped working properly, and you need replace it with a new one that costs $40,000. You borrow this amount from a bank at an annual interest rate of 8 percent compounded on a monthly basis. You will have to pay this back to the bank as equal monthly installments for a period of 5 years. What is the amount of the monthly payment?

12-20   A solar photovoltaic system installed in a facility saves $1200 a year from the electricity it saves. Can you buy another photovoltaic system in 10 years that will cost $15,000 if you deposit these annual savings in a bank with annual interest of 4.5 percent?

12-21   For which option is the inflation-adjusted interest rate greater? (*a*) Interest rate $i = 0.06$, inflation $e = 0.03$, (*b*) interest rate $i = 0.07$, inflation $e = 0.04$. Assume that interest and inflation occur for the same period at constant rates.

12-22   For which option is the inflation-taxation-adjusted interest rate greater? (*a*) Interest rate $i = 0.06$, inflation $e = 0.03$, taxation $t = 0.01$, (*b*) interest rate $i = 0.08$, inflation $e = 0.04$, taxation $t = 0.02$. Assume that interest, inflation, and taxation occur for the same period at constant rates.

12-23 Company A installs a biomass burning boiler system in their facility, which costs $50,000 including labor. Calculations indicate that this system will save $4000 per year from the energy it saves for a period of 15 years. The salvage value of the equipment at the end of 15 years is $5000. Company B takes a different route and deposits $50,000 in a savings account. Taking the inflation-taxation-adjusted interest rate for both applications to be 4 percent, determine which company will have more money after 15 years. Assume that annual savings due to using the biomass system is also deposited into the same bank.

12-24 Which one is not a term used in basic equations for calculating the time value of money?
(*a*) Present value of money   (*b*) Future value of money   (*c*) Uniform series amount
(*d*) Unit product cost   (*e*) Interest rate

12-25 Which one is not a concept or term associated with the time value of money?
(*a*) Present value of money   (*b*) Simple payback period   (*c*) Compounding
(*d*) Inflation   (*e*) Interest

12-26 What is the effect of inflation on the interest rate?
(*a*) Increases   (*b*) Decreases   (*c*) Remains the same

12-27 What is the effect of taxation on the interest rate?
(*a*) Increases   (*b*) Decreases   (*c*) Remains the same

LIFE CYCLE COST ANALYSIS

12-28 What are the methods of evaluating life cycle cost? Describe each of them.

12-29 What is cost-benefit analysis? How is the benefit-cost ratio defined?

12-30 What is the minimum acceptable value of benefit-cost ratio for a project? What does this value represent?

12-31 What is the unit product cost? How does it differ from specific cost?

12-32 Two methods can be used when comparing project options based on life cycle cost analysis. Is there any limitation on using either of these methods? Explain. Which method is more accurate?

12-33 A waste heat recovery heat exchanger (also called a recuperator) is to be installed in a boiler system of a geothermal cogeneration plant. The initial cost of the recuperator is $14,000. It is estimated that the recuperator will save $3000 per year from the fuel it saves. It will take $500 per year to operate and maintain the recuperator. The system will require major maintenance after every 3 years for a cost of $1500. The lifetime of the system is 15 years, after which the recuperator will have a salvage value of $1200. Take the inflation-adjusted interest rate to be 5 percent.
(*a*) Draw the cash flow diagram, indicating all cash flows.
(*b*) Determine the net present value of this project.
(*c*) Determine the benefit-cost ratio.
(*d*) Determine the net value of the project if the time value of money is not considered.

12-34 Reconsider Prob. 12-33. Determine the levelized annual value of this project and the benefit-cost ratio.

12-35 A total of 250 incandescent lamps in a building are to be replaced with 100 compact fluorescent lamps while providing the same amount of lighting. The initial cost of this measure is $800, including the cost of lamps and labor. It is estimated that the high-efficiency fluorescent lighting will save $1300 per year in electricity. The lifetime of new lights is 3 years, and the annual interest rate is 9 percent. Determine the net present value and the benefit-cost ratio of this investment. Also, determine the net value of this investment if the time value of money is not considered.

12-36 Reconsider Prob. 12-35. Determine the levelized annual value and the benefit-cost ratio. Also, determine the levelized annual value of this investment if the time value of money is not considered.

12-37 In order to reduce cooling electricity consumption in an industrial plant, it is proposed to install an absorption refrigeration system running on solar energy. The absorption system costs $300,000, and the operating and maintenance expenses are $8000 per year. It is estimated that this absorption system will save $27,500 per year in electricity. The lifetime of the system is expected to be 20 years with an average annual inflation-adjusted interest rate of 4 percent. The system will have no salvage value after 20 years of its lifetime. Would you support this project from an economic point of view? How would you evaluate this project if the inflation-adjusted interest rate is 2 percent?

12-38   A food manufacturing plant decides to add another type of cookies to its production line. The plant spends $220,000 for the purchase and installation of the equipment. Other estimated costs on an annual basis include the following:

$$Operating\ and\ maintenance\ cost = \$7000$$
$$Electricity\ cost = \$48,000$$
$$Natural\ gas\ cost = \$37,000$$
$$Raw\ material\ cost = \$260,000$$
$$Staff\ cost = \$240,000$$

The lifetime of the equipment is estimated to be 12 years, after which it will have a salvage value of $22,000. The plant is to produce 250,000 kg of cookies per year. Taking the interest rate to be 6 percent, determine the unit product cost.

12-39   Reconsider Prob. 12-38. It is proposed to buy and install renewable energy–based equipment instead of the current option with an initial cost of $240,000. This provides an annual electricity cost savings of $7000 and a natural gas cost savings of $5000. Determine the unit product cost for this new option.

12-40   A manufacturing plant produces towels. The following annual data is obtained:

$$Towel\ production = 4,485,000\ kg$$
$$Electricity\ cost = \$1,456,000$$
$$Natural\ gas\ cost = \$172,500$$
$$Coal\ cost = \$608,700$$
$$Operating\ and\ maintenance\ cost = \$40,000$$
$$Raw\ material\ cost = \$3,280,000$$
$$Staff\ cost = \$3,750,000$$

Taking the interest rate to be 5 percent, determine (*a*) the unit product cost, (*b*) the specific energy cost, and (*c*) the fraction of the total annual cost due to energy consumption.

12-41   A 30-MW geothermal power plant is to be installed on an underground hot water site. The initial cost of installing the power plant is $75 million. The operating and maintenance expenses are $50,000 per year, while the staff cost is $600,000 per year. The plant is expected to operate 8000 h/yr with an average load factor of 0.8. The lifetime of the plant is expected to be 25 years. Taking the annual interest rate to be 3.5 percent, determine the unit product cost.

12-42   A 400-MW combined cycle power plant burning biogas is to be installed with an initial cost of $900 million. The operating and maintenance expenses are $400,000 per year, while the staff cost is $8 million per year. The plant is expected to operate 8200 h/yr with an average load factor of 0.9. The overall efficiency of the plant is 45 percent, and the unit cost of biogas is $1.1/therm (1 therm = 100,000 Btu = 105,500 kJ). The lifetime of the plant is expected to be 30 years. Taking the annual interest rate to be 4.5 percent, determine the unit product cost.

12-43   An electric motor is to be purchased by a facility, and there are three options. Compare the total costs of standard motor, high-efficiency motor, and premium-efficiency motor using the levelized annual cost method. Consider the following data:

All three motors provide the same mechanical output of 22 kW.

The efficiencies of the standard, high-efficiency, and premium-efficiency motors are 85, 88, and 90 percent, respectively. The efficiency is defined as the actual mechanical power output over the electricity input.

The initial costs of the standard, high-efficiency, and premium-efficiency motors are $8000, $10,000, and $12,000, respectively.

The lifetime is 12 years for each motor.

The motor operates 6000 h/yr at an average load factor of 80 percent.

The price of electricity is $0.11/kWh.

The operating and maintenance expenses are $300 per year.

The interest rate is 7 percent.

12-44   You are to buy a hot water heater running on wood pellet for your household. You narrowed down the options to the following two: Heater A costs $2500 and its efficiency is 80 percent. Heater B costs $3000 and its efficiency is 87 percent. The annual pellet cost is estimated to become $600 if heater A is purchased. The lifetime of the heater is 15 years, and the annual interest rate is 12 percent. Which heater should you buy based on an economic comparison of the two heaters?

12-45   Two people are to buy a car with different priorities. Person A wants to buy a powerful SUV that costs $50,000 and gets only 20 mi/gal. Person B is environmentally conscious and wants to buy a fuel-efficient hybrid car that costs $25,000 and gets 35 mi/gal. Each car will be driven for 13,000 mi/yr. The SUV requires $500 per year for maintenance, while the hybrid car requires $300 per year. Person A will sell the car after 4 years, while person B will sell it after 6 years. Each car loses 10 percent of its value every year. The unit cost of gasoline is $3.50/gal, and the annual interest rate is 5 percent. How much more money does the car cost for person A on an annual basis compared to person B? Can you also compare the total costs of the two cars using the net present value method?

12-46   The benefit-cost ratio is calculated for five competing projects as indicated in the following choices. Which project is financially most attractive?
(*a*) 0.5   (*b*) 0.75   (*c*) 1.0   (*d*) 1.25   (*e*) 1.5

12-47   Which one is a suitable unit of unit product cost for a geothermal power plant?
(*a*) $/kWh   (*b*) $/kg   (*c*) $/lbm   (*d*) $/therm   (*e*) $/m³

PAYBACK PERIOD ANALYSIS

12-48   What is the payback period?

12-49   What is the minimum value of the payback period for a project whose expected lifetime is 5 years? Would you support this project if its payback period is 4 years?

12-50   Someone decides to replace their existing windows with energy-efficient ones. This investment has a payback period of 6 years. Would you support this project with this relatively long payback period?

12-51   Is the simple payback period necessarily shorter than the discounted payback period? Why?

12-52   What is the effect of interest rate on the difference between the discounted and simple payback periods?

12-53   A flat-plate solar collector is to be installed on the roof of a house for the hot water needs of the house. The collector costs $650 including material and labor. It is estimated that the collector will save the house $375 per year from the energy it saves. Taking the interest rate to be 7 percent, determine the discounted and simple payback periods.

12-54   In order to generate her own electricity, a homeowner decides to install a photovoltaic panel on the roof of her house. The panel costs $8000, and it does not require any operating and maintenance expenses. It is estimated that the panel will save $110 per month from the electricity bill. Taking the interest rate to be 9 percent, determine the discounted and simple payback periods. What would your answers be if the interest rate were zero percent?

12-55   Frustrated with high electricity bills, the manager of an industrial plant decides to install wind turbines. The total cost of wind turbines is $89,000. The annual electricity savings are estimated to be $9950. Determine the discounted payback period for an annual interest rate of 3 percent. Neglect operating and maintenance costs.

12-56   The Gemasolar solar-power-tower plant has an installed capacity of 19.9 MW and produces 110 GWh of electricity per year. The cost of the Gemasolar plant is $33,000/kW. The unit price of electricity is $0.15/kWh, and the annual interest rate is 1 percent. The operating and maintenance cost is estimated to be $250,000 per year. Determine the discounted and simple payback periods.

12-57   An increase of which parameter increases the difference between the discounted and simple payback periods?
(*a*)   Initial cost   (*b*) O&M costs   (*c*) Annual savings   (*d*) Interest rate   (*e*) Salvage value

12-58   The discounted interest rate is calculated to be *n* years when the investment cost is *P* and the annual savings is *U* (both in $). If the interest rate is zero percent, what is the simple payback period?
(*a*) $n-1$ (*b*) $n+1$ (*c*) n (*d*) U/P (*e*) $P/U - n$

CHAPTER 13

Energy and the Environment

13-1 INTRODUCTION

The conversion of energy from one form to another often affects the environment and the air we breathe in many ways (Fig. 13-1). Fossil fuels such as coal, oil, and natural gas have been powering the industrial development and the amenities of modern life that we enjoy since the 1700s, but this has not been without any undesirable side effects. From the soil we farm and the water we drink to the air we breathe, the environment has often paid a heavy toll. Pollutants emitted during the combustion of fossil fuels contribute to smog, acid rain, and climate change. The environmental pollution has reached such high levels that it has become a serious threat to vegetation, wildlife, and human health. Air pollution has been the cause of numerous health problems, including asthma and cancer. It is estimated that over 60,000 people in the United States alone die each year due to heart and lung diseases related to air pollution.

Considerable amounts of pollutants are emitted as the chemical energy in fossil fuels is converted to thermal, mechanical, or electrical energy via combustion, and thus power plants, industrial processes, motor vehicles, residential and commercial energy consumption, and even stoves take the blame for air pollution. Hundreds of elements and compounds such as benzene and formaldehyde are known to be emitted during the combustion of coal, oil, natural gas, and wood in electric power plants, engines of vehicles, furnaces, and even fireplaces. Some compounds are added to liquid fuels for various reasons (such as methyl tertiary-butyl ether (MTBE) to raise the octane number of the fuel and also to oxygenate the fuel in winter months to reduce urban smog).

The largest sources of air pollution are industrial activities and motor vehicles, and the pollutants released by them are usually grouped as *hydrocarbons* (HC), *sulfur dioxide* (SO_2), *nitrogen oxides* (NO_x), *particulate matter* (PM), and *carbon monoxide* (CO). The HC emissions are a large component of *volatile organic compounds* (VOCs) emissions, and the two terms are generally used interchangeably for motor vehicle emissions.

The increase in environmental pollution at alarming rates and the rising awareness of its dangers made it necessary to control it by legislation and international treaties. In the United States, the Clean Air Act of 1970 (whose passage was aided by the 14-day smog alert in Washington that year) set limits on pollutants emitted by large plants and vehicles. These early standards focused on emissions of HC, NO_x, and CO. The new cars were required to have catalytic converters in their exhaust systems to reduce HC, NO_x, and CO emissions. As a side benefit, the removal of lead from gasoline to permit the use of catalytic converters led to a significant reduction in toxic lead emissions.

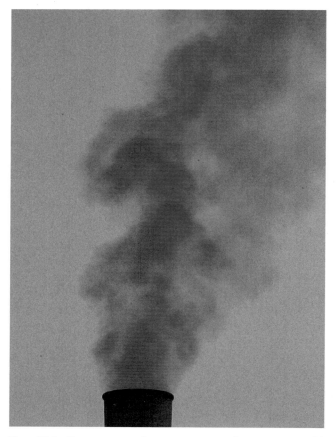

Figure 13-1   Energy conversion processes are often accompanied by air pollution and environmental effects.

Maximum allowable emission limits for SO_2, PM, HC, NO_x, and CO have been declining steadily since 1970. The Clean Air Act of 1990 made the requirements on emissions even tougher, primarily for ozone, CO, nitrogen dioxide, and PM. As a result, today's industrial facilities and vehicles emit a fraction of the pollutants they used to emit a few decades ago.

Children are most susceptible to the damages caused by air pollutants since their organs are still developing. They are also exposed to more pollution since they are more active, and thus they breathe faster. People with heart and lung problems, especially those with asthma, are most affected by air pollutants. This becomes apparent when the air pollution levels in their neighborhoods rise to high levels.

Air pollutant and greenhouse gas emissions can be reduced by replacing fossil fuels with renewable energy sources such as solar, wind, hydroelectric, geothermal, and biomass. However, renewable energy systems also involve some negative environmental impacts. Biomass energy involves ecological impacts of harvesting, transportation, and processing of plants. Some hydroelectric power systems displace people and flood lands as well as intervene with the natural habitat of rivers for fish and other creatures. Disposal of solar panels and collectors after their useful life is a concern for solar systems.

Geothermal systems are known to contribute to hydrogen sulfide (H_2S) emissions. When H_2S is released as a gas into the atmosphere, it eventually changes into SO_2 and sulfuric acid. The removal efficiency of H_2S from geothermal power plants is over 99.9 percent. This is done by converting H_2S into sulfur, which is used in a beneficial way as a fertilizer for soil amendment. The rate of H_2S emission in the United States is now less than 100 kg/h (Kagel et al., 2007).

13-2 AIR POLLUTANTS

The minimum amount of air needed for the complete combustion of a fuel is called the *stoichiometric* or *theoretical air*. When a fossil fuel such as coal, oil, or natural gas is burned completely with air, the combustion products do not contain any air pollutants. Natural gas is usually approximated by methane (CH_4), and the theoretical combustion of CH_4 with air which consists of nitrogen (N_2) and oxygen (O_2) is

$$CH_4 + 2(O_2 + 3.76N_2) \rightarrow CO_2 + 2H_2O + 7.52N_2 \tag{13-1}$$

Gasoline and diesel fuel used in automobile and vehicle engines are commonly approximated by octane (C_8H_{18}). If octane is burned completely with the stoichiometric amount of air, the reaction is

$$C_8H_{18} + 12.5(O_2 + 3.76N_2) \rightarrow 8CO_2 + 9H_2O + (12.5 \times 3.76)N_2 \tag{13-2}$$

If we consider the combustion of a particular composition or assay of coal (80 percent C, 5 percent H_2, 6 percent O_2, 2 percent N_2, 0.5 percent S, and 6.5 percent ash) with the theoretical amount of air, the complete reaction is

$$6.667C + 2.5H_2 + 0.1875O_2 + 0.07143N_2 + 0.01563S + 7.745(O_2 + 3.76N_2) \tag{13-3}$$

$$\rightarrow 6.667CO_2 + 2.5H_2O + 0.01563SO_2 + 29.19N_2$$

In this reaction, we disregarded ash, which is a nonreacting component.

In actual combustion processes, it is common practice to use more air (called excess air) than the stoichiometric amount to increase the chances of complete combustion and/or to control the temperature of the combustion chamber. A combustion process is *complete* if all the carbon in the fuel burns to CO_2, all the hydrogen burns to H_2O, and all the sulfur (if any) burns to SO_2. That is, all the combustible components of a fuel are burned to completion during a complete combustion process. Conversely, the combustion process is *incomplete* if the combustion products contain any unburned fuel or components such as C, H_2, CO, or OH.

Insufficient oxygen is an obvious reason for incomplete combustion, but it is not the only one. Incomplete combustion occurs even when more oxygen is present in the combustion chamber than is needed for complete combustion. This may be attributed to insufficient mixing in the combustion chamber during the limited time that the fuel and the oxygen are in contact. This is particularly true for internal combustion engines. Another cause of incomplete combustion is *dissociation*, which becomes important at high combustion temperatures.

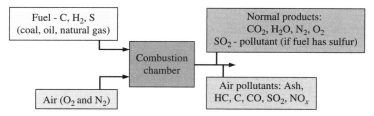

Figure 13-2 Combustion products (normal products and air pollutants) when an HC fuel burns with air.

Oxygen has a much greater tendency to combine with hydrogen than it does with carbon. Therefore, the hydrogen in the fuel normally burns to completion, forming H_2O, even when there is less oxygen than needed for complete combustion. Some of the carbon, however, ends up as CO or just as plain C particles (soot) in the products.

The major air pollutants produced by the combustion of fossil fuels can be summarized as (Fig. 13-2):

- Particulate matter (PM)
- Sulfur dioxide (SO_2)
- Nitrogen oxides (NO_x)
- Hydrocarbons (HC) including carbon soot particles (C)
- Carbon monoxide (CO)

Note that HC, C, and CO emissions are mainly produced due to incomplete combustion, which may be due to insufficient oxygen or insufficient mixing of fuel and air. However, SO_2 is inevitably produced when the fuel contains sulfur. Sulfur is a fuel which reacts with oxygen to form SO_2. NO_x generation is due to high-temperature reactions between atomic nitrogen and oxygen. Particulate matter is commonly called ash, and it is produced mostly when coal or oil is burned.

Average and total emissions of SO_2 and NO_x in U.S. power plants are given in Table 13-1. Approximately 1.17 g of SO_2 are emitted to the environment for each kWh of electricity produced using fossil fuels. This value is decreased to 0.86 g/kWh when all power plants including nuclear and renewable are considered. The corresponding values for NO_x are 0.60 and 0.43 g/kWh, respectively. About 3.46 million tons of SO_2 and 1.74 million tons of NO_x were emitted in the United States in 2012 for electricity production.

There are two general methods of minimizing harmful emissions from a fuel combustion system. The first one is to design and optimize the combustion process such that minimum emissions are generated. The second method involves aftertreatment. In this

TABLE 13-1 **Average and Total Emissions of SO_2 and NO_x in U.S. Power Plants** (EPA, 2012)

	SO_2	NO_x
Average emissions from fossil power plants (g/kWh)	1.17	0.60
Average emissions from all power plants (g/kWh)	0.86	0.43
Annual emissions (tons)	3,463,165	1,736,096

method, the exhaust gases leaving the combustion chamber go through some systems in which harmful components are treated or collected.

There are several environmental controls in a typical coal-fired power plant, as shown in Fig. 13-3. These controls include

- Bottom ash collection from coal before the combustion process
- Selective catalytic reduction (SCR) for NO_x removal
- Electrostatic precipitator for flash removal
- Flue gas desulfurization for SO_2 removal
- Solid waste removal to a pond or landfill

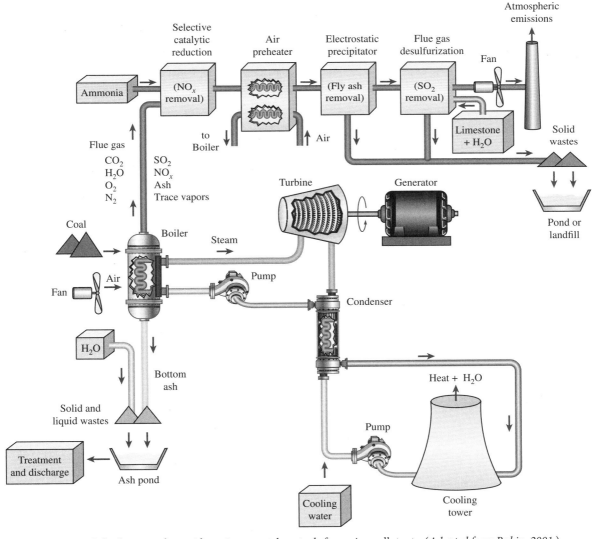

Figure 13-3 A coal-fired power plant with environmental controls for major pollutants. (*Adapted from Rubin, 2001.*)

Particulate Matter

Small solid and liquid particles suspended in air are generally referred to as *particulate matter* (PM). A primary group of PM is ash and soot particles generated from the combustion of coal and oil. Many manufacturing and other industrial processes are responsible for the production of various PMs. Dust particles from roads and construction and agricultural activities as well as iron, silicon, soil, and materials of earth can also be considered PM. Nitrates and sulfates are formed as a result of SO_2 and NO_x emissions and are referred to as secondary PM.

When the particles are less than 10 micrometers (μm or 10^{-6} m), they are called *PM10* and classified as *coarse*. *PM2.5* refers to particles less than 2.5 μm and are referred to as *fine*. Particulate matter is associated with significant health effects—in particular, respiratory diseases such as asthma and bronchitis, cardiovascular and lung diseases, lung tissue damage, vision impairment, and cancer. PM2.5 can penetrate deeply into the respiratory system, and therefore it is more dangerous to health than PM10.

Coal contains considerable amounts of ash, which are noncombustible components of the fuel. The mass percentage of ash in coal varies greatly depending on the source. It is about 10 percent on average for bituminous coal (soft coal with high energy content) and about 5 percent for subbituminous coal (low energy content coal) (EPRI, 1997).

The combustion of coal produces large amounts of ash, which should be removed in a safe and effective manner. In coal-fired power plants, ash particles are carefully collected from the combustion gases. However, some ash and other particulates still remain in the combustion gases and are emitted to the atmosphere. It is estimated that about 1 g of PM is released to the environment for each kWh of electricity produced from a coal-fired power plant.

The collection of ash from coal-fired plants involves some water pollution as water is used to transport ash in the plant. These ash particles are released to the environment when this water along with other water streams is discharged to the lakes or rivers.

In a coal-fired power plant, the flying ash particles (flash) in combustion gases can be removed and safely collected by various means. One common way is by using a device called an *electrostatic precipitator* (ESP), as shown in Fig. 13-4. This prevents flash particles from being emitted to the atmosphere. An ESP is very effective as it removes more than 99 percent of the ash in the combustion gases. In an ESP system, combustion gases flow between two collector plates. An electric field is applied between a discharge electrode wire and the two plates, creating an electric field between the plates. As the combustion gases flow between the plates, ash particles are bombarded with negative ions. The charged particles are then moved toward positively charged plates as a result of electrostatic attraction. Ash particles are collected on the plates. This process is repeated in many plates for maximum removal of flash.

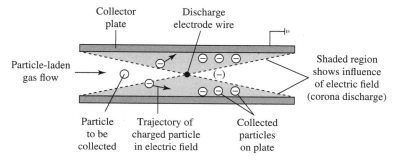

Figure 13-4   Operation of an electrostatic precipitator (ESP). (*Rubin, 2001.*)

EXAMPLE 13-1
Ash Account in a
Coal-Fired Power Plant

Consider a coal with the following composition on a mass basis: 67 percent C, 5 percent H_2, 9 percent O_2, 2 percent N_2, 1.5 percent S, 8.5 percent ash, and 7 percent moisture. The heating value of the coal is 28,400 kJ/kg. This coal is consumed in a power plant at a rate of 2000 kg/day. The plant operates 65 percent of the time annually.

(a) Determine the amount of ash entering the plant per year.

(b) Determine the amount of ash entering the plant per MJ of heat input to the plant.

(c) If only 2 percent of the ash entering the plant is emitted to the atmosphere with combustion gases, determine the mass of ash emitted per kWh of electricity produced in the plant. Assume the overall plant efficiency is 40 percent (40 percent of the heat input to the plant is converted to electricity).

SOLUTION   (a) The plant operates 65 percent of the time. Then, the total number of operating days a year is

$$\text{Operating days} = (0.65)(365 \text{ day/yr}) = 237 \text{ day/yr}$$

The total amount of coal consumption in a year is

$$\text{Annual coal consumption} = (2000 \text{ kg/day})(237 \text{ day/yr}) = 474{,}000 \text{ kg/yr}$$

Noting that 8.5 percent of coal is ash, the amount of ash entering the plant per year is

$$m_{\text{ash}} = (0.085)(474{,}000 \text{ kg/yr}) = \textbf{40,290 kg/yr}$$

(b) The total heat input to the plant in a year is

$$Q_{\text{in}} = m_{\text{coal}}\text{HV}_{\text{coal}} = (2000 \text{ kg/day})(237 \text{ day/yr})(28{,}400 \text{ kJ/kg}) = 1.346 \times 10^{10} \text{ kJ/yr}$$

The amount of ash entering the plant per MJ of heat input to the plant is then

$$m_{\text{ash}} = \frac{40{,}290 \text{ kg/yr}}{1.346 \times 10^7 \text{ MJ/yr}} = \textbf{0.00299 kg/MJ}$$

(c) The amount of electricity produced in the plant is determined using plant efficiency as

$$\eta_{\text{plant}} = \frac{W_{\text{electric}}}{Q_{\text{in}}} \rightarrow W_{\text{electric}} = \eta_{\text{plant}}Q_{\text{in}} = (0.40)(1.346 \times 10^{10} \text{ kJ/yr})\left(\frac{1 \text{ kWh}}{3600 \text{ kJ}}\right) = 1.496 \times 10^6 \text{ kWh/yr}$$

Noting that 2 percent of the ash entering the plant is emitted to the atmosphere, its amount per year is

$$m_{\text{ash,emitted}} = (0.02)(40{,}290 \text{ kg/yr}) = 805.8 \text{ kg/yr}$$

Then the mass of ash emitted per kWh of electricity produced is

$$\frac{m_{\text{ash,emitted}}}{W_{\text{electricity}}} = \frac{805{,}800 \text{ g/yr}}{1.496 \times 10^6 \text{ kWh/yr}} = \textbf{0.539 g/kWh}$$

Therefore, this plant emits 0.539 g of ash per kWh of electricity generated.   ▲

Sulfur Dioxide

Fossil fuels are mixtures of various chemicals, including some amounts of sulfur. The sulfur in the fuel reacts with oxygen to form *sulfur dioxide* (SO_2), which is an air pollutant:

$$S + O_2 \rightarrow SO_2 \tag{13-4}$$

Coal contains the most of sulfur among fossil fuels. Bituminous coal contains 0.5 to 4.0 percent sulfur by mass with an average value of 1.5 percent. Subbituminous coal typically contains much less sulfur with an average of 0.4 percent by mass. Low-sulfur coal is more desirable and has high monetary value. The mass percentage of sulfur in oil may range between 0.3 and 2.3 percent depending on the source and processing in oil refineries (EPRI, 1997). Natural gas also contains varying amounts of sulfur when extracted from the ground but it is removed in gas treatment plants. Therefore, natural gas combustion does not produce any SO_2.

The main source of SO_2 is the electric power plants that burn high-sulfur coal. The Clean Air Act of 1970 limited SO_2 emissions severely, which forced the plants to install SO_2 scrubbers, to switch to low-sulfur coal, or to gasify the coal and recover the sulfur. Volcanic eruptions and hot springs also release sulfur oxides (the cause of the rotten egg smell).

When coal is burned, more than 95 percent of sulfur is oxidized to SO_2. About 2 to 5 percent of sulfur in the coal does not get burned and ends up in the solid ash particles. Less than 1 percent of the SO_2 formed during combustion further reacts with O_2 to form SO_3. The resulting SO_3 reacts with water vapor in the combustion gases to form sulfuric acid (H_2SO_4). SO_2 emitted from the combustion and other systems can react in the atmosphere to form sulfates, which are components of acid rain.

Motor vehicles also contribute to SO_2 emissions since diesel fuel also contain small amounts of sulfur. Recent regulations have reduced the amount of sulfur in diesel fuel. In Europe the sulfur level in diesel fuel was limited to less than 50 ppm (part per million) in 2005. The acceptable sulfur level in diesel fuel dropped from 50 to 10 ppm in the United States in the 1990s. Currently most diesel fuel in Europe and North America is ultra-low-sulfur diesel with a maximum of 10 ppm sulfur. The exhaust of a diesel engine can contain up to 20 ppm of SO_2.

The most effective method of reducing SO_2 emissions is to use low-sulfur coal in electricity generation plants and heating systems. However, this may not be possible if low-sulfur coal is not available or is expensive compared to high-sulfur coal. Also, installing an SO_2 removal system can be more cost-effective than using low-sulfur coal. Today, all coal-fired power plants must be equipped with *flue gas desulfurization* (FGD) systems, which are very effective in SO_2 removal. In this process, a slurry of pulverized limestone (calcium carbonate—$CaCO_3$) mixed with water is used. The process takes place in a vessel called a scrubber. A mixture of limestone and water is sprayed into the combustion gas containing SO_2 according to the reaction (Rubin, 2001):

$$SO_2 + (CaCO_3 + 2H_2O) + \tfrac{1}{2}O_2 \rightarrow CO_2 + (CaSO_4 \cdot 2H_2O) \qquad (13\text{-}5)$$

The product gypsum ($CaSO_4 \cdot 2H_2O$) has commercial value as it is a main constituent in plaster, blackboard chalk, and wallboard. A modern flue gas desulfurization system utilizing limestone ($CaCO_3$) can remove more than 95 percent of SO_2 in combustion gases. Some systems may also use lime (CaO), which is more effective, and can remove up to 99 percent of SO_2.

Sulfur dioxide is a very toxic pollutant with serious health effects, as shown in Fig. 13-5. Short exposure times at higher SO_2 concentrations can have a significant health impact. Longer exposures at low concentrations and shorter exposures at high concentrations may result in similar health effects. When the SO_2 level in the atmosphere is more than 1 ppm, it can lead to airway constriction. The health effects of SO_2 include respiratory, lung, and cardiovascular illnesses. Children and elderly people with asthma and breathing problems are more sensitive to SO_2 exposure. It also negatively affects ecosystems of rivers, lakes, and forests.

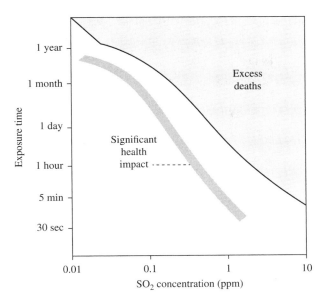

Figure 13-5 Health impact of SO_2 concentration as a function of exposure time. (*Williamson, 1973.*)

Sulfur dioxide is not only a harmful emission but also poisons catalyst materials in catalytic converters. Particulate traps used in reducing carbon soot particles in diesel engines are negatively affected by sulfur components in the exhaust.

Another reaction of sulfur is that it combines with hydrogen at high temperatures to form hydrogen sulfide (H_2S):

$$S + H_2 \rightarrow H_2S \tag{13-6}$$

H_2S is a colorless, poisonous, corrosive, flammable, and explosive gas with a rotten egg smell. Long-term exposure to H_2S is associated with nervous and respiratory system and eye effects.

EXAMPLE 13-2
SO_2 Emission from a
Coal-Fired Power Plant

A coal-fired power plant generates 300 MW of power with an overall efficiency of 35 percent. The capacity factor of the plant is 70 percent. The coal used in the plant has the following composition on a mass basis: 67 percent C, 5 percent H_2, 9 percent O_2, 2 percent N_2, 1.5 percent S, 8.5 percent ash, and 7 percent moisture. The heating value of this coal is 28,400 kJ/kg.

(a) Determine the amount of sulfur entering the plant per year.

(b) Determine the amount of SO_2 produced after the combustion process if all sulfur reacts with oxygen (O_2) during combustion.

(c) A flue gas desulfurization (FGD) system is used to remove 98 percent of SO_2. Determine the annual amounts of SO_2 emitted to the atmosphere, limestone ($CaCO_3$) used, and gypsum ($CaSO_4 \cdot 2H_2O$) produced in the FGD system.

SOLUTION (a) The capacity factor of the plant is given to be 70 percent. That is, the plant operates 70 percent of the time. Then, the annual operating hours are

Operating hours = $(0.70)(365 \times 24 \text{ h/yr}) = 6132 \text{ h/yr}$

The rate of heat input to the plant is determined using plant efficiency as

$$\eta_{plant} = \frac{\dot{W}_{electric}}{\dot{Q}_{in}} \rightarrow \dot{Q}_{in} = \frac{\dot{W}_{electric}}{\eta_{plant}} = \frac{300\ MW}{0.35} = 857.1\ MW$$

The amount of heat input to the plant per year is

$$Q_{in} = \dot{Q}_{in} \times \text{Operating hours} = (857,100\ kJ/s)(6132\ h/yr)\left(\frac{3600\ s}{1\ h}\right) = 1.892 \times 10^{13}\ kJ/yr$$

The corresponding coal consumption by the plant per year is

$$Q_{in} = m_{coal}HV_{coal} \rightarrow m_{coal} = \frac{Q_{in}}{HV_{coal}} = \frac{1.892 \times 10^{13}\ kJ/yr}{28,400\ kJ/kg} = 6.662 \times 10^{8}\ kg/yr$$

The mass percentage of sulfur in coal is given to be 1.5 percent. Then, the amount of sulfur entering the plant per year becomes

$$m_{sulfur} = (0.015)(6.662 \times 10^{8}\ kg/yr) = \mathbf{9.993 \times 10^{6}\ kg/yr}$$

(b) Note that SO_2 is produced according to the reaction $S + O_2 \rightarrow SO_2$. That is, for 1 kmol of S entering the reaction, 1 kmol of SO_2 is produced. The molar masses are 32 kg/kmol for S, 32 kg/kmol for O_2, and 64 kg/kmol for SO_2. Therefore, for 32 kg of sulfur entering the combustion process, 64 kg of SO_2 is produced:

$$m_{SO_2} = m_{sulfur}\frac{M_{SO_2}}{M_{sulfur}} = (9.993 \times 10^{6}\ kg/yr)\frac{64\ kg/kmol}{32\ kg/kmol} = \mathbf{1.999 \times 10^{7}\ kg/yr}$$

(c) Noting that 2 percent of SO_2 produced after combustion is eventually emitted to the atmosphere, its amount per year is

$$m_{SO_2,emitted} = (0.02)(1.999 \times 10^{7}\ kg/yr) = \mathbf{399,800\ kg/yr}$$

The reaction for the flue gas desulfurization system is

$$SO_2 + (CaCO_3 + 2H_2O) + \tfrac{1}{2}O_2 \rightarrow CO_2 + (CaSO_4 \cdot 2H_2O)$$

When treating 1 kmol of SO_2, 1 kmol of limestone ($CaCO_3$) is used and 1 kmol of gypsum ($CaSO_4 \cdot 2H_2O$) is produced. The molar mass of calcium is 40 kg/kmol and that of limestone ($CaCO_3$) is

$$M_{CaCO_3} = 40 + 12 + 48 = 100\ kg/kmol$$

The molar mass of gypsum ($CaSO_4 \cdot 2H_2O$) is

$$M_{gypsum} = 40 + 32 + 2 \times 32 + 2 \times 18 = 172\ kg/kmol$$

The amount of limestone ($CaCO_3$) used in the FGD system is

$$m_{CaSO_4} = m_{SO_2}\frac{M_{CaSO_4}}{M_{SO_2}} = (1.999 \times 10^{7}\ kg/yr)\frac{100\ kg/kmol}{32\ kg/kmol} = \mathbf{6.247 \times 10^{7}\ kg/yr}$$

Noting that the flue gas desulfurization system removes 98 percent of SO_2, the amount of gypsum ($CaSO_4 \cdot 2H_2O$) produced is

$$m_{gypsum} = (0.98)m_{SO_2}\frac{M_{gypsum}}{M_{SO_2}} = (0.98)(1.999 \times 10^{7}\ kg/yr)\frac{172\ kg/kmol}{32\ kg/kmol} = \mathbf{1.053 \times 10^{8}\ kg/yr}$$

Therefore, a 300-MW coal-fired power plant involves approximately 400 tons of SO_2 emission to the atmosphere per year. In the flue gas desulfurization system, approximately 62,000 tons of limestone are consumed and 105,000 tons of gypsum are produced per year. ▲

Nitrogen Oxides

Nitrogen oxide (NO_x) and *nitrogen dioxide* (NO_2) are produced during the combustion of fossil fuels such as coal, oil, and natural gas. Electrical power plants, automobiles, and other vehicles burning fossil fuels are responsible for nitric oxide (NO) and nitrogen dioxide (NO_2) emissions. Together these are called nitrogen oxides or oxides of nitrogen (NO_x). Most of the NO_x emissions from a combustion system to the atmosphere are NO, but it gradually turns into NO_2 by reacting with oxygen in air.

Nitrogen oxides are produced at high combustion temperatures when a fuel is burned with oxygen in air. A number of possible reactions can contribute the formation of NO and NO_2. Some of these reactions are as follows:

$$N_2 + O \rightarrow NO + N \tag{13-7}$$

$$N + O_2 \rightarrow NO + O \tag{13-8}$$

$$N + OH \rightarrow NO + H \tag{13-9}$$

$$NO + O_2 \rightarrow NO_2 + O \tag{13-10}$$

$$NO + H_2O \rightarrow NO_2 + H_2 \tag{13-11}$$

In these reactions, monatomic oxygen (O) reacts with N_2 and monatomic nitrogen (N) reacts with O_2 to form NO. Also, N and OH react to produce NO. Nitrogen is normally stable in air as diatomic (N_2) and does not react with oxygen. However, some nitrogen breaks down at high combustion temperatures as a result of *dissociation* reactions.

$$N_2 \rightarrow 2N \tag{13-12}$$

At high combustion temperatures, O_2 and H_2O also dissociate as follows:

$$O_2 \rightarrow 2O \tag{13-13}$$

$$H_2O \rightarrow OH + \tfrac{1}{2}H_2 \tag{13-14}$$

The monatomic nitrogen N can easily react with O_2 to form NO, N_2 reacts with O to form NO, and NO reacts with O_2 or H_2O to form NO_2.

NO_x can also be formed by the reactions of atomic nitrogen and oxygen particles in fuel. This is particularly the case for coal as it contains about 10 percent oxygen and 1 percent nitrogen by mass. Natural gas also contains small amounts of nitrogen and oxygen. For higher NO_x formation levels, the nitrogen in fuel usually makes a greater contribution to NO_x formation compared to that due to N_2 and O_2 in air.

In an internal combustion engine, the combustion temperatures can reach 2500 or 3000 K, allowing some dissociation reactions, which result in NO_x formation. At higher combustion temperatures, more dissociation reactions cause greater amounts of NO_x production. Very little NO_x is formed at low temperatures.

Nitrogen oxide emissions are greatly reduced in gasoline-burning automobiles by the use of catalytic converters. In diesel engines, catalytic converters are not very effective in reducing NO_x. Instead, *exhaust gas recycling* (EGR) is used effectively to control NO_x generation. In EGR, some exhaust gases (up to 30 percent) are diverted back to the combustion chamber. These gases do not react with air but absorb heat in the chamber, thus reducing the combustion temperature. In some large internal combustion engines such as those used in electricity production and in ships, tiny water droplets are injected into the combustion chamber. Water evaporates, absorbing heat from the surroundings. This reduces combustion temperature and levels of NO_x formation.

An effective method of NO_x control involves design and optimum operation of the combustion process in combustion chambers and burners. This includes controlling combustion temperature, combustion time, and some design characteristics of the burner. The burners specifically designed for controlling NO_x emission are called *low-NO_x burners*.

Another method of NO_x control involves a chemical treatment system to remove NO_x components in combustion gases. In coal-fired power plants, a *selective catalytic reduction* (SCR) unit is installed into the combustion gas stream right after the boiler (Fig. 13-3). The system uses a catalyst material to promote the conversion of NO_x components into nonharmful molecular nitrogen N_2. Ammonia (NH_3) is also injected into the combustion gas stream during the following reactions:

$$NO + NH_3 + \tfrac{1}{4}O_2 \rightarrow N_2 + \tfrac{3}{2}H_2O \tag{13-15}$$

$$NO_2 + 2NH_3 + \tfrac{1}{2}O_2 \rightarrow \tfrac{3}{2}N_2 + 3H_2O \tag{13-16}$$

These reactions require a temperature of about 400°C, and this is the reason the SCR system is installed right after the boiler of the power plant. The SCR systems can typically remove 70 to 90 percent of NO_x in the combustion gases.

Nitrogen oxides react in the atmosphere to form ground-level ozone. It contributes to photochemical smog and acidification of waters and soils. The emission of NO_x causes the formation of nitrates, which are considered PM (mostly PM2.5). Nitrogen dioxide (NO_2) gas is toxic at high concentrations. The primary health effects of NO_2 are related to respiratory illnesses such as asthma and bronchitis.

EXAMPLE 13-3
NO_x Emissions in a
Natural Gas Power Plant

The maximum allowable NO_x emissions for fossil fuel power plants built after 1997 in the United States are specified to be 0.72 g/kWh or 65 ng/J (65 nanograms per J of fuel heat input to the boiler). A natural gas–fired power plant consumes 15 kg/s natural gas and produces 375 MW of power. It is estimated that NO_x emissions from this power plant are 65 ng/J. The heating value of natural gas is 55,000 kJ/kg.

(a) Determine the overall plant efficiency.

(b) Determine the equivalent NO_x emissions in terms of g/kWh (grams per kWh of electricity produced).

(c) If the plant operates 8000 h in a year, determine the amount of NO_x emissions per year.

SOLUTION   (a) The rate of heat input to the plant is

$$\dot{Q}_{in} = \dot{m}_{nat.\ gas} HV_{nat.\ gas} = (15\ \text{kg/s})(55,000\ \text{kJ/kg}) = 825,000\ \text{kJ/s} = 825,000\ \text{kW}$$

The plant efficiency is defined as the power generated divided by the rate of heat input in the boiler. That is,

$$\eta_{plant} = \frac{\dot{W}_{electric}}{\dot{Q}_{in}} = \frac{375,000\ \text{kW}}{825,000\ \text{kW}} = 0.454 = \textbf{45.4\%}$$

(b) The plant efficiency shows that for 1 J of heat input to the plant, 0.4545 J of electricity is generated. Then, the specified NO_x emissions can be expressed in g/kWh unit as

$$NO_x\ \text{emission} = (65 \times 10^{-9}\ \text{g/J heat})\left(\frac{1\ \text{J heat}}{0.4545\ \text{J electricity}}\right)\left(\frac{1000\ \text{J}}{1\ \text{kJ}}\right)\left(\frac{3600\ \text{kJ}}{1\ \text{kWh}}\right) = \textbf{0.515 g/kWh}$$

(*c*) The amount of heat input for annual operating hours of 8000 h is

$$Q_{in} = \dot{Q}_{in} \times \text{Operating hours} = (825{,}000 \text{ kJ/s})(8000 \text{ h/yr})\left(\frac{3600 \text{ s}}{1 \text{ h}}\right) = 2.376 \times 10^{13} \text{ kJ/yr}$$

Noting that the NO_x emissions from this power plant are 65 ng/J of heat input, the amount of NO_x emissions per year is determined as

$$m_{NO_x} = (65 \times 10^{-9} \text{ g/J})(2.376 \times 10^{13} \text{ kJ/yr})\left(\frac{1000 \text{ J}}{1 \text{ kJ}}\right)\left(\frac{1 \text{ kg}}{1000 \text{ g}}\right) = \textbf{1.544} \times \textbf{10}^{\textbf{6}} \text{ \textbf{kg/yr}}$$

Therefore, this natural gas–fired power plant is responsible for 1544 tons of NO_x emissions to the environment a year. ▲

Hydrocarbons

Fossil fuels (coal, oil, natural gas) are primarily made of *hydrocarbons* (HC). In a combustion reaction, some of the fuel cannot find oxygen to react with during the combustion period. As a result, some unburned or partially burned fuel particles leave the exhaust gases as HC components. This will certainly happen when there is deficiency of air (not enough air to burn all the fuel) during combustion. However, having sufficient (i.e., stoichiometric) or excess air does not guarantee that all fuel will be burned at the end of the combustion process. Insufficient mixing between fuel and air, nonhomogeneous mixture, and short time of reaction are some of the causes for incomplete combustion and resulting HC emission.

The terms HC and *volatile organic compounds* (VOC) are usually used interchangeably. However, some of the VOCs found in an automobile exhaust also contain components other than HCs such as aldehydes.

A significant portion of the VOC or HC emissions in vehicles are caused by the evaporation of fuels during refueling or spillage during spit back or by evaporation from gas tanks with faulty caps that do not close tightly. The solvents, propellants, and household cleaning products that contain benzene, butane, or other HC products are also significant sources of HC emissions. Additional sources of HC emissions include venting of crankcase, petroleum refineries, chemical plants, gasoline and diesel fuel distribution and storage facilities, dry cleaners, and other processes involving chemical solvents.

When diesel fuel is burned in an internal combustion engine, combustion gases contain unburned carbon (C) soot particles, which are sometimes seen as a black smog. Carbon soot particles are usually in the form of spheres with diameters between 10 and 80 nm (or 10×10^{-9} and 80×10^{-9} m). The particles are essentially solid carbon with some HC on the surface.

Carbon particles are generated during combustion of an HC fuel in the fuel-rich zones with insufficient air:

$$C_m H_n + w \, (O_2 + 3.76 N_2) \rightarrow a \, CO_2 + b \, H_2O + c \, CO + d \, N_2 + e \, C \text{ (solid)} \quad \text{(13-17)}$$

These soot particles may find oxygen to react with by the time the combustion is finished:

$$C \text{ (solid)} + O_2 \rightarrow CO_2 \quad \text{(13-18)}$$

Therefore, most of the carbon particles generated initially in the combustion chamber will be burned into CO_2, and never get exhausted to the atmosphere. Most diesel automotive engines are equipped with *particulate traps* to collect soot particles. The traps are cleaned/regenerated periodically by burning the collected particles.

In a gasoline automotive engine, up to 6000 ppm of HC components exist in the exhaust after combustion. This represents 1 to 1.5 percent of the fuel. About half of these are unburned fuel components while the rest are partially burned. Fortunately, most of these components are treated in the catalytic converter before exhaust gases end up in the atmosphere.

Compared to gasoline engines, much less HC is found in the engine exhaust of a diesel engine. The reason is that the air-fuel ratio in diesel engines (between 18 and 70) is typically much higher than those in gasoline engines (between 12 and 18). If diesel engines did not use considerable amounts of excess air for combustion, the carbon soot particle emissions would be higher than most legal limits.

Hydrocarbon components and carbon soot particles in air act as irritants and odorants. Some of them are believed to be carcinogenic. Nitrogen oxides and HC are two main sources for the formation of ground-level ozone. Except for CH_4, HC components react in atmosphere to form photochemical smog.

Carbon Monoxide

Carbon monoxide (CO) is a colorless, odorless, poisonous gas. It is mostly emitted by industry and motor vehicles, and it can build to dangerous levels in areas with heavy congested traffic. CO is produced during combustion when there is *fuel-rich mixture* (not enough air or oxygen in the air-fuel mixture). During the combustion of an HC fuel, carbon reacts with oxygen according to the reaction

$$C + \tfrac{1}{2}O_2 \rightarrow CO \tag{13-19}$$

The resulting CO is also a fuel with energy content (HV = 10,100 kJ/kg). It further reacts with oxygen again to form CO_2 to complete the reaction:

$$CO + \tfrac{1}{2}O_2 \rightarrow CO_2 \tag{13-20}$$

If CO cannot find oxygen after the first reaction due to deficiency of oxygen or insufficient mixing, some CO will be part of the exhaust. CO emission can be minimized by avoiding fuel-rich mixtures, better mixing of fuel and air, and increasing combustion time.

In a gasoline automobile, most CO is produced at engine start-up and acceleration, during which the engine runs with fuel-rich mixtures. The percentage of CO in the exhaust of a gasoline engine can be as low as 0.1 percent and as high as 5 percent. Diesel engines generate very small amounts of CO since they use excess air in fuel-air mixtures.

Carbon monoxide deprives the body's organs from getting enough oxygen by binding with the red blood cells that would otherwise carry oxygen. At low levels, CO decreases the amount of oxygen supplied to the brain and other organs and muscles, slows body reactions and reflexes, and impairs judgment. It poses a serious threat to people with heart disease because of the fragile condition of the circulatory system and to fetuses because of the oxygen needs of the developing brain. At high levels, it can be fatal, as evidenced by numerous deaths caused by cars that are warmed up in closed garages or by exhaust gases leaking into the cars.

Hemoglobin is the protein molecule in red blood cells that carries oxygen from the lungs to the body's tissues and returns carbon dioxide from the tissues back to the lungs. Carbon monoxide absorbed in hemoglobin is called carboxyhemoglobin (COHb). Negative physiological effects on the body are observed when the level of COHb is above 100 ppm (Fig. 13-6). Serious effects, including coma, take place after 300 ppm, and long-term exposure above 600 ppm is fatal.

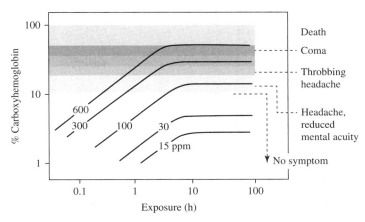

Figure 13-6 Effect of carboxyhemoglobin (COHb: CO absorbed in hemoglobin) on the human body. (*Adapted from Seinfeld, 1975.*)

Ozone, Smog, and Acid Rain

If you live in a metropolitan area such as Los Angeles, you are probably familiar with urban smog (smoke + fog = smog)—the dark yellow or brown haze that builds up in a large, stagnant air mass and hangs over populated areas on calm, hot summer days. *Smog (also called photochemical smog)* is made up mostly of ground-level ozone (O_3), but it also contains numerous other chemicals, including CO, PM such as soot and dust, and volatile organic compounds (VOCs) such as benzene, butane, and other HCs.

The harmful *ground-level ozone* should not be confused with the useful ozone layer high in the stratosphere that protects the earth from the sun's harmful ultraviolet (UV) rays. Ozone at ground level is a pollutant with several adverse health effects. It is harmful to lungs, other biological tissue, and trees. It can react with rubber, plastics, and other materials with serious damage. Agricultural crop losses due to ozone in the United States are estimated to be more than half a billion dollars per year.

The primary sources of NO_x and HCs are motor vehicles and power plants. Hydrocarbons and NO_x react in the presence of sunlight on hot, calm days to form ground-level ozone, which is the primary component of smog (Fig. 13-7). Possible reactions to produce smog and ozone are

$$NO_2 + \text{energy from sunlight} \rightarrow NO + O + \text{smog} \tag{13-21}$$

$$O + O_2 \rightarrow O_3 \tag{13-22}$$

The smog formation usually peaks in late afternoon when the temperatures are highest and there is plenty of sunlight. Although ground-level smog and ozone form in urban areas with heavy traffic or industry, the prevailing winds can transport them several hundred miles to other cities. This shows that pollution knows no boundaries, and it is a global problem.

Ozone irritates eyes and damages the air sacs in the lungs where oxygen and carbon dioxide are exchanged, causing eventual hardening of this soft and spongy tissue. It also causes shortness of breath, wheezing, fatigue, headaches, and nausea and aggravates respiratory problems such as asthma. Every exposure to ozone does a little damage to the lungs, just like cigarette smoke, eventually reducing the individual's lung capacity. Staying indoors

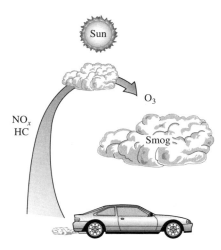

Figure 13-7 Ground-level ozone, which is the primary component of smog, forms when HC and NO_x react in the presence of sunlight in hot calm days.

and minimizing physical activity during heavy smog minimizes damage. Ozone also harms vegetation by damaging leaf tissues. To improve the air quality in areas with the worst ozone problems, reformulated gasoline (RFG) that contains at least 2 percent oxygen was introduced. The use of RFG has resulted in significant reduction in the emission of ozone and other pollutants, and its use is mandatory in many smog-prone areas.

The most effective method of reducing ground-level ozone involves minimizing NO_x and HC emissions. Large reductions of HC emissions from motor vehicles in recent decades have improved conditions. However, significant reductions in NO_x and further reductions in HC emissions are needed to deal with the challenges of controlling ozone levels in the atmosphere.

The other serious pollutant in smog is CO. Smog also contains suspended PM, such as dust and soot emitted by vehicles and industrial facilities. Such particles irritate the eyes and the lungs since they may carry compounds such as acids and metals.

The sulfur oxides and nitric oxides react with water vapor and other chemicals high in the atmosphere in the presence of sunlight to form sulfuric and nitric acids (Fig. 13-8). The acids formed usually dissolve in the suspended water droplets in clouds or fog. These acid-laden droplets, which can be as acidic as lemon juice, are washed from the air onto the soil by rain or snow. This is known as *acid rain*.

Sulfur dioxide reacts with water and dissolved oxygen to produce sulfuric acid (H_2SO_4):

$$H_2O + SO_2 + \tfrac{1}{2}O_2 \rightarrow H_2SO_4 \qquad (13\text{-}23)$$

Another reaction of SO_2 produces sulfurous acid (H_2SO_3):

$$H_2O + SO_2 \rightarrow H_2SO_3 \qquad (13\text{-}24)$$

Both sulfuric acid and sulfurous acid are components of acid rain. Acid droplets can penetrate deeply into lungs during respiration. When SO_2 produced from power plants is transported through long distances due to winds, it may be transformed into sulfate particles and other acidic species. Also, NO_x emissions cause the formation of nitrate species. Sulfates, nitrates, and other acidic species contribute to acid rain.

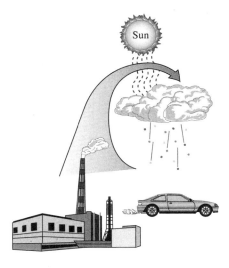

Figure 13-8 Sulfuric acid and nitric acid are formed when sulfur oxides and nitric oxides react with water vapor and other chemicals high in the atmosphere in the presence of sunlight.

The soil is capable of neutralizing a certain amount of acid, but the amounts produced by power plants using inexpensive high-sulfur coal has exceeded this capability, and as a result many lakes and rivers in industrial areas such as New York, Pennsylvania, and Michigan have become too acidic for fish to grow. Forests in those areas also experience a slow death from absorbing the acids through their leaves, needles, and roots. Soil is also adversely affected by acid deposition with consequences such as soil erosion, sedimentation of waterways, and changes in animal habitat. Even limestone and marble structures deteriorate due to acid rain. The magnitude of the problem was not recognized until the early 1970s, and serious measures have been taken since then to reduce SO_2 emissions drastically by installing scrubbers in plants and by desulfurizing coal before combustion.

EXAMPLE 13-4
Rate of Sulfurous Acid Added to the Environment

A diesel engine operates on an average air-fuel ratio of 18 kg air/kg fuel. The engine uses light diesel fuel that contains 750 ppm of sulfur by mass. Assume that all of this sulfur is converted to SO_2 in the combustion chamber and all of the SO_2 produced is converted to sulfurous acid (H_2SO_3) in the atmosphere. If the rate of the air entering the engine is 336 kg/h, determine the mass flow rate of H_2SO_3 added to the environment.

SOLUTION Using the definition of air-fuel ratio, we determine the mass flow rates of fuel and the sulfur in the fuel to be

$$AF = \frac{\dot{m}_{air}}{\dot{m}_{fuel}} \rightarrow \dot{m}_{fuel} = \frac{\dot{m}_{air}}{AF} = \frac{336 \text{ kg air/h}}{18 \text{ kg air/kg fuel}} = 18.67 \text{ kg fuel/h}$$

$$\dot{m}_{sulfur} = (750 \times 10^{-6})\dot{m}_{fuel} = (750 \times 10^{-6})(18.67 \text{ kg/h}) = 0.0140 \text{ kg/h}$$

SO_2 is produced according to the reaction

$$S + O_2 \rightarrow SO_2$$

Molar masses of sulfur and SO_2 are 32 and 64 kg/kmol, respectively. Therefore, for 32 kg of sulfur entering the combustion process, 64 kg of SO_2 are produced. Then,

$$\dot{m}_{SO_2} = \dot{m}_{sulfur} \frac{M_{SO_2}}{M_{sulfur}} = (0.0140 \text{ kg/h}) \frac{64 \text{ kg/kmol}}{32 \text{ kg/kmol}} = 0.0280 \text{ kg/h}$$

The conversion of SO_2 to H_2SO_3 takes place through the reaction

$$H_2O + SO_2 \rightarrow H_2SO_3$$

The molar mass of H_2SO_3 is 82 kg/kmol. Therefore, for 64 kg of SO_2 entering the reaction, 82 kg of H_2SO_3 is produced:

$$\dot{m}_{H_2SO_3} = \frac{M_{H_2SO_3}}{M_{SO_2}} \dot{m}_{SO_2} = \frac{82 \text{ kg/kmol}}{64 \text{ kg/kmol}} (0.0280 \text{ kg/h}) = \textbf{0.0359 kg/h}$$

This diesel engine is responsible for adding 0.0359 kg/h of sulfurous acid to the environment. If the diesel engine used low-sulfur diesel fuel, much less H_2SO_3 would be released to the atmosphere. ▲

13-3 EMISSIONS FROM AUTOMOBILES

Automobiles are the most noticeable source of air pollution and greenhouse gas emissions. Pollutant emissions from automobiles and other internal combustion engine powered vehicles and machines have decreased by more than 90 percent since the 1970s. However, the number of automobiles has greatly increased over the same period, resulting in a significant increase in pollutant and greenhouse gas emissions.

There are four major pollutant emissions from automobiles powered by internal combustion engines burning gasoline or diesel fuel:

- Hydrocarbons (HC). Also called volatile organic compounds (VOC).
- Nitrogen oxides (NO_x).
- Carbon monoxide (CO).
- Carbon soot (C) particles. Primarily emitted from diesel fuel vehicles.

Table 13-2 gives average emissions from different type of motor vehicles in the United States. For passenger cars, HC, CO, and NO_x emissions are 0.79, 8.73, and 0.56 g/mi, as shown in Table 13-2. The corresponding emission values were 4.0, 42.9, and 2.7 g/mi in 1990 (EPA, 2010). These values indicate that HC, CO, and NO_x emissions from automobiles were decreased by about 80 percent over a 20-year period.

The amount of air used for a unit mass of fuel has considerable effect on emissions from automobiles. You may recall that the *air-fuel ratio* is defined as

$$AF = m_{air}/m_{fuel} \tag{13-25}$$

If the theoretical minimum amount of air is used in a combustion process, the air used is called *stoichiometric* or *theoretical air* and the air-fuel ratio is denoted by AF_s. The ratio of stoichiometric air-fuel ratio to actual air-fuel ratio is defined as *equivalence ratio*:

$$\phi = AF_s/AF_a \tag{13-26}$$

TABLE 13-2   **Average Emissions from Automobiles in the United States** (EPA, 2010; EPA, 2008)

	Average Emission (g/mi)	Annual Emission (kg/yr)
Light-duty vehicles (passenger cars)		
HC	0.79	10,700
CO	8.73	117,900
NO$_x$	0.56	7560
PM10	0.0044	59
PM2.5	0.0041	55
CO$_2$	368	4,968,000
Gasoline consumption	0.04149 gal/mi	560 gal/yr
Light-duty trucks		
HC	1.01	13,600
CO	11.02	148,800
NO$_x$	0.81	10,900
PM2.5	0.0049	66
PM2.5	0.0045	61
CO$_2$	514	6,939,000
Gasoline consumption	0.05780 gal/mi	780 gal/yr
Heavy-duty vehicles		
HC	1.14	
CO	9.42	
NO$_x$	2.25	
Motorcycles		
HC	2.29	
CO	14.59	
NO$_x$	1.25	
Diesel, light-duty vehicle		
HC	0.18	
CO	0.90	
NO$_x$	0.42	
Diesel, light-duty trucks		
HC	0.44	
CO	0.76	
NO$_x$	0.72	
Diesel, heavy-duty vehicles		
HC	0.39	
CO	1.75	
NO$_x$	6.87	
Average, gasoline and diesel		
HC	0.89	12,000
CO	9.37	126,500
NO$_x$	1.30	17,600

Notes:
- Annual emissions data are based on an average 13,500 mi of travel per year.
- Vehicles types are defined as follows: light-duty vehicles (passenger cars up to 6000 lbm); light-duty trucks (pickups and minivans up to 8500 lbm); heavy-duty vehicles (8501 lbm or more); motorcycle (highway only).
- Emissions factors are national averages based on the following assumptions: ambient temperature 75°F, daily temperature range 60 to 84°F, average traffic speed 27.6 mph (representative of overall traffic in urban areas).

If the air-fuel mixture has more than theoretical air (excess air), the mixture is called lean (fuel-lean), and the equivalence ratio is less than 1. If the mixture has less than theoretical air (deficiency of air), the mixture is called rich (fuel-rich), and the equivalence ratio is greater than 1.

Figure 13-9 shows the effect of equivalence ratio on the amount of emissions in a gasoline-burning internal combustion engine. HC and CO emissions can be minimized by using fuel-lean mixtures. The reason for increased HC emissions at very lean mixtures is the occurrence of misfires. That is, combustion is not taking place during some engine cycles when the mixture is very lean. All fuel during these particular cycles is released to the atmosphere. NO_x emissions are maximum at slightly lean conditions for which the combustion temperature is high and there is some extra oxygen for monatomic nitrogen with which to react.

Environmental impacts of automobiles are not limited to pollutant and greenhouse gas emissions with the exhaust. For example, disposal of cars with minimum impact on the environment remains a challenge. In the United States, about 75 percent of car materials are recycled and reused. The remaining 25 percent is buried in a landfill. Lead emissions from the use of leaded gasoline in some parts of the world, chlorofluorocarbon (CFC) emissions from automobile air-conditioning, and waste motor oil are among other environmental issues related to automobiles.

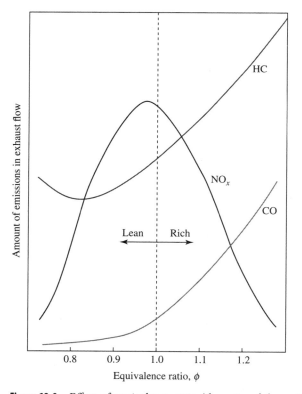

Figure 13-9   Effect of equivalence ratio (the ratio of theoretical air to actual air) on the amount of emissions in a gasoline-burning internal combustion engine. (*Adapted from Heywood, 1988 and Pulkrabek, 2004.*)

EXAMPLE 13-5
Emissions from a Car

An average car emits 0.79 g of HC, 8.73 g CO, and 0.56 g NO_x per mile. How much HC, CO, and NO_x will be emitted by an average car during a 10-year period if it is driven 13,000 mi/yr?

SOLUTION   The amount of each emission during a 10-year period is

$$HC\ emission = Time\ period \times Annual\ drive \times Unit\ HC\ emission$$

$$= (10\ yr)(13{,}000\ mi/yr)(0.79\ g/mi)\left(\frac{1\ kg}{1000\ g}\right)$$

$$= \mathbf{103\ kg}$$

$$CO\ emission = Time\ period \times Annual\ drive \times Unit\ CO\ emission$$

$$= (10\ yr)(13{,}000\ mi/yr)(8.73\ g/mi)\left(\frac{1\ kg}{1000\ g}\right)$$

$$= \mathbf{1135\ kg}$$

$$NO_x\ emission = Time\ period \times Annual\ drive \times Unit\ NO_x\ emission$$

$$= (10\ yr)(13{,}000\ mi/yr)(0.56\ g/mi)\left(\frac{1\ kg}{1000\ g}\right)$$

$$= \mathbf{73\ kg}$$

Therefore, an average car is responsible for about 10 kg of HC, 114 kg of CO, and 7 kg of NO_x emissions to the environment per year. ▲

Catalytic Converters

Catalytic converters are commonly used in automobile engines for aftertreatment and are located in the exhaust system. Hot exhaust gases leaving the combustion chamber are forced to flow into the converter, which is a chemical chamber. Some catalyst materials are used in the chamber to promote treatment reactions. HC and CO that exist in the combustion gases react with O_2 in the converter to form H_2O and CO_2:

$$C_mH_n + w\ O_2 \rightarrow a\ H_2O + b\ CO_2 \qquad (13\text{-}27)$$

$$CO + \tfrac{1}{2}O_2 \rightarrow CO_2 \qquad (13\text{-}28)$$

Here, m, n, w, a, b, and c are some suitable constants and are needed to balance the chemical equation. Another possible reaction is called water-gas shift, which uses water instead of oxygen as the oxidant:

$$CO + H_2O \rightarrow CO_2 + H_2 \qquad (13\text{-}29)$$

Several reactions are possible to convert NO_x to acceptable components such as (Pulkrabek, 2004)

$$NO + CO \rightarrow \tfrac{1}{2}N_2 + CO_2 \qquad (13\text{-}30)$$

$$2NO + CO \rightarrow N_2O + CO_2 \qquad (13\text{-}31)$$

$$NO + H_2 \rightarrow \tfrac{1}{2}N_2 + H_2O \qquad (13\text{-}32)$$

$$2NO + H_2 \rightarrow H_2O + H_2O \tag{13-33}$$

$$2NO + 5CO + 3H_2O \rightarrow 2NH_3 + 5CO_2 \tag{13-34}$$

$$2NO + 5H_2 \rightarrow 2NH_3 + 2H_2O \tag{13-35}$$

The above reactions readily take place in the converter at chamber temperatures of 600 to 700°C. However, due to the existence of catalyst materials in the converter, these reactions can occur 250 to 300°C. Since exhaust temperatures found in automobile and other transport engines are normally between 300 and 500°C, catalytic converters are very effective in reducing harmful emissions.

A well-working catalytic converter reduces HC, CO, and NO_x components by more than 90 percent when the converter temperature is 400°C or above. Converter effectiveness declines when the temperature goes down. For a converter to effectively treat HC and CO components, the air-fuel mixture should have some excess air. Therefore, the converter is most effective for lean air-fuel mixtures. Sometimes gasoline engines operate with rich mixtures, especially during start-up and acceleration, and the catalytic converter becomes very ineffective during these periods. On the other hand, the converter is most effective in NO_x conversion when the mixture is close to stoichiometric (theoretical). At lean mixtures, the converter becomes ineffective for treating NO_x components.

Diesel engines also use catalytic converters, but their design should also include the conversion of carbon soot particles. This is done by using larger flow passages in the converter. Since diesel engines generally use lean mixtures, their converter is not effective in treating NO_x components. The solution for diesel engines is to use exhaust gas recycling (EGR), as mentioned earlier.

Catalytic converters are called *three-way converters* as they mainly work on reducing three major automobile-related pollutants: HC, CO, and NO_x. Some catalytic converters are basically a stainless steel container mounted in the exhaust system, but close to the engine. The inside of the container is a porous ceramic structure with packed spheres. A more common type is a single honeycomb structure in which there are multiple flow passages for the exhaust (Fig. 13-10).

Catalyst materials are located on the surface of ceramic passages. Their task is to accelerate chemical reactions for the treatment of the pollutants. The most common material used as the ceramic material of the converter is aluminum oxide (alumina). Platinum, rhodium, palladium, and iridium are commonly used as catalyst materials. Platinum is particularly effective in HC reaction, palladium with CO reaction, and rhodium with NO_x reactions.

Some components that may exist in fuel such as lead and sulfur and other components coming from engine oil additives such as zinc, phosphorus, antimony, calcium, and magnesium can seriously harm catalytic converters by poisoning catalyst materials. Lead is particularly to be avoided as it poisons catalyst materials quickly. An engine with a catalytic converter must not use leaded gasoline. Leaded gasoline has been phased out in gasoline engines since the 1990s as lead is a poisonous pollutant with significant adverse effects on the liver and kidney. It also accumulates in blood, bone, and soft tissues when ingested. About 0.15 g of lead was present in leaded gasoline, with 10 to 50 percent ending up in the engine exhaust.

Diesel fuel contains some sulfur, which is converted to SO_2 after the combustion process. When the combustion gases containing some SO_2 flow through the catalytic converter, SO_2 may react with O_2, forming SO_3 with the help of some catalyst materials. SO_3 is then converted to H_2SO_4, an ingredient of acid rain. This process degrades the catalytic converter.

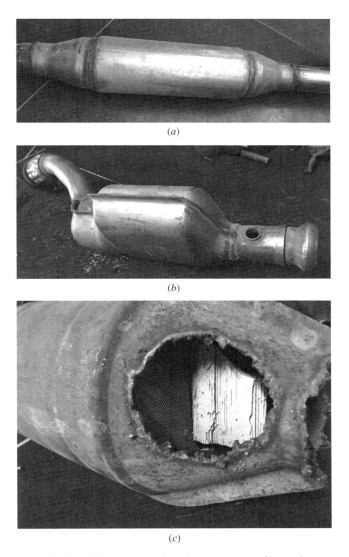

Figure 13-10   Different types of catalytic converters for gasoline engines. (*a*) Honeycomb structure. (*b*) Packed spheres. (*c*) A damaged converter showing inside structure.

13-4 THE GREENHOUSE EFFECT

You have probably noticed that when you leave your car under direct sunlight on a sunny day, the interior of the car gets much warmer than the air outside, and you may have wondered why the car acts like a heat trap. This is because glass at thicknesses encountered in practice transmits over 90 percent of radiation in the visible range and is practically opaque (nontransparent) to radiation in the longer wavelength infrared regions. Therefore, glass allows the solar radiation to enter freely but blocks the infrared radiation emitted by the interior surfaces. This causes a rise in the interior temperature as a result of the thermal energy buildup in the car. This heating effect is known as the *greenhouse effect*, since it is utilized primarily in greenhouses.

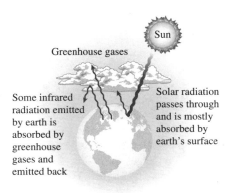

Figure 13-11 The greenhouse effect on earth.

The greenhouse effect is also experienced on a larger scale on earth. The surface of the earth, which warms up during the day as a result of the absorption of solar energy, cools down at night by radiating part of its energy into deep space as infrared radiation. *Carbon dioxide* (CO_2), water vapor (H_2O), and trace amounts of some other gases such as methane (CH_4) and nitrous oxide (N_2O) act like a blanket and keep the earth warm at night by blocking the heat radiated from the earth (Fig. 13-11). Therefore, they are called *greenhouse gases*, with CO_2 being the primary component along with water vapor. Water vapor is usually taken out of this list since it comes down as rain or snow as part of the water cycle, and human activities in producing water (such as the burning of fossil fuels) do not make much difference on its concentration in the atmosphere (which is mostly due to evaporation from rivers, lakes, oceans, etc.). CO_2 is different, however, in that people's activities do make a difference in CO_2 concentration in the atmosphere.

The greenhouse effect makes life on earth possible by keeping the earth warm (about 30°C warmer). However, excessive amounts of these gases disturb the delicate balance by trapping too much energy, which causes the average temperature of the earth to rise and the climate at some localities to change. These undesirable consequences of the greenhouse effect are referred to as *global warming* or *global climate change*.

The global climate change is due in part to the excessive use of fossil fuels such as coal, petroleum products, and natural gas in electric power generation, transportation, buildings, and manufacturing, and it has been a concern in recent decades. The concentration of CO_2 in the atmosphere as of 2019 is about 410 ppm (or 0.41%). This is 30 percent higher than the level a century ago, and it is projected to increase to over 700 ppm by the year 2100. Under normal conditions, vegetation consumes CO_2 and releases O_2 during the photosynthesis process, and thus it keeps the CO_2 concentration in the atmosphere in check. A mature, growing tree consumes about 12 kg of CO_2 a year and exhales enough oxygen to support a family of four. However, deforestation and the huge increase in CO_2 production in recent decades have disturbed this balance.

Greenhouse gas emissions can be reduced by increasing conservation efforts and improving conversion efficiencies, while meeting new energy demands by the use of renewable energy (such as hydroelectric, solar, wind, and geothermal energy) rather than by fossil fuels.

In a 1995 report, the world's leading climate scientists concluded that the earth has already warmed about 0.5°C during the last century, and they estimate that the earth's temperature will rise another 2°C by the year 2100. A rise of this magnitude is feared to cause severe changes in weather patterns with storms and heavy rains and flooding in some places and drought in others, major floods due to the melting of ice at the poles, loss of wetlands and coastal areas due to rising sea levels, variations in water supply, changes in the ecosystem due to the inability of some animal and plant species to adjust to the changes, increases in epidemic diseases due to the warmer temperatures, and adverse side effects on human health and socioeconomic conditions in some areas.

The seriousness of these threats has moved the United Nations to establish a committee on climate change. A world summit in 1992 in Rio de Janeiro, Brazil, attracted world attention to the problem. The agreement prepared by the committee in 1992 to control greenhouse gas emissions was signed by 162 nations. In the 1997 meeting in Kyoto (Japan), the world's industrialized countries adopted the Kyoto Protocol and committed to reduce their CO_2 and other greenhouse gas emissions by 5 percent below the 1990 levels by 2008 to 2012. In December 2011, countries agreed in Durban, South Africa, to forge a new deal forcing the biggest polluting countries to limit greenhouse gas emissions. The Kyoto Protocol was extended to allow 5 more years to finalize a wider agreement. The goal was to produce a new, legally binding accord to cut greenhouse gas emissions that would be completed by 2015 and would come into force by 2020.

In 2015, the United Nations Climate Change Conference was held in Paris (France), resulting in the *Paris Agreement* on the reduction of climate change. The conference included participants from 196 nations. The main result of the conference was the establishment of a goal to limit global warming to less than 2°C compared to preindustrial times. According to the agreement, human-made (also called anthropogenic) greenhouse emissions should be eliminated during the second half of the 21st century.

Major sources of greenhouse gas emissions are the industrial sector and transportation. Each kilowatt-hour of electricity produced in a fossil fuel power plant produces 0.6 to 1 kg (1.3 to 2.2 lbm) of carbon dioxide. Each liter of gasoline burned by a vehicle produces about 2.5 kg of CO_2 (or, each gallon of gasoline burned produces about 20 lbm of CO_2). An average car in the United States is driven about 13,500 mi/yr, and it consumes about 600 gal of gasoline. Therefore, a car emits about 12,000 lbm of CO_2 to the atmosphere a year, which is about four times the weight of a typical car (Fig. 13-12).

Carbon dioxide and other emissions can be reduced significantly by buying an energy-efficient car that burns less fuel over the same distance, and by driving sensibly. Saving fuel also saves money and the environment. For example, choosing a vehicle that gets 30 rather than 20 mi/gal would prevent 2 tons of CO_2 from being released to the atmosphere every year while reducing the fuel cost by $900 per year (under average driving conditions of 13,500 mi/yr and at a fuel cost of $4.00/gal).

CO_2 Production

When a fuel containing carbon is burned, carbon is first converted to CO by the reaction $C + \frac{1}{2}O_2 \rightarrow CO$ and then to CO_2 by the reaction $CO + \frac{1}{2}O_2 \rightarrow CO_2$. When there is a deficiency of O_2 in the combustion process, some CO cannot find O_2 to form into CO_2, and it leaves the combustion chamber as CO. CO is a very undesirable and poisonous emission. CO_2 is not a pollutant. It is undesirable because it is the primary greenhouse gas of concern regarding global warming.

All fossil fuels (coal, oil, natural gas) contain carbon, and burning more of these fuels corresponds to more CO_2 production. The CO_2 emissions can be minimized by replacing fossil fuels with renewable energy sources such as solar, wind, biomass, hydro, and geothermal. The consumption of fossil fuels can also be reduced by energy efficiency measures. The combustion of fuels with a greater carbon percentage by mass produces a greater amount of CO_2 emission.

Figure 13-12   The average car produces several times its weight in CO_2 every year (it is driven 13,500 mi/yr, consumes 600 gal of gasoline, and produces 20 lbm of CO_2 per gallon). (*©mile high traveler/iStockphoto/Getty Images RF.*)

Assuming complete combustion with air with no CO in the products, when 1 kmol of HC fuel (C_mH_n) is burned, m kmol of CO_2 is produced. For example, if octane is burned completely with a stoichiometric amount of air, the complete combustion reaction is

$$C_8H_{18} + 12.5(O_2 + 3.76N_2) \rightarrow 8CO_2 + 9H_2O + (12.5 \times 3.76)N_2$$

That is, when 1 kmol of C_8H_{18} is burned, 8 kmol of CO_2 is produced. In general, the amount of CO_2 production per unit mass of fuel burned can be determined from

$$\frac{m_{CO_2}}{m_{fuel}} = \frac{N_{CO_2}M_{CO_2}}{N_{fuel}M_{fuel}} \tag{13-36}$$

where N is the mole number and M is the molar mass. For example, if pentane (C_5H_{10}) is burned, the amount of CO_2 production will be

$$\frac{m_{CO_2}}{m_{fuel}} = \frac{N_{CO_2}M_{CO_2}}{N_{C_5H_{10}}M_{C_5H_{10}}} = \frac{(5 \text{ kmol})(44 \text{ kg/kmol})}{(1 \text{ kmol})(70 \text{ kg/kmol})} = \frac{220 \text{ kg}}{70 \text{ kg}} = 3.14 \text{ kg } CO_2/\text{kg fuel}$$

Table 13-3 gives the amount of CO_2 production per unit mass of fuel burned for common fuels. When 1 kg of natural gas (approximated as CH_4) is burned, 2.75 kg of CO_2 is produced, as shown in Table 13-3. It appears that the combustion of natural gas involves the least amount of CO_2 production among all fuels. For example, 11 percent less CO_2 is produced per unit mass of fuel when natural gas is burned instead of gasoline (approximated as octane, C_8H_{18}) in a car.

When coal, oil, and natural gas are burned in a combustion unit to provide thermal energy as in a space heating system or boiler of a power plant, the amount of CO_2

TABLE 13-3   The Amount of CO_2 Emission per Unit Mass of Fuel
Burned for Common Fuels

Fuel	Chemical Formula	Molar Mass M, kg/kmol	CO_2 Production, kg CO_2/kg fuel
Carbon	C	12	3.67
Methane	CH_4	16	2.75
Ethane	C_2H_6	30	2.93
Propane	C_3H_8	44	3.00
Butane	C_4H_{10}	58	3.03
Pentene	C_5H_{10}	70	3.14
Isopentane	C_5H_{12}	72	3.06
Benzene	C_6H_6	78	3.38
Hexene	C_6H_{12}	84	3.14
Hexane	C_6H_{14}	86	3.06
Toluene	C_7H_8	92	3.34
Heptane	C_7H_{16}	100	3.08
Octane	C_8H_{18}	114	3.09
Decane	$C_{10}H_{22}$	142	3.10

produced by natural gas is about 40 percent less than that by coal. This is due to the much higher energy content of natural gas (50,000 kJ/kg) per unit mass of the fuel compared to coal (usually between 20,000 and 30,000 kJ/kg). Natural gas also emits less NO_x and PM and no SO_2 compared to coal and gasoline. Thermal and environmental benefits of natural gas compared to coal and oil have made it an attractive fuel for electric power generation and space heating. For these reasons, many consider natural gas to be a clean fuel. However, natural gas is still a fossil fuel, and its combustion results in considerable CO_2 and pollutant emissions.

Methane and N_2O are other notable greenhouse gas emissions as well as air pollutants even though their emission rates are much lower than carbon dioxide. The energy sector (coal mining, natural gas systems, oil systems, combustion systems), agriculture, industrial activities, and waste management are major sources of CH_4 emissions. N_2O is mostly emitted due to agricultural activities (nitrogen fertilization of soils and management of animal waste). Energy use and industrial processes are other sources of N_2O.

Average and total emissions of common greenhouse gases CO_2, CH_4, and NO_x in U.S. power plants are given in Table 13-4. Approximately 743 g of CO_2 are emitted to

TABLE 13-4   Average and Total Emissions of Greenhouse Gases in U.S. Power Plants (EPA, 2012)

	CO_2	CH_4	N_2O
Average emissions from fossil power plants (g/kWh)	743	—	—
Average emissions from all power plants (g/kWh)	516	10.8	7.2
Annual emissions (tons)	2,085,584,291	43,628	14,564

the environment for each kWh of electricity produced using fossil fuels. This value is decreased to 516 g/kWh when nuclear and renewable power plants are also included. The corresponding values for all power plants for CH_4 and N_2O are 10.8 and 7.23 g/kWh, respectively. About 2 billion tons of CO_2 are emitted in the United States per year due to electricity production.

Methane and N_2O emissions can also be expressed as CO_2 equivalent in terms of their effect on global warming. *Global warming potential* (GWP) is defined as the ratio of the global warming effect of a given mass of a substance to that caused by the same mass of CO_2. Carbon dioxide is the reference gas used in measuring GWP, and, by definition, has a GWP of 1. CH_4 is estimated to have a GWP of 28 to 36, and N_2O has a GWP of 265 to 298. It is estimated that about 700 million tons of CO_2 equivalent of CH_4 and 4 million tons of CO_2 equivalent of N_2O are emitted per year in the United States (EIA, 2016).

Other greenhouse gases include CFCs such as R-11, R-12, and R-113, which also cause stratospheric ozone depletion, hydrochlorofluorocarbons (such as HCFC-22), hydrofluorocarbons (HFCs), perfluorocarbons (PFCs), and sulfur hexafluoride (SF_6). The contribution of other greenhouse gases is much less than carbon dioxide. It is estimated that about 85 percent of the global warming emissions in the United States are due to CO_2.

EXAMPLE 13-6
CO_2 Production from the Combustion of Coal

Consider a coal with the following composition on a mass basis: 80 percent C, 5 percent H_2, 6 percent O_2, 2 percent N_2, 0.5 percent S, and 6.5 percent ash (noncombustibles). This coal is burned with a stoichiometric amount of air (Fig. 13-13). Neglecting the ash constituent, determine the amount of CO_2 produced per unit mass of coal.

SOLUTION   We consider 100 kg of coal for simplicity. Noting that the mass percentages in this case correspond to the masses of the constituents, the mole numbers of the constituent of the coal are determined to be

$$N_C = \frac{m_C}{M_C} = \frac{80 \text{ kg}}{12 \text{ kg/kmol}} = 6.667 \text{ kmol}$$

$$N_{H_2} = \frac{m_{H_2}}{M_{H_2}} = \frac{5 \text{ kg}}{2 \text{ kg/kmol}} = 2.5 \text{ kmol}$$

$$N_{O_2} = \frac{m_{O_2}}{M_{O_2}} = \frac{6 \text{ kg}}{32 \text{ kg/kmol}} = 0.1875 \text{ kmol}$$

$$N_{N_2} = \frac{m_{N_2}}{M_{N_2}} = \frac{2 \text{ kg}}{28 \text{ kg/kmol}} = 0.07143 \text{ kmol}$$

$$N_S = \frac{m_S}{M_S} = \frac{0.5 \text{ kg}}{32 \text{ kg/kmol}} = 0.01563 \text{ kmol}$$

Figure 13-13   Schematic for Example 13-6.

Ash consists of the noncombustible matter in coal. Therefore, the mass of ash content that enters the combustion chamber is equal to the mass content that leaves. Disregarding this nonreacting component for simplicity, the combustion equation may be written as

$$6.667C + 2.5H_2 + 0.1875O_2 + 0.07143N_2 + 0.01563S + a_{th}(O_2 + 3.76N_2)$$
$$\rightarrow xCO_2 + yH_2O + zSO_2 + wN_2$$

Performing mass balances for the constituents gives

C balance : $x = 6.667$

H_2 balance : $y = 2.5$

S balance : $z = 0.01563$

O_2 balance : $0.1875 + a_{th} = x + 0.5y + z \rightarrow a_{th} = 6.667 + 0.5(2.5) + 0.01563 - 0.1875 = 7.745$

N_2 balance : $w = 0.07143 + 3.76a_{th} = 0.07143 + 3.76 \times 7.745 = 29.19$

Substituting, the balanced combustion equation without the ash becomes

$$6.667C + 2.5H_2 + 0.1875O_2 + 0.07143N_2 + 0.01563S + 7.745(O_2 + 3.76N_2)$$
$$\rightarrow 6.667CO_2 + 2.5H_2O + 0.01563SO_2 + 29.19N_2$$

The amount of CO_2 production per unit mass of fuel is

$$\frac{m_{CO_2}}{m_{fuel}} = \frac{N_{CO_2}M_{CO_2}}{N_{fuel}M_{fuel}}$$

$$= \frac{(6.667 \text{ kmol})(44 \text{ kg/kmol})}{(6.667 \times 12 + 2.5 \times 2 + 0.1875 \times 32 + 0.07143 \times 28 + 0.01563 \times 32) \text{ kg}}$$

$$= \frac{293.3 \text{ kg}}{93.5 \text{ kg}} = \textbf{3.14 kg CO}_2\textbf{/kg fuel} \quad \blacktriangle$$

EXAMPLE 13-7
CO$_2$ Production from a Car

Calculate the amount of CO_2 emission from a midsize car, in g/mi and g/km, considering that the car gets 35 mi per gal of gasoline. Also, calculate the amount of CO_2 released to the environment per year if the car is driven 12,000 mi/yr. Take the density of gasoline to be 740 kg/m³.

SOLUTION   The mass of fuel per mile of travel is

$$m_{fuel} = \frac{1}{35 \text{ mi/gal}}\left(\frac{3.7854 \text{ L}}{1 \text{ gal}}\right)(740 \text{ g/L}) = 80.03 \text{ g fuel/mi}$$

Approximating gasoline as octane (C_8H_{18}), the complete combustion reaction with stoichiometric air is written as

$$C_8H_{18} + 12.5 \,(O_2 + 3.76N_2) \rightarrow 8CO_2 + 9H_2O + (12.5 \times 3.76)N_2$$

When 1 kmol (114 kg) of C_8H_{18} is burned, 8 kmol (352 kg) of CO_2 is produced. That is, for 1 kg of fuel, 3.088 kg of CO_2 is produced. Also, we note that 1 mi = 1.6093 km and 1 gal = 3.7854 L. Then, the masses of CO_2 per mile of travel and per km of travel are

$$m_{CO_2} \text{ (per mile)} = (80.03 \text{ g fuel/mi})\left(\frac{3.088 \text{ kg CO}_2}{1 \text{ kg fuel}}\right) = \textbf{247 g CO}_2\textbf{/mi}$$

$$m_{CO_2} \text{ (per km)} = (80.03 \text{ g fuel/mi})\left(\frac{3.088 \text{ kg CO}_2}{1 \text{ kg fuel}}\right)\left(\frac{1 \text{ mi}}{1.6093 \text{ km}}\right) = \mathbf{154 \text{ g CO}_2/\text{km}}$$

Noting that the car is driven 12,000 mi/yr, the amount of CO_2 released to the environment per year is determined to be

$$m_{CO_2} \text{ (per year)} = (247 \text{ g CO}_2/\text{mi})(12,000 \text{ mi/yr})\left(\frac{1 \text{ kg}}{1000 \text{ g}}\right) = \mathbf{2966 \text{ kg/yr}}$$

Therefore, an average car is responsible for about 3 tons of CO_2 emissions to the environment per year. ▲

Carbon dioxide concentration in the atmosphere has increased steadily since the Industrial Revolution (1750s), but this increase has been more dramatic in the last 60 years, as shown in Fig. 13-14. The CO_2 concentration in the atmosphere has increased by about 50 percent since the 1750s and has reached above 410 ppm as of 2019.

Coal is the largest source of CO_2 emission, representing 42 percent of total energy-related CO_2 emissions. The share of other sources is oil (33%), natural gas (19%), cement production (6%), and gas flaring (1%). Not all CO_2 emissions end up in the atmosphere. It is estimated that about 40 percent of CO_2 emissions are accumulated in the atmosphere, 30 percent in the oceans, and 30 percent on land (CO_2 Earth, 2016).

The trends in CO_2 emissions are affected by the following parameters:

- Carbon intensity of energy (the amount of energy-related carbon dioxide emissions emitted per unit of energy produced)
- Energy intensity of the economy (energy consumed per dollar of gross domestic product, GDP)
- Output per capita (GDP per person)
- Population

Energy intensity can be decreased by energy efficiency measures. The carbon intensity of energy can be decreased by switching to renewable energy. Both of these will result in a decrease in CO_2 emissions. However, the increases in output per capita and world population contribute higher rates of CO_2 emissions.

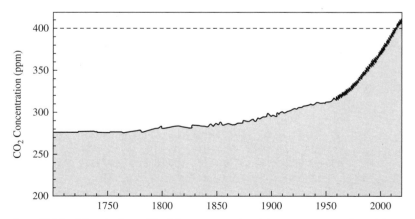

Figure 13-14 Historical trend in CO_2 concentration. Measurements are from Mauna Loa Observatory in Hawaii. (*Scripps Institution of Oceanography, 2019.*)

EXAMPLE 13-8
Reduction in
CO₂ Emission
by a Household

Consider a household that uses 8000 kWh of electricity per year and 1200 gal of fuel oil during a heating season. The average amount of CO_2 produced is 26.4 lbm per gal of fuel oil and 1.54 lbm per kWh of electricity. If this household reduces its oil and electricity usage by 20 percent as a result of implementing some energy conservation measures, determine the reduction in the amount of CO_2 emissions by that household per year.

SOLUTION Noting that this household consumes 8000 kWh of electricity and 1200 gal of fuel oil per year, the amount of CO_2 production this household is responsible for is

Amount of CO_2 produced = (Amount of electricity consumed)(Amount of CO_2 per kWh)
                          + (Amount of fuel oil consumed)(Amount of CO_2 per gal)
= (8000 kWh/yr)(1.54 lbm CO_2/kWh) + (1200 gal/yr)(26.4 lbm CO_2/gal)
= 44,000 lbm CO_2/yr

Then reducing the electricity and fuel oil usage by 20 percent will reduce the annual amount of CO_2 production of this household by

Reduction in CO_2 produced = (0.20)(Current amount of CO_2 production)
= (0.20)(44,000 lbm CO_2/yr)
= **8800 lbm CO_2/yr**

Therefore, any measure that saves energy also reduces the amount of carbon dioxide emitted to the environment. ▲

13-5 STRATOSPHERIC OZONE DEPLETION

Ground-level ozone is a pollutant, but the ozone (O_3) that exist in the stratosphere, which is 10 to 40 km above the earth's surface, is desirable as it protects the earth from harmful *ultraviolet* (UV) *radiation* from the sun, mostly the most intense radiation known as *ultraviolet-B* (UV-B).

The UV radiation lies between the wavelengths 0.01 and 0.40 mm in the solar spectrum. UV rays are to be avoided since they destroy protein and DNA molecules in biological tissue. As a result, they can kill microorganisms and cause serious damage to humans and other living beings. About 12 percent of solar radiation is in the UV range, and it would be devastating if it were to reach the surface of the earth. Fortunately, the ozone (O_3) layer in the upper atmosphere acts as a protective blanket and absorbs most of this UV radiation. The UV rays that remain in sunlight are still sufficient to cause serious sunburns to sun worshippers, and prolonged exposure to direct sunlight is the leading cause of skin cancer with increased risks of cataracts and blindness. It also reduces immune function and harms algae, adversely affecting marine ecosystems.

Discoveries of holes in the ozone layer in the late 1970s and 1980s have prompted the international community to ban the use of ozone-destroying chemicals such as the refrigerant Freon-12 and other CFCs in order to save the earth. The chemical reactions of CFCs in the stratosphere reacting with the ozone can be described by the following reactions (Decher, 1994):

$$CCl_2F_2 + \text{solar energy} \rightarrow Cl + CClF_2 \tag{13-37}$$

$$Cl + O_3 \rightarrow ClO + O_2 \tag{13-38}$$

$$ClO + O \rightarrow Cl + O_2 \tag{13-39}$$

In the first reaction, dichlorodifluoromethane is photodissociated under the effect of intense UV radiation, yielding chlorine (Cl) atom. Cl atom easily react with ozone to produce chlorine monoxide (ClO) and oxygen molecule (O_2), thus breaking ozone. Stratosphere also contains atomic oxygen (O) which reacts with ClO to produce Cl and O_2 in the final reaction. As a result, free Cl atoms are responsible for the breakdown of ozone.

The CFCs have been extensively used as refrigerants in refrigerators, air conditioners, and heat pumps from the 1930s to the 1990s. The CFCs used in the refrigeration systems eventually end up in the atmosphere through leaking during use or after their useful life. Chlorine is also introduced to the atmosphere by salt (NaCl) in the ocean, volcanoes, and industrial processes.

The first member of the CFC family of refrigerants was refrigerant-11 (R-11), developed in 1928 by General Motors' research laboratory. Of several CFCs developed, the research team settled on R-12 as the refrigerant most suitable for commercial use and gave the CFC family the trade name "Freon." Commercial production of R-11 and R-12 was started in 1931 by a company jointly formed by General Motors and E. I. du Pont de Nemours and Co., Inc. The versatility and low cost of CFCs made them the refrigerants of choice. CFCs were also widely used in aerosols, foam insulations, and the electronic industry as solvents to clean computer chips.

R-11 was used primarily in large-capacity water chillers serving air-conditioning systems in buildings. R-12 was used in domestic refrigerators and freezers, as well as automotive air conditioners. R-22 has been used in window air conditioners, heat pumps, air conditioners of commercial buildings, and large industrial refrigeration systems. R-502 (a blend of R-115 and R-22) has been the dominant refrigerant used in commercial refrigeration systems such as those in supermarkets. Fully halogenated CFCs (such as R-11, R-12, and R-115) do the most damage to the ozone layer. The non-fully halogenated refrigerants such as R-22 have about 5 percent of the ozone-depleting capability of R-12. R-22 is a hydro-chlorofluorocarbon (HCFC) and its use was discontinued by 2015.

The ozone crisis has caused a major stir in the refrigeration and air-conditioning industry and has triggered a critical look at the refrigerants in use. It was realized in the 1970s that CFCs allow more UV radiation into the earth's atmosphere by destroying the protective ozone layer. As a result, the use of some CFCs was banned in 1987 by the Montreal Protocol on Substances that Deplete the Ozone Layer, which was signed by 172 countries. An amendment to the Montreal Protocol in 1990 called for the complete phaseout of CFC production by the year 2000 and additional measures for some other ozone-depleting chemicals.

Refrigerants that are friendly to the ozone layer that protects the earth from harmful UV rays have been developed. The once popular refrigerant R-12 has largely been replaced by the chlorine-free R-134a. R-134a is mostly used in household refrigerators and automotive air conditioners. R-134a is a hydrofluorocarbon (HFC) and the first non-ozone-depleting fluorocarbon refrigerant to be commercialized. The most common refrigerants for air-conditioning and heat pump systems are R-410A, R-407C, and R-32, which are also HFC refrigerants. An HFC refrigerant does not contribute to ozone depletion (Çengel et al., 2019).

13-6 NUCLEAR WASTE

A single nuclear power plant using fission reaction in the reactors typically produces thousands of megawatts of electricity. One kilogram of fissioned nuclear material can release 10^{11} kJ of thermal energy, which is roughly two million times the energy released when

1 kg of fossil fuel burns. Note, however, that only about 4 percent of the nuclear fuel is fissile uranium-235. This means a unit mass of fissile nuclear fuel can provide about 80,000 times more energy than a unit mass of fossil fuel.

A nuclear fission reaction breaks the nucleus of an atom, which releases large amounts of energy. In a fission reaction, neutrons react with fissile isotopes of uranium and plutonium. Uranium-233, uranium-235, and plutonium-239 are fissile isotopes. Only uranium-235 is found in nature, but the others can be produced artificially (Decher, 1994). Uranium-235 is used in nuclear reactors.

Some people may regard nuclear power plants as a potential solution to air pollution and greenhouse gas emission. This is because the air pollutants produced by the combustion of fossil fuels such as HC, carbon soot particles, CO, sulfur oxides, and NO_x are not generated after a nuclear fission reaction. In addition, a nuclear reaction does not generate any CO_2, which is the major greenhouse gas. Other people, however, do not consider nuclear energy as an alternative to fossil fuels due to the safety of nuclear power plants and concern over safe and permanent disposal of nuclear waste.

The accident that occurred in 1986 at the Chernobyl nuclear power plant in Ukraine (in former Soviet Union) was the worst nuclear power plant accident in history. Thirty-one people died during the accident, and it affected much of the western Soviet Union and Europe with long-term health effects. The most recent nuclear disaster occurred in Japan at the Fukushima I nuclear power plant in 2011. The accident was initiated by a devastating tsunami caused by an earthquake. This caused some equipment failures and as a result, reactors could not be cooled, resulting in three nuclear meltdowns and large releases of radioactive materials.

Nuclear waste is generated due to fission reactor operation in nuclear power plants. The products of fuel rods after a fission reaction make up most radioactive wastes. Materials located or used in the reactor room absorb neutrons and become radioactive. High-level radioactive waste consists of fission products and transuranic (elements heavier than uranium like plutonium) isotopes.

When a used nuclear fuel is first discharged, intense levels of radioactivity exist. The content of the fuel changes as time passes. However, the total radioactivity does not decrease significantly. After 100 years, the total radioactivity can be as high as 99 percent. Therefore, once a nuclear fuel goes through a fission reaction, it will be radioactive for thousands of years.

Some countries recycle nuclear fuel elements. They put high-level nuclear waste into glass or ceramic logs. In the United States, used fuel elements are not recycled. Instead, they are stored for several years at the reactor site until short-lived isotopes can decay and radioactive heat generation is minimized. Later, they are taken to temporary storage locations for long-term storage. The Yucca Mountain site in Nevada is considered as a potential deep geological repository for the used nuclear fuel.

There are also low-level wastes produced in nuclear power plants. They include gloves, towels, lab coats, filters, and other materials that may contain small amounts of radioactive material. These wastes are carefully placed in drums and stored at a burial site.

Radioactive emissions are harmful to living organisms. The higher the level of radioactivity, the greater the potential health effects on humans. Long-wave radiation causes cell damage. An increased risk of cancer with exposure to nuclear radiation remains a serious concern.

REFERENCES

Amann CA, "Control of the Homogeneous-Charge Passenger Car Engine—Defining the Problem." SAE Paper 801440, 1980.

Çengel YA, Boles MA, and Kanoğlu M, *Thermodynamics: An Engineering Approach*, 9th ed., New York: McGraw Hill, 2019.

CO₂ Earth, *Pro Oxygen* (2016). https://www.co2.earth/global-co2-emissions?itemid=1%3Fitemid=1. Access date: Mar. 9, 2019.

Decher R, *Energy Conversion; Systems, Flow Physics and Engineering*, New York: Oxford University Press, 1994.

EIA, U.S. Energy Information Administration, Washington, DC (2016). http://www.eia.gov/environment/emissions/ghg_report/ghg_nitrous.cfm. Access date: Mar. 9, 2019.

EPA, U.S. Environmental Protection Agency. Office of Transportation and Air Quality, EPA420-F-08-024, 2008. http://www.epa.gov/otaq/. Access date: Sep. 9, 2018.

EPA, National Vehicle and Fuel Emissions Laboratory. U.S. Environmental Protection Agency, 2010.

EPA, Emissions data. eGRID2012 Summary Tables. U.S. U.S. Environmental Protection Agency (2012). https://www.epa.gov/energy/egrid. Access date: Sep. 9, 2018.

EPRI, *PISCES: Power Plant Chemical Assessment Model Version 2.0*. CM-107036-V1, Electric Power Research Institute, Palo Alto, CA, 1997.

Heywood JB, *Internal Combustion Engine Fundamentals*. New York: McGraw Hill, 1988.

Kagel A, Bates D, and Gawell K, *A Guide to Geothermal Energy and the Environment*. Geothermal Energy Association. Washington, D.C, 2007.

Pulkrabek WW, *Engineering Fundamentals of the Internal Combustion Engines*, 2nd ed., Upper Saddle River, NJ: Pearson Prentice Hall, 2004.

Rubin ES, *Introduction to Engineering and the Environment*, New York: McGraw Hill, 2001.

Scripps Institution of Oceanography, 2019. https://scripps.ucsd.edu/programs/keelingcurve/. Access date: Feb. 13, 2019.

Seinfeld JH, *Air Pollution: Physical and Chemical Fundamentals*, New York: McGraw Hill, 1975.

Williamson S, *Fundamentals of Air Pollution*, Reading MA: Addison-Wesley, 1973.

PROBLEMS

AIR POLLUTANTS

13-1   What are the major pollutants emitted by the combustion of fossil fuels?

13-2   What are the environmental concerns due to renewable energy sources?

13-3   What are the causes of incomplete combustion?

13-4   Describe the two general methods of minimizing harmful emissions from a fuel combustion system.

13-5   Give a list of environmental controls in a typical coal-fired power plant.

13-6   What is particulate matter? What is it made of?

13-7   What are PM10 and PM2.5? What are they referred to?

13-8   What are the sources of SO_2 emission?

13-9   List the methods of controlling NO_x emissions.

13-10   Describe the negative effects of NO_x emissions on the environment and human health.

13-11   Compared to gasoline engines, much less HC is found in the engine exhaust of a diesel engine. Why? Explain.

13-12   Under which driving conditions is most CO produced in a gasoline automobile? Why?

13-13   What is the effect of CO when people are exposed to it?

13-14   What is smog? What is it made of?

13-15   How is ground-level ozone formed? Is it the same as stratospheric ozone? Explain.

13-16   What is acid rain? How is it formed?

13-17   Consider a coal with an ash fraction of 6.0 percent. The heating value of the coal is 25,000 kJ/kg. This coal is consumed in a power plant at a rate of 90 kg/h. The plant operates 70 percent of the time annually.

(*a*) Determine the amount of ash entering the plant per year.

(*b*) Determine the amount of ash entering the plant per MJ of heat input to the plant.

13-18 A coal power plant produces 5 million kWh of electricity per year consuming coal with an ash fraction of 5.5 percent. The plant efficiency is 35 percent. The heating value of coal is 23,500 kJ/kg. (*a*) Determine the amount of ash entering the plant per year.
(*b*) If only 3 percent of the ash entering the plant is emitted to the atmosphere with combustion gases, determine the mass of ash emitted per kWh of electricity produced in the plant.

13-19 A residential building is heated by a coal-burning heater. The heater consumes an average of 130 tons of coal in a winter season. The coal has an ash fraction of 7 percent. It is estimated that 95 percent of ash is collected in the heater room and the remaining 5 percent is released to the atmosphere by the exhaust gases. Determine the amount of ash released to the atmosphere by this heater during a 15-year period.

13-20 A coal-fired power plant generates 750 MW power with an overall efficiency of 42 percent. The plant operates 75 percent of time in a year. The coal used in the plant has a sulfur fraction of 1.2 percent, and the heating value of this coal is 11,500 Btu/lbm. Determine the amount of sulfur entering the plant per year, in lbm/yr. Also, determine the amount of SO_2 produced, in lbm/yr, after the combustion process if all sulfur reacts with oxygen (O_2) during combustion.

13-21 A coal-fired power plant generates 500 MW of power and is responsible for 25,000 tons of SO_2 production. A flue gas desulfurization (FGD) system removes 98.5 percent of SO_2. The plant operates 60 percent of the time in a year. Determine the amount of SO_2 released to the atmosphere per kWh of electricity produced.

13-22 A coal-fired power plant generates 500 MW of power and is responsible for 25,000 tons of SO_2 production. A flue gas desulfurization (FGD) system removes 98.5 percent of SO_2. Determine the annual amounts of SO_2 emitted to the atmosphere, limestone ($CaCO_3$) used, and gypsum ($CaSO_4 \cdot 2H_2O$) produced in the FGD system.

13-23 A natural gas–fired power plant produces 900 MW of power with an overall efficiency of 47 percent. It is estimated that NO_x emissions from this power plant are 0.50 g/kWh (grams per kWh of electricity produced).
(*a*) Determine the rate of natural gas consumption, in therm/h. Note that 1 therm = 100,000 Btu = 105,500 kJ.
(*b*) If the plant operates 8200 h in a year, determine the amount of NO_x emissions per year.

13-24 A natural gas–fired boiler consumes natural gas at a rate of 40 m³/h. It is estimated that NO_x production from this boiler is 250 ng/J (nanograms per J of boiler heat input). The heating value of natural gas is 35,000 kJ/m³. Now a selective catalytic reduction (SCR) unit is installed into the combustion gas stream right after the boiler to remove 75 percent of NO_x in the combustion gases. Determine the amount of NO_x removed by the SCR unit if this boiler operates for a period of 500 h in a month.

13-25 A diesel engine with an engine volume of 4.0 L and an engine speed of 2300 rpm operates on an air-fuel ratio of 20 kg air/kg fuel. The engine uses light diesel fuel that contains 500 ppm of sulfur by mass. All of this sulfur is exhausted to the environment where the sulfur is converted to sulfurous acid (H_2SO_3). If the rate of the air entering the engine is 320 kg/h, determine the mass flow rate of sulfur in the exhaust. Also, determine the mass flow rate of sulfurous acid added to the environment if for each kmol of sulfur in the exhaust, 1 kmol sulfurous acid will be added to the environment.

13-26 An average motorcycle emits 2.29 g of HC, 14.59 g CO, and 1.25 g NO_x per mile. How much HC, CO, and NO_x will be emitted by an average motorcycle if it is driven 10,000 km/yr?

13-27 An average car in the United States is driven for 13,500 mi/yr and emits 0.0089 g of PM per mile. Consider a big city with 800,000 cars and determine the amount of PM emissions a year from the cars in that city.

13-28 An environmentally conscious engineer has decided to replace his/her gasoline automobile with a diesel automobile in order to save HC, CO, and NO_x emissions from the car to the environment. Using the emission values in Table 13-2, estimate the amount of HC, CO, and NO_x emissions to be saved per year as a result of this replacement. Assume an annual drive of 15,000 mi.

13-29 Leaded gasoline contains lead that ends up in the engine exhaust. Lead is a very toxic engine emission. The use of leaded gasoline in the United States has been unlawful for most vehicles since the 1980s. However, leaded gasoline is still used in some parts of the world. Consider a city with 150,000 cars using leaded gasoline. The gasoline contains 0.13 g/L of lead, and 50 percent of lead is exhausted to the environment. Assuming that an average car travels 24,000 km/yr with a gasoline consumption of 9.5 L/100 km, determine the amount of lead put into the atmosphere per year in that city.

13-30   Which component is not a pollutant emitted by the combustion of fossil fuels?
(a) NO_x (b) HC   (c) SO_2 (d) CO   (e) CO_2

13-31   Which is not a fuel?
(a) SO_2 (b) HC   (c) Sulfur   (d) CO   (e) CH_4

13-32   Which renewable energy source involves H_2S emission?
(a) Solar   (b) Biomass   (c) Geothermal   (d) Hydro   (e) Wind

13-33   Which is not a cause of incomplete combustion?
(a) Insufficient air   (b) Insufficient mixing   (c) Limited time of combustion
(d) Existence of sulfur in fuel   (e) Dissociation

13-34   Which environmental controls are used in a typical coal-fired power plant?
 I. Bottom ash collection from coal before the combustion process
 II. Exhaust gas recycling to control NO_x
 III. Electrostatic precipitator for flash removal
 IV. Flue gas desulfurization for SO_2 removal
(a) I and III   (b) I, II, and III   (c) I, III, and IV   (d) II, III, and IV
(e) I, II, III, and IV

13-35   In a coal-fired power plant, the flying ash particles (flash) in combustion gases are removed and safely collected by a device/method called
(a) Selective catalytic reduction   (b) Electrostatic precipitator   (c) Catalytic converter
(d) Solid waste removal   (e) Exhaust gas recycling

13-36   The combustion of which fuels involves SO_2 production?
I. Coal   II. Gasoline   III. Diesel fuel   IV. Natural gas
(a) I and III   (b) I, II, and III   (c) I, III, and IV   (d) II, III, and IV
(e) I, II, III, and IV

13-37   The combustion of which fuel involves most SO_2 production?
(a) Coal   (b) Gasoline   (c) Diesel fuel   (d) Natural gas   (e) Biodiesel

13-38   Which pollutant is produced as a result of a dissociation reaction that occurs at high combustion temperatures?
(a) PM   (b) HC   (c) SO_2 (d) CO   (e) NO_x

13-39   Which method/unit is used effectively to control NO_x emissions in diesel engines?
(a) Catalytic converter   (b) Exhaust gas recycling   (c) Selective catalytic reduction
(d) Low-NO_x burner   (e) Water injection to combustion chamber

13-40   Which method/unit is used effectively to control NO_x emissions in gasoline engines?
(a) Catalytic converter   (b) Exhaust gas recycling   (c) Selective catalytic reduction
(d) Low-NO_x burner   (e) Water injection to combustion chamber

13-41   Which method/unit is used effectively to control NO_x emissions in coal-fired power plants?
(a) Catalytic converter   (b) Exhaust gas recycling   (c) Selective catalytic reduction
(d) Low-NO_x burner   (e) Water injection to combustion chamber

13-42   Which pollutants are mainly produced as a result of insufficient air during combustion?
I. SO_2 II. NO_x III. HC   IV. CO
(a) I and III   (b) II and III   (c) III and IV   (d) II, III, and IV   (e) I, II, III, and IV

13-43   Which plants/engines are primarily responsible for carbon soot (C) particle emission?
(*a*) Coal power plants   (*b*) Natural gas power plants   (*c*) Nuclear power plants
(*d*) Diesel engines   (*e*) Gasoline engines

13-44   HC components react in the atmosphere to form photochemical smog except for
(*a*) Methane (CH_4)   (*b*) Ethane (C_2H_6)   (*c*) Carbon soot (C)   (*d*) Benzene (C_6H_6)
(*e*) Octane (C_8H_{18})

13-45   Which is a fuel?
(*a*) NO_x (*b*) H_2O (*c*) SO_2 (*d*) CO   (*e*) CO_2

EMISSIONS FROM AUTOMOBILES

13-46   Why are catalytic converters called three-way converters?

13-47   What is the purpose of catalyst materials in catalytic converters? Which catalyst materials are used?

13-48   Some automobile engines are equipped with a turbocharger located in the exhaust system just before the catalytic converter. What is the effect of a turbocharger on catalytic converter effectiveness?

13-49   Which pollutant is emitted in the greatest amount (in g/mi) from automobiles?
(*a*) NO_x (*b*) CO   (*c*) SO_2 (*d*) PM   (*e*) HC

13-50   Three major automobile-related pollutants are
I. SO_2 II. NO_x III. HC   IV. CO   V. PM
(*a*) I, II, and III   (*b*) I, III, and V   (*c*) I, III, and IV   (*d*) II, III, and V
(*e*) II, III, and IV

13-51   Pollutant emissions from automobiles have decreased by about _____ since 1970s.
(*a*) 20%   (*b*) 35%   (*c*) 50%   (*d*) 65%   (*e*) 90%

13-52   Nitrogen oxide emissions are maximum in a gasoline automobile engine when the air-fuel mixture is
(*a*) Very lean   (*b*) Slightly lean   (*c*) Stoichiometric   (*d*) Slightly rich   (*e*) Very rich

13-53   Three-way catalytic converters mainly work to reduce the following three air pollutants:
I. SO_2 II. NO_x III. HC   IV. CO   V. PM
(*a*) I, II, and III   (*b*) I, III, and V   (*c*) I, III, and IV   (*d*) II, III, and V
(*e*) II, III, and IV

13-54   A well-working catalytic converter reduces HC, CO, and NO_x components by more than _____ when the converter temperature is 400°C or above.
(*a*) 90%   (*b*) 75%   (*c*) 60%   (*d*) 50%   (*e*) 30%

GREENHOUSE EFFECT

13-55   What is the greenhouse effect? Explain.

13-56   Describe the greenhouse effect on earth.

13-57   List greenhouse gases. Is water vapor a greenhouse gas? Explain.

13-58   Is CO_2 a pollutant? Why is it undesirable? Can we avoid CO_2 emission while burning fossil fuels?

13-59   How can the CO_2 emissions be minimized?

13-60   What are the causes of global warming?

13-61   Define global warming potential (GWP). What is the GWP of CO_2?

13-62   Some people consider natural gas a clean fuel. Why? Explain.

13-63   Can electric cars be considered zero emission vehicles? Explain.

13-64   What are the four parameters affecting world CO_2 emissions?

13-65   An average car produces 20 lbm of CO_2 per gal of gasoline consumed. How much CO_2 in kg is produced per liter of gasoline consumed? How much CO_2 will be emitted per year by an average car if it is driven 15,000 km/yr and it consumes 8 L per 100 km of driving?

13-66   An average car produces 20 lbm of CO_2 per gal of gasoline consumed. How much CO_2, in lbm, will be emitted during a 10-year period by an average car if it is driven 12,500 mi/yr and its fuel efficiency is 25 mi/gal?

13-67   Calculate the amount of CO_2 emission from an SUV, in g/mi and g/km, considering that the gasoline consumption of the SUV is 14 L/100 km. Take the density of gasoline to be 740 kg/m^3.

13-68   A household replaces its aging refrigerator that consumes 1500 kWh of electricity a year with an energy-efficient one that consumes only 375 kWh/yr. If the household pays $0.12/kWh for electricity, how much money will they save per year? Considering that an average of 0.7 kg of CO_2 is emitted per kWh of electricity generated, how much CO_2 emission will be avoided per year?

13-69   Consider a household that uses 6500 kWh of electricity and 500 therm (1 therm = 100,000 Btu = 105,500 kJ) of natural gas per year. The average amount of CO_2 produced is 0.71 kg/kWh of electricity and 5.5 kg/therm of natural gas. If this household reduces its electricity and natural gas usage by 15 percent as a result of implementing some energy conservation measures, determine the reduction in the amount of CO_2 emissions by that household per year.

13-70   An industrial facility uses 400,000 kWh of electricity and 200 tons of coal per year. The average amount of CO_2 produced is 0.68 kg/kWh of electricity and 3.5 kg/kg of coal. The unit costs of electricity and coal are $0.09/kWh and $0.4/kg, respectively. If this household reduces its electricity usage by 12 percent and coal usage by 24 percent as a result of implementing some energy conservation measures, determine the cost savings and the reduction in the amount of CO_2 emissions by that facility per year.

13-71   A coal from Colorado which has an ultimate analysis (by mass) of 79.61 percent C, 4.66 percent H$_2$, 4.76 percent O$_2$, 1.83 percent N$_2$, 0.52 percent S, and 8.62 percent ash (noncombustibles) is burned with a stoichiometric amount of air. Neglecting the ash constituent, calculate the mass fractions of the products, the air-fuel ratio, and the mass of carbon dioxide produced per unit mass of coal burned.

13-72   A coal-fired power plant generates 500 MW of power. It is estimated that 3.1 kg of CO_2 is produced per kg of coal burned. The efficiency of the plant is 33 percent and the heating value of coal is 24,000 kJ/kg. The plant operates 65 percent of the time in a year. Determine the amount of CO_2 produced per kWh of electricity produced.

13-73   A natural gas–fired power plant generates 800 MW power. The efficiency of the plant is 45 percent, and the heating value of natural gas is 55,000 kJ/kg. Approximating natural gas as CH$_4$, determine the amount of CO_2 produced per kWh of electricity produced.

13-74   Which is the main greenhouse gas?
(a) N$_2$O   (b) CH$_4$ (c) Water vapor   (d) CO   (e) CO$_2$

13-75   Which is not among common greenhouse gases in the atmosphere?
(a) N$_2$O   (b) CH$_4$ (c) Water vapor   (d) CO   (e) CO$_2$

13-76   When 1 L of gasoline is consumed by a car, how much CO_2 is emitted?
(a) 3.8 kg   (b) 2.5 kg   (c) 1.7 kg   (d) 1.1 kg   (e) 0.6 kg

13-77   Which fuel is the largest source of CO_2 emission?
(a) Coal   (b) Oil   (c) Natural gas   (d) Sulfur   (e) Hydrogen

13-78   The total fossil fuel energy use in the world is expected to increase by 43 percent from 2007 to 2035. What is your estimate for the percentage increase of CO_2 emission during the same period?
(a) 30%   (b) 35%   (c) 38%   (d) 43%   (e) 48%

13-79   The global warming potential (GWP) of CO_2 is
(a) 0   (b) 0.5   (c) 1   (d) 10   (e) 100

STRATOSPHERIC OZONE DEPLETION

13-80 Is the ground-level ozone the same as stratospheric ozone? What are the benefits of stratospheric ozone (O_3)?

13-81 Why are UV rays to be avoided?

13-82 Which refrigerant is the most harmful to stratospheric ozone?
(*a*) R-12   (*b*) R-22   (*c*) R-134a   (*d*) R-410A   (*e*) R-502

13-83 Which refrigerant(s) is ozone-depleting?
I. R-22   II. R-134a   III. R-410A   IV. R-407C
(*a*) Only I   (*b*) Only II   (*c*) I and II   (*d*) I, II, and III   (*e*) II, III, and IV

13-84 Which group or groups of refrigerants do not contribute to ozone depletion?
 I. Chlorofluorocarbon (CFC)
 II. Hydrochlorofluorocarbon (HCFC)
III. Hydrofluorocarbon(HFC)
(*a*) Only I   (*b*) Only II   (*c*) Only III   (*d*) II and III   (*e*) I and III

NUCLEAR WASTE

13-85 Some people regard nuclear power plants as a potential solution to air pollution and greenhouse gas emission. Why? Do you agree? Discuss.

13-86 What are the adverse effects of radioactive emissions?

13-87 A unit mass of fissile nuclear fuel can provide about _____ times more energy than a unit mass of fossil fuel.
(*a*) 1000   (*b*) 10,000   (*c*) 80,000   (*d*) 10^6 (*e*) 10^9

13-88 Which material is used in nuclear reactors?
(*a*) Uranium-233   (*b*) Uranium-235   (*c*) Uranium-239   (*d*) Plutonium-233
(*e*) Plutonium-239

APPENDIX 1

Property Tables (SI Units)

TABLE A-1 Molar Mass, Gas Constant, and Ideal-Gas Specific Heats of Some Substances

Substance	Molar Mass M, kg/kmol	Gas Constant R, kJ/kg·K*	Specific Heat Data at 25°C		
			c_p, kJ/kg·K	c_v, kJ/kg·K	$k = c_p/c_v$
Air	28.97	0.2870	1.005	0.7180	1.400
Ammonia, NH_3	17.03	0.4882	2.093	1.605	1.304
Argon, Ar	39.95	0.2081	0.5203	0.3122	1.667
Bromine, Br_2	159.81	0.05202	0.2253	0.1732	1.300
Isobutane, C_4H_{10}	58.12	0.1430	1.663	1.520	1.094
n-Butane, C_4H_{10}	58.12	0.1430	1.694	1.551	1.092
Carbon dioxide, CO_2	44.01	0.1889	0.8439	0.6550	1.288
Carbon monoxide, CO	28.01	0.2968	1.039	0.7417	1.400
Chlorine, Cl_2	70.905	0.1173	0.4781	0.3608	1.325
Chlorodifluoromethane (R-22), $CHClF_2$	86.47	0.09615	0.6496	0.5535	1.174
Ethane, C_2H_6	30.070	0.2765	1.744	1.468	1.188
Ethylene, C_2H_4	28.054	0.2964	1.527	1.231	1.241
Fluorine, F_2	38.00	0.2187	0.8237	0.6050	1.362
Helium, He	4.003	2.077	5.193	3.116	1.667
n-Heptane, C_7H_{16}	100.20	0.08297	1.649	1.566	1.053
n-Hexane, C_6H_{14}	86.18	0.09647	1.654	1.558	1.062
Hydrogen, H_2	2.016	4.124	14.30	10.18	1.405
Krypton, Kr	83.80	0.09921	0.2480	0.1488	1.667
Methane, CH_4	16.04	0.5182	2.226	1.708	1.303
Neon, Ne	20.183	0.4119	1.030	0.6180	1.667
Nitrogen, N_2	28.01	0.2968	1.040	0.7429	1.400
Nitric oxide, NO	30.006	0.2771	0.9992	0.7221	1.384
Nitrogen dioxide, NO_2	46.006	0.1889	0.8060	0.6171	1.306
Oxygen, O_2	32.00	0.2598	0.9180	0.6582	1.395
n-Pentane, C_5H_{12}	72.15	0.1152	1.664	1.549	1.074
Propane, C_3H_8	44.097	0.1885	1.669	1.480	1.127
Propylene, C_3H_6	42.08	0.1976	1.531	1.333	1.148
Steam, H_2O	18.015	0.4615	1.865	1.403	1.329
Sulfur dioxide, SO_2	64.06	0.1298	0.6228	0.4930	1.263
Tetrachloromethane, CCl_4	153.82	0.05405	0.5415	0.4875	1.111
Tetrafluoroethane (R-134a), $C_2H_2F_4$	102.03	0.08149	0.8334	0.7519	1.108
Trifluoroethane (R-143a), $C_2H_3F_3$	84.04	0.09893	0.9291	0.8302	1.119
Xenon, Xe	131.30	0.06332	0.1583	0.09499	1.667

*The unit kJ/kg·K is equivalent to kPa·m³/kg·K. The gas constant is calculated from $R = R_u/M$, where $R_u = 8.31447$ kJ/kmol·K is the universal gas constant and M is the molar mass.

Source: Specific heat values are obtained primarily from the property routines prepared by The National Institute of Standards and Technology (NIST), Gaithersburg, MD.

TABLE A-2 Boiling and Freezing Point Properties

Substance	Boiling Data at 1 atm		Freezing Data		Liquid Properties		
	Normal Boiling Point, °C	Latent Heat of Vaporization h_{fg}, kJ/kg	Freezing Point, °C	Latent Heat of Fusion h_{if}, kJ/kg	Temperature, °C	Density ρ, kg/m³	Specific Heat c_p, kJ/kg·K
Ammonia	−33.3	1357	−77.7	322.4	−33.3	682	4.43
					−20	665	4.52
					0	639	4.60
					25	602	4.80
Argon	−185.9	161.6	−189.3	28	−185.6	1394	1.14
Benzene	80.2	394	5.5	126	20	879	1.72
Brine (20% sodium chloride by mass)	103.9	—	−17.4	—	20	1150	3.11
n-Butane	−0.5	385.2	−138.5	80.3	−0.5	601	2.31
Carbon dioxide	−78.4*	230.5 (at 0°C)	−56.6		0	298	0.59
Ethanol	78.2	838.3	−114.2	109	25	783	2.46
Ethyl alcohol	78.6	855	−156	108	20	789	2.84
Ethylene glycol	198.1	800.1	−10.8	181.1	20	1109	2.84
Glycerine	179.9	974	18.9	200.6	20	1261	2.32
Helium	−268.9	22.8	—	—	−268.9	146.2	22.8
Hydrogen	−252.8	445.7	−259.2	59.5	−252.8	70.7	10.0
Isobutane	−11.7	367.1	−160	105.7	−11.7	593.8	2.28
Kerosene	204–293	251	−24.9	—	20	820	2.00
Mercury	356.7	294.7	−38.9	11.4	25	13,560	0.139
Methane	−161.5	510.4	−182.2	58.4	−161.5	423	3.49
					−100	301	5.79
Methanol	64.5	1100	−97.7	99.2	25	787	2.55
Nitrogen	−195.8	198.6	−210	25.3	−195.8	809	2.06
					−160	596	2.97
Octane	124.8	306.3	−57.5	180.7	20	703	2.10
Oil (light)					25	910	1.80
Oxygen	−183	212.7	−218.8	13.7	−183	1141	1.71
Petroleum	—	230–384			20	640	2.0
Propane	−42.1	427.8	−187.7	80.0	−42.1	581	2.25
					0	529	2.53
					50	449	3.13
Refrigerant-134a	−26.1	216.8	−96.6	—	−50	1443	1.23
					−26.1	1374	1.27
					0	1295	1.34
					25	1207	1.43
Water	100	2257	0.0	333.7	0	1000	4.22
					25	997	4.18
					50	988	4.18
					75	975	4.19
					100	958	4.22

*Sublimation temperature. (At pressures below the triple-point pressure of 518 kPa, carbon dioxide exists as a solid or gas. Also, the freezing-point temperature of carbon dioxide is the triple-point temperature of −56.5°C.)

TABLE A-3   Saturated Water—Temperature Table

Temp., T, °C	Sat. Press., P_{sat}, kPa	Specific Volume m³/kg		Internal Energy, kJ/kg			Enthalpy, kJ/kg			Entropy, kJ/kg·K		
		Sat. Liquid, v_f	Sat. Vapor, v_g	Sat. Liquid, u_f	Evap., u_{fg}	Sat. Vapor, u_g	Sat. Liquid, h_f	Evap., h_{fg}	Sat. Vapor, h_g	Sat. Liquid, s_f	Evap., s_{fg}	Sat. Vapor, s_g
0.01	0.6117	0.001000	206.00	0.000	2374.9	2374.9	0.001	2500.9	2500.9	0.0000	9.1556	9.1556
5	0.8725	0.001000	147.03	21.019	2360.8	2381.8	21.020	2489.1	2510.1	0.0763	8.9487	9.0249
10	1.2281	0.001000	106.32	42.020	2346.6	2388.7	42.022	2477.2	2519.2	0.1511	8.7488	8.8999
15	1.7057	0.001001	77.885	62.980	2332.5	2395.5	62.982	2465.4	2528.3	0.2245	8.5559	8.7803
20	2.3392	0.001002	57.762	83.913	2318.4	2402.3	83.915	2453.5	2537.4	0.2965	8.3696	8.6661
25	3.1698	0.001003	43.340	104.83	2304.3	2409.1	104.83	2441.7	2546.5	0.3672	8.1895	8.5567
30	4.2469	0.001004	32.879	125.73	2290.2	2415.9	125.74	2429.8	2555.6	0.4368	8.0152	8.4520
35	5.6291	0.001006	25.205	146.63	2276.0	2422.7	146.64	2417.9	2564.6	0.5051	7.8466	8.3517
40	7.3851	0.001008	19.515	167.53	2261.9	2429.4	167.53	2406.0	2573.5	0.5724	7.6832	8.2556
45	9.5953	0.001010	15.251	188.43	2247.7	2436.1	188.44	2394.0	2582.4	0.6386	7.5247	8.1633
50	12.352	0.001012	12.026	209.33	2233.4	2442.7	209.34	2382.0	2591.3	0.7038	7.3710	8.0748
55	15.763	0.001015	9.5639	230.24	2219.1	2449.3	230.26	2369.8	2600.1	0.7680	7.2218	7.9898
60	19.947	0.001017	7.6670	251.16	2204.7	2455.9	251.18	2357.7	2608.8	0.8313	7.0769	7.9082
65	25.043	0.001020	6.1935	272.09	2190.3	2462.4	272.12	2345.4	2617.5	0.8937	6.9360	7.8296
70	31.202	0.001023	5.0396	293.04	2175.8	2468.9	293.07	2333.0	2626.1	0.9551	6.7989	7.7540
75	38.597	0.001026	4.1291	313.99	2161.3	2475.3	314.03	2320.6	2634.6	1.0158	6.6655	7.6812
80	47.416	0.001029	3.4053	334.97	2146.6	2481.6	335.02	2308.0	2643.0	1.0756	6.5355	7.6111
85	57.868	0.001032	2.8261	355.96	2131.9	2487.8	356.02	2295.3	2651.4	1.1346	6.4089	7.5435
90	70.183	0.001036	2.3593	376.97	2117.0	2494.0	377.04	2282.5	2659.6	1.1929	6.2853	7.4782
95	84.609	0.001040	1.9808	398.00	2102.0	2500.1	398.09	2269.6	2667.6	1.2504	6.1647	7.4151
100	101.42	0.001043	1.6720	419.06	2087.0	2506.0	419.17	2256.4	2675.6	1.3072	6.0470	7.3542
105	120.90	0.001047	1.4186	440.15	2071.8	2511.9	440.28	2243.1	2683.4	1.3634	5.9319	7.2952
110	143.38	0.001052	1.2094	461.27	2056.4	2517.7	461.42	2229.7	2691.1	1.4188	5.8193	7.2382
115	169.18	0.001056	1.0360	482.42	2040.9	2523.3	482.59	2216.0	2698.6	1.4737	5.7092	7.1829
120	198.67	0.001060	0.89133	503.60	2025.3	2528.9	503.81	2202.1	2706.0	1.5279	5.6013	7.1292
125	232.23	0.001065	0.77012	524.83	2009.5	2534.3	525.07	2188.1	2713.1	1.5816	5.4956	7.0771
130	270.28	0.001070	0.66808	546.10	1993.4	2539.5	546.38	2173.7	2720.1	1.6346	5.3919	7.0265
135	313.22	0.001075	0.58179	567.41	1977.3	2544.7	567.75	2159.1	2726.9	1.6872	5.2901	6.9773
140	361.53	0.001080	0.50850	588.77	1960.9	2549.6	589.16	2144.3	2733.5	1.7392	5.1901	6.9294
145	415.68	0.001085	0.44600	610.19	1944.2	2554.4	610.64	2129.2	2739.8	1.7908	5.0919	6.8827
150	476.16	0.001091	0.39248	631.66	1927.4	2559.1	632.18	2113.8	2745.9	1.8418	4.9953	6.8371
155	543.49	0.001096	0.34648	653.19	1910.3	2563.5	653.79	2098.0	2751.8	1.8924	4.9002	6.7927
160	618.23	0.001102	0.30680	674.79	1893.0	2567.8	675.47	2082.0	2757.5	1.9426	4.8066	6.7492
165	700.93	0.001108	0.27244	696.46	1875.4	2571.9	697.24	2065.6	2762.8	1.9923	4.7143	6.7067
170	792.18	0.001114	0.24260	718.20	1857.5	2575.7	719.08	2048.8	2767.9	2.0417	4.6233	6.6650
175	892.60	0.001121	0.21659	740.02	1839.4	2579.4	741.02	2031.7	2772.7	2.0906	4.5335	6.6242
180	1002.8	0.001127	0.19384	761.92	1820.9	2582.8	763.05	2014.2	2777.2	2.1392	4.4448	6.5841
185	1123.5	0.001134	0.17390	783.91	1802.1	2586.0	785.19	1996.2	2781.4	2.1875	4.3572	6.5447
190	1255.2	0.001141	0.15636	806.00	1783.0	2589.0	807.43	1977.9	2785.3	2.2355	4.2705	6.5059
195	1398.8	0.001149	0.14089	828.18	1763.6	2591.7	829.78	1959.0	2788.8	2.2831	4.1847	6.4678
200	1554.9	0.001157	0.12721	850.46	1743.7	2594.2	852.26	1939.8	2792.0	2.3305	4.0997	6.4302

(Continued)

TABLE A-3 Saturated Water—Temperature Table (*Continued*)

Temp., T,°C	Sat. Press., P_{sat}, kPa	Specific Volume, m³/kg		Internal Energy, kJ/kg			Enthalpy, kJ/kg			Entropy, kJ/kg·K		
		Sat. Liquid, v_f	Sat. Vapor, v_g	Sat. Liquid, u_f	Evap., u_{fg}	Sat. Vapor, u_g	Sat. Liquid, h_f	Evap., h_{fg}	Sat. Vapor, h_g	Sat. Liquid, s_f	Evap., s_{fg}	Sat. Vapor, s_g
205	1724.3	0.001164	0.11508	872.86	1723.5	2596.4	874.87	1920.0	2794.8	2.3776	4.0154	6.3930
210	1907.7	0.001173	0.10429	895.38	1702.9	2598.3	897.61	1899.7	2797.3	2.4245	3.9318	6.3563
215	2105.9	0.001181	0.094680	918.02	1681.9	2599.9	920.50	1878.8	2799.3	2.4712	3.8489	6.3200
220	2319.6	0.001190	0.086094	940.79	1660.5	2601.3	943.55	1857.4	2801.0	2.5176	3.7664	6.2840
225	2549.7	0.001199	0.078405	963.70	1638.6	2602.3	966.76	1835.4	2802.2	2.5639	3.6844	6.2483
230	2797.1	0.001209	0.071505	986.76	1616.1	2602.9	990.14	1812.8	2802.9	2.6100	3.6028	6.2128
235	3062.6	0.001219	0.065300	1010.0	1593.2	2603.2	1013.7	1789.5	2803.2	2.6560	3.5216	6.1775
240	3347.0	0.001229	0.059707	1033.4	1569.8	2603.1	1037.5	1765.5	2803.0	2.7018	3.4405	6.1424
245	3651.2	0.001240	0.054656	1056.9	1545.7	2602.7	1061.5	1740.8	2802.2	2.7476	3.3596	6.1072
250	3976.2	0.001252	0.050085	1080.7	1521.1	2601.8	1085.7	1715.3	2801.0	2.7933	3.2788	6.0721
255	4322.9	0.001263	0.045941	1104.7	1495.8	2600.5	1110.1	1689.0	2799.1	2.8390	3.1979	6.0369
260	4692.3	0.001276	0.042175	1128.8	1469.9	2598.7	1134.8	1661.8	2796.6	2.8847	3.1169	6.0017
265	5085.3	0.001289	0.038748	1153.3	1443.2	2596.5	1159.8	1633.7	2793.5	2.9304	3.0358	5.9662
270	5503.0	0.001303	0.035622	1177.9	1415.7	2593.7	1185.1	1604.6	2789.7	2.9762	2.9542	5.9305
275	5946.4	0.001317	0.032767	1202.9	1387.4	2590.3	1210.7	1574.5	2785.2	3.0221	2.8723	5.8944
280	6416.6	0.001333	0.030153	1228.2	1358.2	2586.4	1236.7	1543.2	2779.9	3.0681	2.7898	5.8579
285	6914.6	0.001349	0.027756	1253.7	1328.1	2581.8	1263.1	1510.7	2773.7	3.1144	2.7066	5.8210
290	7441.8	0.001366	0.025554	1279.7	1296.9	2576.5	1289.8	1476.9	2766.7	3.1608	2.6225	5.7834
295	7999.0	0.001384	0.023528	1306.0	1264.5	2570.5	1317.1	1441.6	2758.7	3.2076	2.5374	5.7450
300	8587.9	0.001404	0.021659	1332.7	1230.9	2563.6	1344.8	1404.8	2749.6	3.2548	2.4511	5.7059
305	9209.4	0.001425	0.019932	1360.0	1195.9	2555.8	1373.1	1366.3	2739.4	3.3024	2.3633	5.6657
310	9865.0	0.001447	0.018333	1387.7	1159.3	2547.1	1402.0	1325.9	2727.9	3.3506	2.2737	5.6243
315	10,556	0.001472	0.016849	1416.1	1121.1	2537.2	1431.6	1283.4	2715.0	3.3994	2.1821	5.5816
320	11,284	0.001499	0.015470	1445.1	1080.9	2526.0	1462.0	1238.5	2700.6	3.4491	2.0881	5.5372
325	12,051	0.001528	0.014183	1475.0	1038.5	2513.4	1493.4	1191.0	2684.3	3.4998	1.9911	5.4908
330	12,858	0.001560	0.012979	1505.7	993.5	2499.2	1525.8	1140.3	2666.0	3.5516	1.8906	5.4422
335	13,707	0.001597	0.011848	1537.5	945.5	2483.0	1559.4	1086.0	2645.4	3.6050	1.7857	5.3907
340	14,601	0.001638	0.010783	1570.7	893.8	2464.5	1594.6	1027.4	2622.0	3.6602	1.6756	5.3358
345	15,541	0.001685	0.009772	1605.5	837.7	2443.2	1631.7	963.4	2595.1	3.7179	1.5585	5.2765
350	16,529	0.001741	0.008806	1642.4	775.9	2418.3	1671.2	892.7	2563.9	3.7788	1.4326	5.2114
355	17,570	0.001808	0.007872	1682.2	706.4	2388.6	1714.0	812.9	2526.9	3.8442	1.2942	5.1384
360	18,666	0.001895	0.006950	1726.2	625.7	2351.9	1761.5	720.1	2481.6	3.9165	1.1373	5.0537
365	19,822	0.002015	0.006009	1777.2	526.4	2303.6	1817.2	605.5	2422.7	4.0004	0.9489	4.9493
370	21,044	0.002217	0.004953	1844.5	385.6	2230.1	1891.2	443.1	2334.3	4.1119	0.6890	4.8009
373.95	22,064	0.003106	0.003106	2015.7	0	2015.7	2084.3	0	2084.3	4.4070	0	4.4070

Source: Tables A-3 through A-5 are generated using the Engineering Equation Solver (EES) software developed by S. A. Klein and F. L. Alvarado. The routine used in calculations is the highly accurate Steam_IAPWS, which incorporates the 1995 Formulation for the Thermodynamic Properties of Ordinary Water Substance for General and Scientific Use, issued by The International Association for the Properties of Water and Steam (IAPWS). This formulation replaces the 1984 formulation of Haar, Gallagher, and Kell (NBS/NRC Steam Tables, Hemisphere Publishing Co., 1984), which is also available in EES as the routine STEAM. The new formulation is based on the correlations of Saul and Wagner (J. Phys. Chem. Ref. Data, 16, 893, 1987) with modifications to adjust to the International Temperature Scale of 1990. The modifications are described by Wagner and Pruss (J. Phys. Chem. Ref. Data, 22, 783, 1993).

TABLE A-4 Saturated Water—Pressure Table

Press., P, kPa	Sat. Temp., T_{sat},°C	Specific Volume, m³/kg		Internal Energy, kJ/kg			Enthalpy, kJ/kg			Entropy, kJ/kg·K		
		Sat. Liquid, v_f	Sat. Vapor, v_g	Sat. Liquid, u_f	Evap., u_{fg}	Sat. Vapor, u_g	Sat. Liquid, h_f	Evap., h_{fg}	Sat. Vapor, h_g	Sat. Liquid, s_f	Evap., s_{fg}	Sat. Vapor, s_g
1.0	6.97	0.001000	129.19	29.302	2355.2	2384.5	29.303	2484.4	2513.7	0.1059	8.8690	8.9749
1.5	13.02	0.001001	87.964	54.686	2338.1	2392.8	54.688	2470.1	2524.7	0.1956	8.6314	8.8270
2.0	17.50	0.001001	66.990	73.431	2325.5	2398.9	73.433	2459.5	2532.9	0.2606	8.4621	8.7227
2.5	21.08	0.001002	54.242	88.422	2315.4	2403.8	88.424	2451.0	2539.4	0.3118	8.3302	8.6421
3.0	24.08	0.001003	45.654	100.98	2306.9	2407.9	100.98	2443.9	2544.8	0.3543	8.2222	8.5765
4.0	28.96	0.001004	34.791	121.39	2293.1	2414.5	121.39	2432.3	2553.7	0.4224	8.0510	8.4734
5.0	32.87	0.001005	28.185	137.75	2282.1	2419.8	137.75	2423.0	2560.7	0.4762	7.9176	8.3938
7.5	40.29	0.001008	19.233	168.74	2261.1	2429.8	168.75	2405.3	2574.0	0.5763	7.6738	8.2501
10	45.81	0.001010	14.670	191.79	2245.4	2437.2	191.81	2392.1	2583.9	0.6492	7.4996	8.1488
15	53.97	0.001014	10.020	225.93	2222.1	2448.0	225.94	2372.3	2598.3	0.7549	7.2522	8.0071
20	60.06	0.001017	7.6481	251.40	2204.6	2456.0	251.42	2357.5	2608.9	0.8320	7.0752	7.9073
25	64.96	0.001020	6.2034	271.93	2190.4	2462.4	271.96	2345.5	2617.5	0.8932	6.9370	7.8302
30	69.09	0.001022	5.2287	289.24	2178.5	2467.7	289.27	2335.3	2624.6	0.9441	6.8234	7.7675
40	75.86	0.001026	3.9933	317.58	2158.8	2476.3	317.62	2318.4	2636.1	1.0261	6.6430	7.6691
50	81.32	0.001030	3.2403	340.49	2142.7	2483.2	340.54	2304.7	2645.2	1.0912	6.5019	7.5931
75	91.76	0.001037	2.2172	384.36	2111.8	2496.1	384.44	2278.0	2662.4	1.2132	6.2426	7.4558
100	99.61	0.001043	1.6941	417.40	2088.2	2505.6	417.51	2257.5	2675.0	1.3028	6.0562	7.3589
101.325	99.97	0.001043	1.6734	418.95	2087.0	2506.0	419.06	2256.5	2675.6	1.3069	6.0476	7.3545
125	105.97	0.001048	1.3750	444.23	2068.8	2513.0	444.36	2240.6	2684.9	1.3741	5.9100	7.2841
150	111.35	0.001053	1.1594	466.97	2052.3	2519.2	467.13	2226.0	2693.1	1.4337	5.7894	7.2231
175	116.04	0.001057	1.0037	486.82	2037.7	2524.5	487.01	2213.1	2700.2	1.4850	5.6865	7.1716
200	120.21	0.001061	0.88578	504.50	2024.6	2529.1	504.71	2201.6	2706.3	1.5302	5.5968	7.1270
225	123.97	0.001064	0.79329	520.47	2012.7	2533.2	520.71	2191.0	2711.7	1.5706	5.5171	7.0877
250	127.41	0.001067	0.71873	535.08	2001.8	2536.8	535.35	2181.2	2716.5	1.6072	5.4453	7.0525
275	130.58	0.001070	0.65732	548.57	1991.6	2540.1	548.86	2172.0	2720.9	1.6408	5.3800	7.0207
300	133.52	0.001073	0.60582	561.11	1982.1	2543.2	561.43	2163.5	2724.9	1.6717	5.3200	6.9917
325	136.27	0.001076	0.56199	572.84	1973.1	2545.9	573.19	2155.4	2728.6	1.7005	5.2645	6.9650
350	138.86	0.001079	0.52422	583.89	1964.6	2548.5	584.26	2147.7	2732.0	1.7274	5.2128	6.9402
375	141.30	0.001081	0.49133	594.32	1956.6	2550.9	594.73	2140.4	2735.1	1.7526	5.1645	6.9171
400	143.61	0.001084	0.46242	604.22	1948.9	2553.1	604.66	2133.4	2738.1	1.7765	5.1191	6.8955
450	147.90	0.001088	0.41392	622.65	1934.5	2557.1	623.14	2120.3	2743.4	1.8205	5.0356	6.8561
500	151.83	0.001093	0.37483	639.54	1921.2	2560.7	640.09	2108.0	2748.1	1.8604	4.9603	6.8207
550	155.46	0.001097	0.34261	655.16	1908.8	2563.9	655.77	2096.6	2752.4	1.8970	4.8916	6.7886
600	158.83	0.001101	0.31560	669.72	1897.1	2566.8	670.38	2085.8	2756.2	1.9308	4.8285	6.7593
650	161.98	0.001104	0.29260	683.37	1886.1	2569.4	684.08	2075.5	2759.6	1.9623	4.7699	6.7322
700	164.95	0.001108	0.27278	696.23	1875.6	2571.8	697.00	2065.8	2762.8	1.9918	4.7153	6.7071
750	167.75	0.001111	0.25552	708.40	1865.6	2574.0	709.24	2056.4	2765.7	2.0195	4.6642	6.6837

(*Continued*)

TABLE A-4   **Saturated Water—Pressure Table (*Continued*)**

		Specific Volume, m³/kg		Internal Energy, kJ/kg			Enthalpy, kJ/kg			Entropy, kJ/kg·K		
Press., P, kPa	Sat. Temp., T_{sat}, °C	Sat. Liquid, v_f	Sat. Vapor, v_g	Sat. Liquid, u_f	Evap., u_{fg}	Sat. Vapor, u_g	Sat. Liquid, h_f	Evap., h_{fg}	Sat. Vapor, h_g	Sat. Liquid, s_f	Evap., s_{fg}	Sat. Vapor, s_g
800	170.41	0.001115	0.24035	719.97	1856.1	2576.0	720.87	2047.5	2768.3	2.0457	4.6160	6.6616
850	172.94	0.001118	0.22690	731.00	1846.9	2577.9	731.95	2038.8	2770.8	2.0705	4.5705	6.6409
900	175.35	0.001121	0.21489	741.55	1838.1	2579.6	742.56	2030.5	2773.0	2.0941	4.5273	6.6213
950	177.66	0.001124	0.20411	751.67	1829.6	2581.3	752.74	2022.4	2775.2	2.1166	4.4862	6.6027
1000	179.88	0.001127	0.19436	761.39	1821.4	2582.8	762.51	2014.6	2777.1	2.1381	4.4470	6.5850
1100	184.06	0.001133	0.17745	779.78	1805.7	2585.5	781.03	1999.6	2780.7	2.1785	4.3735	6.5520
1200	187.96	0.001138	0.16326	796.96	1790.9	2587.8	798.33	1985.4	2783.8	2.2159	4.3058	6.5217
1300	191.60	0.001144	0.15119	813.10	1776.8	2589.9	814.59	1971.9	2786.5	2.2508	4.2428	6.4936
1400	195.04	0.001149	0.14078	828.35	1763.4	2591.8	829.96	1958.9	2788.9	2.2835	4.1840	6.4675
1500	198.29	0.001154	0.13171	842.82	1750.6	2593.4	844.55	1946.4	2791.0	2.3143	4.1287	6.4430
1750	205.72	0.001166	0.11344	876.12	1720.6	2596.7	878.16	1917.1	2795.2	2.3844	4.0033	6.3877
2000	212.38	0.001177	0.099587	906.12	1693.0	2599.1	908.47	1889.8	2798.3	2.4467	3.8923	6.3390
2250	218.41	0.001187	0.088717	933.54	1667.3	2600.9	936.21	1864.3	2800.5	2.5029	3.7926	6.2954
2500	223.95	0.001197	0.079952	958.87	1643.2	2602.1	961.87	1840.1	2801.9	2.5542	3.7016	6.2558
3000	233.85	0.001217	0.066667	1004.6	1598.5	2603.2	1008.3	1794.9	2803.2	2.6454	3.5402	6.1856
3500	242.56	0.001235	0.057061	1045.4	1557.6	2603.0	1049.7	1753.0	2802.7	2.7253	3.3991	6.1244
4000	250.35	0.001252	0.049779	1082.4	1519.3	2601.7	1087.4	1713.5	2800.8	2.7966	3.2731	6.0696
5000	263.94	0.001286	0.039448	1148.1	1448.9	2597.0	1154.5	1639.7	2794.2	2.9207	3.0530	5.9737
6000	275.59	0.001319	0.032449	1205.8	1384.1	2589.9	1213.8	1570.9	2784.6	3.0275	2.8627	5.8902
7000	285.83	0.001352	0.027378	1258.0	1323.0	2581.0	1267.5	1505.2	2772.6	3.1220	2.6927	5.8148
8000	295.01	0.001384	0.023525	1306.0	1264.5	2570.5	1317.1	1441.6	2758.7	3.2077	2.5373	5.7450
9000	303.35	0.001418	0.020489	1350.9	1207.6	2558.5	1363.7	1379.3	2742.9	3.2866	2.3925	5.6791
10,000	311.00	0.001452	0.018028	1393.3	1151.8	2545.2	1407.8	1317.6	2725.5	3.3603	2.2556	5.6159
11,000	318.08	0.001488	0.015988	1433.9	1096.6	2530.4	1450.2	1256.1	2706.3	3.4299	2.1245	5.5544
12,000	324.68	0.001526	0.014264	1473.0	1041.3	2514.3	1491.3	1194.1	2685.4	3.4964	1.9975	5.4939
13,000	330.85	0.001566	0.012781	1511.0	985.5	2496.6	1531.4	1131.3	2662.7	3.5606	1.8730	5.4336
14,000	336.67	0.001610	0.011487	1548.4	928.7	2477.1	1571.0	1067.0	2637.9	3.6232	1.7497	5.3728
15,000	342.16	0.001657	0.010341	1585.5	870.3	2455.7	1610.3	1000.5	2610.8	3.6848	1.6261	5.3108
16,000	347.36	0.001710	0.009312	1622.6	809.4	2432.0	1649.9	931.1	2581.0	3.7461	1.5005	5.2466
17,000	352.29	0.001770	0.008374	1660.2	745.1	2405.4	1690.3	857.4	2547.7	3.8082	1.3709	5.1791
18,000	356.99	0.001840	0.007504	1699.1	675.9	2375.0	1732.2	777.8	2510.0	3.8720	1.2343	5.1064
19,000	361.47	0.001926	0.006677	1740.3	598.9	2339.2	1776.8	689.2	2466.0	3.9396	1.0860	5.0256
20,000	365.75	0.002038	0.005862	1785.8	509.0	2294.8	1826.6	585.5	2412.1	4.0146	0.9164	4.9310
21,000	369.83	0.002207	0.004994	1841.6	391.9	2233.5	1888.0	450.4	2338.4	4.1071	0.7005	4.8076
22,000	373.71	0.002703	0.003644	1951.7	140.8	2092.4	2011.1	161.5	2172.6	4.2942	0.2496	4.5439
22,064	373.95	0.003106	0.003106	2015.7	0	2015.7	2084.3	0	2084.3	4.4070	0	4.4070

TABLE A-5   Superheated Water

T, °C	v m³/kg	u kJ/kg	h kJ/kg	s kJ/kg·K	v m³/kg	u kJ/kg	h kJ/kg	s kJ/kg·K	v m³/kg	u kJ/kg	h kJ/kg	s kJ/kg·K
	P = 0.01 MPa (45.81°C)*				P = 0.05 MPa (81.32°C)				P = 0.10 MPa (99.61°C)			
Sat.†	14.670	2437.2	2583.9	8.1488	3.2403	2483.2	2645.2	7.5931	1.6941	2505.6	2675.0	7.3589
50	14.867	2443.3	2592.0	8.1741								
100	17.196	2515.5	2687.5	8.4489	3.4187	2511.5	2682.4	7.6953	1.6959	2506.2	2675.8	7.3611
150	19.513	2587.9	2783.0	8.6893	3.8897	2585.7	2780.2	7.9413	1.9367	2582.9	2776.6	7.6148
200	21.826	2661.4	2879.6	8.9049	4.3562	2660.0	2877.8	8.1592	2.1724	2658.2	2875.5	7.8356
250	24.136	2736.1	2977.5	9.1015	4.8206	2735.1	2976.2	8.3568	2.4062	2733.9	2974.5	8.0346
300	26.446	2812.3	3076.7	9.2827	5.2841	2811.6	3075.8	8.5387	2.6389	2810.7	3074.5	8.2172
400	31.063	2969.3	3280.0	9.6094	6.2094	2968.9	3279.3	8.8659	3.1027	2968.3	3278.6	8.5452
500	35.680	3132.9	3489.7	9.8998	7.1338	3132.6	3489.3	9.1566	3.5655	3132.2	3488.7	8.8362
600	40.296	3303.3	3706.3	10.1631	8.0577	3303.1	3706.0	9.4201	4.0279	3302.8	3705.6	9.0999
700	44.911	3480.8	3929.9	10.4056	8.9813	3480.6	3929.7	9.6626	4.4900	3480.4	3929.4	9.3424
800	49.527	3665.4	4160.6	10.6312	9.9047	3665.2	4160.4	9.8883	4.9519	3665.0	4160.2	9.5682
900	54.143	3856.9	4398.3	10.8429	10.8280	3856.8	4398.2	10.1000	5.4137	3856.7	4398.0	9.7800
1000	58.758	4055.3	4642.8	11.0429	11.7513	4055.2	4642.7	10.3000	5.8755	4055.0	4642.6	9.9800
1100	63.373	4260.0	4893.8	11.2326	12.6745	4259.9	4893.7	10.4897	6.3372	4259.8	4893.6	10.1698
1200	67.989	4470.9	5150.8	11.4132	13.5977	4470.8	5150.7	10.6704	6.7988	4470.7	5150.6	10.3504
1300	72.604	4687.4	5413.4	11.5857	14.5209	4687.3	5413.3	10.8429	7.2605	4687.2	5413.3	10.5229
	P = 0.20 MPa (120.21°C)				P = 0.30 MPa (133.52°C)				P = 0.40 MPa (143.61°C)			
Sat.	0.88578	2529.1	2706.3	7.1270	0.60582	2543.2	2724.9	6.9917	0.46242	2553.1	2738.1	6.8955
150	0.95986	2577.1	2769.1	7.2810	0.63402	2571.0	2761.2	7.0792	0.47088	2564.4	2752.8	6.9306
200	1.08049	2654.6	2870.7	7.5081	0.71643	2651.0	2865.9	7.3132	0.53434	2647.2	2860.9	7.1723
250	1.19890	2731.4	2971.2	7.7100	0.79645	2728.9	2967.9	7.5180	0.59520	2726.4	2964.5	7.3804
300	1.31623	2808.8	3072.1	7.8941	0.87535	2807.0	3069.6	7.7037	0.65489	2805.1	3067.1	7.5677
400	1.54934	2967.2	3277.0	8.2236	1.03155	2966.0	3275.5	8.0347	0.77265	2964.9	3273.9	7.9003
500	1.78142	3131.4	3487.7	8.5153	1.18672	3130.6	3486.6	8.3271	0.88936	3129.8	3485.5	8.1933
600	2.01302	3302.2	3704.8	8.7793	1.34139	3301.6	3704.0	8.5915	1.00558	3301.0	3703.3	8.4580
700	2.24434	3479.9	3928.8	9.0221	1.49580	3479.5	3928.2	8.8345	1.12152	3479.0	3927.6	8.7012
800	2.47550	3664.7	4159.8	9.2479	1.65004	3664.3	4159.3	9.0605	1.23730	3663.9	4158.9	8.9274
900	2.70656	3856.3	4397.7	9.4598	1.80417	3856.0	4397.3	9.2725	1.35298	3855.7	4396.9	9.1394
1000	2.93755	4054.8	4642.3	9.6599	1.95824	4054.5	4642.0	9.4726	1.46859	4054.3	4641.7	9.3396
1100	3.16848	4259.6	4893.3	9.8497	2.11226	4259.4	4893.1	9.6624	1.58414	4259.2	4892.9	9.5295
1200	3.39938	4470.5	5150.4	10.0304	2.26624	4470.3	5150.2	9.8431	1.69966	4470.2	5150.0	9.7102
1300	3.63026	4687.1	5413.1	10.2029	2.42019	4686.9	5413.0	10.0157	1.81516	4686.7	5412.8	9.8828
	P = 0.50 MPa (151.83°C)				P = 0.60 MPa (158.83°C)				P = 0.80 MPa (170.41°C)			
Sat.	0.37483	2560.7	2748.1	6.8207	0.31560	2566.8	2756.2	6.7593	0.24035	2576.0	2768.3	6.6616
200	0.42503	2643.3	2855.8	7.0610	0.35212	2639.4	2850.6	6.9683	0.26088	2631.1	2839.8	6.8177
250	0.47443	2723.8	2961.0	7.2725	0.39390	2721.2	2957.6	7.1833	0.29321	2715.9	2950.4	7.0402
300	0.52261	2803.3	3064.6	7.4614	0.43442	2801.4	3062.0	7.3740	0.32416	2797.5	3056.9	7.2345
350	0.57015	2883.0	3168.1	7.6346	0.47428	2881.6	3166.1	7.5481	0.35442	2878.6	3162.2	7.4107
400	0.61731	2963.7	3272.4	7.7956	0.51374	2962.5	3270.8	7.7097	0.38429	2960.2	3267.7	7.5735
500	0.71095	3129.0	3484.5	8.0893	0.59200	3128.2	3483.4	8.0041	0.44332	3126.6	3481.3	7.8692
600	0.80409	3300.4	3702.5	8.3544	0.66976	3299.8	3701.7	8.2695	0.50186	3298.7	3700.1	8.1354
700	0.89696	3478.6	3927.0	8.5978	0.74725	3478.1	3926.4	8.5132	0.56011	3477.2	3925.3	8.3794
800	0.98966	3663.6	4158.4	8.8240	0.82457	3663.2	4157.9	8.7395	0.61820	3662.5	4157.0	8.6061
900	1.08227	3855.4	4396.6	9.0362	0.90179	3855.1	4396.2	8.9518	0.67619	3854.5	4395.5	8.8185
1000	1.17480	4054.0	4641.4	9.2364	0.97893	4053.8	4641.1	9.1521	0.73411	4053.3	4640.5	9.0189
1100	1.26728	4259.0	4892.6	9.4263	1.05603	4258.8	4892.4	9.3420	0.79197	4258.3	4891.9	9.2090
1200	1.35972	4470.0	5149.8	9.6071	1.13309	4469.8	5149.6	9.5229	0.84980	4469.4	5149.3	9.3898
1300	1.45214	4686.6	5412.6	9.7797	1.21012	4686.4	5412.5	9.6955	0.90761	4686.1	5412.2	9.5625

*The temperature in parentheses is the saturation temperature at the specified pressure.

†Properties of saturated vapor at the specified pressure.

(Continued)

TABLE A-5   **Superheated Water (*Continued*)**

T, °C	v m³/kg	u kJ/kg	h kJ/kg	s kJ/kg·K	v m³/kg	u kJ/kg	h kJ/kg	s kJ/kg·K	v m³/kg	u kJ/kg	h kJ/kg	s kJ/kg·K
	P = 1.00 MPa (179.88°C)				*P* = 1.20 MPa (187.96°C)				*P* = 1.40 MPa (195.04°C)			
Sat.	0.19437	2582.8	2777.1	6.5850	0.16326	2587.8	2783.8	6.5217	0.14078	2591.8	2788.9	6.4675
200	0.20602	2622.3	2828.3	6.6956	0.16934	2612.9	2816.1	6.5909	0.14303	2602.7	2803.0	6.4975
250	0.23275	2710.4	2943.1	6.9265	0.19241	2704.7	2935.6	6.8313	0.16356	2698.9	2927.9	6.7488
300	0.25799	2793.7	3051.6	7.1246	0.21386	2789.7	3046.3	7.0335	0.18233	2785.7	3040.9	6.9553
350	0.28250	2875.7	3158.2	7.3029	0.23455	2872.7	3154.2	7.2139	0.20029	2869.7	3150.1	7.1379
400	0.30661	2957.9	3264.5	7.4670	0.25482	2955.5	3261.3	7.3793	0.21782	2953.1	3258.1	7.3046
500	0.35411	3125.0	3479.1	7.7642	0.29464	3123.4	3477.0	7.6779	0.25216	3121.8	3474.8	7.6047
600	0.40111	3297.5	3698.6	8.0311	0.33395	3296.3	3697.0	7.9456	0.28597	3295.1	3695.5	7.8730
700	0.44783	3476.3	3924.1	8.2755	0.37297	3475.3	3922.9	8.1904	0.31951	3474.4	3921.7	8.1183
800	0.49438	3661.7	4156.1	8.5024	0.41184	3661.0	4155.2	8.4176	0.35288	3660.3	4154.3	8.3458
900	0.54083	3853.9	4394.8	8.7150	0.45059	3853.3	4394.0	8.6303	0.38614	3852.7	4393.3	8.5587
1000	0.58721	4052.7	4640.0	8.9155	0.48928	4052.2	4639.4	8.8310	0.41933	4051.7	4638.8	8.7595
1100	0.63354	4257.9	4891.4	9.1057	0.52792	4257.5	4891.0	9.0212	0.45247	4257.0	4890.5	8.9497
1200	0.67983	4469.0	5148.9	9.2866	0.56652	4468.7	5148.5	9.2022	0.48558	4468.3	5148.1	9.1308
1300	0.72610	4685.0	5411.9	9.4593	0.60509	4685.5	5411.6	9.3750	0.51866	4685.1	5411.3	9.3036
	P = 1.60 MPa (201.37°C)				*P* = 1.80 MPa (207.11°C)				*P* = 2.00 MPa (212.38°C)			
Sat.	0.12374	2594.8	2792.8	6.4200	0.11037	2597.3	2795.9	6.3775	0.09959	2599.1	2798.3	6.3390
225	0.13293	2645.1	2857.8	6.5537	0.11678	2637.0	2847.2	6.4825	0.10381	2628.5	2836.1	6.4160
250	0.14190	2692.9	2919.9	6.6753	0.12502	2686.7	2911.7	6.6088	0.11150	2680.3	2903.3	6.5475
300	0.15866	2781.6	3035.4	6.8864	0.14025	2777.4	3029.9	6.8246	0.12551	2773.2	3024.2	6.7684
350	0.17459	2866.6	3146.0	7.0713	0.15460	2863.6	3141.9	7.0120	0.13860	2860.5	3137.7	6.9583
400	0.19007	2950.8	3254.9	7.2394	0.16849	2948.3	3251.6	7.1814	0.15122	2945.9	3248.4	7.1292
500	0.22029	3120.1	3472.6	7.5410	0.19551	3118.5	3470.4	7.4845	0.17568	3116.9	3468.3	7.4337
600	0.24999	3293.9	3693.9	7.8101	0.22200	3292.7	3692.3	7.7543	0.19962	3291.5	3690.7	7.7043
700	0.27941	3473.5	3920.5	8.0558	0.24822	3472.6	3919.4	8.0005	0.22326	3471.7	3918.2	7.9509
800	0.30865	3659.5	4153.4	8.2834	0.27426	3658.8	4152.4	8.2284	0.24674	3658.0	4151.5	8.1791
900	0.33780	3852.1	4392.6	8.4965	0.30020	3851.5	4391.9	8.4417	0.27012	3850.9	4391.1	8.3925
1000	0.36687	4051.2	4638.2	8.6974	0.32606	4050.7	4637.6	8.6427	0.29342	4050.2	4637.I	8.5936
1100	0.39589	4256.6	4890.0	8.8878	0.35188	4256.2	4889.6	8.8331	0.31667	4255.7	4889.1	8.7842
1200	0.42488	4467.9	5147.7	9.0689	0.37766	4467.6	5147.3	9.0143	0.33989	4467.2	5147.0	8.9654
1300	0.45383	4684.8	5410.9	9.2418	0.40341	4684.5	5410.6	9.1872	0.36308	4684.2	5410.3	9.1384
	P = 2.50 MPa (223.95°C)				*P* = 3.00 MPa (233.85°C)				*P* = 3.50 MPa (242.56°C)			
Sat.	0.07995	2602.1	2801.9	6.2558	0.06667	2603.2	2803.2	6.1856	0.05706	2603.0	2802.7	6.1244
225	0.08026	2604.8	2805.5	6.2629								
250	0.08705	2663.3	2880.9	6.4107	0.07063	2644.7	2856.5	6.2893	0.05876	2624.0	2829.7	6.1764
300	0.09894	2762.2	3009.6	6.6459	0.08118	2750.8	2994.3	6.5412	0.06845	2738.8	2978.4	6.4484
350	0.10979	2852.5	3127.0	6.8424	0.09056	2844.4	3116.1	6.7450	0.07680	2836.0	3104.9	6.6601
400	0.12012	2939.8	3240.1	7.0170	0.09938	2933.6	3231.7	6.9235	0.08456	2927.2	3223.2	6.8428
450	0.13015	3026.2	3351.6	7.1768	0.10789	3021.2	3344.9	7.0856	0.09198	3016.1	3338.1	7.0074
500	0.13999	3112.8	3462.8	7.3254	0.11620	3108.6	3457.2	7.2359	0.09919	3104.5	3451.7	7.1593
600	0.15931	3288.5	3686.8	7.5979	0.13245	3285.5	3682.8	7.5103	0.11325	3282.5	3678.9	7.4357
700	0.17835	3469.3	3915.2	7.8455	0.14841	3467.0	3912.2	7.7590	0.12702	3464.7	3909.3	7.6855
800	0.19722	3656.2	4149.2	8.0744	0.16420	3654.3	4146.9	7.9885	0.14061	3652.5	4144.6	7.9156
900	0.21597	3849.4	4389.3	8.2882	0.17988	3847.9	4387.5	8.2028	0.15410	3846.4	4385.7	8.1304
1000	0.23466	4049.0	4635.6	8.4897	0.19549	4047.7	4634.2	8.4045	0.16751	4046.4	4632.7	8.3324
1100	0.25330	4254.7	4887.9	8.6804	0.21105	4253.6	4886.7	8.5955	0.18087	4252.5	4885.6	8.5236
1200	0.27190	4466.3	5146.0	8.8618	0.22658	4465.3	5145.1	8.7771	0.19420	4464.4	5144.1	8.7053
1300	0.29048	4683.4	5409.5	9.0349	0.24207	4682.6	5408.8	8.9502	0.20750	4681.8	5408.0	8.8786

(*Continued*)

TABLE A-5 **Superheated Water (*Continued*)**

T, °C	v m³/kg	u kJ/kg	h kJ/kg	s kJ/kg·K	v m³/kg	u kJ/kg	h kJ/kg	s kJ/kg·K	v m³/kg	u kJ/kg	h kJ/kg	s kJ/kg·K
	P = 4.0 MPa (250.35°C)				P = 4.5 MPa (257.44°C)				P = 5.0 MPa (263.94°C)			
Sat.	0.04978	2601.7	2800.8	6.0696	0.04406	2599.7	2798.0	6.0198	0.03945	2597.0	2794.2	5.9737
275	0.05461	2668.9	2887.3	6.2312	0.04733	2651.4	2864.4	6.1429	0.04144	2632.3	2839.5	6.0571
300	0.05887	2726.2	2961.7	6.3639	0.05138	2713.0	2944.2	6.2854	0.04535	2699.0	2925.7	6.2111
350	0.06647	2827.4	3093.3	6.5843	0.05842	2818.6	3081.5	6.5153	0.05197	2809.5	3069.3	6.4516
400	0.07343	2920.8	3214.5	6.7714	0.06477	2914.2	3205.7	6.7071	0.05784	2907.5	3196.7	6.6483
450	0.08004	3011.0	3331.2	6.9386	0.07076	3005.8	3324.2	6.8770	0.06332	3000.6	3317.2	6.8210
500	0.08644	3100.3	3446.0	7.0922	0.07652	3096.0	3440.4	7.0323	0.06858	3091.8	3434.7	6.9781
600	0.09886	3279.4	3674.9	7.3706	0.08766	3276.4	3670.9	7.3127	0.07870	3273.3	3666.9	7.2605
700	0.11098	3462.4	3906.3	7.6214	0.09850	3460.0	3903.3	7.5647	0.08852	3457.7	3900.3	7.5136
800	0.12292	3650.6	4142.3	7.8523	0.10916	3648.8	4140.0	7.7962	0.09816	3646.9	4137.7	7.7458
900	0.13476	3844.8	4383.9	8.0675	0.11972	3843.3	4382.1	8.0118	0.10769	3841.8	4380.2	7.9619
1000	0.14653	4045.1	4631.2	8.2698	0.13020	4043.9	4629.8	8.2144	0.11715	4042.6	4628.3	8.1648
1100	0.15824	4251.4	4884.4	8.4612	0.14064	4250.4	4883.2	8.4060	0.12655	4249.3	4882.1	8.3566
1200	0.16992	4463.5	5143.2	8.6430	0.15103	4462.6	5142.2	8.5880	0.13592	4461.6	5141.3	8.5388
1300	0.18157	4680.9	5407.2	8.8164	0.16140	4680.1	5406.5	8.7616	0.14527	4679.3	5405.7	8.7124
	P = 6.0 MPa (275.59°C)				P = 7.0 MPa (285.83°C)				P = 8.0 MPa (295.01°C)			
Sat.	0.03245	2589.9	2784.6	5.8902	0.027378	2581.0	2772.6	5.8148	0.023525	2570.5	2758.7	5.7450
300	0.03019	2668.4	2885.6	6.0703	0.029492	2633.5	2839.9	5.9337	0.024279	2692.3	2786.5	5.7937
350	0.04225	2790.4	3043.9	6.3357	0.035262	2770.1	3016.9	6.2305	0.029975	2748.3	2988.1	6.1321
400	0.04742	2893.7	3178.3	6.5432	0.039958	2879.5	3159.2	6.4502	0.034344	2864.6	3139.4	6.3658
450	0.05217	2989.9	3302.9	6.7219	0.044187	2979.0	3288.3	6.6353	0.038194	2967.8	3273.3	6.5579
500	0.05667	3083.1	3423.1	6.8826	0.048157	3074.3	3411.4	6.8000	0.041767	3065.4	3399.5	6.7266
550	0.06102	3175.2	3541.3	7.0308	0.051966	3167.9	3531.6	6.9507	0.045172	3160.5	3521.8	6.8800
600	0.06527	3267.2	3658.8	7.1693	0.055665	3261.0	3650.6	7.0910	0.048463	3254.7	3642.4	7.0221
700	0.07355	3453.0	3894.3	7.4247	0.062850	3448.3	3888.3	7.3487	0.054829	3443.6	3882.2	7.2822
800	0.08165	3643.2	4133.1	7.6582	0.069856	3639.5	4128.5	7.5836	0.061011	3635.7	4123.8	7.5185
900	0.08964	3838.8	4376.6	7.8751	0.076750	3835.7	4373.0	7.8014	0.067082	3832.7	4369.3	7.7372
1000	0.09756	4040.1	4625.4	8.0786	0.083571	4037.5	4622.5	8.0055	0.073079	4035.0	4619.6	7.9419
1100	0.10543	4247.1	4879.7	8.2709	0.090341	4245.0	4877.4	8.1982	0.079025	4242.8	4875.0	8.1350
1200	0.11326	4459.8	5139.4	8.4534	0.097075	4457.9	5137.4	8.3810	0.084934	4456.1	5135.5	8.3181
1300	0.12107	4677.7	5404.1	8.6273	0.103781	4676.1	5402.6	8.5551	0.090817	4674.5	5401.0	8.4925
	P = 9.0 MPa (303.35°C)				P = 10.0 MPa (311.00°C)				P = 12.5 MPa (327.81°C)			
Sat.	0.020489	2558.5	2742.9	5.6791	0.018028	2545.2	2725.5	5.6159	0.013496	2505.6	2674.3	5.4638
325	0.023284	2647.6	2857.1	5.8738	0.019877	2611.6	2810.3	5.7596				
350	0.025816	2725.0	2957.3	6.0380	0.022440	2699.6	2924.0	5.9460	0.016138	2624.9	2826.6	5.7130
400	0.029960	2849.2	3118.8	6.2876	0.026436	2833.1	3097.5	6.2141	0.020030	2789.6	3040.0	6.0433
450	0.033524	2956.3	3258.0	6.4872	0.029782	2944.5	3242.4	6.4219	0.023019	2913.7	3201.5	6.2749
500	0.036793	3056.3	3387.4	6.6603	0.032811	3047.0	3375.1	6.5995	0.025630	3023.2	3343.6	6.4651
550	0.039885	3153.0	3512.0	6.8164	0.035655	3145.4	3502.0	6.7585	0.028033	3126.1	3476.5	6.6317
600	0.042861	3248.4	3634.1	6.9605	0.038378	3242.0	3625.8	6.9045	0.030306	3225.8	3604.6	6.7828
650	0.045755	3343.4	3755.2	7.0954	0.041018	3338.0	3748.1	7.0408	0.032491	3324.1	3730.2	6.9227
700	0.048589	3438.8	3876.1	7.2229	0.043597	3434.0	3870.0	7.1693	0.034612	3422.8	3854.6	7.0540
800	0.054132	3632.0	4119.2	7.4606	0.048629	3628.2	4114.5	7.4085	0.038724	3618.3	4102.8	7.2967
900	0.059562	3829.6	4365.7	7.6802	0.053547	3826.5	4362.0	7.6290	0.042720	3818.9	4352.9	7.5195
1000	0.064919	4032.4	4616.7	7.8855	0.058391	4029.9	4613.8	7.8349	0.046641	4023.5	4606.5	7.7269
1100	0.070224	4240.7	4872.7	8.0791	0.063183	4238.5	4870.3	8.0289	0.050510	4233.1	4864.5	7.9220
1200	0.075492	4454.2	5133.6	8.2625	0.067938	4452.4	5131.7	8.2126	0.054342	4447.7	5127.0	8.1065
1300	0.080733	4672.9	5399.5	8.4371	0.072667	4671.3	5398.0	8.3874	0.058147	4667.3	5394.1	8.2819

(Continued)

TABLE A-5 **Superheated Water (*Continued*)**

T, °C	v m³/kg	u kJ/kg	h kJ/kg	s kJ/kg·K	v m³/kg	u kJ/kg	h kJ/kg	s kJ/kg·K	v m³/kg	u kJ/kg	h kJ/kg	s kJ/kg·K
	P = 15.0 MPa (342.16°C)				P = 17.5 MPa (354.57°C)				P = 20.0 MPa (365.75°C)			
Sat.	0.010341	2455.7	2610.8	5.3108	0.007932	2390.7	2529.5	5.1435	0.005862	2294.8	2412.1	4.9310
350	0.011481	2520.9	2693.1	5.4438								
400	0.015671	2740.6	2975.7	5.8819	0.012463	2684.3	2902.4	5.7211	0.009950	2617.9	2816.9	5.5526
450	0.018477	2880.8	3157.9	6.1434	0.015204	2845.4	3111.4	6.0212	0.012721	2807.3	3061.7	5.9043
500	0.020828	2998.4	3310.8	6.3480	0.017385	2972.4	3276.7	6.2424	0.014793	2945.3	3241.2	6.1446
550	0.022945	3106.2	3450.4	6.5230	0.019305	3085.8	3423.6	6.4266	0.016571	3064.7	3396.2	6.3390
600	0.024921	3209.3	3583.1	6.5796	0.021073	3192.5	3561.3	6.5890	0.018185	3175.3	3539.0	6.5075
650	0.026804	3310.1	3712.1	6.8233	0.022742	3295.8	3693.8	6.7366	0.019695	3281.4	3675.3	6.6593
700	0.028621	3409.8	3839.1	6.9573	0.024342	3397.5	3823.5	6.8735	0.021134	3385.1	3807.8	6.7991
800	0.032121	3609.3	4091.1	7.2037	0.027405	3599.7	4079.3	7.1237	0.023870	3590.1	4067.5	7.0531
900	0.035503	3811.2	4343.7	7.4288	0.030348	3803.5	4334.6	7.3511	0.026484	3795.7	4325.4	7.2829
1000	0.038808	4017.1	4599.2	7.6378	0.033215	4010.7	4592.0	7.5616	0.029020	4004.3	4584.7	7.4950
1100	0.042062	4227.7	4858.6	7.8339	0.036029	4222.3	4852.8	7.7588	0.031504	4216.9	4847.0	7.6933
1200	0.045279	4443.1	5122.3	8.0192	0.038806	4438.5	5117.6	7.9449	0.033952	4433.8	5112.9	7.8802
1300	0.048469	4663.3	5390.3	8.1952	0.041556	4659.2	5386.5	8.1215	0.036371	4655.2	5382.7	8.0574
	P = 25.0 MPa				P = 30.0 MPa				P = 35.0 MPa			
375	0.001978	1799.9	1849.4	4.0345	0.001792	1738.1	1791.9	3.9313	0.001701	1702.8	1762.4	3.8724
400	0.006005	2428.5	2578.7	5.1400	0.002798	2068.9	2152.8	4.4758	0.002105	1914.9	1988.6	4.2144
425	0.007886	2607.8	2805.0	5.4708	0.005299	2452.9	2611.8	5.1473	0.003434	2253.3	2373.5	4.7751
450	0.009176	2721.2	2950.6	5.6759	0.006737	2618.9	2821.0	5.4422	0.004957	2497.5	2671.0	5.1946
500	0.011143	2887.3	3165.9	5.9643	0.008691	2824.0	3084.8	5.7956	0.006933	2755.3	2997.9	5.6331
550	0.012736	3020.8	3339.2	6.1816	0.010175	2974.5	3279.7	6.0403	0.008348	2925.8	3218.0	5.9093
600	0.014140	3140.0	3493.5	6.3637	0.011445	3103.4	3446.8	6.2373	0.009523	3065.6	3399.0	6.1229
650	0.015430	3251.9	3637.7	6.5243	0.012590	3221.7	3599.4	6.4074	0.010565	3190.9	3560.7	6.3030
700	0.016643	3359.9	3776.0	6.6702	0.013654	3334.3	3743.9	6.5599	0.011523	3308.3	3711.6	6.4623
800	0.018922	3570.7	4043.8	6.9322	0.015628	3551.2	4020.0	6.8301	0.013278	3531.6	3996.3	6.7409
900	0.021075	3780.2	4307.1	7.1668	0.017473	3764.6	4288.8	7.0695	0.014904	3749.0	4270.6	6.9853
1000	0.023150	3991.5	4570.2	7.3821	0.019240	3978.6	4555.8	7.2880	0.016450	3965.8	4541.5	7.2069
1100	0.025172	4206.1	4835.4	7.5825	0.020954	4195.2	4823.9	7.4906	0.017942	4184.4	4812.4	7.4118
1200	0.027157	4424.6	5103.5	7.7710	0.022630	4415.3	5094.2	7.6807	0.019398	4406.1	5085.0	7.6034
1300	0.029115	4647.2	5375.1	7.9494	0.024279	4639.2	5367.6	7.8602	0.020827	4631.2	5360.2	7.7841
	P = 40.0 MPa				P = 50.0 MPa				P = 60.0 MPa			
375	0.001641	1677.0	1742.6	3.8290	0.001560	1638.6	1716.6	3.7642	0.001503	1609.7	1699.9	3.7149
400	0.001911	1855.0	1931.4	4.1145	0.001731	1787.8	1874.4	4.0029	0.001633	1745.3	1843.2	3.9317
425	0.002538	2097.5	2199.0	4.5044	0.002009	1960.3	2060.7	4.2746	0.001816	1892.9	2001.8	4.1630
450	0.003692	2364.2	2511.8	4.9449	0.002487	2160.3	2284.7	4.5896	0.002086	2055.1	2180.2	4.4140
500	0.005623	2681.6	2906.5	5.4744	0.003890	2528.1	2722.6	5.1762	0.002952	2393.3	2570.3	4.9356
550	0.006985	2875.1	3154.4	5.7857	0.005118	2769.5	3025.4	5.5563	0.003955	2664.6	2901.9	5.3517
600	0.008089	3026.8	3350.4	6.0170	0.006108	2947.1	3252.6	5.8245	0.004833	2866.8	3156.8	5.6527
650	0.009053	3159.5	3521.6	6.2078	0.006957	3095.6	3443.5	6.0373	0.005591	3031.3	3366.8	5.8867
700	0.009930	3282.0	3679.2	6.3740	0.007717	3228.7	3614.6	6.2179	0.006265	3175.4	3551.3	6.0814
800	0.011521	3511.8	3972.6	6.6613	0.009073	3472.2	3925.8	6.5225	0.007456	3432.6	3880.0	6.4033
900	0.012980	3733.3	4252.5	6.9107	0.010296	3702.0	4216.8	6.7819	0.008519	3670.9	4182.1	6.6725
1000	0.014360	3952.9	4527.3	7.1355	0.011441	3927.4	4499.4	7.0131	0.009504	3902.0	4472.2	6.9099
1100	0.015686	4173.7	4801.1	7.3425	0.012534	4152.2	4778.9	7.2244	0.010439	4130.9	4757.3	7.1255
1200	0.016976	4396.9	5075.9	7.5357	0.013590	4378.6	5058.1	7.4207	0.011339	4360.5	5040.8	7.3248
1300	0.018239	4623.3	5352.8	7.7175	0.014620	4607.5	5338.5	7.6048	0.012213	4591.8	5324.5	7.5111

TABLE A-6 Enthalpy of Formation, Gibbs Function of Formation, and Absolute Entropy at 25°C, 1 atm

Substance	Formula	$\bar{h}_f^\circ$ kJ/kmol	$\bar{g}_f^\circ$ kJ/kmol	$\bar{s}^\circ$ kJ/kmol·K
Carbon	C(s)	0	0	5.74
Hydrogen	H$_2$(g)	0	0	130.68
Nitrogen	N$_2$(g)	0	0	191.61
Oxygen	O$_2$(g)	0	0	205.04
Carbon monoxide	CO(g)	−110,530	−137,150	197.65
Carbon dioxide	CO$_2$(g)	−393,520	−394,360	213.80
Water vapor	H$_2$O(g)	−241,820	−228,590	188.83
Water	H$_2$O(ℓ)	−285,830	−237,180	69.92
Hydrogen peroxide	H$_2$O$_2$(g)	−136,310	−105,600	232.63
Ammonia	NH$_3$(g)	−46,190	−16,590	192.33
Methane	CH$_4$(g)	−74,850	−50,790	186.16
Acetylene	C$_2$H$_2$(g)	+226,730	+209,170	200.85
Ethylene	C$_2$H$_4$(g)	+52,280	+68,120	219.83
Ethane	C$_2$H$_6$(g)	−84,680	−32,890	229.49
Propylene	C$_3$H$_6$(g)	+20,410	+62,720	266.94
Propane	C$_3$H$_8$(g)	−103,850	−23,490	269.91
n-Butane	C$_4$H$_{10}$(g)	−126,150	−15,710	310.12
n-Octane	C$_8$H$_{18}$(g)	−208,450	+16,530	466.73
n-Octane	C$_8$H$_{18}$(ℓ)	−249,950	+6,610	360.79
n-Dodecane	C$_{12}$H$_{26}$(g)	−291,010	+50,150	622.83
Benzene	C$_6$H$_6$(g)	+82,930	+129,660	269.20
Methyl alcohol	CH$_3$OH(g)	−200,670	−162,000	239.70
Methyl alcohol	CH$_3$OH(ℓ)	−238,660	−166,360	126.80
Ethyl alcohol	C$_2$H$_5$OH(g)	−235,310	−168,570	282.59
Ethyl alcohol	C$_2$H$_5$OH(ℓ)	−277,690	−174,890	160.70
Oxygen	O(g)	+249,190	+231,770	161.06
Hydrogen	H(g)	+218,000	+203,290	114.72
Nitrogen	N(g)	+472,650	+455,510	153.30
Hydroxyl	OH(g)	+39,460	+34,280	183.70

Source: From JANAF, *Thermochemical Tables* (Midland, MI: Dow Chemical Co., 1971), *Selected Values of Chemical Thermodynamic Properties*, NBS Technical Note 270-3, 1968, and *API Research Project 44* (Carnegie Press, 1953).

TABLE A-7 **Properties of Some Common Fuels and Hydrocarbons**

Fuel (Phase)	Formula	Molar Mass, kg/kmol	Density,* kg/L	Enthalpy of Vaporization,† kJ/kg	Specific Heat,* c_p kJ/kg·K	Higher Heating Value,‡ kJ/kg	Lower Heating Value,‡ kJ/kg
Carbon (s)	C	12.011	2	—	0.708	32,800	32,800
Hydrogen (g)	H_2	2.016	—	—	14.4	141,800	120,000
Carbon monoxide (g)	CO	28.013	—	—	1.05	10,100	10,100
Methane (g)	CH_4	16.043	—	509	2.20	55,530	50,050
Methanol (ℓ)	CH_4O	32.042	0.790	1168	2.53	22,660	19,920
Acetylene (g)	C_2H_2	26.038	—	—	1.69	49,970	48,280
Ethane (g)	C_2H_6	30.070	—	172	1.75	51,900	47,520
Ethanol (ℓ)	C_2H_6O	46.069	0.790	919	2.44	29,670	26,810
Propane (ℓ)	C_3H_8	44.097	0.500	335	2.77	50,330	46,340
Butane (ℓ)	C_4H_{10}	58.123	0.579	362	2.42	49,150	45,370
1-Pentene (ℓ)	C_5H_{10}	70.134	0.641	363	2.20	47,760	44,630
Isopentane (ℓ)	C_5H_{12}	72.150	0.626	—	2.32	48,570	44,910
Benzene (ℓ)	C_6H_6	78.114	0.877	433	1.72	41,800	40,100
Hexene (ℓ)	C_6H_{12}	84.161	0.673	392	1.84	47,500	44,400
Hexane (ℓ)	C_6H_{14}	86.177	0.660	366	2.27	48,310	44,740
Toluene (ℓ)	C_7H_8	92.141	0.867	412	1.71	42,400	40,500
Heptane (ℓ)	C_1H_{16}	100.204	0.684	365	2.24	48,100	44,600
Octane (ℓ)	C_8H_{18}	114.231	0.703	363	2.23	47,890	44,430
Decane (ℓ)	$C_{10}H_{22}$	142.285	0.730	361	2.21	47,640	44,240
Gasoline (ℓ)	$C_nH_{1.87n}$	100–110	0.72–0.78	350	2.4	47,300	44,000
Light diesel (ℓ)	$C_nH_{1.8n}$	170	0.78–0.84	270	2.2	46,100	43,200
Heavy diesel (ℓ)	$C_nH_{1.7n}$	200	0.82–0.88	230	1.9	45,500	42,800
Natural gas (g)	$C_nH_{3.8n}N_{0.1n}$	18	—	—	2	50,000	45,000

*At 1 atm and 20°C.

†At 25°C for liquid fuels, and 1 atm and 1 normal boiling temperature for gaseous fuels.

‡At 25°C. Multiply by molar mass to obtain heating values in kJ/kmol.

APPENDIX 2

Property Tables (English Units)

TABLE A-1E Molar Mass, Gas Constant, and Ideal-Gas Specific Heats of Some Substances

Substance	Molar Mass, M, lbm/lbmol	Gas Constant R^*		Specific Heat Data at 77°F		
		Btu/ lbm·R	psia·ft³/ lbm·R	c_p, Btu/lbm·R	c_v, Btu/lbm·R	$k = c_p/c_v$
Air	28.97	0.06855	0.3704	0.2400	0.1715	1.400
Ammonia, NH_3	17.03	0.1166	0.6301	0.4999	0.3834	1.304
Argon, Ar	39.95	0.04970	0.2686	0.1243	0.07457	1.667
Bromine, Br_2	159.81	0.01242	0.06714	0.0538	0.04137	1.300
Isobutane, C_4H_{10}	58.12	0.03415	0.1846	0.3972	0.3631	1.094
n-Butane, C_4H_{10}	58.12	0.03415	0.1846	0.4046	0.3705	1.092
Carbon dioxide, CO_2	44.01	0.04512	0.2438	0.2016	0.1564	1.288
Carbon monoxide, CO	28.01	0.07089	0.3831	0.2482	0.1772	1.400
Chlorine, Cl_2	70.905	0.02802	0.1514	0.1142	0.08618	1.325
Chlorodifiuoromethane (R -22), $CHClF_2$	86.47	0.02297	0.1241	0.1552	0.1322	1.174
Ethane, C_2H_6	30.070	0.06604	0.3569	0.4166	0.3506	1.188
Ethylene, C_2H_4	28.054	0.07079	0.3826	0.3647	0.2940	1.241
Fluorine, F_2	38.00	0.05224	0.2823	0.1967	0.1445	1.362
Helium, He	4.003	0.4961	2.681	1.2403	0.7442	1.667
n-Heptane, C_7H_{16}	100.20	0.01982	0.1071	0.3939	0.3740	1.053
n-Hexane, C_6H_{14}	86.18	0.02304	0.1245	0.3951	0.3721	1.062
Hydrogen, H_2	2.016	0.9850	5.323	3.416	2.431	1.405
Krypton, Kr	83.80	0.02370	0.1281	0.05923	0.03554	1.667
Methane, CH_4	16.04	0.1238	0.6688	0.5317	0.4080	1.303
Neon, Ne	20.183	0.09838	0.5316	0.2460	0.1476	1.667
Nitrogen, N_2	28.01	0.07089	0.3831	0.2484	0.1774	1.400
Nitric oxide, NO	30.006	0.06618	0.3577	0.2387	0.1725	1.384
Nitrogen dioxide, NO_2	46.006	0.04512	0.2438	0.1925	0.1474	1.306
Oxygen, O_2	32.00	0.06205	0.3353	0.2193	0.1572	1.395
n-Pentane, C_5H_{12}	72.15	0.02752	0.1487	0.3974	0.3700	1.074
Propane, C_3H_8	44.097	0.04502	0.2433	0.3986	0.3535	1.127
Propylene, C_3H_6	42.08	0.04720	0.2550	0.3657	0.3184	1.148
Steam, H_2O	18.015	0.1102	0.5957	0.4455	0.3351	1.329
Sulfur dioxide, SO_2	64.06	0.03100	0.1675	0.1488	0.1178	1.263
Tetrachloromethane, CCl_4	153.82	0.01291	0.06976	0.1293	0.1164	1.111
Tetrafluoroethane (R-134a), $C_2H_2F_4$	102.03	0.01946	0.1052	0.1991	0.1796	1.108
Trifluoroethane (R-143a), $C_2H_3F_3$	84.04	0.02363	0.1277	0.2219	0.1983	1.119
Xenon, Xe	131.30	0.01512	0.08173	0.03781	0.02269	1.667

*The gas constant is calculated from $R = R_u/M$, where R_u = 1.9859 Btu/lbmol·R = 10.732 psia·ft³/lbmol·R is the universal gas constant and M is the molar mass.

Source: Specific heat values are mostly obtained from the property routines prepared by The National Institute of Standards and Technology (NIST), Gaithersburg, MD.

TABLE A-2E   Boiling and Freezing Point Properties

Substance	Boiling Data at 1 atm		Freezing Data		Liquid Properties		
	Normal Boiling Point, °F	Latent Heat of Vaporization h_{fg}, Btu/lbm	Freezing Point, °F	Latent Heat of Fusion h_{if}, Btu/lbm	Temperature, °F	Density ρ, lbm/ft³	Specific Heat c_p, Btu/lbm·R
Ammonia	−27.9	24.54	−107.9	138.6	−27.9	42.6	1.06
					0	41.3	1.083
					40	39.5	1.103
					80	37.5	1.135
Argon	−302.6	69.5	−308.7	12.0	−302.6	87.0	0.272
Benzene	176.4	169.4	41.9	54.2	68	54.9	0.411
Brine (20% sodium chloride by mass)	219.0	—	0.7	—	68	71.8	0.743
n-Butane	31.1	165.6	−217.3	34.5	31.1	37.5	0.552
Carbon dioxide	−109.2*	99.6 (at 32°F)	−69.8	—	32	57.8	0.583
Ethanol	172.8	360.5	−173.6	46.9	77	48.9	0.588
Ethyl alcohol	173.5	368	−248.8	46.4	68	49.3	0.678
Ethylene glycol	388.6	344.0	12.6	77.9	68	69.2	0.678
Glycerine	355.8	419	66.0	86.3	68	78.7	0.554
Helium	−452.1	9.80	—	—	−452.1	9.13	5.45
Hydrogen	−423.0	191.7	−434.5	25.6	−423.0	4.41	2.39
Isobutane	10.9	157.8	−255.5	45.5	10.9	37.1	0.545
Kerosene	399-559	108	−12.8	—	68	51.2	0.478
Mercury	674.1	126.7	−38.0	4.90	77	847	0.033
Methane	−258.7	219.6	296.0	25.1	−258.7	26.4	0.834
					−160	20.0	1.074
Methanol	148.1	473	−143.9	42.7	77	49.1	0.609
Nitrogen	−320.4	85.4	−346.0	10.9	−320.4	50.5	0.492
					−260	38.2	0.643
Octane	256.6	131.7	−71.5	77.9	68	43.9	0.502
Oil (light)	—	—			77	56.8	0.430
Oxygen	−297.3	91.5	−361.8	5.9	−297.3	71.2	0.408
Petroleum	—	99-165			68	40.0	0.478
Propane	−43.7	184.0	−305.8	34.4	−43.7	36.3	0.538
					32	33.0	0.604
					100	29.4	0.673
Refrigerant-134a	−15.0	93.2	−141.9	—	−40	88.5	0.283
					−15	86.0	0.294
					32	80.9	0.318
					90	73.6	0.348
Water	212	970.5	32	143.5	32	62.4	1.01
					90	62.1	1.00
					150	61.2	1.00
					212	59.8	1.01

*Sublimation temperature. (At pressures below the triple-point pressure of 75.1 psia, carbon dioxide exists as a solid or gas. Also, the freezing-point temperature of carbon dioxide is the triple-point temperature of −69.8°F.)

TABLE A-3E   **Saturated Water-Temperature Table**

Temp., T, °F	Sat. Press., P_{sat}, psia	Specific Volume, ft³/lbm		Internal Energy, Btu/lbm			Enthalpy, Btu/lbm			Entropy, Btu/lbm·R		
		Sat. Liquid, v_f	Sat. Vapor, v_g	Sat. Liquid, u_f	Evap., u_{fg}	Sat. Vapor, u_g	Sat. Liquid, h_f	Evap., h_{fg}	Sat. Vapor, h_g	Sat. Liquid, s_f	Evap., s_{fg}	Sat. Vapor, s_g
32.018	0.08871	0.01602	3299.9	0.000	1021.0	1021.0	0.000	1075.2	1075.2	0.00000	2.18672	2.1867
35	0.09998	0.01602	2945.7	3.004	1019.0	1022.0	3.004	1073.5	1076.5	0.00609	2.17011	2.1762
40	0.12173	0.01602	2443.6	8.032	1015.6	1023.7	8.032	1070.7	1078.7	0.01620	2.14271	2.1589
45	0.14756	0.01602	2035.8	13.05	1012.2	1025.3	13.05	1067.8	1080.9	0.02620	2.11587	2.1421
50	0.17812	0.01602	1703.1	18.07	1008.9	1026.9	18.07	1065.0	1083.1	0.03609	2.08956	2.1256
55	0.21413	0.01603	1430.4	23.07	1005.5	1028.6	23.07	1062.2	1085.3	0.04586	2.06377	2.1096
60	0.25638	0.01604	1206.1	28.08	1002.1	1030.2	28.08	1059.4	1087.4	0.05554	2.03847	2.0940
65	0.30578	0.01604	1020.8	33.08	998.76	1031.8	33.08	1056.5	1089.6	0.06511	2.01366	2.0788
70	0.36334	0.01605	867.18	38.08	995.39	1033.5	38.08	1053.7	1091.8	0.07459	1.98931	2.0639
75	0.43016	0.01606	739.27	43.07	992.02	1035.1	43.07	1050.9	1093.9	0.08398	1.96541	2.0494
80	0.50745	0.01607	632.41	48.06	988.65	1036.7	48.07	1048.0	1096.1	0.09328	1.94196	2.0352
85	0.59659	0.01609	542.80	53.06	985.28	1038.3	53.06	1045.2	1098.3	0.10248	1.91892	2.0214
90	0.69904	0.01610	467.40	58.05	981.90	1040.0	58.05	1042.4	1100.4	0.11161	1.89630	2.0079
95	0.81643	0.01612	403.74	63.04	978.52	1041.6	63.04	1039.5	1102.6	0.12065	1.87408	1.9947
100	0.95052	0.01613	349.83	68.03	975.14	1043.2	68.03	1036.7	1104.7	0.12961	1.85225	1.9819
110	1.2767	0.01617	264.96	78.01	968.36	1046.4	78.02	1031.0	1109.0	0.14728	1.80970	1.9570
120	1.6951	0.01620	202.94	88.00	961.56	1049.6	88.00	1025.2	1113.2	0.16466	1.76856	1.9332
130	2.2260	0.01625	157.09	97.99	954.73	1052.7	97.99	1019.4	1117.4	0.18174	1.72877	1.9105
140	2.8931	0.01629	122.81	107.98	947.87	1055.9	107.99	1013.6	1121.6	0.19855	1.69024	1.8888
150	3.7234	0.01634	96.929	117.98	940.98	1059.0	117.99	1007.8	1125.7	0.21508	1.65291	1.8680
160	4.7474	0.01639	77.185	127.98	934.05	1062.0	128.00	1001.8	1129.8	0.23136	1.61670	1.8481
170	5.9999	0.01645	61.982	138.00	927.08	1065.1	138.02	995.88	1133.9	0.24739	1.58155	1.8289
180	7.5197	0.01651	50.172	148.02	920.06	1068.1	148.04	989.85	1137.9	0.26318	1.54741	1.8106
190	9.3497	0.01657	40.920	158.05	912.99	1071.0	158.08	983.76	1141.8	0.27874	1.51421	1.7930
200	11.538	0.01663	33.613	168.10	905.87	1074.0	168.13	977.60	1145.7	0.29409	1.48191	1.7760
210	14.136	0.01670	27.798	178.15	898.68	1076.8	178.20	971.35	1149.5	0.30922	1.45046	1.7597
212	14.709	0.01671	26.782	180.16	897.24	1077.4	180.21	970.09	1150.3	0.31222	1.44427	1.7565
220	17.201	0.01677	23.136	188.22	891.43	1079.6	188.28	965.02	1153.3	0.32414	1.41980	1.7439
230	20.795	0.01684	19.374	198.31	884.10	1082.4	198.37	958.59	1157.0	0.33887	1.38989	1.7288
240	24.985	0.01692	16.316	208.41	876.70	1085.1	208.49	952.06	1160.5	0.35342	1.36069	1.7141
250	29.844	0.01700	13.816	218.54	869.21	1087.7	218.63	945.41	1164.0	0.36779	1.33216	1.6999
260	35.447	0.01708	11.760	228.68	861.62	1090.3	228.79	938.65	1167.4	0.38198	1.30425	1.6862
270	41.877	0.01717	10.059	238.85	853.94	1092.8	238.98	931.76	1170.7	0.39601	1.27694	1.6730
280	49.222	0.01726	8.6439	249.04	846.16	1095.2	249.20	924.74	1173.9	0.40989	1.25018	1.6601
290	57.573	0.01735	7.4607	259.26	838.27	1097.5	259.45	917.57	1177.0	0.42361	1.22393	1.6475
300	67.028	0.01745	6.4663	269.51	830.25	1099.8	269.73	910.24	1180.0	0.43720	1.19818	1.6354
310	77.691	0.01755	5.6266	279.79	822.11	1101.9	280.05	902.75	1182.8	0.45065	1.17289	1.6235
320	89.667	0.01765	4.9144	290.11	813.84	1104.0	290.40	895.09	1185.5	0.46396	1.14802	1.6120
330	103.07	0.01776	4.3076	300.46	805.43	1105.9	300.80	887.25	1188.1	0.47716	1.12355	1.6007
340	118.02	0.01787	3.7885	310.85	796.87	1107.7	311.24	879.22	1190.5	0.49024	1.09945	1.5897
350	134.63	0.01799	3.3425	321.29	788.16	1109.4	321.73	870.98	1192.7	0.50321	1.07570	1.5789
360	153.03	0.01811	2.9580	331.76	779.28	1111.0	332.28	862.53	1194.8	0.51607	1.05227	1.5683
370	173.36	0.01823	2.6252	342.29	770.23	1112.5	342.88	853.86	1196.7	0.52884	1.02914	1.5580
380	195.74	0.01836	2.3361	352.87	761.00	1113.9	353.53	844.96	1198.5	0.54152	1.00628	1.5478
390	220.33	0.01850	2.0842	363.50	751.58	1115.1	364.25	835.81	1200.1	0.55411	0.98366	1.5378

(Continued)

TABLE A-3E Saturated Water-Temperature Table (*Continued*)

Temp., T, °F	Sat. Press., P_{sat}, psia	Specific Volume, ft³/lbm		Internal Energy, Btu/lbm			Enthalpy, Btu/lbm			Entropy, Btu/lbm·R		
		Sat. Liquid, v_f	Sat. Vapor, v_g	Sat. Liquid, u_f	Evap., u_{fg}	Sat. Vapor, u_g	Sat. Liquid, h_f	Evap., h_{fg}	Sat. Vapor, h_g	Sat. Liquid, s_f	Evap., s_{fg}	Sat. Vapor, s_g
400	247.26	0.01864	1.8639	374.19	741.97	1116.2	375.04	826.39	1201.4	0.56663	0.96127	1.5279
410	276.69	0.01878	1.6706	384.94	732.14	1117.1	385.90	816.71	1202.6	0.57907	0.93908	1.5182
420	308.76	0.01894	1.5006	395.76	722.08	1117.8	396.84	806.74	1203.6	0.59145	0.91707	1.5085
430	343.64	0.01910	1.3505	406.65	711.80	1118.4	407.86	796.46	1204.3	0.60377	0.89522	1.4990
440	381.49	0.01926	1.2178	417.61	701.26	1118.9	418.97	785.87	1204.8	0.61603	0.87349	1.4895
450	422.47	0.01944	1.0999	428.66	690.47	1119.1	430.18	774.94	1205.1	0.62826	0.85187	1.4801
460	466.75	0.01962	0.99510	439.79	679.39	1119.2	441.48	763.65	1205.1	0.64044	0.83033	1.4708
470	514.52	0.01981	0.90158	451.01	668.02	1119.0	452.90	751.98	1204.9	0.65260	0.80885	1.4615
480	565.96	0.02001	0.81794	462.34	656.34	1118.7	464.43	739.91	1204.3	0.66474	0.78739	1.4521
490	621.24	0.02022	0.74296	473.77	644.32	1118.1	476.09	727.40	1203.5	0.67686	0.76594	1.4428
500	680.56	0.02044	0.67558	485.32	631.94	1117.3	487.89	714.44	1202.3	0.68899	0.74445	1.4334
510	744.11	0.02067	0.61489	496.99	619.17	1116.2	499.84	700.99	1200.8	0.70112	0.72290	1.4240
520	812.11	0.02092	0.56009	508.80	605.99	1114.8	511.94	687.01	1199.0	0.71327	0.70126	1.4145
530	884.74	0.02118	0.51051	520.76	592.35	1113.1	524.23	672.47	1196.7	0.72546	0.67947	1.4049
540	962.24	0.02146	0.46553	532.88	578.23	1111.1	536.70	657.31	1194.0	0.73770	0.65751	1.3952
550	1044.8	0.02176	0.42465	545.18	563.58	1108.8	549.39	641.47	1190.9	0.75000	0.63532	1.3853
560	1132.7	0.02207	0.38740	557.68	548.33	1106.0	562.31	624.91	1187.2	0.76238	0.61284	1.3752
570	1226.2	0.02242	0.35339	570.40	532.45	1102.8	575.49	607.55	1183.0	0.77486	0.59003	1.3649
580	1325.5	0.02279	0.32225	583.37	515.84	1099.2	588.95	589.29	1178.2	0.78748	0.56679	1.3543
590	1430.8	0.02319	0.29367	596.61	498.43	1095.0	602.75	570.04	1172.8	0.80026	0.54306	1.3433
600	1542.5	0.02362	0.26737	610.18	480.10	1090.3	616.92	549.67	1166.6	0.81323	0.51871	1.3319
610	1660.9	0.02411	0.24309	624.11	460.73	1084.8	631.52	528.03	1159.5	0.82645	0.49363	1.3201
620	1786.2	0.02464	0.22061	638.47	440.14	1078.6	646.62	504.92	1151.5	0.83998	0.46765	1.3076
630	1918.9	0.02524	0.19972	653.35	418.12	1071.5	662.32	480.07	1142.4	0.85389	0.44056	1.2944
640	2059.3	0.02593	0.18019	668.86	394.36	1063.2	678.74	453.14	1131.9	0.86828	0.41206	1.2803
650	2207.8	0.02673	0.16184	685.16	368.44	1053.6	696.08	423.65	1119.7	0.88332	0.38177	1.2651
660	2364.9	0.02767	0.14444	702.48	339.74	1042.2	714.59	390.84	1105.4	0.89922	0.34906	1.2483
670	2531.2	0.02884	0.12774	721.23	307.22	1028.5	734.74	353.54	1088.3	0.91636	0.31296	1.2293
680	2707.3	0.03035	0.11134	742.11	269.00	1011.1	757.32	309.57	1066.9	0.93541	0.27163	1.2070
690	2894.1	0.03255	0.09451	766.81	220.77	987.6	784.24	253.96	1038.2	0.95797	0.22089	1.1789
700	3093.0	0.03670	0.07482	801.75	146.50	948.3	822.76	168.32	991.1	0.99023	0.14514	1.1354
705.10	3200.1	0.04975	0.04975	866.61	0	866.6	896.07	0	896.1	1.05257	0	1.0526

Source: Tables A-3E through A-5E are generated using the Engineering Equation Solver (EES) software developed by S. A. Klein and F. L. Alvarado. The routine used in calculations is the highly accurate Steam_IAPWS, which incorporates the 1995 Formulation for the Thermodynamic Properties of Ordinary Water Substance for General and Scientific Use, issued by The International Association for the Properties of Water and Steam (IAPWS). This formulation replaces the 1984 formulation of Haar, Gallagher, and Kell (NBS/NRC Steam Tables, Hemisphere Publishing Co., 1984), which is also available in EES as the routine STEAM. The new formulation is based on the correlations of Saul and Wagner (J. Phys. Chem. Ref. Data, 16, 893, 1987) with modifications to adjust to the International Temperature Scale of 1990. The modifications are described by Wagner and Pruss (J. Phys. Chem. Ref. Data, 22, 783, 1993).

TABLE A-4E   Saturated Water—Pressure Table

Press., P, psia	Sat. Temp., T_{sat}, °F	Specific Volume, ft³/lbm		Internal Energy, Btu/lbm			Enthalpy, Btu/lbm			Entropy, Btu/lbm·R		
		Sat. Liquid, v_f	Sat. Vapor, v_g	Sat. Liquid, u_f	Evap., u_{fg}	Sat. Vapor, u_g	Sat. Liquid, h_f	Evap., h_{fg}	Sat. Vapor, h_g	Sat. Liquid, s_f	Evap., s_{fg}	Sat. Vapor, s_g
1	101.69	0.01614	333.49	69.72	973.99	1043.7	69.72	1035.7	1105.4	0.13262	1.84495	1.9776
2	126.02	0.01623	173.71	94.02	957.45	1051.5	94.02	1021.7	1115.8	0.17499	1.74444	1.9194
3	141.41	0.01630	118.70	109.39	946.90	1056.3	109.40	1012.8	1122.2	0.20090	1.68489	1.8858
4	152.91	0.01636	90.629	120.89	938.97	1059.9	120.90	1006.0	1126.9	0.21985	1.64225	1.8621
5	162.18	0.01641	73.525	130.17	932.53	1062.7	130.18	1000.5	1130.7	0.23488	1.60894	1.8438
6	170.00	0.01645	61.982	138.00	927.08	1065.1	138.02	995.88	1133.9	0.24739	1.58155	1.8289
8	182.81	0.01652	47.347	150.83	918.08	1068.9	150.86	988.15	1139.0	0.26757	1.53800	1.8056
10	193.16	0.01659	38.425	161.22	910.75	1072.0	161.25	981.82	1143.1	0.28362	1.50391	1.7875
14.696	211.95	0.01671	26.805	180.12	897.27	1077.4	180.16	970.12	1150.3	0.31215	1.44441	1.7566
15	212.99	0.01672	26.297	181.16	896.52	1077.7	181.21	969.47	1150.7	0.31370	1.44441	1.7549
20	227.92	0.01683	20.093	196.21	885.63	1081.8	196.27	959.93	1156.2	0.33582	1.39606	1.7319
25	240.03	0.01692	16.307	208.45	876.67	1085.1	208.52	952.03	1160.6	0.35347	1.36060	1.7141
30	250.30	0.01700	13.749	218.84	868.98	1087.8	218.93	945.21	1164.1	0.36821	1.33132	1.6995
35	259.25	0.01708	11.901	227.92	862.19	1090.1	228.03	939.16	1167.2	0.38093	1.30632	1.6872
40	267.22	0.01715	10.501	236.02	856.09	1092.1	236.14	933.69	1169.8	0.39213	1.28448	1.6766
45	274.41	0.01721	9.4028	243.34	850.52	1093.9	243.49	928.68	1172.2	0.40216	1.26506	1.6672
50	280.99	0.01727	8.5175	250.05	845.39	1095.4	250.21	924.03	1174.2	0.41125	1.24756	1.6588
55	287.05	0.01732	7.7882	256.25	840.61	1096.9	256.42	919.70	1176.1	0.41958	1.23162	1.6512
60	292.69	0.01738	7.1766	262.01	836.13	1098.1	262.20	915.61	1177.8	0.42728	1.21697	1.6442
65	297.95	0.01743	6.6560	267.41	831.90	1099.3	267.62	911.75	1179.4	0.43443	1.20341	1.6378
70	302.91	0.01748	6.2075	272.50	827.90	1100.4	272.72	908.08	1180.8	0.44112	1.19078	1.6319
75	307.59	0.01752	5.8167	277.31	824.09	1101.4	277.55	904.58	1182.1	0.44741	1.17895	1.6264
80	312.02	0.01757	5.4733	281.87	820.45	1102.3	282.13	901.22	1183.4	0.45335	1.16783	1.6212
85	316.24	0.01761	5.1689	286.22	816.97	1103.2	286.50	898.00	1184.5	0.45897	1.15732	1.6163
90	320.26	0.01765	4.8972	290.38	813.62	1104.0	290.67	894.89	1185.6	0.46431	1.14737	1.6117
95	324.11	0.01770	4.6532	294.36	810.40	1104.8	294.67	891.89	1186.6	0.46941	1.13791	1.6073
100	327.81	0.01774	4.4327	298.19	807.29	1105.5	298.51	888.99	1187.5	0.47427	1.12888	1.6032
110	334.77	0.01781	4.0410	305.41	801.37	1106.8	305.78	883.44	1189.2	0.48341	1.11201	1.5954
120	341.25	0.01789	3.7289	312.16	795.79	1107.9	312.55	878.20	1190.8	0.49187	1.09646	1.5883
130	347.32	0.01796	3.4557	318.48	790.51	1109.0	318.92	873.21	1192.1	0.49974	1.08204	1.5818
140	353.03	0.01802	3.2202	324.45	785.49	1109.9	324.92	868.45	1193.4	0.50711	1.06858	1.5757
150	358.42	0.01809	3.0150	330.11	780.69	1110.8	330.61	863.88	1194.5	0.51405	1.05595	1.5700
160	363.54	0.01815	2.8347	335.49	776.10	1111.6	336.02	859.49	1195.5	0.52061	1.04405	1.5647
170	368.41	0.01821	2.6749	340.62	771.68	1112.3	341.19	855.25	1196.4	0.52682	1.03279	1.5596
180	373.07	0.01827	2.5322	345.53	767.42	1113.0	346.14	851.16	1197.3	0.53274	1.02210	1.5548
190	377.52	0.01833	2.4040	350.24	763.31	1113.6	350.89	847.19	1198.1	0.53839	1.01191	1.5503
200	381.80	0.01839	2.2882	354.78	759.32	1114.1	355.46	843.33	1198.8	0.54379	1.00219	1.5460
250	400.97	0.01865	1.8440	375.23	741.02	1116.3	376.09	825.47	1201.6	0.56784	0.95912	1.5270
300	417.35	0.01890	1.5435	392.89	724.77	1117.7	393.94	809.41	1203.3	0.58818	0.92289	1.5111
350	431.74	0.01912	1.3263	408.55	709.98	1118.5	409.79	794.65	1204.4	0.60590	0.89143	1.4973
400	444.62	0.01934	1.1617	422.70	696.31	1119.0	424.13	780.87	1205.0	0.62168	0.86350	1.4852
450	456.31	0.01955	1.0324	435.67	683.52	1119.2	437.30	767.86	1205.2	0.63595	0.83828	1.4742
500	467.04	0.01975	0.92819	447.68	671.42	1119.1	449.51	755.48	1205.0	0.64900	0.81521	1.4642
550	476.97	0.01995	0.84228	458.90	659.91	1118.8	460.93	743.60	1204.5	0.66107	0.79388	1.4550
600	486.24	0.02014	0.77020	469.46	648.88	1118.3	471.70	732.15	1203.9	0.67231	0.77400	1.4463

(Continued)

TABLE A-4E   **Saturated Water—Pressure Table (*Continued*)**

Press., P, psia	Sat. Temp., T_{sat}, °F	Specific Volume, ft³/lbm		Internal Energy, Btu/lbm			Enthalpy, Btu/lbm			Entropy, Btu/lbm·R		
		Sat. Liquid, v_f	Sat. Vapor, v_g	Sat. Liquid, u_f	Evap., u_{fg}	Sat. Vapor, u_g	Sat. Liquid, h_f	Evap., h_{fg}	Sat. Vapor, h_g	Sat. Liquid, s_f	Evap., s_{fg}	Sat. Vapor, s_g
700	503.13	0.02051	0.65589	488.96	627.98	1116.9	491.62	710.29	1201.9	0.69279	0.73771	1.4305
800	518.27	0.02087	0.56920	506.74	608.30	1115.0	509.83	689.48	1199.3	0.71117	0.70502	1.4162
900	532.02	0.02124	0.50107	523.19	589.54	1112.7	526.73	669.46	1196.2	0.72793	0.67505	1.4030
1000	544.65	0.02159	0.44604	538.58	571.49	1110.1	542.57	650.03	1192.6	0.74341	0.64722	1.3906
1200	567.26	0.02232	0.36241	566.89	536.87	1103.8	571.85	612.39	1184.2	0.77143	0.59632	1.3677
1400	587.14	0.02307	0.30161	592.79	503.50	1096.3	598.76	575.66	1174.4	0.79658	0.54991	1.3465
1600	604.93	0.02386	0.25516	616.99	470.69	1087.7	624.06	539.18	1163.2	0.81972	0.50645	1.3262
1800	621.07	0.02470	0.21831	640.03	437.86	1077.9	648.26	502.35	1150.6	0.84144	0.46482	1.3063
2000	635.85	0.02563	0.18815	662.33	404.46	1066.8	671.82	464.60	1136.4	0.86224	0.42409	1.2863
2500	668.17	0.02860	0.13076	717.67	313.53	1031.2	730.90	360.79	1091.7	0.91311	0.31988	1.2330
3000	695.41	0.03433	0.08460	783.39	186.41	969.8	802.45	214.32	1016.8	0.97321	0.18554	1.1587
3200.1	705.10	0.04975	0.04975	866.61	0	866.6	896.07	0	896.1	1.05257	0	1.0526

TABLE A-5E   Superheated Water

T, °F	v ft³/lbm	u Btu/lbm	h Btu/lbm	s Btu/lbm·R	v ft³/lbm	u Btu/lbm	h Btu/lbm	s Btu/lbm·R	v ft³/lbm	u Btu/lbm	h Btu/lbm	s Btu/lbm·R
	P = 1.0 psia (101.69°F)*				P = 5.0 psia (162.18°F)				P = 10 psia (193.16°F)			
Sat.†	333.49	1043.7	1105.4	1.9776	73.525	1062.7	1130.7	1.8438	38.425	1072.0	1143.1	1.7875
200	392.53	1077.5	1150.1	2.0509	78.153	1076.2	1148.5	1.8716	38.849	1074.5	1146.4	1.7926
240	416.44	1091.2	1168.3	2.0777	83.009	1090.3	1167.1	1.8989	41.326	1089.1	1165.5	1.8207
280	440.33	1105.0	1186.5	2.1030	87.838	1104.3	1185.6	1.9246	43.774	1103.4	1184.4	1.8469
320	464.20	1118.9	1204.8	2.1271	92.650	1118.4	1204.1	1.9490	46.205	1117.6	1203.1	1.8716
360	488.07	1132.9	1223.3	2.1502	97.452	1132.5	1222.6	1.9722	48.624	1131.9	1221.8	1.8950
400	511.92	1147.1	1241.8	2.1722	102.25	1146.7	1241.3	1.9944	51.035	1146.2	1240.6	1.9174
440	535.77	1161.3	1260.4	2.1934	107.03	1160.9	1260.0	2.0156	53.441	1160.5	1259.4	1.9388
500	571.54	1182.8	1288.6	2.2237	114.21	1182.6	1288.2	2.0461	57.041	1182.2	1287.8	1.9693
600	631.14	1219.4	1336.2	2.2709	126.15	1219.2	1335.9	2.0933	63.029	1219.0	1335.6	2.0167
700	690.73	1256.8	1384.6	2.3146	138.09	1256.7	1384.4	2.1371	69.007	1256.5	1384.2	2.0605
800	750.31	1295.1	1433.9	2.3553	150.02	1294.9	1433.7	2.1778	74.980	1294.8	1433.5	2.1013
1000	869.47	1374.2	1535.1	2.4299	173.86	1374.2	1535.0	2.2524	86.913	1374.1	1534.9	2.1760
1200	988.62	1457.1	1640.0	2.4972	197.70	1457.0	1640.0	2.3198	98.840	1457.0	1639.9	2.2433
1400	1107.8	1543.7	1748.7	2.5590	221.54	1543.7	1748.7	2.3816	110.762	1543.6	1748.6	2.3052
	P = 15 psia (212.99°F)				P = 20 psia (227.92°F)				P = 40 psia (267.22°F)			
Sat.	26.297	1077.7	1150.7	1.7549	20.093	1081.8	1156.2	1.7319	10.501	1092.1	1169.8	1.6766
240	27.429	1087.8	1163.9	1.7742	20.478	1086.5	1162.3	1.7406				
280	29.085	1102.4	1183.2	1.8010	2 1.739	1101.4	1181.9	1.7679	10.713	1097.3	1176.6	1.6858
320	30.722	1116.9	1202.2	1.8260	22.980	1116.1	1201.2	1.7933	11.363	1112.9	1197.1	1.7128
360	32.348	1131.3	1221.1	1.8496	24.209	1130.7	1220.2	1.8171	11.999	1128.1	1216.9	1.7376
400	33.965	1145.7	1239.9	1.8721	25.429	1145.1	1239.3	1.8398	12.625	1143.1	1236.5	1.7610
440	35.576	1160.1	1258.8	1.8936	26.644	1159.7	1258.3	1.8614	13.244	1157.9	1256.0	1.7831
500	37.986	1181.9	1287.3	1.9243	28.458	1181.6	1286.9	1.8922	14.165	1180.2	1285.0	1.8143
600	41.988	1218.7	1335.3	1.9718	31.467	1218.5	1334.9	1.9398	15.686	1217.5	1333.6	1.8625
700	45.981	1256.3	1383.9	2.0156	34.467	1256.1	1383.7	1.9837	17.197	1255.3	1382.6	1.9067
800	49.967	1294.6	1433.3	2.0565	37.461	1294.5	1433.1	2.0247	18.702	1293.9	1432.3	1.9478
1000	57.930	1374.0	1534.8	2.1312	43.438	1373.8	1534.6	2.0994	21.700	1373.4	1534.1	2.0227
1200	65.885	1456.9	1639.8	2.1986	49.407	1456.8	1639.7	2.1668	24.691	1456.5	1639.3	2.0902
1400	73.836	1543.6	1748.5	2.2604	55.373	1543.5	1748.4	2.2287	27.678	1543.3	1748.1	2.1522
1600	81.784	1634.0	1861.0	2.3178	61.335	1633.9	1860.9	2.2861	30.662	1633.7	1860.7	2.2096
	P = 60 psia (292.69°F)*				P = 80 psia (312.02°F)				P = 100 psia (327.81°F)			
Sat.	7.1766	1098.1	1177.8	1.6442	5.4733	1102.3	1183.4	1.6212	4.4327	1105.5	1187.5	1.6032
320	7.4863	1109.6	1192.7	1.6636	5.5440	1105.9	1187.9	1.6271				
360	7.9259	1125.5	1213.5	1.6897	5.8876	1122.7	1209.9	1.6545	4.6628	1119.8	1206.1	1.6263
400	8.3548	1140.9	1233.7	1.7138	6.2187	1138.7	1230.8	1.6794	4.9359	1136.4	1227.8	1.6521
440	8.7766	1156.1	1253.6	1.7364	6.5420	1154.3	1251.2	1.7026	5.2006	1152.4	1248.7	1.6759
500	9.4005	1178.8	1283.1	1.7682	7.0177	1177.3	1281.2	1.7350	5.5876	1175.9	1279.3	1.7088
600	10.4256	1216.5	1332.2	1.8168	7.7951	1215.4	1330.8	1.7841	6.2167	1214.4	1329.4	1.7586
700	11.4401	1254.5	1381.6	1.8613	8.5616	1253.8	1380.5	1.8289	6.8344	1253.0	1379.5	1.8037
800	12.4484	1293.3	1431.5	1.9026	9.3218	1292.6	1430.6	1.8704	7.4457	1292.0	1429.8	1.8453
1000	14.4543	1373.0	1533.5	1.9777	10.8313	1372.6	1532.9	1.9457	8.6575	1372.2	1532.4	1.9208
1200	16.4525	1456.2	1638.9	2.0454	12.3331	1455.9	1638.5	2.0135	9.8615	1455.6	1638.1	1.9887
1400	18.4464	1543.0	1747.8	2.1073	13.8306	1542.8	1747.5	2.0755	11.0612	154 2.6	1747.2	2.0508
1600	20.438	1633.5	1860.5	2.1648	15.3257	1633.3	1860.2	2.1330	12.2584	1633.2	1860.0	2.1083
1800	22.428	1727.6	1976.6	2.2187	16.8192	172 7.5	1976.5	2.1869	13.4541	172 7.3	1976.3	2.1622
2000	24.417	1825.2	2096.3	2.2694	18.3117	1825.0	2096.1	2.2376	14.6487	1824.9	2096.0	2.2130

*The temperature in parentheses is the saturation temperature at the specified pressure.

†Properties of saturated vapor at the specified pressure.

(*Continued*)

TABLE A-5E   Superheated Water (Continued)

T, °F	v ft³/lbm	u Btu/lbm	h Btu/lbm	s Btu/lbm·R	v ft³/lbm	u Btu/lbm	h Btu/lbm	s Btu/lbm·R	v ft³/lbm	u Btu/lbm	h Btu/lbm	s Btu/lbm·R
	P = 120 psia (341.25°F)				P = 140 psia (353.03°F)				P = 160 psia (363.54°F)			
Sat.	3.7289	1107.9	1190.8	1.5883	3.2202	1109.9	1193.4	1.5757	2.8347	1111.6	1195.5	1.5647
360	3.8446	1116.7	1202.1	1.6023	3.2584	1113.4	1197.8	1.5811				
400	4.0799	1134.0	1224.6	1.6292	3.4676	1131.5	1221.4	1.6092	3.0076	1129.0	1218.0	1.5914
450	4.3613	1154.5	1251.4	1.6594	3.7147	1152.6	1248.9	1.6403	3.2293	1150.7	1246.3	1.6234
500	4.6340	1174.4	1277.3	1.6872	3.9525	1172.9	1275.3	1.6686	3.4412	1171.4	1273.2	1.6522
550	4.9010	1193.9	1302.8	1.7131	4.1845	1192.7	1301.1	1.6948	3.6469	1191.4	1299.4	1.6788
600	5.1642	1213.4	1328.0	1.7375	4.4124	1212.3	1326.6	1.7195	3.8484	1211.3	1325.2	1.7037
700	5.6829	1252.2	1378.4	1.7829	4.8604	1251.4	1377.3	1.7652	4.2434	1250.6	1376.3	1.7498
800	6.1950	1291.4	1429.0	1.8247	5.3017	1290.8	1428.1	1.8072	4.6316	1290.2	1427.3	1.7920
1000	7.2083	1371.7	1531.8	1.9005	6.1732	1371.3	1531.3	1.8832	5.3968	1370.9	1530.7	1.8682
1200	8.2137	1455.3	1637.7	1.9684	7.0367	1455.0	1637.3	1.9512	6.1540	1454.7	1636.9	1.9363
1400	9.2149	1542.3	1746.9	2.0305	7.8961	1542.1	1746.6	2.0134	6.9070	1541.8	1746.3	1.9986
1600	10.2135	1633.0	1859.8	2.0881	8.7529	1632.8	1859.5	2.0711	7.6574	1632.6	1859.3	2.0563
1800	11.2106	1727.2	1976.1	2.1420	9.6082	1727.0	1975.9	2.1250	8.4063	1726.9	1975.7	2.1102
2000	12.2067	1824.8	2095.8	2.1928	10.4624	1824.6	2095.7	2.1758	9.1542	1824.5	2095.5	2.1610
	P = 180 psia (373.07°F)				P = 200 psia (381.80°F)				P = 225 psia (391.80°F)			
Sat.	2.5322	1113.0	1197.3	1.5548	2.2882	1114.1	1198.8	1.5460	2.0423	1115.3	1200.3	1.5360
400	2.6490	1126.3	1214.5	1.5752	2.3615	1123.5	1210.9	1.5602	2.0728	1119.7	1206.0	1.5427
450	2.8514	1148.7	1243.7	1.6082	2.5488	1146.7	1241.0	1.5943	2.2457	1144.1	1237.6	1.5783
500	3.0433	1169.8	1271.2	1.6376	2.7247	1168.2	1269.0	1.6243	2.4059	1166.2	1266.3	1.6091
550	3.2286	1190.2	1297.7	1.6646	2.8939	1188.9	1296.0	1.6516	2.5590	1187.2	1293.8	1.6370
600	3.4097	1210.2	1323.8	1.6897	3.0586	1209.1	1322.3	1.6771	2.7075	1207.7	1320.5	1.6628
700	3.7635	1249.8	1375.2	1.7361	3.3796	1249.0	1374.1	1.7238	2.9956	1248.0	1372.7	1.7099
800	4.1104	1289.5	1426.5	1.7785	3.6934	1288.9	1425.6	1.7664	3.2765	1288.1	1424.5	1.7528
900	4.4531	1329.7	1478.0	1.8179	4.0031	1329.2	1477.3	1.8059	3.5530	1328.5	1476.5	1.7925
1000	4.7929	1370.5	1530.1	1.8549	4.3099	1370.1	1529.6	1.8430	3.8268	1369.5	1528.9	1.8296
1200	5.4674	1454.3	1636.5	1.9231	4.9182	1454.0	1636.1	1.9113	4.3689	1453.6	1635.6	1.8981
1400	6.1377	1541.6	1746.0	1.9855	5.5222	1541.4	1745.7	1.9737	4.9068	1541.1	1745.4	1.9606
1600	6.8054	1632.4	1859.1	2.0432	6.1238	1632.2	1858.8	2.0315	5.4422	1632.0	1858.6	2.0184
1800	7.4716	1726.7	1975.6	2.0971	6.7238	1726.5	1975.4	2.0855	5.9760	1726.4	1975.2	2.0724
2000	8.1367	1824.4	2095.4	2.1479	7.3227	1824.3	2095.3	2.1363	6.5087	1824.1	2095.1	2.1232
	P = 250 psia (400.97°F)				P = 275 psia (409.45°F)				P = 300 psia (417.35°F)			
Sat.	1.8440	1116.3	1201.6	1.5270	1.6806	1117.0	1202.6	1.5187	1.5435	1117.7	1203.3	1.5111
450	2.0027	1141.3	1234.0	1.5636	1.8034	1138.5	1230.3	1.5499	1.6369	1135.6	1226.4	1.5369
500	2.1506	1164.1	1263.6	1.5953	1.9415	1162.0	1260.8	1.5825	1.7670	1159.8	1257.9	1.5706
550	2.2910	1185.6	1291.5	1.6237	2.0715	1183.9	1289.3	1.6115	1.8885	1182.1	1287.0	1.6001
600	2.4264	1206.3	1318.6	1.6499	2.1964	1204.9	1316.7	1.6380	2.0046	1203.5	1314.8	1.6270
650	2.5586	1226.8	1345.1	1.6743	2.3179	1225.6	1343.5	1.6627	2.1172	1224.4	1341.9	1.6520
700	2.6883	1247.0	1371.4	1.6974	2.4369	1246.0	1370.0	1.6860	2.2273	1244.9	1368.6	1.6755
800	2.9429	1287.3	1423.5	1.7406	2.6699	1286.5	1422.4	1.7294	2.4424	1285.7	1421.3	1.7192
900	3.1930	1327.9	1475.6	1.7804	2.8984	1327.3	1474.8	1.7694	2.6529	1326.6	1473.9	1.7593
1000	3.4403	1369.0	1528.2	1.8177	3.1241	1368.5	1527.4	1.8068	2.8605	1367.9	1526.7	1.7968
1200	3.9295	1453.3	1635.0	1.8863	3.5700	1452.9	1634.5	1.8755	3.2704	1452.5	1634.0	1.8657
1400	4.4144	1540.8	1745.0	1.9488	4.0116	1540.5	1744.6	1.9381	3.6759	1540.2	1744.2	1.9284
1600	4.8969	1631.7	1858.3	2.0066	4.4507	1631.5	1858.0	1.9960	4.0789	1631.3	1857.7	1.9863
1800	5.3777	1726.2	1974.9	2.0607	4.8882	1726.0	1974.7	2.0501	4.4803	1725.8	1974.5	2.0404
2000	5.8575	1823.9	2094.9	2.1116	5.3247	1823.8	2094.7	2.1010	4.8807	1823.6	2094.6	2.0913

(Continued)

TABLE A-5E   Superheated Water (*Continued*)

T, °F	v ft³/lbm	u Btu/lbm	h Btu/lbm	s Btu/ lbm·R	v ft³/lbm	u Btu/lbm	h Btu/lbm	s Btu/ lbm·R	v ft³/lbm	u Btu/lbm	h Btu/lbm	s Btu/ lbm·R
	P = 350 psia (431.74°F)				*P* = 400 psia (444.62°F)				*P* = 450 psia (456.31°F)			
Sat.	1.3263	1118.5	1204.4	1.4973	1.1617	1119.0	1205.0	1.4852	1.0324	1119.2	1205.2	1.4742
450	1.3739	1129.3	1218.3	1.5128	1.1747	1122.5	1209.4	1.4901				
500	1.4921	1155.2	1251.9	1.5487	1.2851	1150.4	1245.6	1.5288	1.1233	1145.4	1238.9	1.5103
550	1.6004	1178.6	1282.2	1.5795	1.3840	1174.9	1277.3	1.5610	1.2152	1171.1	1272.3	1.5441
600	1.7030	1200.6	1310.9	1.6073	1.4765	1197.6	1306.9	1.5897	1.3001	1194.5	1302.8	1.5737
650	1.8018	1221.9	1338.6	1.6328	1.5650	1219.4	1335.3	1.6158	1.3807	1216.9	1331.9	1.6005
700	1.8979	1242.8	1365.8	1.6567	1.6507	1240.7	1362.9	1.6401	1.4584	1238.5	1360.0	1.6253
800	2.0848	1284.1	1419.1	1.7009	1.8166	1282.5	1417.0	1.6849	1.6080	1280.8	1414.7	1.6706
900	2.2671	1325.3	1472.2	1.7414	1.9777	1324.0	1470.4	1.7257	1.7526	1322.7	1468.6	1.7117
1000	2.4464	1366.9	1525.3	1.7791	2.1358	1365.8	1523.9	1.7636	1.8942	1364.7	1522.4	1.7499
1200	2.7996	1451.7	1633.0	1.8483	2.4465	1450.9	1632.0	1.8331	2.1718	1450.1	1631.0	1.8196
1400	3.1484	1539.6	1743.5	1.9111	2.7527	1539.0	1742.7	1.8960	2.4450	1538.4	1742.0	1.8827
1600	3.4947	1630.8	1857.1	1.9691	3.0565	1630.3	1856.5	1.9541	2.7157	1629.8	1856.0	1.9409
1800	3.8394	1725.4	1974.0	2.0233	3.3586	1725.0	1973.6	2.0084	2.9847	1724.6	1973.2	1.9952
2000	4.1830	1823.3	2094.2	2.0742	3.6597	1823.0	2093.9	2.0594	3.2527	1822.6	2093.5	2.0462
	P = 500 psia (467.04°F)				*P* = 600 psia (486.24°F)				*P* = 700 psia (503.13°F)			
Sat.	0.92815	1119.1	1205.0	1.4642	0.77020	1118.3	1203.9	1.4463	0.65589	1116.9	1201.9	1.4305
500	0.99304	1140.1	1231.9	1.4928	0.79526	1128.2	1216.5	1.4596				
550	1.07974	1167.1	1267.0	1.5284	0.87542	1158.7	1255.9	1.4996	0.72799	1149.5	1243.8	1.4730
600	1.15876	1191.4	1298.6	1.5590	0.94605	1184.9	1289.9	1.5325	0.79332	1177.9	1280.7	1.5087
650	1.23312	1214.3	1328.4	1.5865	1.01133	1209.0	1321.3	1.5614	0.85242	1203.4	1313.8	1.5393
700	1.30440	1236.4	1357.0	1.6117	1.07316	1231.9	1351.0	1.5877	0.90769	1227.2	1344.8	1.5666
800	1.44097	1279.2	1412.5	1.6576	1.19038	1275.8	1408.0	1.6348	1.01125	1272.4	1403.4	1.6150
900	1.57252	1321.4	1466.9	1.6992	1.30230	1318.7	1463.3	1.6771	1.10921	1316.0	1459.7	1.6581
1000	1.70094	1363.6	1521.0	1.7376	1.41097	1361.4	1518.1	1.7160	1.20381	1359.2	1515.2	1.6974
1100	1.82726	1406.2	1575.3	1.7735	1.51749	1404.4	1572.9	1.7522	1.29621	1402.5	1570.4	1.7341
1200	1.95211	1449.4	1630.0	1.8075	1.62252	1447.8	1627.9	1.7865	1.38709	1446.2	1625.9	1.7685
1400	2.1988	1537.8	1741.2	1.8708	1.82957	1536.6	1739.7	1.8501	1.56580	1535.4	1738.2	1.8324
1500	2.4430	1529.4	1855.4	1.9291	2.0340	1628.4	1854.2	1.9085	1.74192	1627.5	1853.1	1.8911
1800	2.6856	1724.2	1972.7	1.9834	2.2369	1723.4	1971.8	1.9630	1.91643	1722.7	1970.9	1.9457
2000	2.9271	1822.3	2093.1	2.0345	2.4387	1821.7	2092.4	2.0141	2.08987	1821.0	2091.7	1.9969
	P = 800 psia (518.27°F)				*P* = 1000 psia (544.65°F)				*P* = 1250 psia (572.45°F)			
Sat.	0.56920	1115.0	1199.3	1.4162	0.44604	1110.1	1192.6	1.3906	0.34549	1102.0	1181.9	1.3623
550	0.61586	1139.4	1230.5	1.4476	0.45375	1115.2	1199.2	1.3972				
600	0.67799	1170.5	1270.9	1.4866	0.51431	1154.1	1249.3	1.4457	0.37894	1129.5	1217.2	1.3961
650	0.73279	1197.6	1306.0	1.5191	0.56411	1185.1	1289.5	1.4827	0.42703	1167.5	1266.3	1.4414
700	0.78330	1222.4	1338.4	1.5476	0.60844	1212.4	1325.0	1.5140	0.46735	1198.7	1306.8	1.4771
750	0.83102	1246.0	1369.1	1.5735	0.64944	1237.6	1357.8	1.5418	0.50344	1226.4	1342.9	1.5076
800	0.87678	1268.9	1398.7	1.5975	0.68821	1261.7	1389.0	1.5670	0.53687	1252.2	1376.4	1.5347
900	0.96434	1313.3	1456.0	1.6413	0.76136	1307.7	1448.6	1.6126	0.59876	1300.5	1439.0	1.5826
1000	1.04841	1357.0	1512.2	1.6812	0.83078	1352.5	1506.2	1.6535	0.65656	1346.7	1498.6	1.6249
1100	1.13024	1400.7	1568.0	1.7181	0.89783	1396.9	1563.1	1.6911	0.71184	1392.2	1556.8	1.6635
1200	1.21051	1444.6	1623.8	1.7528	0.96327	1441.4	1619.7	1.7263	0.76545	1437.4	1614.5	1.6993
1400	1.36797	1534.2	1736.7	1.8170	1.09101	1531.8	1733.7	1.7911	0.86944	1528.7	1729.8	1.7649
1600	1.52283	1626.5	1851.9	1.8759	1.21610	1624.6	1849.6	1.8504	0.97072	1622.2	1846.7	1.8246
1800	1.67606	1721.9	1970.0	1.9306	1.33956	1720.3	1968.2	1.9053	1.07036	1718.4	1966.0	1.8799
2000	1.82823	1820.4	2091.0	1.9819	1.46194	1819.1	2089.6	1.9568	1.16892	1817.5	2087.9	1.9315

(Continued)

TABLE A-5E   **Superheated Water (*Continued*)**

T, °F	v ft³/lbm	u Btu/lbm	h Btu/lbm	s Btu/ lbm·R	v ft³/lbm	u Btu/lbm	h Btu/lbm	s Btu/ lbm·R	v ft³/lbm	u Btu/lbm	h Btu/lbm	s Btu/ lbm·R
	$P = 1500$ psia (596.26°F)				$P = 1750$ psia (617.17°F)				$P = 2000$ psia (635.85°F)			
Sat.	0.27695	1092.1	1169.0	1.3362	0.22681	1080.5	1153.9	1.3112	0.18815	1066.8	1136.4	1.2863
600	0.28189	1097.2	1175.4	1.3423								
650	0.33310	1147.2	1239.7	1.4016	0.26292	1122.8	1207.9	1.3607	0.20586	1091.4	1167.6	1.3146
700	0.37198	1183.6	1286.9	1.4433	0.30252	1166.8	1264.7	1.4108	0.24894	1147.6	1239.8	1.3783
750	0.40535	1214.4	1326.9	1.4771	0.33455	1201.5	1309.8	1.4489	0.28074	1187.4	1291.3	1.4218
800	0.43550	1242.2	1363.1	1.5064	0.36266	1231.7	1349.1	1.4807	0.30763	1220.5	1334.3	1.4567
850	0.46356	1268.2	1396.9	1.5328	0.38835	1259.3	1385.1	1.5088	0.33169	1250.0	1372.8	1.4867
900	0.49015	1293.1	1429.2	1.5569	0.41238	1285.4	1419.0	1.5341	0.35390	1277.5	1408.5	1.5134
1000	0.54031	1340.9	1490.8	1.6007	0.45719	1334.9	1482.9	1.5796	0.39479	1328.7	1474.9	1.5606
1100	0.58781	1387.3	1550.5	1.6402	0.49917	1382.4	1544.1	1.6201	0.43266	1377.5	1537.6	1.6021
1200	0.63355	1433.3	1609.2	1.6767	0.53932	1429.2	1603.9	1.6572	0.46864	1425.1	1598.5	1.6400
1400	0.72172	1525.7	1726.0	1.7432	0.61621	1522.6	1722.1	1.7245	0.53708	1519.5	1718.3	1.7081
1600	0.80714	1619.8	1843.8	1.8033	0.69031	1617.4	1840.9	1.7852	0.60269	1615.0	1838.0	1.7693
1800	0.89090	1716.4	1963.7	1.8589	0.76273	1714.5	1961.5	1.8410	0.66660	1712.5	1959.2	1.8255
2000	0.97358	1815.9	2086.1	1.9108	0.83406	1814.2	2084.3	1.8931	0.72942	1812.6	2082.6	1.8778
	$P = 2500$ psia (668.17°F)				$P = 3000$ psia (695.41°F)				$P = 3500$ psia			
Sat.	0.13076	1031.2	1091.7	1.2330	0.08460	969.8	1016.8	1.1587				
650									0.02492	663.7	679.9	0.863.2
700	0.16849	1098.4	1176.3	1.3072	0.09838	1005.3	1059.9	1.1960	0.03065	760.0	779.9	0.9511
750	0.20327	1154.9	1249.0	1.3686	0.14840	1114.1	1196.5	1.3118	0.10460	1057.6	1125.4	1.2434
800	0.22949	1195.9	1302.0	1.4116	0.17601	1167.5	1265.3	1.3676	0.13639	1134.3	1222.6	1.3224
850	0.25174	1230.1	1346.6	1.4463	0.19771	1208.2	1317.9	1.4085	0.15847	1183.8	1286.5	1.3721
900	0.27165	1260.7	1386.4	1.4761	0.21640	1242.8	1362.9	1.4423	0.17659	1223.4	1337.8	1.4106
950	0.29001	1289.1	1423.3	1.5028	0.23321	1273.9	1403.3	1.4716	0.19245	1257.8	1382.4	1.4428
1000	0.30726	1316.1	1458.2	1.5271	0.24876	1302.8	1440.9	1.4978	0.20687	1289.0	1423.0	1.4711
1100	0.33949	1367.3	1524.4	1.5710	0.27732	1356.8	1510.8	1.5441	0.23289	1346.1	1496.9	1.5201
1200	0.36966	1416.6	1587.6	1.6103	0.30367	1408.0	1576.6	1.5850	0.25654	1399.3	1565.4	1.5627
1400	0.42631	1513.3	1710.5	1.6802	0.35249	1507.0	1702.7	1.6567	0.29978	1500.7	1694.8	1.6364
1600	0.48004	1610.1	1832.2	1.7424	0.39830	1605.3	1826.4	1.7199	0.33994	1600.4	1820.5	1.7006
1800	0.53205	1708.6	1954.8	1.7991	0.44237	1704.7	1950.3	1.7773	0.37833	1700.8	1945.8	1.7586
2000	0.58295	1809.4	2079.1	1.8518	0.48532	1806.1	2075.6	1.8304	0.41561	1802.9	2072.1	1.8121
	$P = 4000$ psia				$P = 5000$ psia				$P = 6000$ psia			
650	0.02448	657.9	676.1	0.8577	0.02379	648.3	670.3	0.8485	0.02325	640.3	666.1	0.8408
700	0.02871	742.3	763.6	0.9347	0.02678	721.8	746.6	0.9156	0.02564	708.1	736.5	0.9028
750	0.06370	962.1	1009.2	1.1410	0.03373	821.8	853.0	1.0054	0.02981	788.7	821.8	0.9747
800	0.10520	1094.2	1172.1	1.2734	0.05937	986.9	1041.8	1.1581	0.03949	897.1	941.0	1.0711
850	0.12848	1156.7	1251.8	1.3355	0.08551	1092.4	1171.5	1.2593	0.05815	1018.6	1083.1	1.1819
900	0.14647	1202.5	1310.9	1.3799	0.10390	1155.9	1252.1	1.3198	0.07584	1103.5	1187.7	1.2603
950	0.16176	1240.7	1360.5	1.4157	0.11863	1203.9	1313.6	1.3643	0.09010	1163.7	1263.7	1.3153
1000	0.17538	1274.6	1404.4	1.4463	0.13128	1244.0	1365.5	1.4004	0.10208	1211.4	1324.7	1.3578
1100	0.19957	1335.1	1482.8	1.4983	0.15298	1312.2	1453.8	1.4590	0.12211	1288.4	1424.0	1.4237
1200	0.22121	1390.3	1554.1	1.5426	0.17185	1372.1	1531.1	1.5070	0.13911	1353.4	1507.8	1.4758
1300	0.24128	1443.0	1621.6	1.5821	0.18902	1427.8	1602.7	1.5490	0.15434	1412.5	1583.8	1.5203
1400	0.26028	1494.3	1687.0	1.6182	0.20508	1481.4	1671.1	1.5868	0.16841	1468.4	1655.4	1.5598
1600	0.29620	1595.5	1814.7	1.6835	0.23505	1585.6	1803.1	1.6542	0.19438	1575.7	1791.5	1.6294
1800	0.33033	1696.8	1941.4	1.7422	0.26320	1689.0	1932.5	1.7142	0.21853	1681.1	1923.7	1.6907
2000	0.36335	1799.7	2068.6	1.7961	0.29023	1793.2	2061.7	1.7689	0.24155	1786.7	2054.9	1.7463

TABLE A-6E   Enthalpy of Formation, Gibbs Function of Formation, and Absolute Entropy at 77°F, 1 atm

Substance	Formula	$\overline{h}_f^\circ$ Btu/lbmol	$\overline{g}_f^\circ$ Btu/lbmol	$\overline{s}^\circ$ Btu/ lbmol·R
Carbon	C(s)	0	0	1.36
Hydrogen	$H_2(g)$	0	0	31.21
Nitrogen	$N_2(g)$	0	0	45.77
Oxygen	O_2	0	0	49.00
Carbon monoxide	CO(g)	−47,540	−59,010	47.21
Carbon dioxide	$CO_2(g)$	−169,300	−169,680	51.07
Water vapor	$H_2O(g)$	−104,040	−98,350	45.11
Water	$H_2O(\ell)$	−122,970	−102,040	16.71
Hydrogen peroxide	$H_2O_2(g)$	−58,640	−45,430	55.60
Ammonia	$NH_3(g)$	−19,750	−7,140	45.97
Methane	$CH_4(g)$	−32,210	−21,860	44.49
Acetylene	$C_2H_2(g)$	+97,540	+87,990	48.00
Ethylene	$C_2H_4(g)$	+22,490	+29,306	52.54
Ethane	$C_2H_6(g)$	−36,420	−14,150	54.85
Propylene	$C_3H_6(g)$	+8,790	+26,980	63.80
Propane	$C_3H_8(g)$	−44,680	−10,105	64.51
n-Butane	$C_4H_{10}(g)$	−54,270	−6,760	74.11
n-Octane	$C_8H_{18}(g)$	−89,680	+7,110	111.55
n-Octane	$C_8H_{18}(\ell)$	−107,530	+2,840	86.23
n-Dodecane	$C_{12}H_{26}(g)$	−125,190	+21,570	148.86
Benzene	$C_6H_6(g)$	+35,680	+55,780	64.34
Methyl alcohol	$CH_3OH(g)$	−86,540	−69,700	57.29
Methyl alcohol	$CH_3OH(\ell)$	−102,670	−71,570	30.30
Ethyl alcohol	$C_2H_5OH(g)$	−101,230	−72,520	67.54
Ethyl alcohol	$C_2H_5OH(\ell)$	−119,470	−75,240	38.40
Oxygen	O(g)	+107,210	+99,710	38.47
Hydrogen	H(g)	+93,780	+87,460	27.39
Nitrogen	N(g)	+203,340	+195,970	36.61
Hydroxyl	OH(g)	+16,790	+14,750	43.92

Source: From JANAF, *Thermochemical Tables* (Midland, MI: Dow Chemical Co., 1971), *Selected Values of Chemical Thermodynamic Properties*, NBS Technical Note 270-3, 1968, and *API Research Project 44* (Carnegie Press, 1953).

TABLE A-7E Properties of Some Common Fuels and Hydrocarbons

Fuel (Phase)	Formula	Molar Mass, lbm/lbmol	Density,* lbm/ft³	Enthalpy of Vaporization,† Btu/lbm	Specific Heat,* c_p Btu/ lbm·°F	Higher Heating Value,‡ Btu/lbm	Lower Heating Value,‡ Btu/lbm
Carbon (s)	C	12.011	125	—	0.169	14,100	14,100
Hydrogen (g)	H_2	2.016	—	—	3.44	60,970	51,600
Carbon monoxide (g)	CO	28.013	—	—	0.251	4,340	4,340
Methane (g)	CH_4	16.043	—	219	0.525	23,880	21,520
Methanol (ℓ)	CH_4O	32.042	49.3	502	0.604	9,740	8,570
Acetylene (g)	C_2H_2	26.038	—	—	0.404	21,490	20,760
Ethane (g)	C_2H_6	30.070	—	74	0.418	22,320	20,430
Ethanol (ℓ)	C_2H_6O	46.069	49.3	395	0.583	12,760	11,530
Propane (ℓ)	C_3H_8	44.097	31.2	144	0.662	21,640	19,930
Butane (ℓ)	C_4H_{10}	58.123	36.1	156	0.578	21,130	19,510
1-Pentene (ℓ)	C_5H_{10}	70.134	40.0	156	0.525	20,540	19,190
Isopentane (ℓ)	C_5H_{12}	72.150	39.1	—	0.554	20,890	19,310
Benzene (ℓ)	C_6H_6	78.114	54.7	186	0.411	17,970	17,240
Hexene (ℓ)	C_6H_{12}	84.161	42.0	169	0.439	20,430	19,090
Hexane (ℓ)	C_6H_{14}	86.177	41.2	157	0.542	20,770	19,240
Toluene (ℓ)	C_7H_8	92.141	54.1	177	0.408	18,230	17,420
Heptane (ℓ)	C_7H_{16}	100.204	42.7	157	0.535	20,680	19,180
Octane (ℓ)	C_8H_{18}	114.231	43.9	156	0.533	20,590	19,100
Decane (ℓ)	$C_{10}H_{22}$	142.285	45.6	155	0.528	20,490	19,020
Gasoline (ℓ)	$C_nH_{1.87n}$	100–110	45–49	151	0.57	20,300	18,900
Light diesel (ℓ)	$C_nH_{1.8n}$	170	49–52	116	0.53	19,800	18,600
Heavy diesel (ℓ)	$C_nH_{1.7n}$	200	51–55	99	0.45	19,600	18,400
Natural gas (g)	$C_nH_{3.8n}N_{0.1n}$	18	—	—	0.48	21,500	19,400

*At 1 atm and 68°F.

†At 77°F for liquid fuels, and 1 atm and 1 normal boiling temperature for gaseous fuels.

‡At 77°F. Multiply by molar mass to obtain heating values in Btu/lbmol.

Index

Note: Page numbers followed by *f* denote figures; by *t*, tables.

CONVERSION FACTORS

Dimension	Metric	Metric/English
Acceleration	$1 \text{ m/s}^2 = 100 \text{ cm/s}^2$	$1 \text{ m/s}^2 = 3.2808 \text{ ft/s}^2$ $1 \text{ ft/s}^2 = 0.3048^* \text{ m/s}^2$
Area	$1 \text{ m}^2 = 10^4 \text{ cm}^2 = 10^6 \text{ mm}^2 = 10^{-6} \text{ km}^2$	$1 \text{ m}^2 = 1550 \text{ in}^2 = 10.764 \text{ ft}^2$ $1 \text{ ft}^2 = 144 \text{ in}^2 = 0.09290304^* \text{ m}^2$
Density	$1 \text{ g/cm}^3 = 1 \text{ kg/L} = 1000 \text{ kg/m}^3$	$1 \text{ g/cm}^3 = 62.428 \text{ lbm/ft}^3 = 0.036127 \text{ lbm/in}^3$ $1 \text{ lbm/in}^3 = 1728 \text{ lbm/ft}^3$ $1 \text{ kg/m}^3 = 0.062428 \text{ lbm/ft}^3$
Energy, heat, work, internal energy, enthalpy	$1 \text{ kJ} = 1000 \text{ J} = 1000 \text{ N·m} = 1 \text{ kPa·m}^3$ $1 \text{ kJ/kg} = 1000 \text{ m}^2\text{/s}^2$ $1 \text{ kWh} = 3600 \text{ kJ}$ $1 \text{ cal}^\dagger = 4.184 \text{ J}$ $1 \text{ IT cal}^\dagger = 4.1868 \text{ J}$ $1 \text{ cal}^\dagger = 4.1868 \text{ kJ}$	$1 \text{ kJ} = 0.94782 \text{ Btu}$ $1 \text{ Btu} = 1.055056 \text{ kJ}$ $\quad = 5.40395 \text{ psia·ft}^3 = 778.169 \text{ lbf·ft}$ $1 \text{ Btu/lbm} = 25{,}037 \text{ ft}^2\text{/s}^2 = 2.326^* \text{ kJ/kg}$ $1 \text{ kJ/kg} = 0.430 \text{ Btu/lbm}$ $1 \text{ kWh} = 3412.14 \text{ Btu}$ $1 \text{ therm} = 10^5 \text{ Btu} = 1.055 \times 10^5 \text{ kJ (natural gas)}$
Force	$1 \text{ N} = 1 \text{ kg·m/s}^2 = 10^5 \text{ dyne}$ $1 \text{ kgf} = 9.80665 \text{ N}$	$1 \text{ N} = 0.22481 \text{ lbf}$ $1 \text{ lbf} = 32.174 \text{ lbm·ft/s}^2 = 4.44822 \text{ N}$
Heat flux	$1 \text{ W/cm}^2 = 10^4 \text{ W/m}^2$	$1 \text{ W/m}^2 = 0.3171 \text{ Btu/h·ft}^2$
Heat transfer coefficient	$1 \text{ W/m}^2\text{·}^\circ\text{C} = 1 \text{ W/m}^2\text{·K}$	$1 \text{ W/m}^2\text{·}^\circ\text{C} = 0.17612 \text{ Btu/h·ft}^2\text{·}^\circ\text{F}$
Length	$1 \text{ m} = 100 \text{ cm} = 1000 \text{ mm} = 10^6 \text{ } \mu\text{m}$ $1 \text{ km} = 1000 \text{ m}$	$1 \text{ m} = 39.370 \text{ in} = 3.2808 \text{ ft} = 1.0926 \text{ yd}$ $1 \text{ ft} = 12 \text{ in} = 0.3048^* \text{ m}$ $1 \text{ mile} = 5280 \text{ ft} = 1.6093 \text{ km}$ $1 \text{ in} = 2.54^* \text{ cm}$
Mass	$1 \text{ kg} = 1000 \text{ g}$ $1 \text{ metric ton} = 1000 \text{ kg}$	$1 \text{ kg} = 2.2046226 \text{ lbm}$ $1 \text{ lbm} = 0.45359237^* \text{ kg}$ $1 \text{ ounce} = 28.3495 \text{ g}$ $1 \text{ slug} = 32.174 \text{ lbm} = 14.5939 \text{ kg}$ $1 \text{ short ton} = 2000 \text{ lbm} = 907.1847 \text{ kg}$
Power, heat transfer rate	$1 \text{ W} = 1 \text{ J/s}$ $1 \text{ kW} = 1000 \text{ W} = 1341 \text{ hp}$ $1 \text{ hp}^\ddagger = 745.7 \text{ W}$	$1 \text{ kW} = 3412.14 \text{ Btu/h}$ $\quad = 737.56 \text{ lbf·ft/s}$ $1 \text{ hp} = 550 \text{ lbf·ft/s} = 0.7068 \text{ Btu/s}$ $\quad = 42.41 \text{ Btu/min} = 2544.5 \text{ Btu/h}$ $\quad = 0.74570 \text{ kW}$ $1 \text{ boiler hp} = 33{,}475 \text{ Btu}$ $1 \text{ Btu/h} = 1.055056 \text{ kJ/h}$ $1 \text{ ton of refrigeration} = 200 \text{ Btu/min}$
Pressure	$1 \text{ Pa} = 1 \text{ N/m}^2$ $1 \text{ kPa} = 10^3 \text{ Pa} = 10^{-3} \text{ MPa}$ $1 \text{ atm} = 101.325 \text{ kPa} = 1.01325 \text{ bars}$ $\quad = 760 \text{ mm Hg at } 0^\circ\text{C}$ $\quad = 1.03323 \text{ kgf/cm}^2$ $1 \text{ mm Hg} = 0.1333 \text{ kPa}$	$1 \text{ Pa} = 1.4504 \times 10^{-4} \text{ psia}$ $\quad = 0.020886 \text{ lbf/ft}^2$ $1 \text{ psi} = 144 \text{ lbf/ft}^2 = 6.894757 \text{ kPa}$ $1 \text{ atm} = 14.696 \text{ psia} = 29.92 \text{ in Hg at } 30^\circ\text{F}$ $1 \text{ in Hg} = 3.387 \text{ kPa}$
Specific heat	$1 \text{ kJ/kg·}^\circ\text{C} = 1 \text{ kJ/kg·K} = 1 \text{ J/g·}^\circ\text{C}$	$1 \text{ Btu/lbm·}^\circ\text{F} = 4.1868 \text{ kJ/kg·}^\circ\text{C}$ $1 \text{ Btu/lbmol·R} = 4.1868 \text{ kJ/kmol·K}$ $1 \text{ kJ/kg·}^\circ\text{C} = 0.23885 \text{ Btu/lbm·}^\circ\text{F}$ $\quad = 0.23885 \text{ Btu/lbm·R}$

*Exact conversion factor between metric and English units.

†Calorie is originally defined as the amount of heat needed to raise the temperature or 1 g of water by 1°C, but it varies with temperature. The international steam table (IT) calorie (generally preferred by engineers) is exactly 4.1868 J by definition and corresponds to the specific heat of water at 15°C. The thermochemical calorie (generally preferred by physicists) is exactly 4.184 J by definition and corresponds to the specific heat of water at room temperature. The difference between the two is about 0.06 percent, which is negligible. The capitalized Calorie used by nutritionists is actually a kilocalorie (1000 IT calories).